MULTIPHASE FLOW IN PERMEABLE MEDIA
A Pore-Scale Perspective

Hydrocarbon production, gas recovery from shale, CO_2 storage and water management have a common scientific underpinning: multiphase flow in porous media. This book provides a fundamental description of multiphase flow through porous rock, with emphasis on the understanding of displacement processes at the pore, or micron, scale. Fundamental equations and the principal concepts using energy, momentum and mass balance are developed. The latest advances in high-resolution three-dimensional imaging and associated modelling are explored. The treatment is pedagogical, developing sound physical principles to predict flow and recovery through complex rock structures while providing a review of the recent literature. This systematic approach makes it an excellent reference for those who are new to the field. It provides the scientific background necessary for a quantitative assessment of multiphase subsurface flow processes, and is ideal for hydrology and environmental engineering students, as well as professionals in the hydrocarbon, water and carbon storage industries.

MARTIN J. BLUNT is professor of Petroleum Engineering at Imperial College London, visiting professor at Politecnico di Milano and editor-in-chief of the journal *Transport in Porous Media*. He publishes widely on multiphase flow in porous media applied to oil recovery, groundwater flows and carbon dioxide storage. He is a distinguished member of the Society of Petroleum Engineers (SPE), having won the 2011 SPE Lester C. Uren Award, and the 2012 Darcy Medal for lifetime achievement from the Society of Core Analysts. The book is inspired by recent research and based on courses taught to thousands of students and professionals from around the world.

MULTIPHASE FLOW IN PERMEABLE MEDIA

A Pore-Scale Perspective

MARTIN J. BLUNT

Imperial College London

CAMBRIDGE
UNIVERSITY PRESS

CAMBRIDGE
UNIVERSITY PRESS

University Printing House, Cambridge CB2 8BS, United Kingdom

One Liberty Plaza, 20th Floor, New York, NY 10006, USA

477 Williamstown Road, Port Melbourne, VIC 3207, Australia

314–321, 3rd Floor, Plot 3, Splendor Forum, Jasola District Centre,
New Delhi – 110025, India

79 Anson Road, #06–04/06, Singapore 079906

Cambridge University Press is part of the University of Cambridge.

It furthers the University's mission by disseminating knowledge in the pursuit of
education, learning, and research at the highest international levels of excellence.

www.cambridge.org
Information on this title: www.cambridge.org/9781107093461
DOI: 10.1017/9781316145098

First published 2017
3rd printing 2019

Printed in the United Kingdom by TJ International Ltd. Padstow Cornwall

A catalogue record for this publication is available from the British Library.

Library of Congress Cataloging-in-Publication Data
Names: Blunt, Martin J.
Title: Multiphase flow in permeable media: a pore-scale perspective / Martin J. Blunt,
Imperial College London
Description: Cambridge : Cambridge University Press, 2017.
Identifiers: LCCN 2016026616 | ISBN 9781316145098
Subjects: LCSH: Multiphase flow. | Fluid dynamics. | Porous materials.
Classification: LCC TA357.5.M84 B58 2017 | DDC 532/.56–dc23
LC record available at https://lccn.loc.gov/2016026616

ISBN 978-1-107-09346-1 Hardback

This work is dedicated to my family, and particularly to Catherine, who is no longer here to read it. I am sorry that this is far less than you could have achieved, but it is the best I can manage.

Contents

The plate section can be found between pages 140 and 141

Preface

Among the principal challenges of the twenty-first century are how to secure access to clean water for drinking and agriculture, and how to provide sufficient energy for a growing and hopefully more prosperous population, while coping with the threat of climate change. The economics and social aspects of these challenges have a common scientific underpinning: flow in porous media. The majority of the world's freshwater resides underground in aquifers; most of our energy comes from oil and gas extracted from porous rock; and one promising method to reduce atmospheric emissions of carbon dioxide is to collect it from major sources, such as fossil-fuel burning power stations, and inject it deep underground into saline aquifers or depleted hydrocarbon reservoirs. Indeed, global-scale carbon dioxide storage is necessary if we are to avoid dangerous climate change. In any event, the understanding and management of fresh water, oil and gas recovery and carbon dioxide storage all rely on quantifying how fluids flow through porous rocks.

The emphasis in this book is on multiphase flow and applications in hydrocarbon recovery, geological carbon dioxide storage and contaminant transport, as mentioned above. However, the basic principles are relevant to many other applications in science and engineering, including in fuel cells, membranes and biological systems.

The material will focus on the physics of fluid displacement at the pore scale in geological systems, meaning that we will be concerned with the scale of the interstices between solid grains, or the size of void spaces in the rock that is typically of the order of microns (μm). It is the behaviour at this scale that plays a key role in determining the overall movement of the fluids and how much can be recovered.

In combination with the huge practical applications that drive the science, two recent advances have transformed our understanding of how fluids move at the pore scale: the development of methods to image rocks, the pore space and the fluids within them, in three dimensions at a resolution from nanometres to centimetres;

and the availability for good public-domain software for solving flow and transport problems.

While I intend the book to be a valuable reference resource for researchers and professional scientists and engineers working on porous media problems, it is primarily developed as a textbook to teach and explain the basic concepts of multiphase flow in porous media. As a result, the review of the literature is not intended to be comprehensive, although I do provide references to some of the more important papers and more recent work that provide fresh insights and describe new developments in the field. To emphasize that this is a textbook, rather than a research reference, at the end of the book I provide some examination exercises, with solutions online, based on questions set to my own students, as an aid to understanding and to provide practice in making calculations. It is essential to be able to employ the methods and ideas presented here to provide a quantitative analysis of fluid flow and transport.

The main physical concepts employed are conservation of energy, momentum and mass. Conservation of energy, with specific consideration of the interfacial energy between fluids, and between fluids and solid, is used to define fluid configurations in the pore space and the pressures at which one fluid phase will displace another. It is also the principle used to derive the Young and Young-Laplace equations which form the foundations for the first half of the book. Conservation of momentum, for viscous fluids, leads to the Navier-Stokes equations and their macroscopic counterpart in porous media, Darcy's law. These concepts are introduced half-way through the book, once we have established the behaviour of fluids in capillary equilibrium. Finally, conservation of mass, and consequently of volume for incompressible fluids, is used to derive equations for flow controlled by both viscous and capillary forces. The aim is to apply universal concepts to important problems associated with flow in porous media, rather than attempt a comprehensive overview of the myriad of complex and fascinating phenomena that have been studied in the literature. I trust, however that armed with these foundations, a reader will be able to approach novel problems with some confidence and understanding.

This book is related to the subject called, in the oil industry, digital rock analysis or digital rock physics, although in my opinion these are misleading terms. The word 'digital' implies a separation from experiment, or even an opposition to it, which is unfortunate: it is not possible to make predictions or interpretations of porous media behaviour without data. These data include images of the pore-space geometry and measurements of rock and fluid properties, such as interfacial tension and contact angle. This book provides the underlying scientific understanding necessary to interpret and make use of the results of pore-scale models. It is not sufficient simply to run a complex black-box code that computes average properties in some sort of experimental vacuum, or without a good grasp of what physics is incorporated into the model and the likely resultant behaviour. This book will not provide a shopping list of properties that can be predicted on rock samples, nor will it provide a critique of different modelling approaches. What is needed, and what I hope to provide here, is information on how to complement new tools in imaging and numerical analysis with a sound understanding of the fundamental concepts.

Acknowledgements

This book is based on lectures on multiphase flow and reservoir engineering that I have given to both undergraduate and graduate students at Imperial College London, to MSc students at Politecnico di Milano and to academics and professionals around the world as part of short courses. I would like to thank all the students in these classes for their attention and curiosity. I hope that you have learnt as much from me as I have from you.

I would also like to thank the many PhD students and other colleagues with whom I have worked over the years and whose research I will try to synthesize and explain in this book.

I am very grateful to Dr Ali Raeini, who calculated the displacement statistics in Chapters 4 and 5; Dr Bagus Muljadi, who reintroduced me to LaTeX; Hasan Nourdeen, who prepared the example solutions in Chapter 9; Angus Morrison and Bhavik Lodhia for drafting many of the figures; Dr Kamajit Singh for the cover image, and Harris Rana for obtaining permissions to use the figures of others.

Symbols

Here I provide the symbols and units used for the most commonly employed terms in the book. Unfortunately, with such a long presentation it is not possible to maintain unique symbols for every quantity. Where there is an ambiguity, I indicate the chapter where different meanings are used: if no chapter is shown, it is used consistently throughout the book. Where no unit is shown, the quantity is dimensionless.

Symbol	Meaning	Chapter	SI unit
a	Molecular size	1	m
	Aspect ratio	3	
	Permeability scaling	6	
	Specific surface area		1/m
A	Area		m^2
b	Length of wetting phase in a corner		m
B	Bond number	3	
c	Concentration		
C	Capillary pressure multiplier	3, 4	
	Land trapping constant	4.6	
	Rate constant for imbibition	9	m/s$^{1/2}$
Ca	Capillary number		
C_s	Spreading coefficient	8	N/m
d	Spatial dimension (3)		
D	Capillary pressure factor	3	
	Fractal dimension	3.4	

(*cont.*)

Symbol	Meaning	Chapter	SI unit
	Grain diameter	6	m
	Capillary dispersion	9	m^2/s
f	Throat radius distribution		m^{-1}
	Fraction of oil-wet elements	5	
f_w	Water fractional flow	9	
F	Free energy	1, 3	J
F_d	Dimensionless factor for capillary pressure	3	
F_w	Capillary fractional flow	9	
g	Acceleration due to gravity		m/s^2
	Flow conductance	6	
G	Shape factor		
	Normalized throat radius distribution	3	
h	Height		m
H	Hessian matrix	2	m^{-2}
I	Wettability index		
k	Boltzmann's constant	1	J/K
k_r	Relative permeability		
K	Permeability		m^2
l	Pore-scale length		m
L	Length		m
M	Molecular mass	1	kg/mol
M	Minkowski functional	2	
n	Number		
N_A	Avogadro's number	1	
N_p	Oil produced	9	m^3
N_{pD}	Pore volumes produced	9	
p	Percolation probability		
P	Pressure		Pa
P_c	Capillary pressure		Pa
q	Darcy velocity		m/s
Q	Flow rate		m^3/s
r	Radius of curvature		m
R	Radius		m
R_{vc}	Ratio of capillary to viscous forces		
S	Surface	2	m^2
S	Saturation		

t	Time		s
	Percolation conductance exponent	6.4	
T	Temperature		K
U	Cohesion energy	1	J
v	Velocity		m/s
V	Volume		m^3
W	Finger width in percolation		m
x	Distance		m
z	Depth		m
	Coordination number	2, 3.4	

Greek symbols

α	Trapping constant	4.6	
β	Betti number	2	
	Corner half angle		radians
	Filling percolation exponent	3.4, 6.4	
β_R	Corner resistance factor		
η	Effective saturation	6	
θ	Contact angle		radians
κ	Curvature		m^{-1}
λ	Mobility		1/(Pa.s)
μ	Viscosity		Pa.s
ν	Correlation length exponent		
ξ	Correlation length		m
ρ	Density		kg/m^3
σ	Interfacial tension		N/m
τ	Percolation trapping exponent		
	Relaxation time scale	6	s
τ'	Burst exponent in percolation theory		
$\mathcal{T}$	Viscous stress tensor		Pa
ϕ	Porosity		
χ	Euler characteristic	2	
ω	Dispersion scaling	9	m/s$^{1/2}$

Superscripts and subscripts

adv	Advancing		
A	Advancing		
AH	Amott-Harvey	5	
AM	Arc meniscus		

(cont.)

Symbol	Meaning	Chapter	SI unit
b	Backbone	3.4	
c	Corner		
	Critical (in percolation)	3.4	
$crit$	Critical		
D	Drainage	3, 4	
	Dimensionless	9	
$entry$	Entry (for capillary pressure)		
g	Gas		
H	Hinging		
I	Imbibition		
l	Layer		
L	Left		
m	Macroscopic		
max	Maximum		
min	Minimum		
nw	Non-wetting		
o	Oil		
r	Residual		
R	Receding		
	Right	9	
s	Solid		
t	Throat or total		
tot	Total		
TM	Terminal meniscus		
$USBM$	USBM index	5	
w	Wetting or water		
wc	Irreducible or connate		
$*$	Threshold for spanning		
	To indicate when $P_c = 0$	9	

1
Interfacial Curvature and Contact Angle

1.1 Interfacial Tension

In this book, we will be concerned with the arrangement and displacement of multiple fluid phases in the pore space. The pore space could be the micron-sized interstices between grains in a rock or soil, the gaps between fibres in a tissue or blood vessels. The fluids could be water, oil, natural gas, carbon dioxide, blood plasma or electrolyte solution. Here we will present the basic equations before introducing the complex and fascinating geometry of the porous medium itself.

If we have two fluid phases, without a porous medium present, then the fluids arrange to minimize the surface area between them. For instance, a small droplet of water in air will be spherical, since this minimizes the area of contact with the air for a fixed volume, as shown in Fig. 1.1. For larger drops, the shape is perturbed by gravitational forces, which we ignore here for simplicity.

We define an interfacial tension, σ, which is the energy per unit area of the surface between the phases, or the change in free energy (in an open system with a defined pressure, this is the Gibbs free energy) for a change in area: $\sigma = dF/dA$. The two phases involved may be fluid or solid. It is the energy penalty of breaking the intermolecular interactions between the two phases and themselves only, and instead creating an interface between them; this energy is largest if one of the phases has strong intermolecular bonding with itself (such as a crystalline solid or metal) and lowest between two similar fluids (such as oil and a hydrocarbon gas at high pressure).

Note that sometimes, erroneously, the term surface tension is used: strictly this refers to the energy per unit area of a surface between a fluid or solid and its vapour in thermodynamic equilibrium, with no other components present. While surface tension is precisely defined, and is a useful concept for the discussion below, it is of little relevance in the majority of cases we will consider, where we have complex mixtures of many chemical components, at least two fluid phases and solid all present: in these cases the more general, and correct, expression interfacial tension should be used.

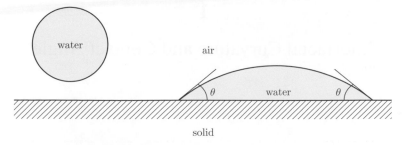

Figure 1.1 An illustration of fluid arrangements. In free space (left) – and ignoring the effects of gravity – a small volume of water, as the wetting phase, forms a spherical droplet in air: this is the configuration which minimizes the surface energy, since the area of the surface between the water and the air is the smallest possible for a fixed volume. In contact with a solid surface the droplet spreads (right). The interface between the water and the air now has the shape of the top slice of a sphere. The water contacts the solid at an angle θ: this is the contact angle.

The likely magnitude of the interfacial tension can be estimated from the strength of molecular forces. If the intermolecular cohesion energy for each molecule is U inside its own phase, then a molecule at a surface sees approximately half that, or $U/2$. If the molecular size is a, then the surface tension (that is, if there is no second phase to interact with) is given approximately by $\sigma \approx U/(2a^2)$ (de Gennes, 1985; de Gennes et al., 2003). For non-polar liquids, whose intermolecular interactions are largely van der Waals forces, $U \approx kT$, where k is the Boltzmann constant, 1.3806×10^{-23} J/K. As an aside, this book will use SI units throughout – it is a nonsense to attempt a serious scientific discussion using an inconsistent unit system. Strange and baffling units are a peculiar and persistent feature of the oil industry, but don't need to be used! Returning to the science, with confidence that dimensional quantities can now be understood by more than a narrow and narrow-minded cohort of specialist engineers, we may consider octane as an example oil phase whose molecular size can be derived from the density and molecular mass to be approximately 6.5×10^{-10} m. The steps are as follows: if the density is ρ and the molecular mass is M, then the effective molecular size can be estimated as $a = \left(\frac{M}{\rho N_A}\right)^{1/3}$, where N_A is Avogadro's number, 6.022×10^{23}. For octane with $\rho = 703$ kg/m^3 and $M = 0.114$ kg/mol, we obtain $a = 6.5 \times 10^{-10}$ m. Then at 20°C ($T = 293$ K; $kT = 4.04 \times 10^{-21}$ J) we find $\sigma \approx 5$ mJ/m^2. This is only an order of magnitude estimate: the real value is around 22 mJ/m^2 because the cohesion energy of octane is larger than estimated here. For water, where the molecules have hydrogen bonds between them, the surface tension is greater, around 73 mJ/m^2, while mercury, since it is metallic, has a much higher value in the range 425–485 mJ/m^2 dependent on temperature.

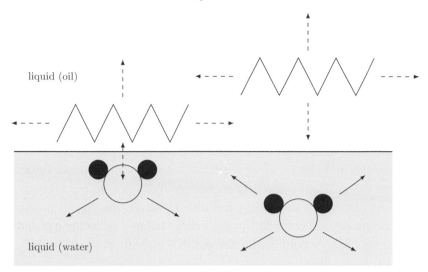

liquid (oil)

liquid (water)

Figure 1.2 A schematic diagram of the forces at an interface between two phases: water and oil. The oil molecules experience relatively weak van der Waals inter-molecular forces (the dotted arrows), while the water is more strongly hydrogen bonded (solid arrows). At the surface, some of the hydrogen bonding between the water is removed, but replaced by van der Waals forces with the oil.

It is also possible to estimate interfacial tensions from surface tensions if one of the phases is non-polar with only van der Waals attraction. Let phase 1 be non-polar with a surface tension σ_1. The second phase (solid or liquid) is more strongly bonded with a larger surface tension σ_2. To a first approximation, we can assume that at the surface the van der Waals interactions are similar between the two phases as within phase 1.

Consider, for example, the interfacial tension between a non-polar liquid (oil) and a polar liquid (water) in Fig. 1.2. At the interface we remove approximately half the hydrogen bonding of the water and replace it with a weaker van der Waals attraction. For oil, the degree of molecular interaction is similar in both cases. Therefore, to consider the energy penalty of the surface, it is equivalent to cre-ating a water surface with a vacuum and then adding van der Waals forces with oil, whose energy per unit area is similar to the surface tension of oil. Then we find

$$\sigma_{12} \approx \sigma_2 - \sigma_1. \tag{1.1}$$

Notice the signs: σ_1 is taken away from the expression for σ_{12} in Eq. (1.1) since it is energetically favourable to add the van der Waals interactions. For instance, for an octane/water interface this would give an interfacial tension of around $73 - 22 = 51$ mJ/m^2 using the values above, which is equal to the measured value. This approximation will be important when we consider the flow of three liquid phases in Chapter 8 as it can be used to explain the spreading behaviour of one liquid phase between two others. It will also be important – indeed crucial – in our discussion of wettability and contact angle later in this section, as this is controlled by a balance of interfacial tensions.

1.2 Young-Laplace Equation

The arrangement of fluid phases in contact with each other and a solid is controlled locally by an energy balance. Displacements, when one fluid pushes out another from the pore space, occur when they are energetically favourable, while positions of equilibrium represent local energy minima. However, the porous medium does introduce constraints and complexity: this energy balance occurs at a pore-by-pore level; we do not observe fluid configurations that globally conserve energy, across the entire porous medium, or at some macro-scale, averaged over many pores. As shown later, there is instead a sequence of local pore-scale configurations representing energy minima determined by the order in which displacements occurred. A series of local displacements, controlled by how fluids are injected into the pore space, provides fluid configurations that are far from global equilibrium, while maintaining, step by step, positions of local equilibrium.

Consider two fluid (liquid or gas) phases in contact with a solid. We denote one of the phases as non-wetting, in that it has less preference to coat, or reside next to, the solid than the other fluid phase, called the wetting phase. Since one fluid preferentially wets the surface, while the other is repelled, the interface between the phases is curved: consider familiar situations, such as a bead of water on a table-top, or the meniscus of a fluid in a narrow tube. This is illustrated in Fig. 1.1, where a droplet of water (the wetting phase in this example) comes to rest on a solid surface surrounded by air (the non-wetting phase). As shown below, the curvature of the interface leads to a pressure difference between the phases. The phase bulging out into the other has a higher pressure. In a porous medium, or in a capillary tube, it is the non-wetting phase that protrudes into the wetting phase and has a higher pressure; it does not want to be next to the solid and requires a higher pressure than the wetting phase to be forced through the porous medium. As we discuss later, it is a balance of surface forces that determines the angle θ at the contact between water, air and the solid.

The first relationship to be derived, the Young-Laplace equation, relates the pressure difference between the phases to the curvature of the interface. It is found from an energy balance, considering a change in volume of one of the phases. An energy balance relates the work done against this pressure difference to the change in surface energy. To determine how fluids are configured in the pore space and how they move, we can apply energy balance directly; indeed this is done to derive configurations and displacements in complex situations (see, for instance, Mayer and Stowe, 1965; Princen, 1969a, b, 1970; and the later discussion in Chapters 3 and 4). However, it is more convenient first to have a clear conceptual picture of how fluids are arranged backed up with equations which allow a rapid quantitative analysis.

The work done against the pressure difference is equal to the change in interfacial energy:

$$(P_{nw} - P_w) \, dV = \sigma \, dA, \tag{1.2}$$

where dV is the infinitesimal change in volume (the increase in volume of the non-wetting phase relative to the wetting phase) and dA is the corresponding change in surface area. The subscripts w and nw refer to the non-wetting and wetting phase, respectively; here we consider an increase in the non-wetting phase volume, where the work against the pressure difference is matched, in equilibrium, with a corresponding increase in surface energy.

We call $P_{nw} - P_w$ the local capillary pressure P_c and hence:

$$P_c = \sigma \frac{dA}{dV}. \tag{1.3}$$

It is easy to compute the derivative dA/dV for simple geometries. If we consider a sphere of radius r, and a small change in radius dr, then $A = 4\pi r^2$, $V = \frac{4}{3}\pi r^3$ and so $dA/dr = 8\pi r$ and $dV/dr = A = 4\pi r^2$, from which we find:

$$P_c^{sphere} = \frac{2\sigma}{r}. \tag{1.4}$$

If, instead, we consider a cylindrical geometry with fixed length, l, $V = \pi r^2 l$ and $A = 2\pi r l$, then there is only curvature in one direction and for a change in radius, $dA/dV = 1/r$ and

$$P_c^{cylinder} = \frac{\sigma}{r}. \tag{1.5}$$

In general, the interface between the two phases can be curved in two directions with different radii of curvature. A full discussion of curvature lies outside the scope of this book – here the main results are stated, which should make intuitive sense. Consider any smooth surface, and some arbitrary point on this surface. Now take a plane that cuts the surface through this point; the intersection of the plane with the surface is a smooth curve. At the chosen point the curvature is the inverse of the radius of a circle that fits the curve, as illustrated in Fig. 1.3. We can vary the orientation of the plane, and the measured curvature will vary. We define the principal radii of curvature r_1 and r_2 as the minimum and maximum radii measured as the orientation of the plane is varied. The planes where r_1 and r_2 are defined will be orthogonal.

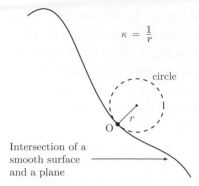

$$\kappa = \frac{1}{r}$$

circle

r

O

Intersection of a
smooth surface
and a plane

Figure 1.3 A description of curvature. The intersection between a smooth surface
and a plane is a curve, as shown. At an arbitrary point O, a circle can be drawn
whose circumference passes through O and whose radius is chosen to fit the curvature of the curve at that point. The radius of curvature is r and the curvature
$\kappa = 1/r$.

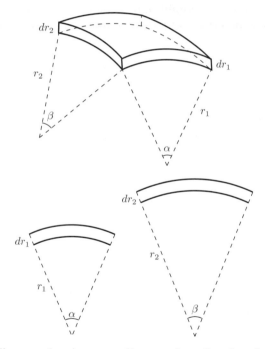

dr_2

r_2

β

dr_1

r_1

α

dr_2

dr_1

r_2

r_1

α

β

Figure 1.4 A diagram showing a small expansion of an interface between two
liquids whose principal radii of curvature are r_1 and r_2 (top). The lower figure
illustrates two perpendicular cuts through the interface showing the radii of curvature. Considering the change in volume and interfacial area for a small change
dr in both radii allows the Young-Laplace equation (1.6) to be derived.

Now consider an interface between two fluid phases, which may be curved in two
directions. In Fig. 1.4 we consider an infinitesimal piece of an interface, subtended
by an infinitesimal angle α with radius of curvature r_1 in one direction, and an angle
β with radius of curvature r_2 in the other. The area of this portion of the interface is

$\alpha\beta r_1 r_2$. Now consider an infinitesimal but equal change in both radii of curvature $dr = dr_1 = dr_2$ with the angles kept constant (the radii increments have to be equal to make sense geometrically). The new area of interface is $\alpha\beta (r_1 + dr) (r_2 + dr)$, which for small dr represents a change in area $dA = \alpha\beta (r_1 + r_2) dr$. The change in volume is $dV = \alpha\beta r_1 r_2 dr$, from which we derive using Eq. (1.3):

$$P_c = \sigma \left(\frac{1}{r_1} + \frac{1}{r_2} \right) = \kappa\sigma, \tag{1.6}$$

where κ is the total curvature of the interface. Note that the radii of curvature can be either positive or negative: bulging out or convex curvature is positive, while concave or inwards curvature is negative. Traditionally, for oil/water systems, P_c is defined as $P_o - P_w$, where the subscripts o and w label oil and water, respectively. If κ is positive, then oil is at a higher pressure than water; if κ is negative, water has the higher pressure.

If the fluids are at rest – in equilibrium – then the pressure within each phase is constant. This means that the capillary pressure is also fixed, and hence the interface between the fluids has a constant curvature. Therefore, in any system with stationary fluids the radii r_1 and r_2 may vary in space, but only if they maintain the same value of $\kappa = 1/r_1 + 1/r_2$.

1.3 The Young Equation and Contact Angle

The second fundamental equation finds an expression for the contact angle in terms of the interfacial tensions between the solid and the two fluids, and the interfacial tension of the fluid interface itself. The easiest, and standard, way to derive this is to represent, or interpret, the interfacial tension as a force: it has the units of energy per unit area, which is equivalent to a force per unit length. Interfacial energy is an energy penalty: the surface is energetically unfavourable and either wants to minimize its area (which is why a small drop of water, for instance, is spherical) or to coat the surface with a fluid with which it has a lower interfacial tension. Either way, this can be viewed as a tugging on an interface, or a real tension, to cover or reduce the surface. With this conceptual picture, the derivation is easy.

Consider the fluid arrangement shown in Fig. 1.5, where the contact between two fluids and a locally flat solid is shown. Treating the interfacial tensions as forces, we apply a horizontal force balance to obtain:

$$\sigma_{nws} = \sigma_{ws} + \sigma \cos\theta, \tag{1.7}$$

where σ_{nws} is the interfacial tension between the non-wetting phase and the solid, while σ_{ws} is the interfacial tension between the wetting phase and the solid. θ is the contact angle: the angle that the two-fluid interface makes at the solid surface; see Fig. 1.1.

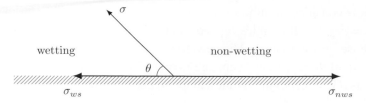

Figure 1.5 Two fluid phases in contact with a solid. A horizontal force balance yields the Young equation (1.7). The exact point of contact is shown enlarged from Fig. 1.1, so that the curvature of the fluid interface cannot be discerned.

This is the Young equation that relates the interfacial tensions to the contact angle. By convention, the angle is measured through the denser fluid phase. If we consider oil (hydrocarbon) and water, then water (considered here to be the wetting phase) is denser and so the contact angle is measured through the water. For a gas/oil system, the contact angle would be measured through the oil phase. For mercury/air (or mercury/vacuum), the contact angle is measured through the mercury. Eq. (1.7) can be rearranged to find the contact angle:

$$\cos\theta = \frac{\sigma_{nws} - \sigma_{ws}}{\sigma}. \tag{1.8}$$

The vertical force balance is accommodated by sub-atomic perturbations in the solid at the point of fluid contact: this produces a downwards force to balance the vertical component from the fluid/fluid interfacial tension (Morrow, 1970); if we were to consider three fluids in contact, we would indeed have both a proper vertical and horizontal force balance accommodated by the arrangement of the phases, presented later in Chapter 8.

The Young equation makes qualitative physical sense: consider, for instance, oil (hydrocarbon liquid) and water on quartz (sand or sandstone surface). Quartz is a crystalline solid with strong inter-atomic bonding; breaking these bonds results in a high surface energy. Water is a polar liquid and so can allow some electrostatic bonding with the solid; this is energetically favourable, leading to a lower interfacial tension between water and quartz than between oil and quartz, where only van der Waals interactions are possible. This makes water the wetting phase and the contact angle is less than $90°$ in Eq. (1.8).

It is possible that the force balance in Eq. (1.7) cannot be obeyed, since $\sigma_{nws} > \sigma_{ws} + \sigma$. In this case we have complete wetting where the wetting phase spontaneously covers the entire surface (Dussan, 1979; Ngan and Dussan, 1989; Adamson and Gast, 1997); the effective contact angle $\theta = 0$. The wetting phase resides as a thin film of molecular thickness across the solid surface. As we show later, in a porous medium, the excess wetting phase will also collect in grooves and crannies in the rock, where it can maximize its contact with the solid.

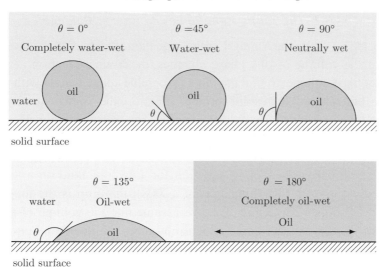

Figure 1.6 An illustration of different contact angles measured through water in an oil/water system. Here a droplet of oil is placed on a solid mineral surface, surrounded by water. The contact angle is measured through the water. We ignore buoyancy effects. A contact angle of zero is complete wetting (top left): water spreads over the surface and oil makes no direct contact with the solid. A contact angle less than 90° is water-wet; 90° is neutrally wet (top right), while a contact angle greater than 90° is oil-wet (lower figures). Complete wetting of oil occurs for a contact angle of 180°: oil spreads across the surface (lower right). Adapted from Morrow (1990).

The contact angle θ may take any value between 0 and π (0 to 180°) dependent on the interfacial tensions. For clarity, we have described the two phases as wetting and non-wetting: by definition the wetting phase preferentially contacts the surface and has a contact angle $\theta < 90°$; if the contact angle is greater than 90° the denser phase is non-wetting. As discussed further in the next chapter, for oil and water systems on mineral surfaces, the contact angle may indeed vary from strongly water-wet (contact angles near 0) to strongly oil-wet (contact angles close to 180°), as shown schematically in Fig. 1.6.

Eq. (1.8) is useful as it relates the surface properties of the solid in interaction with the fluids, to the fluid interfacial tension and contact angle alone. However, it is rarely used directly to compute or predict contact angle, as it is difficult to measure interfacial tensions involving solids directly. As a consequence, in the discussion that follows, we will assume that there is a contact angle, an angle that can be measured directly from observing fluids on a solid surface, as shown in Fig. 1.6, whose value is related to the balance of the three interfacial tensions between the solid and the fluids and between the fluids themselves.

1.3.1 The Young Equation as an Energy Balance

Strictly speaking, while the Young-Laplace equation is an energy balance for work done when the volume of one phase changes, the Young equation is an energy balance at a fixed volume, to find the most favourable contact angle with a constant volume of fluid residing on a solid surface. Therefore the Young equation can also be derived directly from conservation of energy. The reason why this is not usually performed is because the algebra is cumbersome and the derivation is difficult to generalize for any fluid configuration.

Here we will derive the Young equation for a drop of fluid on a flat surface, following the approach of Whyman et al. (2008): the drop is the top slice of a hemisphere, as illustrated in Fig. 1.7. We assume that the volume of the droplet is fixed and find the contact angle which minimizes the interfacial energy of the system. We assume that the droplet is small, so that we need not consider the perturbative effects of gravitational forces on its shape. Some straightforward algebra and geometry leads to the following expressions needed in the analysis.

The volume of the drop is:

$$V = \frac{\pi r^3}{3}(1 - \cos\theta)^2(2 + \cos\theta), \qquad (1.9)$$

where the contact angle is θ and r is the radius of curvature of the drop. The area of the drop (the area of the fluid/fluid interface) is:

$$A = 2\pi r^2(1 - \cos\theta). \qquad (1.10)$$

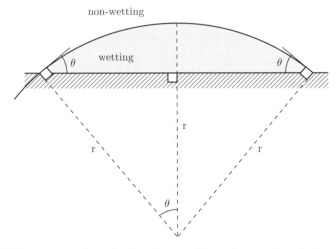

Figure 1.7 A cross-section of a droplet on a flat surface used to derive the Young equation from energy principles: we find the contact angle θ that minimizes the interfacial energy subject to a fixed fluid volume. The shape of the drop is the top slice of a hemisphere.

The change in energy of the system — strictly speaking, if we consider a closed system, with a fixed saturation or phase volumes, at fixed temperature, this is the change in Helmholtz free energy (Morrow, 1970; Hassanizadeh and Gray, 1993a) — is given by the change in interfacial energy when placing the drop on the fluid surface:

$$\Delta F = \sigma A - \pi r^2 \sin^2 \theta \Delta \sigma_s, \qquad (1.11)$$

where $r \sin \theta$ is the radius of the droplet on the surface, while $\Delta \sigma_s = \sigma_{nws} - \sigma_{ws}$ is the difference between the interfacial tensions with the surface and the non-wetting and wetting phases.

To find the minimum energy configuration, we need to write the energy in terms of the fixed volume V and contact angle. From Eqs. (1.9) and (1.10) we rewrite Eq. (1.11) as:

$$\Delta F = \left[\frac{9\pi V^2}{(1 - \cos \theta)(2 + \cos \theta)^2} \right]^{1/3} (2\sigma - \Delta \sigma_s)(1 + \cos \theta). \qquad (1.12)$$

The equilibrium contact angle is that which gives the minimum free energy ($d\Delta F/d\theta = 0$ and $d^2 \Delta F/d\theta^2 < 0$). We find, differentiating Eq. (1.12):

$$\frac{d\Delta F}{d\theta} = 2 \sin \theta \left[\frac{9\pi V^2}{(1 - \cos \theta)^4 (2 + \cos \theta)^5} \right]^{1/3} (\Delta \sigma_s - \sigma \cos \theta). \qquad (1.13)$$

The non-trivial solution for zero derivative is:

$$\Delta \sigma_s = \sigma_{nms} - \sigma_{ws} = \sigma \cos \theta, \qquad (1.14)$$

which is equivalent to the Young equation (1.7), as before.

1.3.2 Interfacial Tension, Roughness and Wettability

Throughout this book we will use the term wettability: this refers to the distribution of contact angles throughout the pore-space. This will be presented in further detail in Chapter 2. However, here we will anticipate the discussion by exploring the implications of changes in interfacial tension and roughness on contact angle. This in turn – as we show later – will affect how fluids are arranged in the pore space, the nature of displacement and the movement and recovery of oil or carbon dioxide (or other phases).

Imagine that we introduce a surfactant to an oil/water system. The surfactant molecules reside preferentially on the fluid/fluid interface and lower the interfacial tension. As presented later, in Chapter 6.4.6, a significant lowering of interfacial tension changes the pore-scale balance of forces and can aid oil recovery directly. Here though, we consider a very low surfactant concentration and a modest reduction in σ.

If the surfactant does not change the interfacial tensions with the solid (it does not preferentially adhere to it), from Eq. (1.7) it is apparent that if σ decreases, then, to maintain the force balance, $\cos\theta$ must increase. This means that the system tends towards complete wetting, and may be completely wetting if the force balance can no longer be maintained. If $\theta < 90°$ then the contact decreases towards zero; if $\theta > 90°$ the system has a larger contact angle and may become completely non-wetting to water (oil becomes completely wetting). Hence, a surfactant alters the wettability of the surface: not through a direct change in the interaction of the fluids with the solid, but through perturbing the balance of forces at the three-phase contact line by lowering the fluid/fluid interfacial tension.

In deep aquifers and oilfields, the aqueous phase is not pure water, but a highly saline solution containing many salts in solution. Now consider another possible change to interfacial properties, this time caused by altering the composition of the brine. This can be achieved in oilfield operations by injecting a brine with a different composition to the resident water. In this case, as a first approximation, we assume that brine composition is unlikely to have a significant effect on the interfacial tension between the fluids: at the high temperatures and pressures typical of storage aquifers and oilfields, the interfacial tensions between oil and brine are approximately half those encountered at ambient (surface conditions), normally in the range 15–25 mN/m. Note the units that are now used here. While interfacial tension is indeed an energy per unit area and so has, hitherto, been quoted in units of J/m^2, it is more usual to use equivalent units based on the concept of interfacial *tension* and use N/m. The brine composition can alter the interfacial energy between the solid and the fluids, particularly between the solid and the water. The brine composition and pH can alter the surface charge on the solid and the arrangement of ions in solution. The exact behaviour is complex and generally hard to predict from simple principles: for a review of how surface forces affect wettability see Hirasaki (1991). However, this is the fundamental concept behind low salinity waterflooding, or – more generally – controlled salinity waterflooding.

The composition of the brine injected to displace oil in a hydrocarbon reservoir can be tailored to alter the wettability so that displacement is most favourable. This is achieved through altering the energy of the solid surfaces in the system and hence changing the contact angle: normally the aim is to alter the surface to more water-wet (low contact angle) conditions. A description of how this changes contact angle is beyond the scope of this book, and is in any event not completely understood. For instance, lowering the salinity does tend to reduce the interfacial tension between oil and water slightly, and this on its own – as discussed above for surfactants – tends to drive the system towards complete wetting. However, rather than continue this discussion, in the chapters that follow, we will show how contact angle impacts fluid distribution, displacement and oil recovery. Then, in principle, we will be able to determine the contact angle that is optimal for waterflooding.

The modification of injected water is an important topic in the oil industry, since it offers a relatively low-cost method to improve recovery. Low salinity waterflooding is now undergoing field trials, complemented by core-scale experimentation and modelling. For a review of this technology see Morrow and Buckley (2011); however, the basic idea of using agents in the water to modify contact angle, usually towards a more water-wet state, and thereby improve recovery, has been investigated in the oil industry for over 50 years; see, for instance, Harvey and Craighead (1965).

Now let's consider a third situation where the apparent wettability can be altered: a rough surface. Here we consider the so-called Wenzel regime, where the wetting phase resides on a surface that is not smooth; see, for instance, de Gennes (1985), Hazlett (1993), and de Gennes et al. (2003). This possibility is shown schematically in Fig. 1.8, with the effective contact angle defined for an equivalent smooth surface. The wetting phase (say water) fills the grooves, crevices and corners of this rough surface. $f \geq 1$ is defined such that for a surface of apparent (smooth) area A, the real area, accounting all the roughness down to the atomic scale, is fA. Then, if we repeat the energy balance for a droplet on a surface as before, we replace $\Delta\sigma$ representing the change in solid surface energy by $f\Delta\sigma$ in Eq. (1.14) to find:

$$f \Delta\sigma_s = f(\sigma_{nws} - \sigma_{ws}) = \sigma \cos\theta. \qquad (1.15)$$

The effect of surface roughness is similar to a reduction in interfacial tension: as f increases, $\cos\theta$ has to increase in magnitude: it drives the contact angle towards a completely wetting state. In porous media with roughness on all length scales, f is likely to be large – often over 100 – and so the surface appears completely wetting, at least to the extent that the roughness is filled with the wetting phase.

In reality, the situation is more complex. Firstly, we need to define carefully the length scale on which we define an equivalent smooth surface: in natural

Figure 1.8 Wetting and contact angle in the Wenzel wetting regime. Here the wetting phase (shaded) fills roughness on the solid surface (grey). If we define a contact angle on an equivalent smooth surface, as shown by the dashed line, energy balance can be used to find a relationship between the interfacial tensions and this contact angle, Eq. (1.15). The apparent contact angle, θ, can be much lower than the value at the atomic level.

porous media there are frequently no apparently flat surfaces at any scale. Secondly, the fluid configurations are established by displacement and so a purely equilibrium calculation in complex geometries is no longer necessarily complete. A full description of this requires the introduction of porous media in the next chapter. However, energetic considerations do explain why the wetting phase (water) always resides in the smaller regions of the pore space, the grooves, scratches and other micro-scale roughness of the solid surface of a rock, even if the true contact angle on an atomically smooth surface is not zero.

1.3.3 Capillary Rise

Before proceeding, we will consider one simple example of the use of the Young-Laplace equation, which we will also derive directly from a consideration of energy balance: capillary rise (Young, 1805; Leverett, 1941). If we have a cylindrical tube of radius r containing two fluids with a contact angle θ, then the radius of curvature of the fluid interface $r_1 = r_2 = r/\cos\theta$. The capillary pressure is given from Eq. (1.6) as

$$P_c = \frac{2\sigma\cos\theta}{r}. \qquad (1.16)$$

If instead we had two parallel plates a distance $2r$ apart, the capillary pressure would be half the value in Eq. (1.16) since there is only curvature in one direction $(r_2 = \infty)$:

$$P_c = \frac{\sigma\cos\theta}{r}. \qquad (1.17)$$

One way to determine contact angle is to measure the capillary rise in a narrow cylindrical capillary tube of known radius; if the interfacial tension is also known, then Eq. (1.16) can be used to find the contact angle: Fig. 1.9.

The capillary pressure is related to the height of the rise of the meniscus in the tube. Pressure, P, in a fluid – with no motion – increases with depth such that $dP/dz = \rho g$, where z is depth, ρ is the density and g is the acceleration due to gravity (9.81 m/s^2). If we have two fluids – usually air and water – in contact across a flat interface (say, in a large beaker) at a depth defined as $z = 0$, then the capillary pressure, or the pressure difference between the air and water, increases with height, $h \equiv -z$ as $P_c = \Delta\rho g h$, where $\Delta\rho = \rho_w - \rho_g$ is the density difference between water, ρ_w, and air, ρ_g (normally the air density is neglected in the analysis). Then, instead of Eq. (1.16), we may write:

$$\Delta\rho g h = \frac{2\sigma\cos\theta}{r}, \qquad (1.18)$$

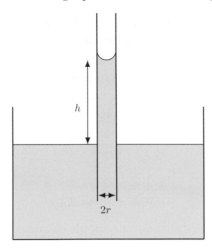

Figure 1.9 A diagram showing a capillary rise experiment. A capillary tube is placed in a beaker of wetting fluid (shaded). The capillary rise, h, can be derived from the Young-Laplace equation or directly from energy balance. If the radius of the tube, r, and the interfacial tension, σ, are known, this is a method to measure contact angle; if the fluid instead is known to be completely wetting ($\theta = 0$), then the experiment can be used to determine the interfacial tension from Eq. (1.18).

or the capillary rise, h is:

$$h = \frac{2\sigma \cos \theta}{\Delta\rho g r}. \tag{1.19}$$

We can derive Eq. (1.19) directly from energy balance and the Young equation (1.7). The wetting phase will rise until it is no longer energetically favourable to go further. Imagine that the wetting phase meniscus resides at a height, h, and consider the change in energy for an infinitesimal change to $h + dh$. The tube radius is r and so there is a change $2\pi r dh$ in the area covered by the wetting phase, corresponding to a difference $2\pi r \Delta\sigma_s dh$ in surface energy (which is energetically favourable). This is counter-balanced by an increase in potential energy, mgh, where m is the mass (or, strictly speaking, the change in mass): in this context it is the change in mass from replacing the slither of air between h and $h + dh$ with the wetting phase: $m = \pi r^2 \Delta\rho dh$. In equilibrium, the two energies balance and

$$2\pi r \Delta\sigma_s dh = \pi r^2 \Delta\rho g h dh. \tag{1.20}$$

Rearranging Eq. (1.20) gives:

$$h = \frac{2\Delta\sigma_s}{\Delta\rho g}; \tag{1.21}$$

then using Eq. (1.14) for $\Delta\sigma_s$ we recover Eq. (1.19) above.

We can extend this treatment of other cases of interest with different pore shapes; however, this rapidly becomes a somewhat dry exercise in geometry. Eq. (1.16) is a useful guide to the relationship between capillary pressure and the radius of a pore. If we have pores of more complex cross-section, the capillary pressure is given, approximately, by Eq. (1.16), but with r now representing the inscribed radius of the cross-section: in the next chapter, once we have introduced the pore space of natural materials, we will show how to extend the concept of energy balance to these cases.

III. *An Essay on the Cohesion of Fluids.*   *By* Thomas Young,
M. D. For. Sec. R. S.

Read December 20, 1804.

1.3.4 Historical Interlude: Thomas Young and the Marquis de Laplace

Eq. (1.6) is called the Young-Laplace equation, after the two physicists, Thomas Young and the Marquis de Laplace, who independently developed the ideas presented here (Young, 1805; Laplace, 1805). Sometimes, confusingly, Eq. (1.6) is called the Laplace equation alone, perhaps as a reaction to some perceived British bias towards Young: after all, he does get another equation to his name: Eq. (1.7). Rather than wade into a fruitless French vs. English historical debate, I will simply note that it is not correct to call this the Laplace equation, since this unnecessarily creates confusion: what everyone else calls the Laplace equation is different ($\nabla^2 \phi = 0$ for some potential ϕ).

The picture above shows the top of the first page of Young (1805) where he describes a series of capillary rise experiments, albeit without either diagrams or equations. Young had impressive physical insight, which he also used to describe elasticity (Young's modulus) and diffraction (Young's double slit experiment); he even got close to deciphering hieroglyphics.

Laplace was a brilliant scientist and mathematician who made significant contributions to potential theory, astronomy and statistics. He was friendly with Napoleon and was his minister of the interior, where he insisted on the implementation of metric (now SI) units (Casado, 2012): we will follow his instructions in this book! He acquired his anachronistic title during the restoration of the Bourbon monarchy.

2

Porous Media and Fluid Displacement

2.1 Pore-Space Images

As mentioned in the Introduction, the discipline of transport in porous media has been transformed by our ability to image, in three dimensions, the pore space of materials at different resolutions, from the nanometre to centimetre scales. Arguably the most versatile instruments use X-rays to construct three-dimensional images at micron resolution. The principles and methodology used to construct an image are similar to that used in CT scanning for medical examinations, and indeed adapted medical scanners are also used routinely to scan rock cores that are a few cm in diameter and 1–2 m long (similar in size, or at least length, to the patients the scanner was designed for). However, in these cases the resolution is around 1 mm, which is insufficient to see, directly, the pore spaces of most rocks. The limitation though is not the wavelength of the X-rays themselves: typical X-ray energies are in the range 10–160 keV – with corresponding wavelengths 0.1–0.01 nm. To obtain higher-resolution images, it is necessary to scan a smaller sample which is placed close to the X-ray source.

The first micron-resolution images of porous rock were obtained by Flannery et al. (1987) using both X-rays from a synchrotron (X-rays are emitted from electrons moving at almost the speed of light as they are accelerated around a ring by strong magnets) and a bench-top instrument with its own X-ray source (here electrons are accelerated and impact on a metal target that then emits X-rays). Since then the technology both to acquire and process these images has improved enormously; most major companies and many universities now have good laboratory instruments to produce three-dimensional micron-resolution images, combined, in many cases, with access to central synchrotron facilities. We will make use of this technology to illustrate the concepts in this book. For reference, Fig. 2.1 shows schematics of the types of apparatus used for X-ray tomography. However, a discussion of imaging and image analysis, or how the image is taken and how it is

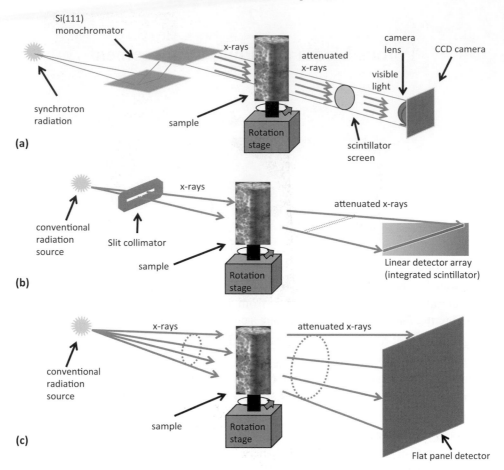

Figure 2.1 Schematic of the apparatus used for X-ray tomography. (a) The configuration of source, sample and receiver (the CCD camera) at a synchrotron beam-line. (b) The apparatus for a micro-CT system with a fan-beam. (c) A micro-CT apparatus with a cone-beam configuration. From Wildenschild and Sheppard (2013).

processed, lies outside the scope of this work: readers are referred to the excellent scholarly reviews of this topic by Cnudde and Boone (2013), Wildenschild and Sheppard (2013) and Schlüter et al. (2014); here it will be assumed that we can acquire a representative image of the pore space.

X-ray scanning is useful since there is a clear contrast between the rock, which absorbs X-rays strongly, and the fluids, which are more X-ray transparent. This allows us to distinguish between solid rock and pore space readily and, with careful experimental design, between the fluid phases as well. As examples, Fig. 2.2 shows a selection of images which show both porous media themselves and the chemical, biological and flow processes occurring within them. Here the greyscale, which

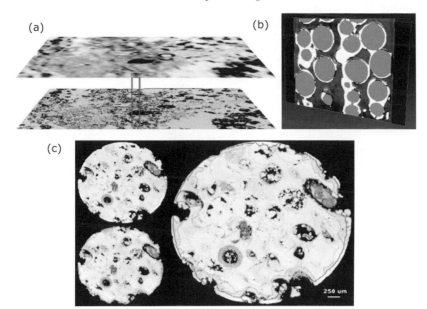

Figure 2.2 X-ray images of different porous materials. (a) The use of images at multiple scales. An example of the mapping of micro-porosity from a registered slice of a three-dimensional micro-CT image (2.85 μm voxel size; upper region) to higher resolution information (0.25 μm pixels) in a scanning electron microscope image (lower region). (b) A two-dimensional slice through a three-dimensional image of a bead pack showing a biofilm (white), solids (light grey) and water (darker grey). (c) Images of reactive transport in a carbonate rock; initial state (upper left) and final state (lower left) after 48 hours in the presence of CO_2 at 1 MPa pressure. The right figure shows the final state at higher resolution (2.8 μm voxel size) indicating the dissolution of micro-porosity. From Wildenschild and Sheppard (2013).

represents the local absorption coefficient of the incident radiation (determined largely by the atomic number of the material), is sufficient to discern the structure. The resolution is determined by the sample size, beam quality and the detector specifications; for cone-beam set-ups (in laboratory-based instruments), resolution is also controlled by the proximity of the sample to the beam.

Current micro-CT scanners will produce images of around $1,000^3$–$2,000^3$ voxels. The rock samples are normally a few mm across, constraining the resolution to a few microns; sub-micron resolution, nano-CT, is possible using a specially designed instrument and smaller samples. Some scanners can now acquire $3,000^3$ images with further improvement likely in the future, but at present most have an approximately 1,000-fold range from resolution to sample size. Recent developments have enabled both rock and fluids to be imaged, and at high temperatures and pressures reproducing conditions deep underground (Iglauer et al., 2011; Silin et al., 2011; Andrew et al., 2013; Wildenschild and Sheppard, 2013). A selection of images, with associated analysis, can be found online, for example Prodanović (2016) and Blunt and Bijeljic (2016).

We show exemplar images of sandstones and carbonates in Fig. 2.3. Some of these images will be used as the basis of calculations later in the book, but are shown now simply for illustrative purposes. It can be seen that the pore space is highly irregular, with rough surfaces and tortuous pathways through the rock.

Our focus will be on fluid flow in the pores, and to emphasize this, Fig. 2.4 shows just the pore space of three of the samples shown previously in Fig. 2.3.

To contrast different rock types, Fig. 2.5 shows the structure of a predominantly quartz sandstone (Doddington) with relatively large pores, while Fig. 2.6 shows details of the structure of Ketton limestone with pores both between grains and within the grains themselves (micro-porosity). This is a higher-resolution image than in Fig. 2.3, displaying further detail of the individual grains, with the cement and grain fragments between them.

Electron microscopy techniques can also be used to image samples both in two dimensions (using standard methods; see Fig. 2.2(a)) and in three dimensions through taking serial cuts through a sample (called FIB-SEM for focussed ion beam scanning electron microscopy, or BIB-SEM, where BIB stands for back-scattered ion beam). Here voxel sizes as small as 5 nm can be achieved; this resolution is necessary to observe the full range of micro-porosity in carbonates and to image the pore space and organic material in shales. These methods, however, destroy the sample and therefore cannot be used in conjunction with dynamic flow experiments.

Fig. 2.7 shows further detail of Ketton limestone taken using scanning electron microscopy (SEM) showing that the individual grains in Fig. 2.6 are micro-porous at the sub-micron scale with different texture in different parts of the rock fabric.

The state of the art in multi-scale imaging is illustrated in Figs. 2.8 and 2.9. In Fig. 2.8, X-ray tomography can resolve the pore space of a tight sandstone down to the nm scale to detect the interstices between platelets in the inter-granular clay. In Fig. 2.9, a two-dimensional back-scattered electron microscopy (BSE) image of a shale is shown with 10^{10} pixels: regions of the picture can be selected for three-dimensional imaging at the nm scale using a FIB-SEM technique, showing the pores within the organic material. Also shown is a QEMSCAN (Quantitative Evaluation of Minerals by SCANning electron microscopy) that can be used to identify the mineralogy of the sample. In both these figures, it is the nm-scale pores that control the flow properties, yet there is significant structure up to the mm or cm scales, requiring images that encompass six or seven orders of magnitude in resolution. A vital component of such a multi-scale analysis is the ability to register images taken using various techniques at different resolutions in three dimensions (Latham et al., 2008), shown in Fig. 2.2(a).

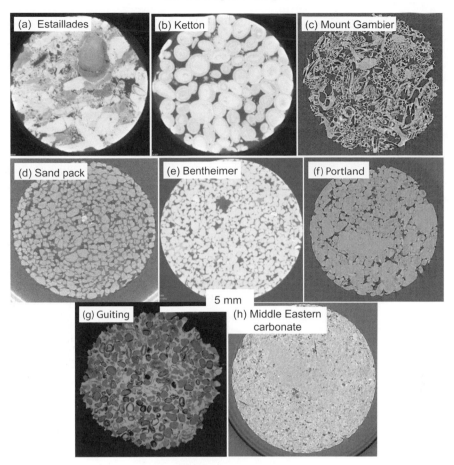

Figure 2.3 Two-dimensional cross-sections of three-dimensional micro-CT images of different samples. The pore space is shown dark. (a) Estaillades carbonate. The pore space is highly irregular with unresolved micro-porosity. (b) Ketton limestone, an oolitic quarry limestone of Jurassic age. The grains are smooth spheres with large pore spaces. The grains themselves contain micro-pores that are not resolved. (c) Mount Gambier limestone is of Oligocene age from Australia. This is a high-porosity, high-permeability sample with an open pore space. (d) A sand pack of angular grains. (e) Bentheimer sandstone, a quarry stone used in buildings. (f) Portland limestone. This is another oolitic limestone of Jurassic age that is well-cemented with some shell fragments. Portland is another building material used, for instance, in the Royal of School of Mines at Imperial College where the author works. (g) Guiting is another Jurassic limestone, but the pore space contains many more shell fragments with evidence of dissolution and precipitation. (h) Carbonate from a deep highly saline Middle Eastern aquifer. From Blunt et al. (2013).

(a) Estaillades   (b) Ketton

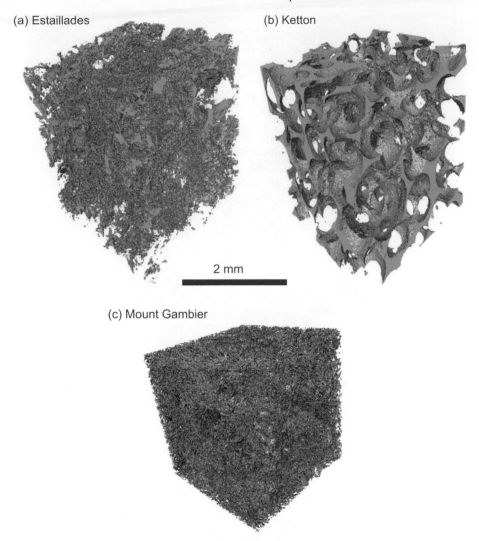

(c) Mount Gambier

Figure 2.4 Pore-space images of three quarry carbonates shown in Fig. 2.3: (a) Estaillades; (b) Ketton; (c) Mount Gambier. The images shown in cross-section in Fig. 2.3(a)–(c) have been binarized into pore and grain. A central $1,000^3$ (Estaillades and Ketton) or 350^3 (Mount Gambier) section has been extracted. The images show only the pore space. From Blunt et al. (2013).

2.1.1 Statistical and Process-Based Pore-Space Reconstruction

To complement direct imaging of rock samples, statistical methods have been developed to generate representations of the pore space which capture key features of the porous medium, such as porosity and connectivity. The advantage of this approach is that it does not suffer from any limitations, in theory, on resolu-

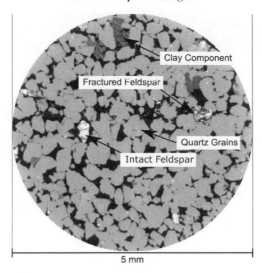

Figure 2.5 A two-dimensional cross-section of the three-dimensional micro-CT image of Doddington sandstone identifying the principal mineralogical components. Doddington is mainly composed of large quartz grains with a small amount of clay. From Andrew (2014).

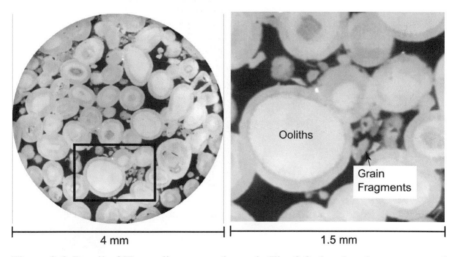

Figure 2.6 Detail of Ketton limestone shown in Fig. 2.3 showing the structure of the grains and the grain fragments between them, indicating the irregularity of the pore space. From Andrew (2014).

tion or size. Moreover, many realizations of the pore space can be created to study variations in properties.

There are two types of approach to the reconstruction of rock images. The first simulates explicitly a packing of grains combined with compaction and cementation. Such process-based modelling attempts to mimic the sedimentary and

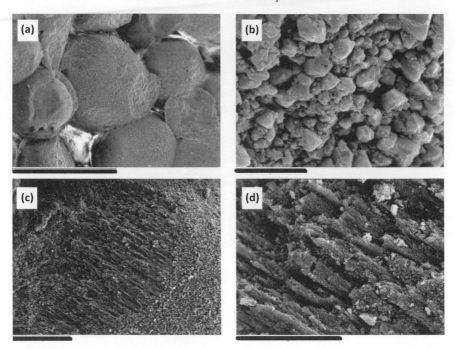

Figure 2.7 Scanning electron microscope images of Ketton limestone showing that the grains illustrated in Fig. 2.6 are micro-porous with different textures. The dark black lines are scale bars representing lengths of (a) 800 μm, (b) 10 μm, (c) 100 μm and (d) 30 μm. From Andrew (2014).

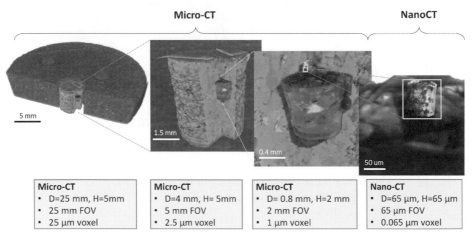

Figure 2.8 An illustration of imaging across multiple scales. Here a tight sandstone with clay cement is imaged at both the core (cm) scale and at sub-micron resolution to resolve the pore space between kaolinite particles. FOV stands for field of view, while D and H refer to the diameter and height of the samples, respectively. Adapted from Roth et al. (2016). (A black and white version of this figure will appear in some formats. For the colour version, please refer to the plate section.)

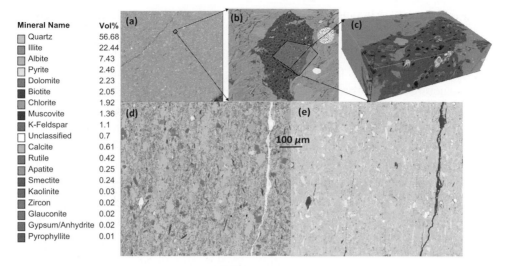

Mineral Name	Vol%
Quartz	56.68
Illite	22.44
Albite	7.43
Pyrite	2.46
Dolomite	2.23
Biotite	2.05
Chlorite	1.92
Muscovite	1.36
K-Feldspar	1.1
Unclassified	0.7
Calcite	0.61
Rutile	0.42
Apatite	0.25
Smectite	0.24
Kaolinite	0.03
Zircon	0.02
Glauconite	0.02
Gypsum/Anhydrite	0.02
Pyrophyllite	0.01

Figure 2.9 Images of a shale gas sample from the Sichuan Basin China show-ing the state of the art in nm-scale imaging. (a) A back-scattered electron microscopy (BSE) image approximately 1 mm^2 in area with a pixel size of 10 nm: there are 10^{10} pixels in the image. (b) A magnified portion of the image of size 34.2×29.5 μm showing organic material (dark). (c) A three-dimensional FIB-SEM image showing nano-pores within the organic matter. Each voxel is approximately 6 nm cubed; the physical size is $8.8 \times 5.6 \times 2.5$ μm. (d) A QEM-SCAN (Quantitative Evaluation of Minerals by SCANning electron microscopy) image showing details of the mineralogy using the key on the left. The image has an area of 699×636.9 μm with a pixel size of 1 μm. (e) The corresponding BSE image. Courtesy of iRock Technologies, Beijing. (A black and white version of this figure will appear in some formats. For the colour version, please refer to the plate section.)

diagenetic processes by which the rock was formed. This technique was employed by Bakke and Øren (1997) to reconstruct sandstones and compute flow proper-ties in the pore space using a grain size distribution inferred from the analysis of good-quality two-dimensional thin section images. It is also possible to benchmark and tune the method to capture the structure in three-dimensional images as well (Hilfer and Zauner, 2011).

Subsequent developments have allowed the packings of grains of different size and shape to be simulated (see, for instance, Guises et al. (2009)), mimicking a wide range of sedimentary rocks, including different types of intergranular poros-ity, cement and clays (Øren and Bakke, 2002). The method can be extended to carbonates, capturing the wide range of length scales between the largest pores (vugs) and micro-porosity (Biswal et al., 2007, 2009). Predictions of flow prop-erties using this approach are close to those computed directly on an image of the same rock (Øren et al., 2007). In theory, these models have infinite resolution, since they are composed of components of known size and shape. However, normally the resultant packing is then discretized to allow analysis of the pore structure and to perform flow simulations.

The second approach uses statistical methods to generate a discretized image directly, borrowing ideas from geostatistics used to create realizations of the field-scale structure and flow properties of a reservoir (Strebelle, 2002; Caers, 2005). The simplest methods reproduce the single- and two-point correlation function of the pore space, determined from the analysis of benchmark (or training) images (where an indicator value of 0 is given to the void and 1 to the solid). The training image is a high-quality two-dimensional thin section from which a three-dimensional representation of the pore space is constructed. This was first applied by Adler (1990) using optical thin sections of Fontainebleau sandstone to determine correlation functions that were then reproduced in a three-dimensional reconstruction of the pore space. However, the long-range connectivity of the pore space that controls the flow properties is not well represented by low-order correlation functions derived from two-dimensional cuts through the pore space.

The desire to preserve connectivity has prompted the use of other statistical measures, such as the path lengths, or the probability that the voids span a region of space (Hilfer, 1991; Coker et al., 1996; Hazlett, 1997; Manwart et al., 2000). This may be combined with the reproduction of more complex patterns delineating typical pore shapes (pore architecture models, Wu et al. (2006)) or multiple-point statistics to capture representative patterns in the void geometry (Strebelle, 2002; Okabe and Blunt, 2004, 2005). In principle, any information can be introduced into a statistical reconstruction to preserve essential geometric and flow properties (Yeong and Torquato, 1998). More importantly, three-dimensional benchmark images can now be used to condition statistical reconstructions (Latief et al., 2010).

A fuller discussion of these approaches, and hybrid methods that use a combination of statistical and object-based reconstruction, are described in detail elsewhere (Adler, 2013). We will use both reconstructed and directly imaged pore-space models as examples in our subsequent description of displacement and flow.

Many reconstruction algorithms have been tested on Fontainebleau sandstone, a widely used exemplar of a sandstone comprised of grains of approximately equal size. Example images are shown in Fig. 2.10. The three-dimensional micro-CT image shows clearly the grains and the connected pore space (Biswal et al., 1999). For comparison, a synthetic representation generated using a model of grain packing is shown: while the angular nature of some of the grains may be lost, the connectivity of the pore space is preserved. Also shown are images generated using statistical techniques. Here more complex pore shapes can be reproduced, but the connectivity of the pore space is not necessarily preserved.

Fig. 2.11 shows the largest publicly available synthetic image generated to date, based on a continuum model of grain packing and diagenesis in Fontainebleau that

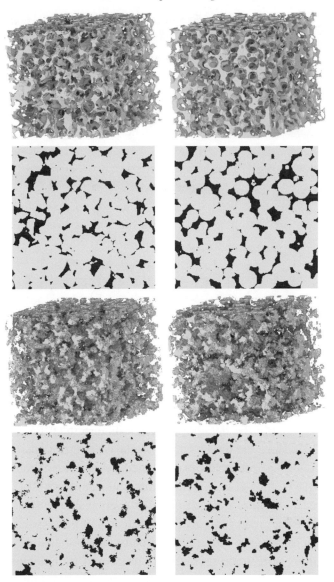

Figure 2.10 Top left: a three-dimensional micro-CT image of Fontainebleau sandstone with 300^3 voxels of size 7.5 μm. Shown below it is a two-dimensional cross-section showing the pore (black) and grains (grey). Top right is a synthetic image generated using a process-based model of the sedimentation of spherical grains followed by diagenesis. While the grain shapes are smoother than in reality, the connectivity of the pore space is preserved. In the lower figures, two statistical techniques are used to capture the spatial correlation of the void geometry in the measured image. While complex pore and grain shapes can be generated, the image has a poorer connectivity. From Biswal et al. (1999).

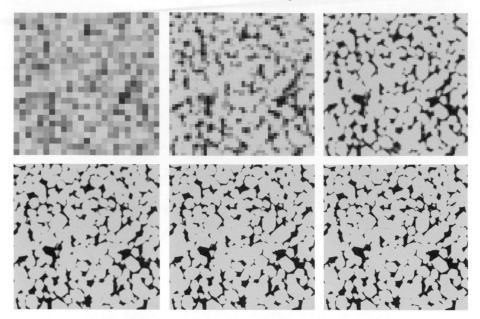

Figure 2.11 Two-dimensional cross-sections of synthetic three-dimensional images of Fontainebleau sandstone of size 3 mm square. The resolution of the pictures (from top left to lower right) is 120 μm, 60 μm, 30 μm, 15 μm, 7.5 μm and 3.75 μm. Images of a cube of size 1.5 cm down to a resolution of 0.458 μm are available to download from Hilfer et al. (2015b). Figure from Latief et al. (2010).

captures more accurately the details of grain structure than the earlier work presented in Fig. 2.10. Here the total system size is 1.5 cm, described at a resolution down to 458 nm, or $32,768^3$ grid blocks – well beyond the capability of any one imaging technique (Latief et al., 2010; Hilfer and Zauner, 2011). However, such detail is required to capture the geometry of even a simple sandstone: as illustrated already, an even greater range of length scales of pore-space description are required for most carbonates with micro-porosity, or for shales.

Other benchmark pore-scale representations of rock samples have been introduced in the literature, using either particle-based simulation, direct imaging or statistical methods. For instance, Andrä et al. (2013a) have presented four exemplar systems: Fontainebleu sandstone, Berea sandstone, a carbonate and a sphere pack. These were used to test different methods to segment images into pore and grain, and to assess the predictive power of the resultant pore-scale models (Andrä et al., 2013b); experimental and modelling results using these rocks will be presented later.

Pore-space reconstruction does allow some of the limitations of direct imaging to be overcome, but even so this does not address how then to quantify flow and transport in these models. This requires some degree of simplification of the void geometry and connectivity, presented below, once some important, if basic, definitions have been introduced.

2.1.2 Definition of a Porous Medium, Representative Volumes, Porosity and Saturation

Having mentioned porous media, it is necessary to give the term a precise definition. A porous medium is any substance that contains both a solid matrix and void space: the pores. The pores may contain fluid phases (gas or liquid). The porosity, ϕ, is the fraction of the total volume of the porous medium that is pore space. The total volume is that of both the solid and voids: we never talk of the volume of a porous medium being just the contribution from the solid portion. Note the symbol used: it is a Greek phi (f), ϕ, not a variety of Norwegian O! And while we are being pedantic: we have one porous medium (medium is singular) but two or more porous media (media is the irregular plural of medium).

We can distinguish two types of porosity: the total porosity (as already defined) and the effective or connected porosity, ϕ_e (the fraction of the volume of the porous medium containing void space that connects across the system). For most practical applications we are interested in the effective porosity when considering flow: pore spaces that are disconnected cannot flow and the fluids in them cannot be displaced.

The total porosity need not necessarily be the porosity apparent from, for instance, a pore-space image. As an example, in Fig. 2.6 approximately half the void space is visible in a micron-resolution image of Ketton limestone, comprising voids between the grains or ooliths, while the other half is micro-porosity or submicron sized pores within the grains themselves, invisible to micro-CT scanning but revealed in higher-resolution images, such as Fig. 2.7.

In the remainder of this book, it will be assumed that the effective and total porosities are the same. This is a reasonable assumption for most geological media, where the processes by which the rock is formed are a combination of sedimentation (deposition of grains) with cementation and diagenesis (changes in pore structure or composition, normally mediated by chemical processes where the reactants are transported in an aqueous phase in the pore space). In all cases, the pore space can only become blocked via the transport of reactants in the connected porosity, and by logical deduction it is difficult to conceive, in all but exceptional cases, how a pore can become completely disconnected as a consequence.

This line of reasoning is similar to the problem of a blocked sink: it never completely clogs; instead the flow declines asymptotically to zero. The reason for this is similar: there has to be some flow to transport more debris to block the pipes further; as the flow stops, so does the rate of clogging. In any event, most measurements of porosity rely on the pore space being invaded by another fluid and so, axiomatically, only measure the connected porosity.

The discussion above leads to the definition of a permeable medium. A permeable medium is a porous medium with connected porosity ($\phi_e > 0$) which will, as a consequence, allow flow through it. Again it will be assumed that all the porous media considered are also permeable. Sometimes a permeable medium is defined as one that allows significant flow, classifying some rocks with sufficiently small or poorly connected pores as being impermeable. I will not entertain this type of discussion, as it hinges on an arbitrary definition of what significant means. Indeed, it is a dangerous trap: shales, containing pores of nanometre size, and with a very low porosity, were traditionally classed as being impermeable, but now economic quantities of oil and gas are recovered from those shales containing organic material (the source rock for oil and gas fields). While hydraulic fracturing (or fracking, a term, incidentally, which was never in common use in the oil industry) permits large-scale production, the shale oil and gas still need to flow through the small pores spaces to the fractures: the properties of the porous medium have not changed; it was the previous categorization that was misleading.

The next definition is saturation, S: the fraction of the porosity occupied by a given phase. The sum of all the phase saturations is, by definition, 1. A fully saturated porous medium is one where there is only a single fluid phase (usually the aqueous phase or water) present, with saturation 1. For convenience, in the following, we will talk about water, oil and gas phases with the subscripts w, o and g, respectively: $S_w + S_o + S_g = 1$. However, strictly speaking, we mean an aqueous phase for water: a phase chemically predominately composed of water, but also containing other dissolved materials, often salts and hydrocarbon pollutants. For oil, we normally (but not always) refer to a liquid hydrocarbon-rich phase, containing a mixture of chemical components. For gas, we refer to a gaseous phase that may also be comprised of a mixture of components. When we consider systems at high temperatures and pressures, we may also treat super-critical gases (phases that, with smooth changes in temperature and/or pressure, can transform to a distinct liquid or gas phase, but without phase change): an example will be carbon dioxide, CO_2, injected into deep saline aquifers.

For a spatially dependent variable f defined in the pore space, an averaged quantity $F \equiv \bar{f}$ over a volume V can be determined from

$$F = \frac{1}{V} \int f \, dV, \tag{2.1}$$

where the integral is over both the void and solid (where $f = 0$). For porosity, $F \equiv \phi$ and $f = 1$ in the pore space; for the saturation of phase p, $F \equiv \phi S_p$ and $f = 1$ where p is present in the void.

Implicit in any description of the pore space – from an image or a statistical reconstruction – is the hypothesis that the averaged properties that we measure or compute are representative of any similar-sized or larger region of the same rock. Imagine that we computed F in Eq. (2.1) for different-sized volumes centred on the same location. If the averaging volume is, for instance, contained within a single grain, $F = 0$ and is not representative of the porous medium as a whole: it is necessary to consider a volume that encompasses several pores. The representative elementary volume, REV, is the minimum volume needed to obtain an accurate computation for an averaged quantity: while it is easy to define and appreciate conceptually, it is difficult to define precisely, since there is no unambiguous assessment of what is meant by sufficient accuracy, while the REV depends both on the rock itself and the quantity being averaged. For the statistical pore-space descriptions mentioned above, the size of the REV may be estimated from the correlation length of the local porosity or the connected path length. In the quarry samples we study, a volume that spans a few grains is generally sufficient to define porosity, whereas a larger volume is necessary for saturation, since this is controlled by the dynamics of the process by which fluids are displaced, as discussed in subsequent chapters. For flow properties, particularly when multiple phases are present, the REV is even larger, since now it is dependent on the connectivity of the pore space and the fluids within it.

As an example, Fig. 2.12 shows the average porosity, calculated using Eq. (2.1), on the statistical image of Fontainebleau sandstone shown in Fig. 2.11. The porosity has a value determined to within less than 0.5% when averaged over a length of around 3 mm or more (a volume of 27 mm^3), or the size of the pictures in Fig. 2.11: however, this is a simple example on a uniform sandstone — the REV may be larger for more complex media, such as carbonates, or for other properties.

The REV is a central concept in the treatment of flow in porous media, since we switch from quantities defined within the pore space – often Boolean indicators, such as the presence of void, or a particular phase – to averaged properties, where the average encompasses a region of the porous medium containing both solid and void. However, this important idea will only be mentioned briefly here, as it is discussed in detail elsewhere; see, for instance, Bear (1972). In order to proceed without undue concern over how we can assign macroscopic quantities, as a general rule, it is necessary to have a sample that is around 10–20 pores across to provide a reasonable average: how we define a pore is described in the following section.

One other, related, concept is that of heterogeneity. Let's assume that we have a well-defined REV. We can define $F(\mathbf{x})$ as an averaged quantity over the REV centred at location $\mathbf{x}$. A heterogeneous system is one where F is a function of location. In the quarry samples we have presented, macroscopic quantities tend to be similar between samples of the same rock type (F is approximately constant

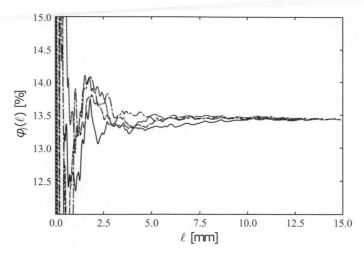

Figure 2.12 The porosity averaged over cubes of different length l using Eq. (2.1) for the statistical realization of Fontainebleau sandstone shown in Fig. 2.11: the lines are for averages centred on different parts of the image. In this example the representative elementary volume for porosity is a few mm in extent. From Hilfer and Lemmer (2015).

and the medium is homogeneous); however, for reservoir samples, we may see very strong variations in properties over all length scales, from the pore upwards, dependent on the sedimentary and diagenetic history of the system. This somewhat complicates the definition of averaged quantities, since F may contain significant spatial variation over lengths comparable with the averaging volume for the REV.

2.2 Pore-Scale Networks and Topological Description

This book concerns flow and transport in porous media, or how fluids move. In theory, if we have an image of the pore space and a determination of contact angle we can, using the Young-Laplace equation (1.6), calculate configurations of fluid equilibrium and – as we describe later – changes in fluid configurations as one phase displaces another. Indeed, direct methods to track fluid interfaces in a sequence of positions of capillary equilibrium have been developed, of which the most widely used is the level set method (see, for instance, Sussman et al. (1994); Spelt (2005); Prodanović and Bryant (2006); Jettestuen et al. (2013)). These approaches are useful to elucidate displacement processes at the pore scale. However, in complex cases, this would require images resolved down to the molecular scale with an equivalently high-resolution analysis of the balance of interfacial forces to specify contact angle. In reality, this is impossible and, in any event, unnecessary. We need only to capture the essence of the pore-space geometry and displacement physics to understand, interpret and predict displacement.

We will construct a simplified geometrical representation of the pore space which preserves the relevant geometric information needed to understand flow and transport. To motivate this approach we will first present an analogy.

2.2.1 Transport Networks

Imagine that we acquired a high-resolution shallow seismic image of the ground below London, which we could interpret and segment to identify all the underground tunnels. This is our analogy to the pore-space images discussed above. Now imagine that someone wants to plan a journey by underground (the tube or subway), from, for instance, Heathrow Airport to visit the author at Imperial College London. South Kensington is the nearest station to Imperial College. Of course, you would study the famous map of the London Underground, Fig. 2.13. Apart from a representation of the River Thames (for orientation) the remarkable feature of the map is that it is entirely topological, in that it represents the train lines and their connections without geographic reference to where the lines and stations are in real space. This makes the map very easy to read and is, for the specific purposes of planning a journey, much more useful than a standard geographically accurate map, or indeed our subsurface image. It would be ludicrous to suggest that the seismic image should be used instead because it is more accurate and that the tube map is useless because it is less detailed. Of course, in theory the

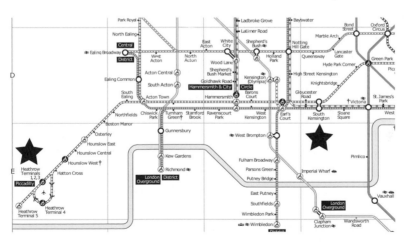

Figure 2.13 A section of the famous London Underground map. This is a topological representation of the train lines useful for planning journeys across the city: specifically we consider travel from Heathrow airport to South Kensington (shown by the stars). By analogy, we will use a network model representation to help understand and interpret transport in complex pore spaces: the pores are stations (or junctions), while throats are the lines between them. Adapted from https://tfl.gov.uk/maps/track/tube.

seismic image could be used to construct the tube map (with some difficulty and additional information to identify train lines). The seismic image does indeed contain more information than the map, but not, for this example, in a manner that is particularly useful. On the other hand, were we to design trains for the underground and needed to ensure that the carriages were not too wide to fit in the tunnels and platforms, the seismic map would be extremely valuable and the tube map largely useless.

The map analogy can be taken a step further, to appreciate the wide range scales present in natural porous media. In my transport example, a tube map alone is insufficient: you will need a more detailed map to direct you from the underground station to Imperial College, and once at Imperial to the author's office. Again, the obvious approach is intuitive when it comes to personal transport, and the same concepts can – and should – be extended for fluid transport as well: you use a street map of London, or more specifically a local map of South Kensington, combined with a map of Imperial's central campus. For each map – underground, street and campus – we resolve finer scales of detail. The same approach is pertinent for porous media, with images and analysis required at different scales to understand flow and displacement.

As evident in Figs. 2.7, 2.8 and 2.9, for instance, in a carbonate, sandstone or shale, large pores or vugs may be present which are 100s μm across, combined with micro-porosity, clays or organic material with structure down to 0.01 μm. Let us take a representative sample, which is, for instance, around 1 cm across (this is the smallest scale on which traditional flow and displacement experiments are performed). Hence, to resolve the smallest pore spaces in even this small rock sample requires an analysis over six orders of magnitude in length (from 10^{-8} to 10^{-2} m). To revert to the analogy of the visitor to London, to locate a single office requires a map with a resolution of a few metres: if we now increase this by six orders of magnitude we arrive at 1,000s km. So our challenge in porous media is to understand flow and transport over a range of lengths equivalent to a visitor flying to London from, say, Chicago, and then locating a single room in the city. But for the visitor the approach is straightforward: a world map showing flight routes, a map of the London underground, a street map of a local area of London and a map of the university. We need to do the same for porous media, using a conceptualization of the pore space at multiple scales as required for different applications. What we cannot achieve, either in terms of images or associated computation, is one super-image at nanometre resolution that captures every single feature in a sample centimetres across. Even if this were possible technologically (and it is not), it would be as absurd as carrying a room-by-room resolution map of the whole world for a visit to London. Now, while much of this information may be available (say with Google images) we would still need additional analysis and interpretation to use it sensibly.

The problem we face with porous media is, however, considerably more difficult than our map-reading analogy. At present there is only one planet that we need to navigate; however, there are lots of rock samples that we may wish to study. For porous media, six orders of magnitude in scale simply allows us to describe one centimetre-sized core sample. To represent flows at the field scale – or tens of kilometres – requires another million-fold change in size, and, needless to say, the rock is typically heterogeneous at all these scales. This problem of determining large-scale flow from measurements (or predictions) on smaller rock samples, traditionally called upscaling in reservoir engineering, is beyond the scope of this book, although we will later use pore-scale information to interpret field-scale recovery. Furthermore, this is only for one field, and we would like, in principle, to develop a predictive description of flow applicable to any macroscopic displacement. Then last, we need to consider the dimensionality of the problem. Broadly speaking, navigating (at least at ground level) is a two-dimensional problem: a map with six orders of magnitude between the overall size and the smallest feature resolved will require $10^6 \times 10^6 = 10^{12}$ pixels. Our rock images are three-dimensional: a million-fold change in scale requires $10^6 \times 10^6 \times 10^6 = 10^{18}$ voxels. Finding your way around London is much easier than moving through the pore space of a reservoir rock!

The message I would like to convey is that it is useful to derive an abstraction of an image that is pertinent for a particular application. For most flow in porous media problems, especially those associated with hydrocarbon reservoir engineering, their sheer boggling complexity often leads engineers to presume that more detail must be better, or more accurate, even if this comes at the cost of understanding. Instead it is more illuminating to develop a simplification of the problem which allows it to be appreciated, understood and interpreted. This simplification needs to retain information about the connectivity of the pore space, as well as geometric information that will allow an accurate estimation of capillary pressure for different fluid distributions.

An example of such a simplification is a network for Berea sandstone shown in Fig. 2.14. To make the link between the pore space and the network even more explicit, Fig. 2.15 shows a two-dimensional cross-section of a three-dimensional image of Bentheimer sandstone with the network superimposed on the pore space. Bentheimer will serve as an exemplar throughout this book. This concept of an equivalent network will be developed further when we describe pore-scale displacement and flow.

2.2.2 Network Construction

For porous media applications, we conceive of the void space as being composed of wide regions, the pores, that are connected together by narrower regions, the

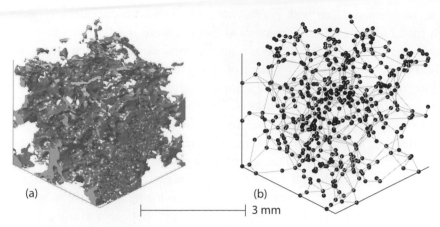

Figure 2.14 On the right is a network model representation of the pore-space on the left. The synthetic pore-space image is of a Berea sandstone 3 mm across with a voxel size of approximately 3 μm generated using a process-based reconstruction of grain packing, cementation and compaction (Bakke and Øren, 1997). The network model is an approximate topological representation of the pore space, indicating what wide voids (pores) are connected to which others through narrower restrictions (throats). It bears the same relationship as the London Underground map does to a real map (or subsurface image); see Fig. 2.13. The pores and throats are shown as balls and sticks, respectively, for illustrative purposes only – they can have assigned to them volumes, inscribed radii, shapes, flow conductances, capillary pressures and other properties based on the underlying image. Here the pores are located at the correct physical locations from the image. From Valvatne and Blunt (2004).

throats. This network model representation of a porous medium was pioneered by Fatt (1956a, b, c); at that time as he was unable to compute flow properties numerically, so he constructed a physical network of electrical resistors instead. The current represented flow, while the voltage difference was equivalent to the pressure drop. Subsequently, networks have been used principally to understand and explain pore-scale displacement processes using idealized lattices that were not necessarily based on a real rock structure. However, in principle, and in practice, the same approach can be used to predict properties of specific rocks if we start from an accurate representation of the pore space.

A network can be constructed from an underlying image, or from a statistical realization of the pore space of a rock. The principal approaches that have been used will be explained briefly, which should help to understand the concepts, before providing a precise definition of what a network is, based on modern methods of image analysis. This is not intended as a detailed literature review, but is provided to explain the idea of a network model from different perspectives.

One ambiguity with the determination of a network for a particular rock is that it has is no universally accepted, mathematically rigorous definition. If we return

Figure 2.15 A two-dimensional cross-section of a three-dimensional image of Bentheimer sandstone. In the pore space (shown dark) we have superimposed the pores (balls) and throats (cylinders) of the equivalent network. This illustrates how a network model captures the connectivity of the pore space. From Bultreys et al. (2016b). (A black and white version of this figure will appear in some formats. For the colour version, please refer to the plate section.)

to our transport analogy, it is intuitively obvious that a good map of the underground shows all the train lines, indicates junctions and clearly shows connections between lines. It is related to the topology of the subsurface arrangement of tunnels, but is not designed to capture this precisely. The same pragmatic approach is needed for porous media: we do not necessarily need to capture precisely particular topological or geometric features of the pore space; instead we simply require a characterization which will allow us to describe the key features of flow and fluid displacement accurately.

The first representation of a disordered porous medium was provided by Finney (1970). In his work, almost eight thousand ball bearings were massaged together in a weather balloon to create a random close packing of equally sized spheres. The centres of these spheres were then painstakingly measured and different properties of the packing were analysed. This provided a description of a porous medium created experimentally, before numerical methods to simulate such a process were available. Those regions of space furthest from the centres of the ball bearings (representing grains) define a pore, while the more constricted voids connecting these pores are throats. This concept can be more precisely defined using a Voronoi tessellation of the pore space. A Voronoi polyhedron is centred on each grain centre and is defined as the region of space closer to a particular grain centre than any other. The vertices of these polyhedra are the pores – they are regions equidistant from two or more grain centres – while the edges are the throats. For a random close packing of spheres, each pore is connected to four others: this is called the coordination number.

Bryant and co-workers (Bryant and Blunt (1992); Bryant et al. (1993a, b)) used this tessellation to define the structure of the pore space and thence to calculate flow properties. They also mimicked compaction of a rock by moving the grain centres closer together in one direction, and diagenesis (the precipitation of solid material) by swelling the grains. In both cases the grains were allowed to inter-penetrate. This process shrank the pores and throats and, as compaction and diagenesis proceeded, filled some of the connections between pores, reducing the coordination number. This model successfully represented rocks that are composed, to a reasonable approximation, of spherical grains of equal size, such as Fontainebleau sandstone, Fig. 2.10, and was the first time that predictions of flow and transport properties could be made on a realistic rock structure: later in the book we will present some of the main results when we discuss flow. However, in general, rocks have a more complex sedimentary history and pore structure, making the wider application of this method difficult.

The pioneering research of Bakke and Øren (1997) and Øren et al. (1998) extended this approach using a numerical model of the sedimentary and diagenetic processes by which the rock was formed, as discussed previously in Chapter 2.1.1. Rather than use a direct image, or the Finney packing, a numerical model of the settling, compaction and cementation of grains was developed. The identification of pores and throats was conceptually similar to the work of Bryant et al. (1993a): the centres of the grains are known, while points furthest from the grain centres are likely to represent pores. From this a physically based network can be constructed. This approach allowed a variety of networks to be generated representing different rocks from which successful predictions were made.

The network shown in Fig. 2.14 for Berea sandstone was generated using this process-based method. The pores and throats themselves are given angular cross-sections, derived from the representation of the rock itself. The cross-sections have the same inscribed radius r and shape factor G, the ratio of the cross-sectional area A to the square of the perimeter p (Mason and Morrow, 1991)

$$G = \frac{A}{p^2}. \tag{2.2}$$

This concept is illustrated in Fig. 2.16, where different idealized shapes based on a real cross-section are shown. Most of the pores and throats had an irregular triangular shape. Preserving the inscribed radius allows accurate calculations of threshold capillary pressures, while the corners account for the wetting phase that resides within them while non-wetting phase occupies the centres.

However, this technique is still limited to circumstances where the formation of the rock itself can be simulated, or at least where a quantification of the grain shapes

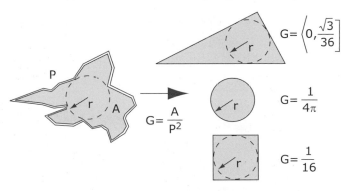

Figure 2.16 Different pore and throat shapes used in network modelling. The figure on the left represents the real cross-section derived from an image or other representation of the pore space. This is idealized by a simple shape that has the same inscribed radius r to allow accurate computation of threshold capillary pressures. The shape also has the same shape factor G, which accounts for the wetting phase retained in the corners when the non-wetting phase occupies the centre of the pore or throat. Most pores and throats generated using this method have cross-sections that are scalene triangles. From Valvatne and Blunt (2004).

and sizes is available. It is normally assumed that the medium has a consolidated granular structure, which may not be an accurate characterization for many rocks of biogenic origin or which have undergone significant diagenesis, involving dissolution and re-precipitation. The representation of many carbonates, for instance, becomes challenging, although sophisticated grain and morphological characterizations of carbonates do, in principle, enable this method to be extended for a wide range of rocks (Mousavi et al., 2012). Moreover, while it is possible to apply grain-identification algorithms to images and then use this as the starting point for a process-based model (Thompson et al., 2008), the method does not directly use three-dimensional images, which are now widely available.

For general application to pore-space images, a more topological approach is required. Medial axis skeletonization was developed for porous media applications by Thovert et al. (1993), Spanne et al. (1994) and Lindquist et al. (1996); it is illustrated schematically in Fig. 2.17. The idea is to reduce a three-dimensional pore-space to a topologically equivalent graph of interconnected lines. The lines represent the medial axis, or a series of points that are most distant from the solid surface. To construct this skeleton the pore space is shrunk voxel by voxel: any void voxel next to the solid is transformed into solid. This process is performed iteratively until a cross-section through a pore becomes a single voxel – the centre of this voxel represents a point that is furthest from the original solid surface. In three dimensions a network of lines – the skeleton – can be constructed. This method, when given an ideal infinite-resolution image, will extract the topology directly.

While the medial axis does, in theory at least, preserve the topology of the pore space, in real applications the method suffers from three problems. Firstly, when

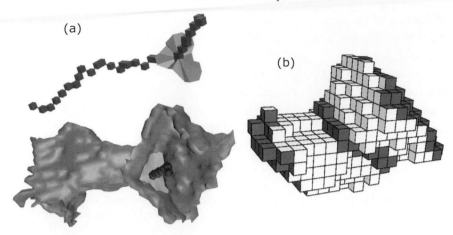

Figure 2.17 (a) (Top) A medial axis path on a discretized image of the pore space showing the computed throat surface. (Bottom) Here a triangulated rendering of the channel (throat) surface in the vicinity of this throat has been added to the top image. (b) A digitized rendering of the pore voxels within a distance of 6 voxels on each side of this throat surface. For clarity of presentation, grain voxels are not displayed. From Prodanović et al. (2007).

applied to a discrete image of voxels, the identification becomes ambiguous and may depend on the order in which the pore space is shrunk (for a fuller discussion, see Silin and Patzek (2006)). This is a significant problem in real images where there is an inevitable trade-off between resolution and sample size, and many of the smaller throats are resolved with only one or two voxels. Secondly, the medial axis may contain features which are not relevant for fluid flow, such as dead ends in the skeleton whenever there is any rugosity or irregularity in the pore walls. Thirdly, as we present later, fluid displacement is generally discussed as proceeding through a network of large pore spaces connected by narrower throats. The skeleton captures the locations of the throats, but the identification of pores is less straightforward. A pore can be located at any junction in the skeleton. However, it is not evident what volume should be assigned to this pore and how to accommodate branches that are in close proximity: in practical application these need to be merged into a single effective pore. Furthermore, a pore defined as a junction in the medial axis is not necessarily in the widest region of the pore space. This leads to a genuine ambiguity in network extraction: is the purpose of the network to preserve strictly the topology – in which case the medial axis must be retained – or is it more a convenient geometric simplification to locate the narrowest and widest regions of the void space with only an approximate representation of connectivity?

Despite these apparent practical limitations and potential ambiguities, medial axis skeletonization, with appropriate guidelines to identify pores, has been successfully used to extract network structures of a variety of rocks. Then, using this simplification of the pore space, accurate predictions of flow and transport properties have been made (Arns et al., 2001; Sok et al., 2002; Prodanović et al., 2007; Yang et al., 2015).

An alternative approach to the problem of computing a network relies instead on a robust method to identify pores, and therefore places the emphasis on finding wide and narrow regions of the pore space, rather than computing a topologically equivalent skeleton. The first application of this concept was by Hazlett (1995), who placed spheres to locate pores in an image of Berea sandstone and from this to calculate, semi-analytically, fluid configurations for different displacement processes. The idea has also been employed to determine pore sizes and connectivity (Arns et al., 2005). To extract a network, Silin and Patzek (2006) took spheres centred in each void voxel; these were grown until they touched the solid: they are called maximal balls. A sphere that is larger than any other sphere that it overlaps uniquely defines a pore. This is denoted an ancestor (see Fig. 2.18). Any sphere completely contained within another is ignored. A smaller maximal ball that overlaps with the ancestor is assigned to the same family. We can then define a cascade of smaller overlapping maximal balls and assign them again to that family. However, we will eventually find one ball that is a child of two families, as shown in Fig. 2.18. This then defines a throat: the restriction that separates two wider regions of the pore space. In theory, the lines connecting the centres of the maximal balls ordered in this way defines the medial axis skeleton (with some caveats, see Silin and Patzek (2006)).

By construction, this method identifies wide pores reliably, but tends to assign a cascade of very small pores and throats of approximately the voxel size in pore-space roughness and corners that do not necessarily contribute to the connectivity of the pore space. Furthermore, the method still suffers from the same ambiguities associated with the discrete nature of pore-space images: the spheres are in fact voxelized approximations, and the method can find multiple connections with spheres of apparently equal size between the same pores. The other problem is the distinction between a pore and a throat, with some largely arbitrary criterion used to partition void voxels between them. Again, however, networks can be constructed using this method that make successful predictions of flow properties; see Dong and Blunt (2009) and Blunt et al. (2013), when combined with the simplification of pore and throat geometry illustrated in Fig. 2.16.

It is possible to combine the strengths of a maximal-ball-type approach with medial axis skeletonization. Al-Raoush and Willson (2005) used the largest spheres in the pore space to define pores, while using the medial axis method to locate throats. This technique successfully captured the pore geometry of unconsolidated media imaged using micro-CT scanning and provided a robust algorithm for pore identification.

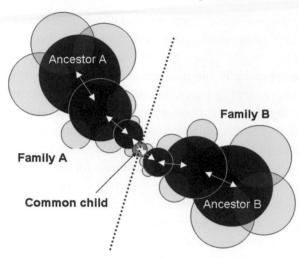

Figure 2.18 A description of the maximal ball method to extract a pore network. Spheres are grown in the pore space until they touch the solid. Spheres which are larger than any others that they intersect define a pore: shown are two such spheres, denoted ancestor A and B. The two families A and B are defined from smaller overlapping spheres with a common ancestor. Shown is a child sphere with both A and B as ancestors: this defines a throat. The black balls define the connection between the two pores, while the lighter coloured balls do not lie on the skeleton of the pore space, but define its volume. From Dong and Blunt (2009).

The final method I will mention is based on techniques used in image analysis (see, for instance, Schlüter et al. (2014)). This uses a seeded watershed algorithm to segment the void space into pores (Wildenschild and Sheppard, 2013; Rabbani et al., 2014; Prodanović et al., 2015; Taylor et al., 2015). First, a distance map is created in the pore space: a concept that was first used in an efficient network extraction algorithm by Jiang et al. (2007b). This is the distance from the centre of a void voxel to the nearest solid surface, d: it is equivalent to the radius of a maximal ball. Where this distance is larger than in any neighbouring points we locate a pore: again this is equivalent to the ancestor maximal ball described previously. The seeded nature of the algorithm allows the user to identify voxels that are definitely located in particular pores by hand. The pore space is then partitioned into regions associated with each pore. This is where a watershed algorithm is used: this is a method used widely in image processing to identify phases and segment images. In hydrology the watershed divides regions where water flows, eventually, into different drainage basins (rivers, lakes or seas): typically it is located along the ridge of a hill. We now apply this concept in three dimensions to a pore space image. The distance to the solid surface represents depth: regions of large distance reside downhill, while small distances represent high ground. Continuing the analogy, water will flow downhill and collect in a local maximum in the distance map – the centres of the pores. Any void voxels where the water will flow to a particular pore are assigned to that pore. Algorithmically, the pore regions are identified by starting at a pore centre and iteratively moving to neighbouring voxels with a continuous decrease in the distance map. A throat represents the watershed, where we have a surface of minimum distances that increase as we move in two directions associated with two distinct pores. This idea is conceptually similar to the approach developed by Baldwin et al. (1996), who used a thinning algorithm to identify pores, bounded by throats defined as cross-sections of the pore space with a local minimum in hydraulic radius (or flow conductance).

In this abstraction, all the volume of the pore space is assigned to pores. A throat is a surface at a restriction in the pore space that defines the boundary between two pore regions, as illustrated in Fig. 2.19: this is not necessarily a planar surface. In the remainder of the book we will use this conceptualization of the pore space to describe fluid displacement: a pore is centred on a wide region in the void space, while a throat represents the restriction between pores. In a discrete image, the throats are a series of adjoining voxels traversing a region of the pore space: a surface can be defined by joining up the centres of these voxels. In any digitalized image this approach still results in some ambiguities, and is less reliable when we have sheet-like pores where the distance map is not a good characterization of the structure (Wildenschild and Sheppard, 2013). Furthermore, it is possible for the same two pores to be connected through more than one throat. Even with these complexities, this definition of a network does serve as a reasonably consistent, robust and useful characterization of the pore space.

We can now present a more formal definition of pores and throats based on a continuous image of the pore space, Fig. 2.19. We define the distance map d anywhere in the void space $\mathbf{x}$, from which a gradient $\mathbf{g}$ can be computed:

$$\nabla d(\mathbf{x}) = \mathbf{g}. \tag{2.3}$$

A local maximum, minimum or saddle point is defined where $\mathbf{g} = \mathbf{0}$ indicating that the gradient is everywhere locally zero. To identify the nature of this extremum, the Hessian matrix is introduced (this is the multi-dimensional version of a second derivative):

$$H_{ij} = \frac{\partial^2 d}{\partial x_i \partial x_j}, \tag{2.4}$$

employing component notation. The eigenvalues, λ, are found from

$$H_{ij}e_j = \lambda e_i, \tag{2.5}$$

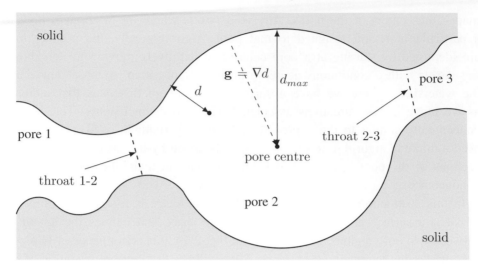

Figure 2.19 A definition of pores and throats based on a distance map. Here we illustrate the concept in two dimensions, but it is applied to three-dimensional pore-space images. d is the distance from any point in the void space to the nearest solid surface. A pore centre is defined where this distance is a local maximum, d_{max}: this is its inscribed radius. A pore is a region of the void space where we follow $\mathbf{g} = \nabla d$, Eq. (2.3), until we reach the centre of that pore, as indicated by the dashed arrow. A throat is a surface (or line in two dimensions) that divides or bounds two pore regions.

where e_i are the corresponding eigenvectors. A local maximum in the distance map occurs when the eigenvalues in Eq. (2.5) are all negative: these locations define the position of pore centres.

A streamline can be defined as a line that begins at some point in the pore space and then follows a direction parallel to $\mathbf{g}$. A streamline ends when $\mathbf{g} = \mathbf{0}$. Let $\mathbf{x}_i$ represent the centre of pore i: a point $\mathbf{x}_j$ in the pore space is associated with pore i if the streamline starting at this point ends at $\mathbf{x}_i$. Since $\mathbf{g}$ defines a direction uphill, we can never arrive at a local minimum (and indeed a local minimum cannot exist for geometrical reasons), but it is possible to find a saddle point, where one of the eigenvalues is positive and another negative. This locates the centre of a throat. The throat is a surface of points whose streamlines end at the throat centre: on either side of this surface the streamlines will progress up the distance map to two different pores. For a two-dimensional image the concept is easy to visualize, as we can identify distance as depth in a landscape, and consider the lowest part of a valley as the pore centres, any region that leads downhill to the valley as part of the pore, while the mountain ridges are the throats.

Fig. 2.20 shows an example of the application of this concept. Here the throats in the pore space of a Berea sandstone are shown; these throats separate distinct pore regions. The pore space of Berea sandstone itself is shown in Fig. 2.14 together

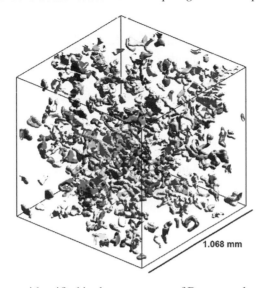

1.068 mm

Figure 2.20 Throats identified in the pore space of Berea sandstone using a watershed algorithm. The throats are surfaces dividing different pores; see Fig. 2.19. In a digital image, these throats are adjoining voxels cutting across the pore space with different pores on either side. From Rabbani et al. (2014). (A black and white version of this figure will appear in some formats. For the colour version, please refer to the plate section.)

with its associated network (albeit generated using a different algorithm): Fig. 2.20 indicates the digitalized surfaces that define the throats. It is evident that when dealing with real images containing hundreds of millions or even billions of voxels, the network may be complex and difficult to visualize, but the underlying concept is simple. Every region of the void space is associated with a pore, while the throats are surfaces separating the pores. When we discuss displacement later, this distinction will be important. The wetting phase tends to reside in the narrower regions of the pore space and the non-wetting phase in the wider regions. Hence movement of the non-wetting phase is impeded by the throats – the narrow restrictions – and is easy through the pores, while, in contrast, the wetting phase advance is limited by the pores or the wide regions. The change in fluid volume associated with fluid movement is associated with the pores. This is a convenient geometric representation, but does not strictly identify the medial axis nor necessarily preserve topological parameters of the pore space. However, for our applications to flow and fluid displacement it is arguably the most useful way to apportion the pore space.

2.2.3 Generalized Network Models

In network modelling it is often unhelpful to revert to an overly explicit picture of links (throats) and junctions (pores). Instead, to repeat, a network representation is a partitioning of the void space which attempts to preserve, approximately, the connectivity and other useful geometric information. We have a physical picture now of how an image is sub-divided into pores, with the throats representing the boundaries between two pore regions. Other information can then be assigned to the elements (pores or throats) as necessary to make reliable predictions of flow and transport properties, or at least in order to interpret the behaviour. Examples of these properties will be the area and inscribed radius of a throat (the radius of the largest maximal ball centred in the surface of the throat), and the volume and inscribed radius (the radius of the ancestor maximal ball) for a pore. These quantities will determine the radii of curvature and hence the local capillary pressure for displacement, as well as the conductance to flow and changes in fluid volume. Again, by analogy, consider an interactive underground map, where assigned to a specific station would be information on train times, street maps to local attractions, journey times to the next station or live information on imminent departures and alerts of disruption. The same is true of a network model: as required, additional information can be given to the elements to allow computations of different properties.

We can consider one further level of abstraction. We use networks to describe and quantify fluid displacement. This in turn is controlled by the capillary pressures required for one phase to push out another. We could calculate at every voxel in an image the radii of curvature of an interface between two fluids that passes through the centre of that voxel obeying the Young equation at the solid surface. From this the capillary pressure could be computed using the Young-Laplace equation. The throats define a surface of voxels where this capillary pressure is a local maximum (smallest region of the pore space with the smallest radii of curvature), while the centre of the pores would represent a local minimum in the capillary pressure. We could perform the same watershed segmentation as above, but this time on the capillary pressure, dividing the void into regions where the capillary pressure continually decreases to a minimum value (the pore centre) divided by surfaces of local maxima (the throats). This then would allow an accurate calculation of displacement pressures and resultant fluid configurations and removes the largely unhelpful dependence on some simplified geometrical representation. However, this has not been performed, except for the simple case where the contact angle is zero and the methodology is essentially equivalent to the maximal ball approach. The reason for this is that the threshold capillary pressures depend on contact angle, which, as we show later, varies across the rock and is dependent on flow direction, as well as the previous sequence of fluid displacement. The calculation is sufficiently complex, and solves the problem directly on the pore space, that the simplification of a network, however accurate, is no longer needed. This approach is in essence a direct computation of the Young-Laplace equation in pore space (see, for instance,

Table 2.1 *Properties of networks: those for Ketton, Estaillades and Mount Gambier are shown in Fig. 2.21, while the Berea network is illustrated in Fig. 2.14. The networks have been extracted from images with approximately* $1,000^3$ *voxels at a resolutions between 2.7 and 3.8 μm. Note the range of coordination number, which quantifies the connectivity of the pore structure, and the number of elements (pores and/or throats), which can be used to estimate a typical distance between pores.*

Rock	Pores	Throats	Volume (mm^3)	Coordination number
Ketton	1,916	3,503	50.5	3.66
Estaillades	83,072	120,867	20.7	2.91
Mount Gambier	66,279	94,678	27.0	2.86
Bentheimer	28,601	54,741	27.0	3.83
Berea	12,349	26,146	27.0	4.23

Prodanović and Bryant (2006)). Hence, we use the geometrical characterization to mimic the likely behaviour of capillary pressure since it is controlled by the radius of curvature which in turn is related to the pore and throat sizes, defined by their inscribed radii.

Fig. 2.21 shows images and associated networks for different rocks types using the maximal ball method: the pore space itself was shown in Fig. 2.4. The important concept is that the network readily displays the connectivity of the pore space and it is this connectivity that controls fluid flow, as we describe later. To quantify this, Table 2.1 shows the number of pores and throats, and the average coordination number (the average number of throats bounding each pore) for the networks shown in Fig. 2.4. The Mount Gambier pore space, with a rich geological structure, is quite open with some large well-connected pores: the overall coordination number is low because there are also many smaller dead-end pores with only a single adjoining throat. For comparison, as mentioned above, most rocks formed by sedimentation of approximately equally sized grains have a coordination number of 4, which may be reduced during compaction and cementation (diagenesis): this is seen for the two sandstone examples. Estaillades is much more poorly connected, which will restrict fluid flow; indeed, many of the pores have only one or two connections and the average coordination number is less than 3. Ketton is better coordinated with far fewer pores per unit volume, indicative of its large grain size. In reality, the connectivity for Estaillades and Ketton is better than predicted here, since the image on which the network extraction is based cannot resolve micro-porosity (see, for instance, Fig. 2.7). However, this micro-porosity contains very small pores and throats that may not allow significant flow. Below, though, we show how this micro-porosity can be accommodated using a dual-scale network.

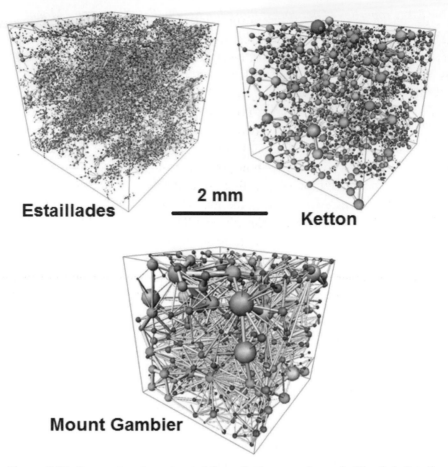

Figure 2.21 Pore networks extracted from the images shown in Fig. 2.4: Estaillades, Ketton and Mount Gambier. The pore space is represented as a lattice of wide pores (shown as spheres) separated by narrower throats (shown as cylinders, even though strictly they are surfaces separating pores). The size of the pore or throat indicates the inscribed radius. Adapted from Blunt et al. (2013). (A black and white version of this figure will appear in some formats. For the colour version, please refer to the plate section.)

In general a network element need not even be a single region of the void space, but could represent the average properties of a smaller-scale network, which cannot be directly resolved in the image. This is necessary to describe carbonates with a huge range of pore size: with a network model, many billions of pores are present in even a small cm-sized piece of rock. One way to accommodate this is through a multi-scale network, analogous to the hierarchy of maps described at the beginning of this section (Jiang et al., 2007b, c; Mehmani and Prodanović, 2014a, b; Prodanović et al., 2015). A network is constructed at the smallest scale to describe micro-porosity, based on a sub-micron resolution image. Then the average properties are captured in an upscaled element that is pasted onto a larger-scale network describing the larger pores, resolvable at the micron level. The details of this methodology may be complex, but do allow both micro- and macro-porosity to be described in a single model, including micro-porosity both inside grains and between them (Mehmani and Prodanović, 2014b). This approach sacrifices the detail of the micro-porous network at a huge saving in computational efficiency. With care, as in the London Underground map, this simplification retains the information most pertinent for transport. An example of this approach is shown in Fig. 2.22, where micro-porosity is first explicitly described in a small-scale local network, using – in this example – the maximal ball algorithm (Bultreys et al., 2015a). Then the averaged properties (the capillary pressure and flow conductance as a function of the volume of the wetting and non-wetting phases in the network) are assigned to an effective network element, providing additional connectivity to the network of larger pores and throats. Fig. 2.23 provides a picture of this concept applied to an image of Estaillades limestone, whose structure was shown in Fig. 2.3 with the pore space, imaged at a resolution of a few microns, illustrated in Fig. 2.4. The same concepts can be used to describe pore sizes spanning three orders of magnitude or more in shale samples (Mehmani and Prodanović, 2014a).

The network representation is used mainly in this book to aid the understanding of multiphase fluid displacement; however, it can and will be used to make predictions when the displacement physics and pore-space geometry are known. In some applications though it is easier and more accurate (within the limits of current computer power) to simulate flow and transport directly on the images: as we show later in Chapter 6, we need to consider each application on its merits and choose the correct tool to solve the problem. A one-method-to-rule-them-all approach is not appropriate.

To recap: the connected void space of a rock can be partitioned into pores. The centre of a pore is defined as the point locally furthest from the solid surface: moving the point in any direction decreases the distance to the surface (it is the centre of an ancestor maximal ball, or a local maximum in the distance map). In terms of capillary pressure, a fluid meniscus passing this point will encounter a minimum in pressure, or a maximum in the average radius of curvature. The volume associated with a pore is the region of the void space where the distance to the surface continually increases as we move towards the pore centre. A throat is defined as a surface that bounds two pore regions. It is the surface where the distance to the solid is lowest, and increases in two directions as we move towards the two adjacent pore centres. A fluid meniscus moving through a throat will experience a local maximum in capillary pressure, or a minimum radius of curvature.

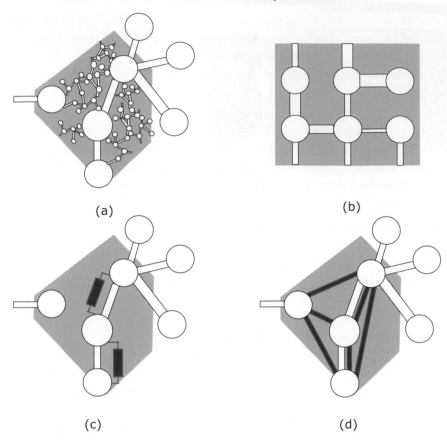

(a)   (b)

(c)   (d)

Figure 2.22 An illustration of a generalized multi-scale network model to accommodate both micro- and macro-porosity. (a) The network of macro pores, based on
network extraction of an image with micron resolution, is shown combined with
local micro-porous networks. Simulation of displacement on this network is difficult since billions of pores may be needed to represent even a small rock sample.
(b) A possible upscaling approach using a regular lattice of large pores with averaged properties assigned to a continuous matrix of micro-porosity. (c) A better
approach is to preserve the structure of the macro-porous network with additional connections and average properties assigned to the micro-pores. (d) The
end result is additional connectivity between the large pores, with upscaled properties assigned to these new connections based on the small-scale micro-porous
networks shown in (a). From Bultreys et al. (2015a).

2.2.4 Topological Descriptors of the Pore Space

While the emphasis here is on a network characterization of the pore space, since
this will be used to both interpret and predict displacement processes, there are
other useful measures of void-space topology which quantify the connectedness
of the rock. We start with a definition of what are called Minkowski functionals

Figure 2.23 A visualization of a dual-porosity network model for Estaillades limestone. Here, in contrast to the network shown in Fig. 2.21, we also accommodate micro-porosity. On the left is the macro-network derived from a pore-space image similar to that shown in Fig. 2.4: the image size is approximately 3.1 mm. The middle image shows the micro-porous structure with different colours for different clusters: in blue is the largest connected cluster. On the right is an illustration of the two-scale network superimposed on the underlying image, showing pores and throats at both the macro and micro scales. From Bultreys et al. (2015a). (A black and white version of this figure will appear in some formats. For the colour version, please refer to the plate section.)

or Quermass integrals, which define different characteristics of any binary image (here a pore space segmented into pore and solid): there are four of these functionals for a three-dimensional object. This will be a brief introduction to this topic; further details can be found elsewhere (Mecke and Arns, 2005; Arns et al., 2009; Vogel et al., 2010).

The zeroth-order Minkowski functional, M_0, is the volume of the pore space. For an image of size V, the total porosity $\phi = M_0/V$. The remaining three functionals are defined on the surface between solid and grain. The first-order functional M_1 is the total area of this surface

$$M_1 = \int dS, \tag{2.6}$$

where S is the surface between pore and grain. We can define a specific surface area $M_{1V} \equiv a = M_1/V$, which has the units of 1/length and whose value is the inverse of a typical pore size.

Curvature κ was introduced in Chapter 1 in the context of the interface between fluid phases. Here we will use the same idea (see, for instance, Fig. 1.3) applied instead to S in Eq. (2.6). The second Minkowski functional is the average curvature of the boundary between solid and void,

$$M_2 = \frac{1}{2} \int \left(\frac{1}{r_1} + \frac{1}{r_2} \right) dS = \frac{1}{2} \int \kappa \, dS, \tag{2.7}$$

where r_1 and r_2 are the principal radii of curvature, defining the curvature κ from Eq. (1.6).

The third-order, and final, functional is the total curvature of the interface, defined by

$$M_3 = \int \frac{1}{r_1 r_2} dS. \tag{2.8}$$

Note that unlike the other quantities, M_3 is dimensionless.

The first three functionals all quantify intuitive features of the pore space related to the porosity, the surface area (which will control adsorption or dissolution reactions) and the curvature (characterizing the shape of the voids, or the grains that surround them: a packing of solid spherical grains will have an interfacial curvature M_2 of opposite sign to a porous medium formed, for instance, by the dissolution of approximately spherical voids). M_3 appears, at first sight, to be simply a mathematical object of less utility. It is, however, associated with possibly the most important property of the pore space, namely its connectedness.

The third-order Minkowski functional is related to the Euler characteristic, χ, which is the well-known relationship between the number of edges, faces and vertices of a polyhedron,

$$\chi = V - E + F - O, \tag{2.9}$$

where V is the number of vertices, E is the number of edges, F is the number of faces and O is the number of distinct objects (this is often ignored in the more common application of Eq. (2.9) to a single solid object: for instance, any convex shape has $V - E + F = 2$, where, in this context, $O = 1$ and hence $\chi = 1$). χ can be computed on a pore-space image. Take, for example, an image comprising cubic voxels. A single isolated void voxel has $V = 8$, $E = 12$, $F = 6$ and $O = 1$ (there is one object) giving $\chi = 1$. It is simply a counting exercise to determine χ for an arbitrary image, summing faces, edges and vertices on a voxel-by-voxel basis.

However, Eq. (2.9) is not obviously related to a measure of connectivity. To do this, it is first necessary to generalize this relation to arbitrary shapes defining the pore space (Serra, 1986; Vogel and Roth, 2001; Vogel, 2002):

$$\chi = \beta_0 - \beta_1 + \beta_2, \tag{2.10}$$

where β represents a Betti number. β_0 is the number of discrete holes in the porous medium. In most cases, all the pores are interconnected and $\beta_0 = 1$: in general, β_0 is the number of distinct regions of the pore space that are not connected to each other. β_1 is the number of handles or loops in the structure: these are called redundant loops as they can be eliminated without changing the connectivity or breaking up the pore space (quantified by increasing the value of β_0). β_2 is the number of isolated regions of solid completely surrounded by pore space. In a consolidated medium we would expect $\beta_2 = 0$:

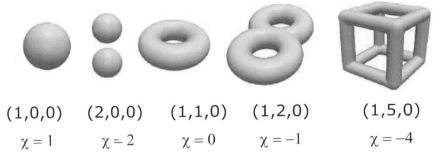

(1,0,0)   (2,0,0)   (1,1,0)   (1,2,0)   (1,5,0)

$\chi = 1$ $\chi = 2$ $\chi = 0$ $\chi = -1$ $\chi = -4$

Figure 2.24 Betti numbers $(\beta_0, \beta_1, \beta_2)$ and Euler values χ for simple three-dimensional solid objects: single sphere, two disjoint spheres, torus, double torus and cubic frame. The same concepts can be used if these objects describe instead the pore space. From Wildenschild and Sheppard (2013).

we do not have disconnected grains that can rattle around in the pore space. To help illustrate this concept, Fig. 2.24 shows the Betti numbers and the Euler characteristics for some simple shapes. χ is a topological measure of the pore space, in that it is invariant under continuous deformation of the void geometry.

We can also relate χ to M_3 as follows (Vogel et al., 2010):

$$\chi = \frac{1}{4\pi} M_3. \tag{2.11}$$

β_1, or the number of loops in the structure, measures how many connections can be broken while maintaining connectivity. This will be important when we describe multiphase flow, since the presence of one phase can disconnect the other: if the pore space contains many loops, it is more likely that the fluid phases connect and can flow.

Hence β_1 is the key connectedness parameter. While this appears to be rather difficult to compute for a general pore-space image, it can instead be found from first finding χ using Eq. (2.9) (by counting edges, faces and vertices in a discretized image) and then using Eq. (2.10) to find $\beta_1 = -\chi + \beta_0 + \beta_2$. Furthermore, in most cases the pore space is all interconnected (and in any event this is the only part of the pore space of interest for flow and transport) with no completely isolated grains, giving $\beta_0 = 1$ and $\beta_2 = 0$.

One remarkable feature is that *any* additive property of a structure can be written as a linear function of the Minkowski functionals. This concept has been reviewed in the context of pore-scale imaging by Schladitz (2011).

In a completely interconnected network, graph theory can be used to find

$$\chi = n_p - n_t = n_p(1 - z/2), \tag{2.12}$$

Table 2.2 *Euler numbers per unit volume for different porous media. Adapted from Herring et al. (2013).*

	Intensive Euler characteristic χ_V, 10^9 m^{-3}
Bentheimer sandstone	−47.3
Sintered glass beads	− 4.29
Loose pack of glass beads	− 4.58
Crushed tuff	− 1.06

where n_p is the number of pores, n_t is the number of throats and $z = 2n_t/n_p$ is the average coordination number (Vogel and Roth, 2001). As evident in Table 2.1, for the samples studied $z > 2$, and so χ has a large and negative value (β_1 is large and positive).

Intensive quantities or densities, $M_{nV} = M_n/V$ can be defined for the four Minkowski functionals with $\chi_V = \chi/V$. Table 2.2 shows example Euler characteristics per unit volume, χ_V, measured on micro-CT scans of different porous media (Herring et al., 2013). For Bentheimer sandstone, $\chi_V \approx -\beta_{1V} = -4.73 \times 10^{10}$ m^{-3}. In contrast, applying Eq. (2.12) using the data in Table 2.1 gives $\chi_V \approx -9.7 \times 10^{11}$ m^{-3}. The discrepancy is likely a combination of two factors. The network model probably over-states the true connectivity of the pore space through incorrectly assigning throats between small pores in roughness and corners of the void space. On the other hand, the image used to find χ directly had a voxel size of 10 μm, as opposed to 3 μm for the image used to construct the network, and so some connectivity through smaller throats may not have been resolved.

This topological information can be used to determine more accurately the connectivity of a network in construction algorithms (Vogel and Roth, 2001; Jiang et al., 2012). In any case, for a structure where each pore has at least two bounding throats we expect $\chi_V \ll 0$ with a characteristic magnitude of $1/l^3$, where l is a typical pore size. For the bead pack in Table 2.2 if we take l to be the bead diameter of 0.85 mm, we predict $\chi_V = -1.6 \times 10^9$ m^{-3} which is of the same order of magnitude as the measured values.

These concepts can be extended to determine the connectivity of the pore space for different threshold pore sizes, ignoring all portions of the pore space with a distance function d less than some value, Fig. 2.19: in this case, we can define a critical inscribed radius at which the pore space begins to disconnect (when $\chi = 0$). This is illustrated in Fig. 2.25, which shows the four Minkowski functionals computed as a function of minimum pore size (essentially d) for micro-CT images of different

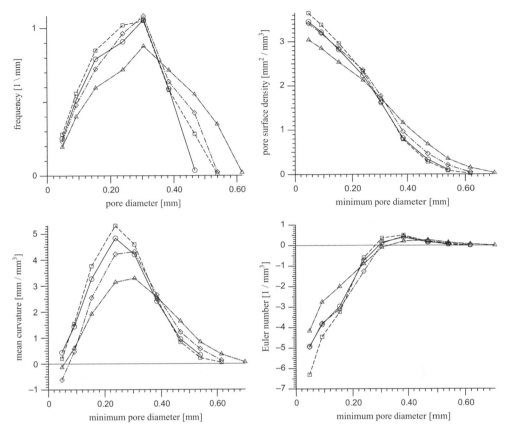

Figure 2.25 Minkowski functions, M_{nV}, computed on a micro-CT image of a sand for different minimum pore diameters. For pore volume density (M_{0V}, upper left figure), the derivative is plotted to illustrate the pore size distribution. Results from four layers are shown with the different symbols. The average pore diameter is around 0.3 mm. The pore surface area increases as the minimum size of the pores considered decreases, as more features of the grain surfaces are detected. The curvature is positive for pore diameters around the average pore size: here a positive curvature represents approximately spherical grains bulging into the void space. However, the mean curvature when smaller features are considered is close to zero, as both small protuberances and chasms in the pore space are included. The Euler number becomes more negative as small features, allowing more connections, are resolved. The Euler number is around zero, implying limited connectivity, when the minimum size is the average pore size. If we only allow larger pores to be considered, the connectivity of the void space is poor. From Vogel et al. (2010).

layers of sand. This analysis provides a quantification of connectivity when channels of different size are considered. It describes, for instance, the connectivity of a non-wetting phase that only occupies the larger regions of the pore space.

When the smallest features are included, which corresponds to a non-wetting phase occupying most of the voids, the pore space is well connected with a negative value of χ_V, a large specific surface area and an average curvature, M_{2V}, close to zero.

The Euler characteristic is thus a valuable measure of the connectivity of networks, distinguishing between branched pathways and ones with many loops. If we return to our analogy of the London underground map, Fig. 2.13, a good network, particularly one where it is possible to travel even if some lines are closed (corresponding to having different phases in the void space of a rock), is not necessarily one with many lines at some stations, but one with many redundant loops, allowing several different ways to move between two stations: this implies a large and negative Euler characteristic. This is related to the average coordination number through Eq. (2.12).

Thus, while the terminology may appear baffling, the concepts have proved a powerful tool for the characterization of images (Brun et al., 2010), to aid and quality control the construction of network models (Vogel and Roth, 2001; Jiang et al., 2012) and to provide statistical measures to match in reconstruction algorithms as described earlier (Vogel, 2002; Arns et al., 2009). Furthermore, related, but distinct properties, such as the local porosity and percolation probability, can also be calculated: the latter is a quantitative non-additive measure of connectivity which provides a robust indicator of the flow properties of the pore space (Hilfer, 2002).

Moreover, the same topological analysis can be performed for fluids residing in the pore space and the interfaces between them (Herring et al., 2013, 2015). Now M_0 is related to the fluid saturation, M_1 to the surface area between the phases (which will control, for instance, fluid/fluid reaction and dissolution), M_2, the curvature, to the local capillary pressure, and M_3 to the connectedness of the phases, and hence their ability to flow. These concepts will be explored in the subsequent chapters.

2.3 Wettability and Displacement

2.3.1 Thermodynamic Description of Displacement Processes

As mentioned above, from energy balance and a knowledge of pore geometry – however complex this may be – and contact angle, we can determine fluid configurations in any system of interest. The general approach is to consider the change in free energy for an infinitesimal movement of the fluids in a porous medium (Mayer and Stowe, 1965; Princen, 1969a, b, 1970). Let A_{1s} be the interfacial area between phase 1 and the solid, A_{2s} is the area between phase 2 and the solid while A_{12} is the area between the two fluids. We use the same definition of subscripts to define the interfacial tensions σ_{12}, σ_{1s} and σ_{2s}. This is illustrated for a circular tube in Fig. 2.26. There is a pressure difference between the phases $P_1 - P_2$ and work is performed against this pressure to make a small change in the volume of the phases in some local region of space (normally in a pore or a throat and its adjacent pores):

$$dF = (P_1 - P_2)dV + \sigma_{12}dA_{12} + \sigma_{1s}dA_{1s} + \sigma_{2s}dA_{2s}, \qquad (2.13)$$

where dF is the change in free energy. Since the solid must be covered by one fluid phase or the other, $dA_{1s} = -dA_{2s}$, and so,

$$dF = (P_1 - P_2)dV + \sigma_{12}dA_{12} - \sigma_{1s}dA_{2s} + \sigma_{2s}dA_{2s}, \qquad (2.14)$$

then applying Eq. (1.7), where we assume that the contact angle θ is measured through phase 1 (which we consider therefore to be the denser phase),

$$dF = (P_1 - P_2)dV + \sigma_{12}\left(dA_{12} + dA_{2s}\cos\theta\right). \qquad (2.15)$$

The final step is to invoke the Young-Laplace equation (1.6), where conventionally if phase 1 is the denser, wetting phase (water), $P_1 - P_2 = -P_c = -\kappa\sigma_{12}$, where κ is the curvature to find

$$dF = -\sigma_{12}\left[\kappa dV - (dA_{12} + dA_{2s}\cos\theta)\right]. \qquad (2.16)$$

The relationship between dV, dA_{12} and dA_{2s} is determined from the geometry of the system under consideration, with menisci between the fluids which have a constant curvature κ in capillary equilibrium. This perturbation of the system is favourable for $dF \le 0$ and we reach equilibrium when $dF = 0$.

We are interested principally in a displacement sequence: that is, how one phase replaces another in a porous medium on a pore-by-pore basis. At each step, we do indeed have a position of capillary equilibrium, obeying the Young-Laplace and Young equations, or, more generally, the energy balance in Eq. (2.16), but the real interest is in the transition from one position of local equilibrium to another, when this occurs, and in which order. The important aspect of this discussion is that we only see a sequence of positions of *local* equilibrium: we do not impose a global energy minimization, since to secure this the fluids have to move to this configuration and they cannot do this is if, at any step, the free energy has to increase.

In a displacement, the configuration of fluid in a pore or throat undergoes a discrete change. The threshold capillary pressure at which this occurs will be given by Eq. (2.15) or (2.16) but written, not for an infinitesimal perturbation of area and volume, but for a finite jump,

$$\Delta F = -\sigma\left[\kappa\Delta V - (\Delta A_{12} + \Delta A_{2s}\cos\theta)\right], \qquad (2.17)$$

writing the fluid/fluid interfacial tension $\sigma_{12} \equiv \sigma$.

There is a further simplification, which we will use in an example application below. We consider displacement in a single pore or throat and assume that this

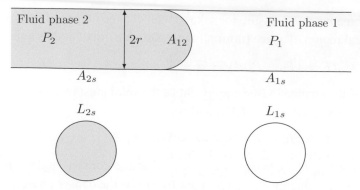

Figure 2.26 A schematic of energy balance in a circular channel. We have two
fluid phases, 1 and 2, with an interface between them of area A_{12}. The area of the
solid in contact with phase 1 is A_{1s} and in contact with phase 2 is A_{2s}. The lower
figures show cross-sections through the channel with the lengths of the interfaces
L_{1s} and L_{2s} indicated. There is a pressure difference between the phases $P_1 -$
P_2: if we change the volumes of either phase, we do work against this pressure
difference that is balanced by the change in interfacial energies associated with
changes in the areas A, Eq. (2.13). For a system of constant cross-section, we can
consider changes in the length of the interfaces: L_{12}, L_{1s} and L_{2s}, Eq. (2.18), as
shown in the lower figure. In this simple example we can use energy balance to
derive the familiar expression for capillary pressure in a circular tube, Eq. (1.16).

element has a constant cross-section (or at least a curvature perpendicular to the
cross-section that is much smaller compared to curvatures of fluid menisci in the
plane of the cross-section); see Fig. 2.26. We can rewrite Eq. (2.17) in terms of a
change in volume per unit length (ΔA) and the change in area per unit length (or
length of the interface ΔL) as,

$$\Delta F_L = -\sigma \left[\kappa \, \Delta A - (\Delta L_{12} + \Delta L_{2s} \cos\theta) \right], \qquad (2.18)$$

where now ΔF_L is a change in free energy per unit length.

Let us consider one definite example, similar to when we treated capillary rise
in Chapter 1.3.3: we will use Eq. (2.18) to find the threshold capillary pressure for
filling a cylindrical capillary tube of radius r. In this case we consider the change
in free energy from a configuration with the tube full of wetting phase (phase 1) to
one full of non-wetting phase (phase 2). In this case $\Delta A = \pi r^2$ (the cross-sectional
area of the tube), $\Delta L_{12} = 0$ (there is no change in the area of the interface between
the two fluids) and $\Delta L_{2s} = 2\pi r$ (the circumference). Hence we find,

$$\Delta F = -\sigma \left(\pi r^2 \kappa - 2\pi r \cos\theta \right). \qquad (2.19)$$

If $\Delta F < 0$ the displacement is favourable and will occur spontaneously, if $\Delta F >$
0 it will not happen. The threshold, or entry condition, when the event can first
proceed is when $\Delta F = 0$, for which

$$\kappa = \frac{2\cos\theta}{r},$$ (2.20)

or in terms of the capillary pressure, $P_c \equiv P_2 - P_1$, Eq. (2.20) is from Eq. (1.6)

$$P_c = \frac{2\sigma\cos\theta}{r},$$ (2.21)

which is the same as Eq. (1.16) derived previously.

This is a somewhat simple example, but the same approach can be used, in general, to study any configuration. We will return to this in later chapters to derive threshold capillary pressures.

2.3.2 Displacement Sequences

The concept of energy balance allows us to determine threshold pressures, or curvatures, from one fluid configuration to another on a pore-by-pore basis. Macroscopically, this leads to a sequence of saturation changes.

Let's consider a concrete example: the injection of CO_2 into a deep saline aquifer for long-term storage (preventing the CO_2 entering the atmosphere and contributing to climate change). The CO_2 is injected through a well and – at high pressure – is forced into the pore space of the rock. Now imagine a small portion of the porous aquifer some way from the well. Carbon dioxide will, at some time, begin to fill the pore space, displacing brine. In what order will it fill the pores, what will be the corresponding sequence of fluid configurations and capillary pressures and how much of the pore space will it invade? Then, after injection, the CO_2, since it has a lower density than the surrounding brine, will tend to rise. However, it does not rise as some sort of large bubble underground; the CO_2 has to displace brine as it moves upwards, and be itself displaced, pore by pore, by brine at its trailing (lower) edge. Again, what is the order of filling by brine and how much of the CO_2 is left behind?

For slow displacement, the pore-scale filling is controlled by the local capillary pressure, which in turn is determined from a local conservation of energy. For every region of the pore space there is a threshold capillary pressure at which one phase is first able to displace another. This threshold pressure is derived from energy balance with a known contact angle. These capillary pressures determine the order of filling and the arrangement of fluids; at the macro-scale they determine how readily each phase flows and how much is displaced.

In terms of the network model conceptualization introduced in the previous section, we will determine the sequence in which pores and throats are filled (or partially filled) with the invading phase, and the configuration of fluids within each pore and throat during the displacement. From this we can compute how much is recovered (removed from the pore space) and how the fluids flow.

While this is a general introduction to what follows, there are three subtleties associated with contact angle that we need to address first: wettability changes, surface roughness and the direction of displacement.

2.3.3 Wettability and Wettability Change

As alluded to previously, most clean rocks are naturally water-wet. Clean means that the surface of the solid in contact with the fluids is the same chemically as the bulk composition: the surfaces are not coated by other material; this case is discussed later. Quartz (silicon dioxide, a major constituent of many sandstones) or calcite (calcium carbonate, a major component of many carbonate rocks, or limestones), for instance, are ionically bonded solids with strong inter-atomic inter-actions. This makes the solid interfacial tensions large, but also allows stronger bonding with water through electrostatic attraction between hydrogen or oxygen and the surface than with hydrocarbons, which do not allow ionic or hydrogen bonding. Hence the surface energies of a solid surface coated with water tend to be less than the same solid in contact with gas or oil, corresponding to less inter-molecular attraction. Hence, when another fluid is introduced into a water-saturated porous medium, be it air, oil or CO_2, this new fluid is likely to be the non-wetting phase and the invasion proceeds as though the contact angle through the water is close to zero.

Since the solid interface represents a high-energy surface, it is favourable for species to reside on the surface if they can lower this energy. It is not flippant to say that clean surfaces get dirty quickly, however frustrating this is when doing housework: they do because it is favourable for dust and other detritus to settle on a high-energy interface. The same happens in porous media, whose large surface area can attract different materials present in either the aqueous phase or in the injected non-wetting phase.

The compounds that attach to the solid are called surface active: they tend to have both polar and non-polar portions in their molecular structure. The polar ends adhere to the surface, while the non-polar fraction presents an oil-wet or water-repellent surface. In groundwater applications, these can be components present in pollutants from, for instance, oil spills or decaying organic matter, including surfac-tants generated by biological activity. As a result, the solid surfaces are not clean, but weathered, indicating that the exposed solid surface is no longer crystalline, but also contains a mixture of organic compounds which lower the surface energy between the solid and an oleic (oil or oily) phase, hence rendering the system less water-wet.

In oil reservoirs the same phenomenon occurs, but it can have a more dramatic impact on wettability. In crude oil, asphaltenes (so-called since they represent the high molecular weight tarry fraction of oil used, after distilling, as asphalt, for making roads) are surface active components. Asphaltenes are a problem in oilfield operations, as they may precipitate at low pressures, clogging the pore space of the reservoir and flow lines. They also, and this is the emphasis of this discussion, will alter the wettability of the rock surface under reservoir flow conditions (Saraji et al., 2010). These compounds have a complex molecular structure, with a large variety of specific forms, but all contain some polar components (such as nitrogen, oxygen or sulphur) with benzene rings and long hydrocarbon chains (which are non-polar). Example asphaltene structures are shown in Fig. 2.27. In general, we do not know the exact chemical formulation of these molecules for a specific crude oil, nor precisely how the molecules will adhere to the solid surface and the resultant change in contact angle. However, we have a good empirical understanding that these substances cause a wettability change, and the physical mechanisms by which this occurs, which can involve direct ionic bonding, precipitation and Coulomb interactions between the brine, oil and solid surface (see, for instance, Buckley and Liu (1998), Buckley et al. (1989) and the review by Morrow (1990)). Here we will simply assume that a wettability change can take place, and explore the consequences for fluid configurations and displacement.

As an example, Fig. 2.28 shows a compilation of measured contact angles for crude oil/brine systems on a flat mineral surface. This surface was either quartz (seen in sandstones) or calcite (in carbonates). These are data collected on 30 oils from the United States (Morrow, 1990; Treiber et al., 1972). A strong wettability alteration is possible, but not all surfaces become oil-wet. In general, but not always, a stronger wettability change is seen on the calcite surfaces.

Fig. 2.29 indicates the configuration of oil and water in a single pore or throat, after oil has migrated into an oil reservoir and filled most of the pore space. This is one of the most important conceptual figures in the book, derived from the original work of Kovscek et al. (1993). The first feature to note is the way in which a pore is represented in cross-section. Referring back to the images shown earlier, it is evident that real pore spaces have irregular shapes; they are certainly not triangles. However, a triangle is shown for illustrative purposes since it preserves one important qualitative feature of real systems, namely that one phase (oil) can reside in the centre of the pore space, while the other (water) is retained in the corners and roughness close to the surface; see Fig. 2.16. We would not capture this feature if instead we assumed that the pores had a circular cross-section. Out of the plane of the diagram, the shape and size of the cross-section will vary: the inscribed radius of the cross-section will be a local maximum at the centre of a pore and a minimum in a throat.

Porous Media and Fluid Displacement

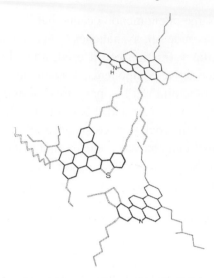

Figure 2.27 Example chemical structures of asphaltene molecules present in crude oil: the carbon chain and polar atoms (N and S) are indicated. There is a wide variety of specific structures, but the compounds all tend to be of high molecular weight (up to around 1,000 atomic mass units), with some polar components, which are attracted to a clean solid surface, benzene rings (shown dark in the figure), and long hydrocarbon chains, which are non-polar and tend to stick out, creating an apparently oil-wet (or at least modified wettability) surface. Reprinted with permission from Groenzin and Mullins (2000). Copyright (2000) American Chemical Society.

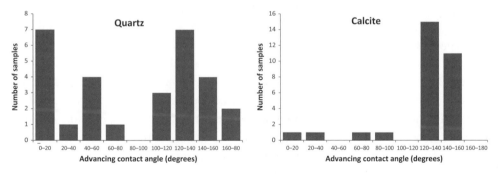

Figure 2.28 Measured contact angles for water displacing oil, the advancing contact angle, measured on 30 crude oil/brine systems on quartz and calcite surfaces. Replotted from Morrow (1990) based on the work of Treiber et al. (1972).

When oil invades the pore space, it enters as the non-wetting phase, while the water preferentially stays close to the solid surfaces. Therefore, the water clings in the smaller pores and throats, narrow crevices, corners and roughness, while the oil occupies the centres of the larger voids. This is shown here, again just to illustrate the concept, as the water residing in the corners of the triangle, while oil occupies

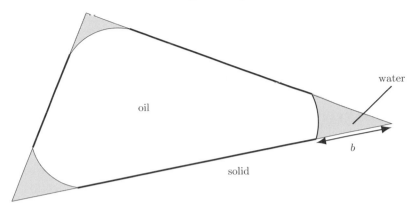

Figure 2.29 The arrangement of oil and water in a single pore after primary oil migration, showing surfaces of altered wettability (in bold). The picture shows a cross-section through a pore – the scale for many reservoir rocks will be of the order of 1–10 μm. Of course, a real pore does not have an exact triangular shape; see Fig. 2.16. The important feature is that the pore space has some corners or roughness, such that oil resides in the centre and water in the corners: the pore is not completely circular with only one phase residing in a cross-section. Any shape of pore is allowed, except a circle; as we showed in the previous section, real pore shapes are highly irregular and do, inevitably, contain corners and crevices in which the wetting phase will reside. Where oil directly contacts the surface, shown by the thick line, surface active compounds will adhere to the surface, altering the wettability, while the water-filled corners will remain water-wet. b is the length of corner in contact with water that remains water-wet: this is typically around 1 μm or less. During waterflooding, where water displaces oil, the contact angle for water invasion will be altered: in many cases the effective contact angle will be greater than 90° over the surfaces indicated by the thick line, meaning that we now have oil-wet regions of the rock, or at least regions with a larger contact angle than before.

the centre. We assume that this water is interconnected: the corners are continuous allowing the water to flow, albeit very slowly, through them.

The oil can also come into direct contact with the solid. When this occurs, the surface active components of the crude oil can adhere to the solid surface. The oil resides in the pore space over geological time scales (millions of years) and so there is plenty of time to reach a position of equilibrium, with the solid coated by high molecular weight asphaltenes, altering the wettability of the surface. In experiments which reproduce this wettability change, the process takes up to 40 days, so millions of years are more than sufficient (Buckley and Liu, 1998).

The asphaltene deposition engenders a change in contact angle. The exact magnitude of this alteration is dependent on the crude oil (the higher the ashpaltene content, the greater the change in contact angle), the brine itself and the temperature and pressure (the amount of deposition is determined by thermodynamic equilibrium of the whole fluid-rock system; there tends to be less wettability alteration at higher temperature), as well as the mineralogy of the rock, giving a wide range of values, as indicated in Fig. 2.28. The change also depends on the capillary pressure; a higher value provides more force for the oil to rupture any protective water films and contact the surface directly at a molecular level (Hirasaki, 1991).

The result is that the solid surfaces in an oil reservoir are no longer strongly water-wet (that is, with a contact angle close to 0) for subsequent waterflooding (displacement of oil by water). This does not necessarily mean that the rock is oil-wet either, as the contact angle is governed by the wide range of factors listed previously. The wettability change in gas reservoirs (where there are no asphaltenes in the hydrocarbon phase) and in most aquifers (even after CO_2 injection) is likely to be much smaller: generally these systems remain water-wet.

We can classify the wettability of an oil/water system by its contact angles, as shown previously in Fig. 1.6. If the contact angle is less than $90°$, it is water-wet; if the contact angle is greater than $90°$ then it is oil-wet. If the contact angle is close to $90°$, and exactly how close is a matter of largely arbitrary definition – typically to within $10°–35°$ – then the system is considered to be neutrally wet.

This categorization is simple, but also generally rather simplistic. Firstly, it is not yet routine to measure contact angle *in situ* for porous media (although see recent developments in this area, such as Armstrong et al. (2012); Andrew et al. (2014b); Schmatz et al. (2015), and the discussion below), and so the apparent contact angle is usually inferred, either from macroscopic indicators of wettability (described later in Chapter 5) or from measurements on smooth surfaces, Fig. 2.28, which do not represent the roughness encountered in the rock (see the next section). Secondly, the contact angle is not a constant value, but will vary dependent on the exact nature of the solid surface and the surface active materials that coat it, the direction of displacement, the local capillary pressure, and the surface roughness. Hence, it is generally inappropriate to attempt to assign a single contact angle, or even some implausibly narrow range, to characterize real systems. Despite these caveats, it is useful to have some categorization of overall wettability based on contact angle and, for reference, the definitions used in Iglauer et al. (2015) are shown in Table 2.3.

A more relevant characterization of wettability is based on the range of contact angle, rather than a fixation on a single value. Brown and Fatt (1956) were the first researchers to coin the term fractionally wet to refer to mixtures of oil-wet and water-wet grains. However, this definition is not consistent with what we now know about wettability changes in reservoirs. Salathiel (1973) used the term mixed-wet to refer to cases where portions of the pore space are water-wet while others are oil-wet, as a result of an altered contact angle in oil-filled pores. Here I will use mixed-wettability in a more restrictive sense to describe rocks that spontaneously imbibe both oil and water (described in Chapters 4 and 5), which means that the

Table 2.3 *Definitions of wettability based on contact angle measurements. From Iglauer et al. (2015).*

Wettability state	Contact angles (degrees)
Complete wetting or spreading of water	0
Strongly water-wet	0–50
Weakly water-wet	50–70
Neutrally wet	70–110
Weakly non-wetting to water	110–130
Strongly non-wetting to water	130–180
Completely non-wetting to water	180

water-wet and oil-wet regions of the pore space are both connected across the rock. Each pore is itself strictly speaking fractionally wet, since it has regions that remain water-wet with surfaces of altered wettability; however, for simplicity, we define the wettability of a pore only by the contact angle of the surfaces contacted by oil.

To recap: when oil first invades the pore space during primary oil migration, the rock is water-wet and oil is the non-wetting phase. However, once oil has resided next to the solid surface for geological times, the surfaces are rarely, if ever, strongly water-wet. Most rocks display some degree of wettability alteration and are best characterized with a broad range of contact angles for subsequent water injection. The exact nature of the wettability change and the distribution of contact angle are, at present, not possible to predict reliably. We can make assessments based on the oil and brine composition, the temperature and pressure of the reservoir and the mineralogy of the rock (Buckley and Liu, 1998). Our understanding is also improving through a quantification of the surface charge and potential, which has a major impact on wettability (Hiorth et al., 2010; Jackson and Vinogradov, 2012); however, this still does not allow a pore-by-pore prediction of contact angle.

2.3.4 Surface Roughness and Contact Angle Hysteresis

While most experimental measurements of contact angle are performed on flat surfaces, ideally ones that are smooth at the atomic level, the natural systems we will consider are not smooth at any scale. This is evident in Figs. 2.3 and 2.7, for instance, where irregularities in the pore-solid interface are evident from the nanometre to micron level. A contact angle still exists, but at the molecular scale of the contact between the fluids and the solid surface; however, locally this surface need not be aligned with the larger-scale features of the pore space, leading to an apparent contact angle as measured at a larger scale, where the fine features of the roughness are not resolved. This value may be very different from the real value, or at least that on a smooth surface.

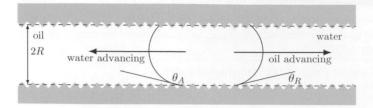

Figure 2.30 A schematic demonstrating contact angle hysteresis for displacement in a pore with a rough surface (shaded). The intrinsic contact angle is approximately 90°. However, depending on the local pore-space geometry, the apparent effective angle, as measured if the rough solid were replaced by an equivalent smooth surface (the dashed line), can be very different. For oil invasion, the displacement is limited by the smallest effective contact angle (the greatest capillary pressure, or the largest pressure required in the oil to push out water from the centre of the pore space), shown as θ_R, the receding contact angle, whose value here is close to 0. For water invasion, the process is limited by the lowest capillary pressure (corresponding to the highest water pressure) and hence the largest angle, indicated by θ_A, which in this example approaches 180° with a negative curvature and capillary pressure (the meniscus bulges into the oil): the medium appears to be non-wetting to both fluid phases; there is no obvious wetting and non-wetting phase. $2R$ is the average separation between the rough surfaces: we define the contact angles such that the capillary pressure for displacement is given by $\sigma \cos \theta / R$, ignoring any interfacial curvature out of the plane of the diagram. For oil to displace water, a capillary pressure $\sigma \cos \theta_R / R$ is required, whereas for water to displace oil the capillary pressure is $\sigma \cos \theta_A / R$.

Fig. 2.30 illustrates the difference between the local (molecular) contact angle and the apparent, macroscopic value. Now, for our applications, it is this apparent macroscopic value that is of interest: we would like to calculate the critical or threshold capillary pressure at which one phase can advance and displace another. This is controlled, directly, through the Young-Laplace equation to the curvature of the fluid interface. On a rough surface, the apparent contact angle or the angle needed to give the same curvature were the surface to be in approximately the same location, but smooth, as indicated by the dashed line in Fig. 2.30, may be very different from the local value; furthermore it can take on a range of values, dependent on the exact nature of the local pore geometry. This means, as we describe below, that the effective angle is dependent on the direction of fluid displacement, leading to contact angle hysteresis.

Displacement, as any process driven by energetic considerations, is limited by the most difficult step, or the one portion of the fluid movement that requires the highest pressure in the invading fluid. Hence, if we consider, for instance, the non-wetting phase displacing the wetting phase over a rough surface, as in Fig. 2.30, the capillary pressure necessary for the displacement, meaning that the non-wetting phase can move through the pore space shown, will be that for the most difficult step, the highest capillary pressure, since this will require the greatest imposed pressure on the non-wetting phase. This is when the effective angle is as small as possible. There will be other configurations with a larger angle which are more favoured (in that they require a smaller pressure in the non-wetting phase) but the contact will move rapidly in these cases; it will be impeded by the configuration which gives the smallest angle, or the highest pressure needed to traverse the surface.

In contrast, if we consider the wetting phase displacing the non-wetting phase, then the most difficult, or limiting, step in the process is that which requires the highest wetting phase pressure or the lowest capillary pressure. This then will be given by the fluid configuration with the largest effective contact angle.

We see a displacement limited by different effective or apparent angles, or strictly speaking different limiting curvatures of the fluid interface, dependent on the direction of displacement. This is contact angle and capillary pressure hysteresis. The contact angle when the non-wetting phase displaces the wetting phase is called the receding angle, θ_R (the wetting phase, assuming that this is the denser phase through which the contact angle is measured, is receding). For the wetting phase displacing the non-wetting phase, the advancing contact angle is θ_A: $\theta_A \geq \theta_R$.

In the example illustrated in Fig. 2.30, it is not evident that we have a clear wetting and non-wetting phase. While the intrinsic angle θ_i is approximately $90°$, and the receding angle θ_R is close to zero, indicating strongly wetting conditions, the advancing angle $\theta_A > 90°$, and so the medium appears to be non-wetting to both fluid phases.

It is difficult to quantify the effect of surface roughness on effective contact angle, since it depends exactly on the nature of the solid. However, a convenient characterization can be obtained from the work of Morrow (1976), who measured advancing and receding contact angles as a function of intrinsic angle on rough surfaces, as shown in Fig. 2.31: note that there is significant hysteresis for intrinsic angles around $90°$.

This argument is different from that presented in Chapter 1, where we used energy balance on a rough surface to derive a modified form of the Young equation (1.15), which implies that a very rough surface is driven to either very strongly wetting or non-wetting conditions. The difference is that here we are explicitly considering displacement. As we show later in Chapter 3, roughness does indeed lead to more apparently wetting conditions since the wetting phase (water) will tend to collect in the crevices and is not completely displaced by the non-wetting phase (oil). Then when water moves back across the surface, it encounters water – and water on water has a contact angle of zero – rather than the solid surface directly.

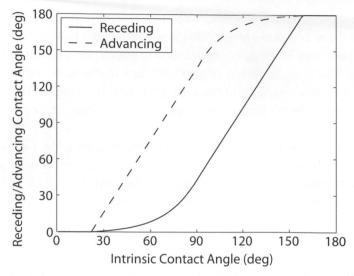

Figure 2.31 The advancing and receding contact angles as a function of the intrinsic contact angle, as measured on rough surfaces. Note that there is significant contact angle hysteresis, or a difference between the advancing and receding angles, when the intrinsic angle is close to 90°. From Valvatne and Blunt (2004) based on Morrow (1976).

This is evident in Fig. 2.31, where, for small intrinsic contact angles, both the receding and advancing angles are close to zero. In reverse, for a strongly oil-wet system (contact angles close to 180°) we move towards complete wetting by oil. However, for intermediate contact angles, less wetting phase resides in the crevices of the pore space and the displacing fluid encounters a bare solid surface, giving the large degree of contact angle hysteresis illustrated schematically in Fig. 2.30.

There are four reasons for contact angle hysteresis, listed here for completeness. We have, so far, covered the first two items, which tend to be the most significant in natural systems.

(1) **Wettability alteration** caused by the sorption of surface-active compounds to the solid surfaces.
(2) **Surface roughness.** Here, as discussed above, there is no change in intrinsic angle at the molecular level, but the effective contact angle, on an equivalent smooth surface, which gives the critical curvature for displacement, is different and depends on the flow direction.
(3) **Chemical heterogeneity.** Consider a surface that contains disconnected regions of different wettability at the microscopic, pore scale. Just as in the case of a rough surface, displacement is limited by the fluid configuration which requires the highest pressure in the invading fluid. Movement across the surface is impeded by those patches with the least favourable wettability. If

we consider oil displacing water, then the oil moves easily across the oil-wet zones (the capillary pressure is negative and the oil pressure is lower than that of the water), but is limited by the water-wet patches. Overall, oil only moves across the surface at a higher pressure than the water with an apparent contact angle determined by the water-wet portions: $\theta_R < 90°$. In reverse, were water to move across the surface, displacement across the water-wet zones is easy (the water pressure is lower than the oil pressure) but a higher water pressure (a negative capillary pressure) is needed to get across the oil-wet regions, giving $\theta_A > 90°$.

(4) **Flow rate.** Even on chemically homogeneous, molecularly smooth surfaces, we can see contact angle hysteresis, which is a function of the speed of movement of the contact line between the two fluids and the surface. This is controlled by the contribution of viscous forces to the fluid movement and is important when we consider very rapid displacements (see, for instance, Thompson and Robbins (1989)). We will return to this later when we consider rapid downwards movement of a wetting phase, in Chapter 6.5.1. However, for the very slow advance of fluids that we encounter in most systems of interest in this book, these effects are small.

2.3.5 Effective Contact Angle and Curvature

Significant contact angle hysteresis is frequently encountered. We are not principally interested in the molecular contact, or the angle which the two-fluid interface makes with the solid at the molecular or atomic level, but the apparent angle necessary to make an accurate determination of local capillary pressure and fluid configuration. In this sense, the contact angle has a somewhat less precise definition, since its value will depend on the resolution at which we image the rock, or the simplified model of the pore-space geometry employed to calculate capillary pressure. For instance, in Fig. 2.30, $2R$ indicates the average separation between the two surfaces shown. We could define a contact angle such that the capillary pressure for an interface residing between these surfaces is given by $\sigma \cos \theta / R$ (see Eq. (1.17) where we ignore any curvature out of the plane of the diagram). This contact angle – as we have shown – will depend on the direction of fluid movement. However, had we instead described the pore space more accurately, but not in sufficient detail to resolve all the roughness, the apparent angle would be different, since we could no longer use Eq. (1.17) to determine capillary pressure, but would require a more detailed characterization that incorporated the geometry and a spatially and directionally dependent contact angle. If we return to our network conceptualization of the pore space, we see that a more complicated or superficially more accurate description of the void geometry does not necessarily lead to more precise calculations of displacement processes, unless the rock-fluid interaction (here the contact angle) is described at the same degree of detail.

For calculations of capillary pressure, the curvature of the fluid interface, Eq. (1.6), is needed, which can now be obtained through direct pore-scale imaging (Armstrong et al., 2012). The contact angle itself can be measured, as shown in Figs. 2.32 and 2.33 (Andrew et al., 2014b; Schmatz et al., 2015). The method can be automated, allowing contact angles to be recorded from large images rapidly (Klise et al., 2016). This work builds on previous techniques which froze rock samples containing fluids and then imaged the distributions using scanning electron microscopy (Robin et al., 1995; Durand and Rosenberg, 1998). In the cases shown, the solid surface appears smooth in the image; however, in Fig. 2.33 the grains of the Ketton limestone are micro-porous and rough, and an effective angle is seen, which ignores these complexities. The micro-porosity itself is shown in Fig. 2.7 and it is assumed that the porous grains remain fully saturated with brine in this experiment. Since we have a rough surface, the texture of the grains themselves may affect the apparent contact angle.

These effective angles could be used in calculations were we, for instance, to describe the pore space of the rock shown in Fig. 2.33 at the resolution of the X-ray scan combined with the contact angles measured at the same resolution. Although this requires detailed imaging and sophisticated experimental methods, it does allow the computation of multiphase fluid configurations in rock samples at representative reservoir conditions.

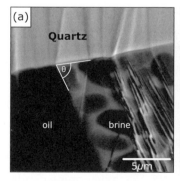 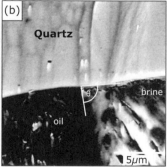

Figure 2.32 Contact angles measured between oil and brine in a sandstone. The images are acquired using a BIB-SEM technique with 8 μm between slices and a resolution within a slice of much less than 1 μm. Here we see weakly water-wet conditions, with contact angles close to, but less than, 90° on a quartz surface that has been aged in crude oil. From Schmatz et al. (2015).

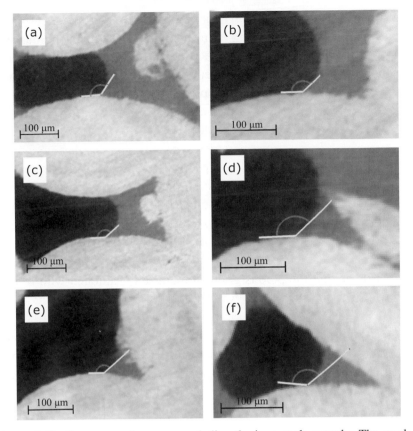

Figure 2.33 Contact angles measured directly in a rock sample. The rock is Ketton limestone, shown in Fig. 2.3, while the fluids are CO_2 at super-critical conditions (10 MPa pressure and a temperature of 50°C) and brine. The distribution of fluid at the end of brine injection is shown, when the CO_2 is disconnected in the pore space. The voxel size is approximately 2 μm. The three-phase contact line between the two fluids and the solid is identified. A plane perpendicular to the contact line is then taken at selected locations and the contact angle is measured. The angle through CO_2 is shown: the contact angle (measured through the water) is the complement, and has a value between approximately 39° and 53° for the six locations shown. The solid is itself micro-porous with roughness at a scale that is not resolved in this image, Fig. 2.7. From Andrew et al. (2014b).

This direct approach has been used successfully to predict fluid configurations and multiphase properties using contact angles measured using a combination of micro-CT scanning, high-resolution SEM images and imaging to determine the chemical composition of the rock surface (Idowu et al., 2015). Other techniques, such as atomic force microscopy, can also assess wettability at the sub-pore scale and from this deduce effective wetting properties at a larger scale (Hassenkam et al., 2009). This is less complex than deducing the molecular-scale contact angle, requiring sub-micron imaging throughout the sample of interest, and using this for larger-scale calculations, accounting for every detail of the micro-porous structure of the rock.

This concludes our discussion of porous media, networks and wettability. In subsequent chapters we will describe displacement processes in detail.

3

Primary Drainage

We will now describe fluid displacement, or how one fluid replaces another in a porous medium. We will start with a pore space that is entirely saturated with the wetting phase and allow a non-wetting phase to enter. This is a primary drainage process. The word primary (meaning first) indicates that this is the first time the non-wetting phase enters the pore space: we start with a wetting phase saturation of 1. Drainage, in general, refers to the displacement of a wetting phase by a non-wetting phase.

There are three common examples of primary drainage in natural systems, listed below.

(1) The migration of oil and gas from source rock (shale) to a hydrocarbon reservoir. Here the hydrocarbon migrates upwards under buoyancy (it is less dense than the brine in the pore space) until it encounters a barrier to movement, under which it then collects. This accumulation becomes the reservoir. Locally, oil invades the pore space driven by the buoyancy force caused by the density difference between oil and water.

(2) The injection of CO_2 into a saline aquifer. Here CO_2 is forced into the pore space of the rock, displacing brine, for long-term underground storage. The driving force, providing the local capillary pressure, comes from the CO_2 pressure in the injection well, which is higher than in the resident brine. Furthermore, as in oil migration, the CO_2 is less dense than brine and will rise in a storage aquifer and again collects under impediments to flow. This is also a primary drainage process.

(3) Mercury injection capillary pressure (MICP) measurements. These are measurements routinely performed on small (cm-sized) cylindrical rock samples (called plugs) to assess the pore-size distribution, as discussed below. The rock is cleaned and dried and placed in a vacuum that acts here as the wetting phase. Mercury is the non-wetting phase. It has a very high surface tension, since it

is metallic (see Chapter 1), and a correspondingly high interfacial tension with the solid, since not only the bonds in the solid are broken, but the bonding in the metal too. Indeed, the interfacial tension between mercury and the solid is higher than the solid and a vacuum (where only the solid bonds are broken), resulting in a contact angle that is greater than 90° when measured through the mercury: this can be seen from the Young equation (1.7).

3.1 Entry Pressures and Fluid Configurations

Consider an experiment where the non-wetting phase initially surrounds a porous medium filled with the wetting phase. The pressure in the non-wetting phase is then increased very slowly. From the Young-Laplace equation (1.6), the non-wetting phase will first have sufficient pressure to invade the widest portions of the pore space, where the radius of curvature of the interface between the phases is as large as possible, and will then progressively fill smaller regions. If we have very slow displacement, the fluid configurations will progress from one equilibrium arrangement to another with displacement occurring when the capillary pressure increases.

We now zoom into the rock and study displacement at the micro-scale. If we consider that a pore or throat has a circular cross-section of radius r with no curvature along its length, then the capillary pressure for an interface residing across the pore is, from Eq. (1.16):

$$P_c = \frac{2\sigma \cos \theta_R}{r},$$ (3.1)

where now we have used the receding contact angle θ_R for drainage, assuming that the denser phase initially saturates the pore space. We call this interface a terminal meniscus (TM) since it straddles a pore or throat, blocking flow through its centre. Throats of non-circular cross-section will be treated later, once we have discussed the displacement sequence encountered in drainage.

Primary drainage is a displacement process controlled by the threshold entry pressures for invasion: that is the pressure necessary for the non-wetting phase to enter different regions of the porous medium. We have considered in Chapter 1 positions of capillary equilibrium and the corresponding pressure differences between the phases. However, we are interested in *displacement* where one fluid moves to push out another. In this context we are not strictly looking at an equilibrium configuration of fluids – where everything is at rest – but how the fluids move. This has a number of implications. Firstly, as explored in Chapter 2, and explicit in Eq. (3.1), the contact angles we use in any calculations will depend on the direction of flow: here for drainage we will use the (water) receding angle θ_R. Secondly, we need to clarify precisely what we mean by a threshold capillary pressure. The

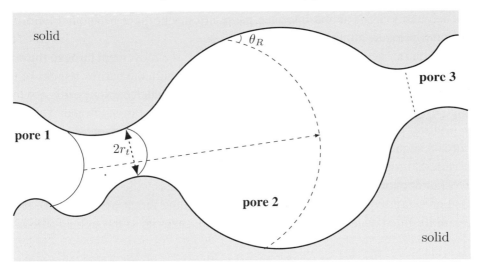

Figure 3.1 An illustration of non-wetting phase advance from one pore to another through a throat. The pore structure is as illustrated in Fig. the throats are indicated by dotted lines. As the capillary pressure increases, the non-wetting phase advances further towards the throat and the radius of curvature of the interface decreases, as shown by the solid arcs. The governing contact angle is θ_R (in this example with a value around 30°) since the wetting phase is receding. The maximum capillary pressure is reached at the throat whose inscribed radius is r_t. Once the interface, the TM, has passed the throat, it moves rapidly through the adjacent pore at a lower capillary pressure, indicated by the dashed line.

concept is illustrated in Fig. 3.1, where we show the non-wetting phase entering a throat. Let us assume that the capillary pressure is given by Eq. (3.1) with a fixed contact angle. This equation gives the capillary pressure not just in a throat, but applies regardless of where the fluid/fluid interface resides. As the capillary pressure increases, the radius of curvature r must decrease: this is accommodated by the interface, the terminal meniscus, moving further into the narrow region of the pore space, away from the pore centre and towards the throat. Since we assume a very low flow rate, or a very gradual increase in imposed capillary pressure, this process occurs slowly. The capillary pressure increases progressively until the interface straddles the throat: this is the configuration with the largest capillary pressure. Indeed, since we have defined a throat as the location where the pore space is locally at its narrowest (see page 43), we assume that this is where the capillary pressure reaches a local maximum value. The threshold capillary pressure is that required for this interface to move across the throat – hence it depends on the receding contact angle.

Once we have exceeded this capillary pressure, by even an infinitesimal amount, the interface moves into wider regions of the pore space, where the local capillary pressure is lower. At the pore scale the capillary pressure therefore increases and then decreases again as the interface passes from one pore to another, with the maximum pressure attained at the throat.

Based on our network concept of the pore space, the movement through the pore space in primary drainage is limited by progress through the narrow regions of the pore space. As a consequence, the capillary pressure needed to fill a pore is not controlled by the size of the pore itself, but by the size of the throat through which the non-wetting phase must pass to access that pore. The capillary pressure therefore records a sequence of throat invasions: the non-wetting phase first passes through the widest throats and then it fills progressively narrower ones. Since pore filling can occur at a lower capillary pressure (the pore space is wider and the non-wetting phase pressure required for invasion is lower than the throat by definition), pores are rapidly filled once the non-wetting phase squeezes itself through an adjoining throat.

This process can be observed experimentally in micro-model experiments, where displacement occurs in a two-dimensional lattice of pores and throats. Fig. 3.2 illustrates pore and throat filling in drainage from the classic work of Lenormand et al. (1983). Here we see filling of pores and throats as the non-wetting phase advances through the network, as well as the filling of throats between two previously filled pores. The same behaviour has also been observed in three-dimensional porous media. Fig. 3.3 illustrates a primary drainage experiment in a packing of sintered glass beads imaged using confocal microscopy with refractive index matched fluids. The progressive invasion of the pore space can be observed: at the end of the displacement, the wetting phase is confined to small pores and crevices.

In a macroscopic experiment to measure capillary pressure, the rock sample is a few centimetres (cm) across, and contains millions of pores and throats. An increasing external pressure difference between the non-wetting and wetting phases is imposed. Alternatively, injection may proceed at a low flow rate and the pressure difference between the phases is recorded; however, in most situations this pressure difference increases monotonically over time. In a field setting, buoyancy or viscous forces impose a rising pressure difference between the phases over time, as CO_2 is injected, or more oil rises from the source rock.

If the macroscopically imposed pressure difference between the phases increases, what happens when, locally, the capillary pressure decreases? Let us consider the particular case of mercury injection, where the wetting phase is a vacuum, held at a constant zero pressure. The mercury pressure is the capillary pressure. If we consider very slow flow, so ignore any pressure differences in the mercury as it moves into the rock, then the externally imposed mercury pressure is the capillary pressure as the fluid interface advances towards a throat. However, once the throat has been passed, the interface resides in a wider region of the pore

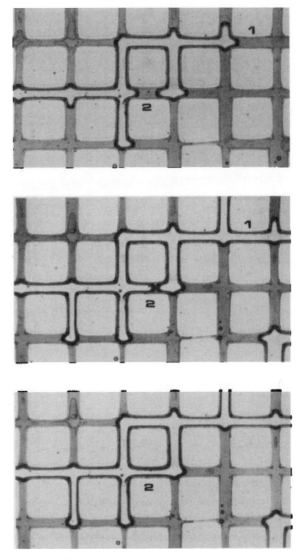

Figure 3.2 Micro-model observations of drainage in a two-dimensional lattice etched in glass: the distance between the pores is 4 mm. The non-wetting phase, white, displaces a wetting phase, grey, from left to right. Advance through the throat labelled 1 (middle figure) is succeeded by the filling of a throat between two previously filled pores, 2 (bottom figure). This filling is only possible because the wetting phase can escape through the corners of the pore space. From Lenormand et al. (1983).

space with a lower local capillary pressure. The vacuum pressure is still zero, so the mercury close to the interface is now at a lower pressure than before. In other regions of the pore space, however, the mercury is still creeping gradually towards smaller throats, and therefore has a pressure equal to the externally imposed value.

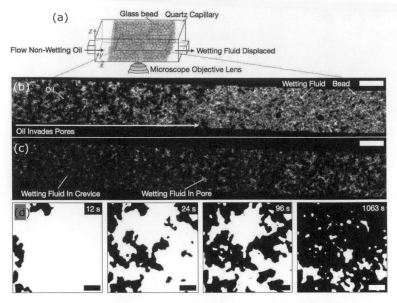

Figure 3.3 (a) Primary drainage in a bead pack imaged using confocal microscopy. (b) Optical section through part of the medium, taken as the oil displaces the wetting fluid at a low flow rate. Bright areas show the pore space saturated with the wetting phase. (c) After the invasion of around nine pore volumes of the oil, some wetting fluid remains trapped in the crevices and pores of the medium, as indicated. (d) Time sequence of micrographs, with the pore space subtracted; these are binary images showing oil in black. The upper and lower arrows in the last frame show wetting fluid trapped in a crevice or in a pore, respectively. Scale bars in (b) and (c)–(d) are 500 μm and 200 μm, respectively. From Datta et al. (2014b).

Fluid flows from high to low pressure, and so – regardless of how slowly the mercury is injected from an external source – within the pore space itself there is a pressure gradient. Mercury will retract from regions of high pressure and flow towards regions of low pressure. The consequence is a rapid filling of the wide pores and a re-arrangement of fluid to achieve a new position of capillary equilibrium.

This rapid filling of multiple pores is called a Haines jump and was first described in the context of water flow in soils, where the invasion of air to displace water was observed to be different from the reverse process of wetting a dry soil (Haines, 1930). The dynamics of this process is complex, involving a combination of viscous (pressure difference) and capillary forces. The retraction of the non-wetting phase is not, at the pore scale, a drainage process: technically it is an imbibition event, since this is displacement of the non-wetting phase by the wetting phase and will be discussed further in Chapter 4.1.1.

If the capillary pressure continues to increase, new positions of equilibrium will be reached at higher pressures, where the non-wetting phase proceeds towards progressively narrower throats, as illustrated in Fig. 3.3. Whenever a new maximum is reached in both the local and the imposed capillary pressure, all the non-wetting phase that may have receded (having been displaced by the wetting phase) during a Haines jump will have re-invaded these regions and advanced further.

3.1.1 Wetting Layers

When the non-wetting phase moves through a pore or throat, it does not entirely remove the wetting phase from that region of the pore space. Instead, the wetting phase is retained in the cracks, crevices and corners of the pore space, leaving the non-wetting phase in the centres, as shown in Fig. 2.29, and observed experimentally; see Fig 3.3. We call these wetting layers. This is distinct from a wetting film: a film is typically of molecular thickness and while it will impact the surface properties of the solid it coats (such as the effective interfacial tension) it does not allow any significant movement of fluid. Over the time and length scales of interest to the displacements in this book, namely oil recovery or CO_2 storage, with displacements spanning kilometres over years to decades, molecular films do not allow any appreciable flow. Wetting *layers* are different: they are of macroscopic thickness, have bulk or close-to-bulk properties and do allow flow, albeit slow flow, over reasonable time scales. They are generally of a thickness that is related to the pore size, typically microns in reservoir rocks.

The configuration of fluid is illustrated in Fig. 3.4. We assume that the non-wetting phase with its associated terminal meniscus has passed through this part of the pore space and now resides elsewhere. However, interfaces between the fluids are still seen, in capillary equilibrium, in the corners. These menisci do not block the centres of pores and throats and are called arc menisci (AM) to distinguish them from the terminal menisci (TM) defined previously. We simplify the complex geometry of a real pore structure with an idealized element (in this case a scalene triangle). We preserve the inscribed radius to capture the entry capillary pressure, but now we also need to reproduce the volume and conductance of the fluid in the corners, as well as the local capillary pressure. If the radius of curvature of the AM is r and we ignore the curvature out of the plane of the diagram, then the capillary pressure is given by:

$$P_c = \frac{\sigma}{r}. \tag{3.2}$$

Note that there is no factor of 2 in this equation, unlike Eq. (3.1), for instance, for a TM, that generally has two significant, and approximately equal, radii of curvature

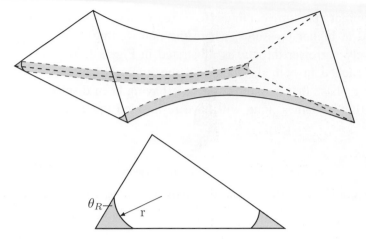

Figure 3.4 The fluid configuration in parts of the pore space invaded by non-wetting phase. The wetting phase (shaded) resides in corners and roughness in the pore space as a layer, with a contact angle θ_R and a local capillary pressure given by Eq. (3.2) with a radius of curvature r as shown. The fluid interfaces, called arc menisci, AM, do not block the pore space. The complex geometry of the pore space is represented by a simplified model. The lower figure shows the throat (narrowest part of the pore space) in cross-section, represented by a scalene triangle. The wetting phase can only reside in corners where the half angle β obeys Eq. (3.3): the top corner of the triangle, with $\beta \approx 45°$ cannot support the wetting phase for $\theta_R \geq 45°$.

to span a pore or throat. In capillary equilibrium, the local capillary pressure is constant throughout the sample, and so we have a fixed average radius of curvature of the AM determined by the imposed capillary pressure, Eq. (1.6): this curvature decreases as the capillary pressure increases, forcing the fluid further into the corners.

Not every corner contains a wetting phase: to be able to maintain a positive and increasing capillary pressure, then for a corner of half angle β (see Fig. 3.4) we require:

$$\beta + \theta_R < \pi/2. \tag{3.3}$$

The presence of the wetting phase clinging to the corners of the pore space is the manifestation of the Wenzel regime for a rough surface mentioned in Chapter 2.3, page 13. In any crack or crevice in the pore space, the wetting phase can reside if Eq. (3.3) is satisfied. The non-wetting phase will still be able to encounter the surface directly, and so this is not, at the micro-scale, complete wetting, but some wetting phase is always present in the pore space, for any finite capillary pressure.

In the micro-model experiments shown previously (Fig. 3.2), escape of the wetting fluid out of corners in the pore space allows the non-wetting phase to displace

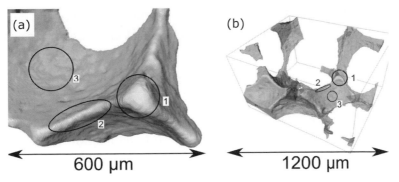

Figure 3.5 Fluid interfaces imaged directly in Ketton limestone. In (a) only the non-wetting phase, CO_2, is shown. There are three types of interface. The first, labelled 1, is a terminal meniscus, TM, where the two principal radii of curvature are approximately equal, straddling a pore or throat. The second is an arc meniscus, AM, where one of the principal radii of curvature is much smaller than the other, indicating a wetting layer in the corner or roughness of the pore space. The third type has a negative curvature and is the direct contact of CO_2 with the solid grains which bulge out into the pore space. (b) shows in pink the location of the wetting phase, allowing a visual confirmation of the types of interface present. The colours on the contour of the CO_2 phase indicate the curvature, which is approximately the same for both the TM and AM, suggesting local capillary equilibrium. From Andrew et al. (2015). (A black and white version of this figure will appear in some formats. For the colour version, please refer to the plate section.)

fluid from throats between two pores already filled with wetting phase: the wetting phase is not trapped.

Fluid menisci have been observed directly in three-dimensional rock samples. An example is shown in Fig. 3.5, where CO_2 as the non-wetting phase has been imaged in Ketton limestone. Three types of interface are shown. The first is a terminal meniscus with significant curvature in two directions, straddling a pore or throat. Indeed, a TM is identified from the image by measuring the radii of curvature of the interface: a TM is defined as a region of the interface where the two principal radii are approximately equal. An arc meniscus, AM, is located where one principal radius of curvature is much smaller than the other, and represents an interface in a corner of the pore space, but not directly blocking the centre of a pore or throat. The third type of interface has a negative curvature and represents the contact of the CO_2 directly with the solid: in this case the Ketton grains bulge out into the pore space (see Fig. 2.6). The total curvature κ, see Eq. (1.6), is approximately the same for all the fluid/fluid interfaces, both TM and AM, indicating local capillary equilibrium.

At this stage, the key features of the displacement to note are three-fold: (i) that displacement is limited by the throats; (ii) there is a sequence of configurations of capillary equilibrium uniquely defined by the throats when the capillary pressure reaches a local maximum value; and (iii) that after invasion of the non-wetting phase, the wetting phase resides in the corners and crevices of the pore space, maintaining connectivity.

3.1.2 Entry Pressures for Irregular Throats

If the pore or throat has an irregular cross-section, we cannot use such a simple expression for the entry pressure as Eq. (3.1) (Mason and Morrow, 1984, 1991; Ma et al., 1996; Lago and Araujo, 2001). We apply the approach developed in Chapter 2.3.1 and start with Eq. (2.18) for the change in free energy for a displacement using the receding contact angle for drainage and labelling phase 1 as wetting, w, and phase 2 and non-wetting nw,

$$\Delta F_L = -\sigma \left[\kappa \Delta A - (\Delta L_{wnw} + \Delta L_{nws} \cos \theta_R) \right], \qquad (3.4)$$

where ΔA is the change in area occupied by wetting phase, ΔL_{wnw} is the change of length of any arc menisci, while ΔL_{nws} is the change in contact length of the non-wetting phase with the solid. We have assumed a throat of constant angular cross-section and ignore any curvature perpendicular to this cross-section.

We will calculate the change in free energy for a displacement from a configuration with the cross-section full of wetting phase, to one with non-wetting phase in the centre and wetting layers in the corners, as shown in Fig. 3.6, following the approach of Mason and Morrow (1991) as applied by Øren et al. (1998) in pore-scale modelling. The threshold entry capillary pressure, or the corresponding interfacial curvature of the arc menisci, AM, in the corners, $\kappa = 1/r$, is found when ΔF_L in Eq. (3.4) is zero:

$$\frac{\Delta A}{r} - \Delta L_{wnw} - \Delta L_{nws} \cos \theta_R = 0 \qquad (3.5)$$

The application of Eq. (3.5) to find r and hence the capillary entry pressure now becomes an exercise in geometry. Note that the curvature is independent of the interfacial tension.

First, some useful relations are defined. For a corner of half angle β, where the radius of curvature of the interface is r and the contact angle is θ_R, the length b of wetting phase in the corner, see Fig. 2.29, is given by

$$b = r \frac{\cos \theta_R + \beta)}{\sin \beta} = r \left(\cos \theta_R \cot \beta - \sin \theta_R \right) \equiv S_b r, \qquad (3.6)$$

where S_b is a coefficient defined by Eq. (3.6). Since we only consider curvature in this plane, $\kappa = 1/r$. The length of the AM is,

$$L_{AM} = 2r \left(\frac{\pi}{2} - \theta_R - \beta \right) \equiv S_L r, \qquad (3.7)$$

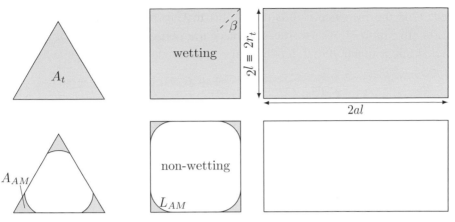

Figure 3.6 Configuration changes during primary drainage. We compute the change in free energy for a displacement where the cross-section of a throat changes from being full of wetting fluid (top) to containing non-wetting phase in the centre and wetting phase in layers (bottom). We consider three example cases: an equilateral triangle, a square and a rectangle. A_t is the total area of the throat, A_{AM} is the area of the arc meniscus, L_{AM} is the length of the meniscus and β is the half angle of the corner. For the rectangular example, side length $2l$, where $l = r_t$ the inscribed radius, and the aspect ratio $a > 1$, we assume that no wetting phase remains in the corners: $\theta_R \geq 45°$.

while the area of the wetting phase in the corner is a somewhat more complex expression:

$$A_{AM} = r^2 \left(\frac{\cos \theta_R \cos(\theta_R + \beta)}{\sin \beta} - \frac{\pi}{2} + \theta_R + \beta \right) \equiv S_A r^2, \qquad (3.8)$$

where again S_L and S_A are geometrical coefficients.

Now consider a throat with n corners, and for simplicity we assume that the contact angle and half-angles β are all the same. Let the total cross-sectional area of this throat be A_t and each side have length l, then in Eq. (3.5) we can define $\Delta A = A_t - n A_{AM}$, $\Delta L_{nws} = nl - 2nb$ and $\Delta L_{wnw} = n L_{AM}$ using Eqs. (3.6)–(3.8). Substituting these relations into Eq. (3.5) we obtain a quadratic equation for r:

$$A_t - n S_A r^2 - n S_L r^2 - nl \cos \theta_R r - 2n S_b \cos \theta_R r^2 = 0, \qquad (3.9)$$

where the physically valid root for r is given by the radius which fits in the throat.

With some algebra, a solution may be obtained, and indeed generalized for throats where the half-angles in each corner are different (Ma et al., 1996; Øren et al., 1998). We find for $P_c = \sigma/r$,

$$P_c = \frac{\sigma \left(1 + 2\sqrt{\pi G} \right) \cos \theta_R F_d(\theta_R, G)}{r_t}, \qquad (3.10)$$

where r_t is the inscribed radius of the throat, G is the shape factor of the cross-section (the ratio of area to the square of the perimeter), Eq. (2.2), and F_d is a dimensionless function that depends on contact angle and the shape factor:

$$F_d(\theta_R, G) = \frac{1 + \sqrt{1 + 4GD/\cos^2 \theta_R}}{1 + 2\sqrt{\pi G}}, \tag{3.11}$$

where D depends on the coefficients S_b, S_L and S_A. If we have AM present in every corner, we can write

$$D = \pi - \frac{2}{3}\theta_R + 3 \sin \theta_R \cos \theta_R - \frac{\cos^2 \theta_R}{4G}. \tag{3.12}$$

For a discussion of the implications of this calculation, we can simplify these expressions and write

$$P_c = \frac{C_D \sigma \cos \theta_R}{r_t}, \tag{3.13}$$

where the dimensionless coefficient is defined through Eqs. (3.11) and (3.12). For reference $C_D = 2$ if we have a throat of circular cross-section, Eq. (3.1).

We will now present a simple case to illustrate the concept. If we consider a strongly water-wet rock with $\theta_R = 0$, then from Eq. (3.12), $D = \pi - 1/4G$ and $F_D = 1$ in Eq. (3.11). Hence from Eq. (3.13):

$$C_D = 1 + 2\sqrt{\pi G}. \tag{3.14}$$

We can now address the two cases shown in Fig. 3.6. For a square of side $2l$ where $r_t = l$, we have $G = 4l^2/64l^2 = 1/16$ and $C_D = 1 + 0.5\sqrt{\pi} \approx 1.89$. For an equilateral triangle of side $2l$, $A_t = \sqrt{3}l^2$ and $G = \sqrt{3}/36$, giving $C_D \approx 1.78$. If we take the extreme case of a slit-like throat, where $G \to 0$ ($\beta \to 0$) then $C_D \to 1$. In general $2 \geq C_D \geq 1$ with the largest entry pressures for circular throats tending to half the entry pressure (for the same inscribed radius) as the throat becomes more elongated with smaller corner angles.

The full variation of F_d is provided for reference in Fig. 3.7 (Øren et al., 1998). It is evident that this correction factor is principally a function of contact angle and is close to 1 for strongly wetting conditions.

One final example, from Fig. 3.6, helps see the variation in capillary pressure with shape. If Eq. (3.3) is not obeyed, then there is no wetting phase in the corners after primary drainage. This simplifies the calculations, as Eq. (3.5) becomes:

$$\frac{A_t}{r} - \Delta L_{nws} \cos \theta_R = 0, \tag{3.15}$$

where $\Delta A = A_t$ the total area of the throat, as it becomes completely filled. Fundamentally, this equates the capillary pressure times the cross-sectional area to the

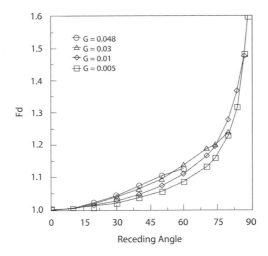

Figure 3.7 The dimensionless correction factor F_d in Eq. (3.10) used to compute the threshold entry pressure in drainage as a function of receding contact angle θ_R. Curves for different shape factors are shown. From Øren et al. (1998).

change in surface energy from the contact around the perimeter of the throat. For a rectangle of aspect ratio a where $a > 1$ and minor side length $2l$, $A_t = 4al^2$ and $\Delta L_{nws} = 2l(1 + a)$, which is simply the perimeter length. Then we find,

$$\frac{\sigma}{r} \equiv P_c = \frac{(1 + a)}{a} \frac{\sigma \cos \theta_R}{r_t}, \tag{3.16}$$

where we have used $r_t = l$. In Eq. (3.13), $C_D = (1 + a)/a$: for a square, $a = 1$, $C_D = 2$ and we have the same expression for capillary pressure as a circular tube of the same inscribed radius: for $a \to \infty$, or a slit, $C_D \to 1$. The same trend, as we have noted above, albeit with more algebraically complex expressions, is seen when the wetting phase is retained in the corners.

This analysis has been confined to situations where the cross-section of the throat is constant, such that the terminal meniscus, TM, has constant area. In reality, the throat represents a local minimum in the inscribed radius of the pore space (see the discussion in the previous chapter, page 43), and so the full three-dimensional energy balance includes the change in interfacial energy as the TM passes into the wider pore space: it is this that results in Haines jumps, since new local configurations of capillary equilibrium can now be reached in the wider regions of the pore space.

3.2 Macroscopic Capillary Pressure in Drainage

We can define a macroscopic capillary pressure, P_{cm}, as a function of saturation, traditionally the saturation of the wetting phase, S_w. As the capillary pressure increases, the wetting phase saturation decreases, as it is displaced from the pore space. This macroscopic pressure may be different from (for primary drainage, we mean larger than) the local capillary pressure difference across a meniscus during a Haines jump. P_{cm} only represents the local capillary pressure when there is no, or infinitesimal, flow, either imposed externally, or within the rock due to internal pressure differences, and the system is hence everywhere in equilibrium. This is normally only achieved in very low flow-rate experiments when a new local maximum in the capillary pressure is reached. We will assume that the flow is infinitesimally slow and neglect any changes in pressure across the sample caused by the movement of fluid. This is an important simplification that will be explored properly later when we introduce flow in Chapter 6.

The measured macroscopic value of capillary pressure is a monotonically decreasing function of wetting phase saturation, in contrast to the local capillary pressure which controls a specific filling event; it fluctuates as the terminal menisci pass through alternating regions of low and high entry pressure.

Example capillary pressure curves, $P_{cm}(S_w)$, are shown for Doddington sandstone, Berea sandstone and Ketton limestone in Fig. 3.8. They were measured using mercury injection at a commercial laboratory; images of these rocks have been shown in Figs. 2.5, 2.14 and 2.6, respectively. The reason for displaying two curves for Berea sandstone, taken from different blocks of stone, is to demonstrate that the differences between samples from the same quarry are much smaller than between different rock types.

These capillary pressures indicate the sizes of the throats invaded in the pore space, Eq. (3.1), where $r = r_t$ the inscribed radius of the throat, and we assume that we have largely circular throats: in general we should use a more accurate expression, Eq. (3.10), but we do not know, from a capillary pressure measurement, the shapes of each throat in the rock. Mercury is the non-wetting phase with a receding contact angle variously estimated at around 130° and 140°. Implicit in Eq. (3.1) is that the contact angle is measured through the wetting phase. In what follows, I will take $\theta_R = 50°$ (corresponding to a mercury contact angle of 130°) and an interfacial tension of 485 mN/m, giving $2\sigma \cos \theta_R = 624$ mN/m. Therefore, a capillary pressure of 1 MPa corresponds to a throat radius of around 0.62 μm.

There is an entry capillary pressure, P_c^{entry}, when the non-wetting phase is first able to pass into the porous medium. This represents the invasion of the largest throats adjacent to the surface of the sample and their neighbouring pores. In the capillary pressure graphs, this is the pressure when the curve first deviates from a wetting saturation of 1. In Doddington this is around 0.02 MPa, or a radius of 29 μm (Fig. 3.8), around 0.04 MPa with a corresponding radius of approximately 15 μm for the two Berea samples, while for Ketton it is at 0.015 MPa or a radius of 40 μm. As the capillary pressure is progressively increased, the non-wetting

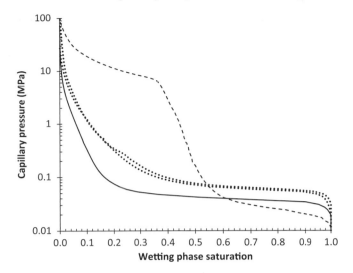

Figure 3.8 The primary drainage capillary pressure for Doddington sandstone (solid line), two different samples of Berea sandstone (dotted lines) and Ketton limestone (dashed line). Pore-space images can be seen in Figs. 2.5, 2.14 and 2.6, respectively. The wetting phase saturation in this case is the saturation of the vacuum that mercury displaces. These measurements were performed by Weatherford Laboratories in East Grinstead, UK, on samples provided by Imperial College London.

phase pressure rises and the non-wetting phase is able to access more and more of the pore space. P_c^* defines the pressure when the non-wetting phase first spans the porous medium, meaning that it is connected across the sample. As we discuss later, in a sufficiently large statistically homogeneous sample, P_c^{entry} (with careful definition) and P_c^* should be the same. As the pressure increases further, so too will the saturation of the non-wetting phase and it will become increasingly well connected, filling the centres of the larger pores.

In the Doddington sample, with large pores, most of the wetting phase is displaced with a capillary pressure of around 0.1 MPa, corresponding to radii of approximately 6.2 μm and larger; for Berea, displacement continues to larger pressures, requiring $P_c \approx 1$ MPa to displace most of the void space, since the pore sizes are smaller. There is still further displacement at higher capillary pressure, representing the wetting phase becoming more and more pressed into the corners of the pore space and, for Berea, the invasion of clays. Since the wetting phase is a vacuum it is possible to displace everything at a sufficiently high pressure and fill the pore space of the rock entirely with mercury.

For Ketton, we also see displacement through wide throats with radii that are several microns across, corresponding to the void spaces between the ooliths, shown in Fig. 2.6. However, this only displaces approximately half the pore space:

the remainder of the pores and throats reside in much smaller micro-porosity within the ooliths that are only accessed at pressures of around 10 MPa, corresponding to a radius of only 0.062 μm, or 62 nm. This is broadly consistent with the images of micro-porosity shown in Fig. 2.7.

For completeness, we also show measured primary drainage capillary pressure curves for some of the other samples whose images were provided in Chapter 2. Fig. 3.9 shows capillary pressures measured on four of the samples illustrated in Fig. 2.3: Bentheimer sandstone and three carbonates: Estaillades, Mount Gambier, and Guiting. Here a wide range of pore structure is revealed, from the sandstone sample with a relatively homogeneous arrangement of large pores, similar to Doddington (Fig. 3.8), to Guiting, where the capillary pressure is much higher, implying that only micro-porosity (pores with a radius of less than around 1 μm) are connected. Mount Gambier has a very open pore space with a low entry pressure, implying that some very large pores are present, while Estaillades shows a dual-porosity behaviour similar to Ketton, with two stages of invasion for macro- and micro-pores. Further discussion is reserved until after we have presented a quantitative analysis, below.

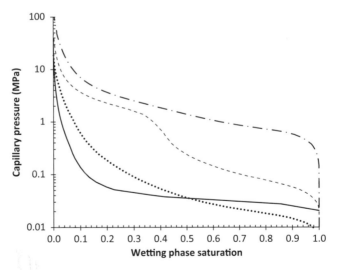

Figure 3.9 The primary drainage capillary pressure for Bentheimer sandstone (solid line), Mount Gambier (dotted line), Estaillades (dashed line) and Guiting limestone (dot-dashed line showing the highest capillary pressures). Pore-space images are shown in Fig. 2.3. The measurements on the carbonates were performed by Weatherford Laboratories in East Grinstead, UK, on samples provided by Imperial College London. The capillary pressure for Bentheimer was measured at Imperial College London (Reynolds and Krevor, 2015).

3.3 Bundle of Tubes Model and the Throat Size Distribution

Capillary pressure can be used to estimate a throat size distribution by assuming a particular model of the pore structure. The simplest model of a porous medium is to consider a bundle of parallel cylindrical tubes, of equal length L, as shown in Fig. 3.10. Imagine a very slow increase in pressure, such that the tubes only fill one at a time from a reservoir to the left of the tubes, while the wetting phase escapes on the right. In this case the microscopic and macroscopic capillary pressures are the same: $P_c = P_{cm}$.

The tubes fill in order of radius, with the largest filled first at a capillary pressure given by Eq. (3.1). If the largest tube has a radius $r_t = r_{max}$ then:

$$P_c^{entry} = P_c^* = \frac{2\sigma \cos \theta_R}{r_{max}}. \tag{3.17}$$

If the smallest tube has a radius $r_t = r_{min}$, then when $P_c = 2\sigma \cos \theta / r_{min}$ all the tubes are filled and $S_w = 0$; there is no further displacement.

Imagine now that we have a distribution of tube radius, such that between a radius r and $r + dr$ there are $f(r)dr$ tubes. Then, if we just filled a tube of radius $r_t = r$, the capillary pressure is $2\sigma \cos \theta_R / r$. The wetting phase volume is:

$$V_w = \int_{r_{min}}^{r} \pi r^2 L f(r) dr, \tag{3.18}$$

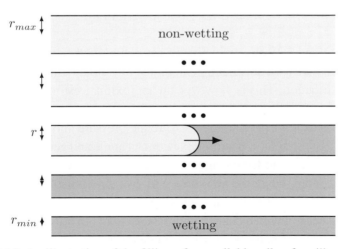

Figure 3.10 An illustration of the filling of a parallel bundle of capillary tubes in primary drainage. The tubes are arranged in order of size, which also represents the filling sequence: the largest are filled first. r represents the radius of the tube that has just been filled. Using this model it is possible to find a relationship between the capillary pressure and throat size distribution, Eq. (3.23).

from which the saturation, S_w, is:

$$S_w = \frac{\int_{r_{min}}^{r} r^2 f(r)dr}{\int_{r_{min}}^{r_{max}} r^2 f(r)dr}, \tag{3.19}$$

noting that πL cancels in the numerator and denominator.

The change in saturation with radius is:

$$r\frac{dS_w}{dr} \equiv \frac{dS_w}{d\ln r} = \frac{r^3 f(r)}{\int_{r_{min}}^{r_{max}} r^2 f(r)dr} \equiv G(r), \tag{3.20}$$

where $G(r)$ is the normalized distribution of volume on a logarithmic scale of radius: a fraction $G(r)/r\,dr$ of the pore volume is contained in tubes of radius between r and $r + dr$. From Eq. (3.20)

$$\int_{r_{min}}^{r_{max}} \frac{G(r)}{r}dr = 1. \tag{3.21}$$

Hence, from the measured capillary pressure as a function of saturation $P_c(S_w)$ we can infer the distribution of throat radius. We can either use Eq. (3.20), or starting from Eq. (3.1) with $r \equiv r_t$ and differentiating

$$\frac{dP_c}{dS_w} = \frac{dP_c}{dr}\frac{dr}{dS_w} = -\frac{P_c}{G(r)}, \tag{3.22}$$

from which we find

$$G(r) = -P_c\frac{dS_w}{dP_c} \equiv -\frac{dS_w}{d\ln P_c}. \tag{3.23}$$

In Figs. 3.11 and 3.12 the derivative of saturation with respect to the logarithm of capillary pressure (or equivalently calculated threshold radius) is used to find the throat size distribution. This is, however, an approximate estimation and is not necessarily an accurate reflection of the true distribution: after all, the porous medium is not really a bundle of parallel tubes. To relate pressure and throat radius, we do have to make assumptions concerning the contact angle and geometry, which, while plausible, do not account for changes in effective contact angle or different throat shapes. Furthermore, while the pressures will give a reasonably good indication of the radii of curvature of the interfaces passing through restrictions in the pore space, the volume associated with a change in pressure is not obviously related to our network representation of pores and throats. By our definition, throats have no volume and so any change in saturation is associated with the pores adjacent to the throat. Yet this is ambiguous, since each throat connects two pores, while each pore may be bounded by several throats. More significantly, Haines jumps, when the non-wetting phase rapidly advances, are not confined to a single pore. Imagine that the

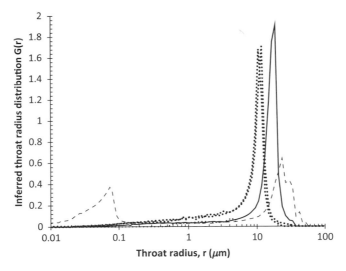

Figure 3.11 The derivative of the saturation with respect to the logarithm of the capillary pressure, Eq. (3.23). The capillary pressure curves are shown in Fig. 3.8. The solid line is for Doddington sandstone, the dotted lines for two Berea sandstone samples and the dashed line for Ketton limestone. Of the sandstones, Doddington tends to have larger pore sizes. Ketton has large throats whose radii exceed those in the sandstones, but also has micro-porosity in the oolithic grains with radii of less than 1 μm, as shown in Fig. 2.7.

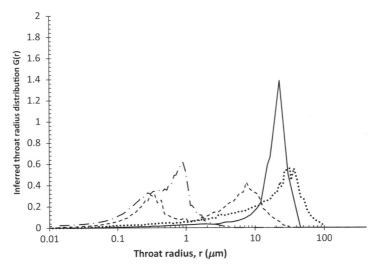

Figure 3.12 The derivative of the saturation with respect to the logarithm of the capillary pressure, Eq. (3.23). The capillary pressure curves are shown in Fig. 3.9. The solid line is for Bentheimer sandstone, the dashed line for Estaillades, the dotted line for Mount Gambier and the dot-dashed line for Guiting. The sandstone has a narrow throat size distribution centred around 20 μm. Mount Gambier has the largest throats, but has a wider distribution. Estaillades shows a bimodal distribution indicative of macro- and micro-porosity. Guiting is largely micro-porosity with throat sizes smaller than 1 μm.

non-wetting phase passes through a narrow throat into a pore that has other wider throats bounding it. This is possible, since these wider throats may not be accessible for invasion until non-wetting phase has moved through narrower restrictions. Then the non-wetting phase enters a pore, passes through adjacent throats and subsequently fills further pores. This may engender a cascade of pore filling until the meniscus comes to rest in a restriction narrower than any others encountered hitherto. All of this volume will be associated with the throat that was initially filled. This effect of multiple pore filling during a Haines jump is more marked early in the displacement, when most of the pores are unfilled. This explains why the measured capillary pressures have a gradual increase with decreasing wetting phase saturation initially, before rising more steeply. When the capillary pressure is high, most of the pores may be filled. In this case we are not seeing the filling of pores and the accessing of throats as the pressure increases further, but more a smooth change as wetting phase is forced further into the corners of the pore space, as illustrated in Fig. 3.4. In this case we see some small volume associated with apparently very small throats, which are not throats at all, but irregularities in the pore walls.

Despite all these caveats, the apparent throat size distribution $G(r)$ does serve as a useful characterization of the pore space of rocks and is routinely measured. It is able to distinguish between rocks with larger or smaller throat sizes (the capillary pressure is larger for smaller pore spaces) and to quantify the heterogeneity or variability in pore size. A flat capillary pressure indicates that most of the throats have similar size, as seen most markedly in the Bentheimer and Doddington sandstones. A wider distribution, indicative of a modest amount of clay, is seen for Berea sandstone. Mount Gambier also has a broader distribution of throat size, with some very large throats, as evident from its open pore structure (see Fig 2.3). Guiting also has a broad throat size distribution, but with radii that are approximately two orders of magnitude smaller than for Mount Gambier.

Two of the samples studied, Estaillades and Ketton, have a dual porosity distribution with wide throats between calcite grains and much smaller throats within them. We see features spanning almost four orders of magnitude variation in size.

3.3.1 Prediction of Capillary Pressure from Images

While the pore-space images allow a qualitative interpretation of the capillary pressure curves that we have presented, it is possible to predict the capillary pressure based on a more quantitative analysis of the pore space. As an example, Fig. 3.13 shows the capillary pressure for Bentheimer sandstone compared to predictions using network modelling. Here, we have used the maximal ball method to extract a network from the image shown in Fig. 2.3 whose properties are listed in Table 2.1; see also Fig. 2.15: the base image has a voxel size of 3 μm with $1,000^3$ voxels. We

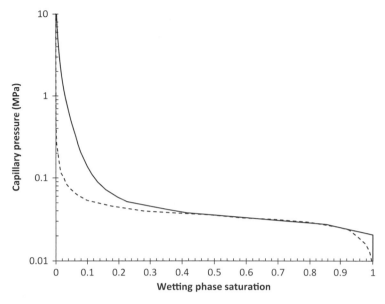

Figure 3.13 A comparison of the measured primary drainage capillary pressure for Bentheimer sandstone, solid line, shown in Fig. 3.9 compared to network model predictions based on the image shown in Fig. 2.3, dashed line.

have given each pore and throat a cross-section with an inscribed radius and shape factor derived from the image (see Fig. 2.16) and computed displacement using the network model of Valvatne and Blunt (2004). This is not the only way to generate such predictions: for instance, the maximal ball method can be used directly to compute displacement during primary drainage (by finding the pore space accessed by balls of different size), which can be modified to account for different contact angles (Silin and Patzek, 2006; Kneafsey et al., 2013). The displacement may also be modelled using a morphological approach to capture the advancing meniscus (see, for instance, Hilpert and Miller (2001)) or through using the pore size distribution computed from an erosion-dilation algorithm (Yang et al., 2015). The purpose of this section is not to critique modelling approaches – indeed, an independent comparison of predictions with experiments for Bentheimer sandstone of better quality has already been presented by Patzek (2001) – but to interpret displacement processes and to illustrate some inevitable limitations of image analysis and associated pore-scale modelling.

The measured and predicted curves show a reasonable agreement for the mid-range of saturation. This implies that the images capture the majority of the throat sizes fairly accurately. However, there are two discrepancies. The first occurs at the highest wetting phase saturation, when the non-wetting phase first accesses the system. The experimental measurements see a higher entry pressure followed by a sharp decrease in saturation with a small increase in pressure: in contrast, the model predictions show a lower entry pressure and an initial rapid rise in pressure with saturation. The experimental sample is around 4 mm in diameter and 10 mm long, or almost five times larger than the model (volume of 126 mm³ as against 27 mm³; see Table 2.1). As discussed below, in a small system, we see a noticeable shift in saturation from the invasion of a few large throats and pores connected to the inlet. For a larger sample, a detectable shift in saturation is only seen once the non-wetting phase connects across the system; the invasion of a few large inlet pores and throats has a negligible impact on the saturation. As the system size becomes larger, the apparent entry pressure tends to increase to a value close to the value P_c^* necessary for the non-wetting phase to span the rock. However, in this case the size difference between the model and the experimental sample is modest, and so this is not a significant effect.

The second, and more important difference, is at low saturation. The model cannot capture features in the pore space below the voxel size, or 3 μm in this case. The invasion of smaller throats, or significant displacement of the wetting phase in tight corners, cannot be captured accurately. This is an inevitable limitation of using images of finite resolution and is even more marked in samples with microporosity, where a multi-scale network is required, as described in Chapter 2.2.3.

Fig. 3.14 shows the calculated throat size distributions from the measurements and model, using Eq. (3.23). The network approach tends to give a narrower distribution shifted to slightly smaller throats. There is also an underestimation of the contribution of the smallest throats whose radii lie below the resolution of the image.

Fig. 3.14 also plots the throat size distribution defined using Eq. (3.20) as

$$G(r) = \frac{r^3 f(r)}{\int_{r_{min}}^{r_{max}} r^2 f(r) dr}, \tag{3.24}$$

where $f(r)dr$ is the number of throats between radius r and $r + dr$. Here $f(r)$ is found simply from counting the number of throats of different size in the network. Putting aside our caveats concerning image resolution, this is the real throat size distribution, since it does not assume that the porous medium is a bundle of parallel cylindrical tubes but is based directly on the pore space geometry.

The true distribution of throat radii is wider than that inferred from the capillary pressure, with more large and small throats. We do not detect some of the wider throats, as these are often filled during a Haines jump, and which therefore contribute a saturation change associated with a smaller throat that limited the ingress of the non-wetting phase. We also underestimate the number of smaller throats, since these tend to be filled with a smaller saturation change than one proportional to r^2 for cylindrical tubes. Instead, these small throats, filled late in the displacement, are frequently adjacent to pores that have been filled already, giving them a

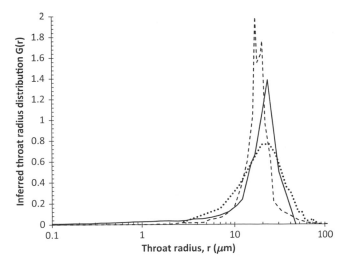

Figure 3.14 The computed distribution of throat radius in Bentheimer using Eq. (3.23) based on the experimental measurements, solid line, and the network model predictions, dashed line (see Fig. 3.13). Also shown as the dotted line is the throat size distribution calculated directly on the network using Eq. (3.24).

tiny contribution to the saturation change, even when we account for their small radius.

This exercise indicates that network analysis combined with experiment can be used to interpret the geometry of the pore space; however, the traditional analysis of capillary pressure does not and cannot alone capture precisely the true distribution of throat size.

3.4 Invasion Percolation

The network representation of a porous medium allows an elegant and rapid way to predict the sequence of filling during primary drainage: the throats fill in order of size, with the largest filled first. This is akin to a percolation process (Broadbent and Hammersley, 1957). In percolation we start with a lattice of sites (the pores) that are connected by bonds (throats). Initially all the sites and bonds are empty; they are then filled one-at-a-time at random. We may consider bond percolation where we fill bonds at random and ignore the sites, and site percolation where sites are filled and bonds are ignored. At any time a fraction p of the elements (sites or bonds) are filled. p is called the percolation probability or fraction. The percolation threshold $p = p_c$ is reached when there is a connected path of filled sites or bonds across the network. For bond percolation this assumes that all the sites are filled (or at least allow connections of the invading phase), while for site percolation we assume that all the bonds are filled.

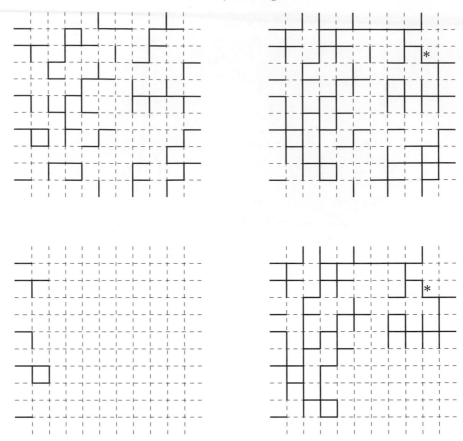

Figure 3.15 An illustration of percolation on a square lattice with 10×10 sites and $2 \times 10 \times 11$ bonds. The upper figures show ordinary percolation where bonds are filled at random. The upper left figure shows the filled bonds when $p = 0.35$. There is no connected path across the network. The upper right figure shows the situation at the percolation threshold, $p = p_c = 0.5$, where there is just a connected pathway: the final connecting bond filled is indicated by the *. The lower pictures show the same thresholds, but for invasion percolation, where only bonds connected to the inlet are filled.

Percolation is illustrated in Fig. 3.15 on a square lattice: this is clearly a simplification of the complex three-dimensional networks of the pore spaces of real rocks described previously, but allows a simple and instructive visualization of the concepts. In this example, bonds (throats) are filled at random. The left figure shows when 35% the bonds are filled. At this point the filled bonds do not connect across the system. When 50% of the bonds fill, on the right, there is first a connected path from left to right across the lattice.

The critical fraction of filled sites or bonds necessary to connect, p_c, is a function of the type of lattice, whether we consider sites or bonds, and the spatial dimension in which the lattice is embedded. For the square lattice in Fig. 3.15, $p_c = 0.5$ for bonds (throats) and around 0.59 for sites (pores). In three dimensions, a cubic lattice has a threshold of approximately 0.245 for bonds and 0.316 for sites (Wang et al., 2013). For bond percolation, the threshold can be estimated from

$$p_c \approx \frac{d}{(d-1)z},$$ (3.25)

where d is the dimension of space and z is the coordination number of the lattice: in our context this is the average number of throats bounding each pore (Stauffer and Aharony, 1994).

At the percolation threshold, the pattern of filled elements is self-similar, or fractal, in that one piece of the pattern resembles the whole: a section of the pattern looks statistically similar regardless of the scale at which it is viewed, from the pore level to the size of the system.

Displacement in a porous medium where the pore and throat size vary, but without any long-range spatial correlation, is a percolation-type process: filling proceeds in a sequence of size (or threshold capillary pressure) for elements distributed statistically randomly in the system. Since drainage is limited by the invasion of throats, we have a bond percolation process. However, we cannot simply fill the throats strictly in order of size: they are only filled if they are adjacent to a pore (and throat) already filled with the non-wetting phase, which is also connected all the way to the inlet of the sample. In this case we have an invasion percolation process as first described by Lenormand and Bories (1980) and Wilkinson and Willemsen (1983). This is the same as ordinary bond percolation, but where any filled bonds not connected to the inlet are removed, shown in the lower pictures in Fig. 3.15.

The invasion percolation threshold is the same as in ordinary percolation, except that not all possible throats are filled: only the connected ones are. This threshold can be related to the size of the smallest throat, r_c, that needs to be filled in order to connect across the system. This defines the capillary pressure P_c^* from Eq. (3.1) as:

$$P_c^* = \frac{2\sigma \cos \theta_R}{r_c}.$$ (3.26)

Returning to the throat size distribution with $f(r)dr$ throats with radius between r and $r + dr$, the percolation probability p is equivalent to:

$$p = \frac{\int_r^{r_{max}} f(r)dr}{\int_{r_{min}}^{r_{max}} f(r)dr},$$ (3.27)

where r is the radius of the largest throat just filled. At the percolation threshold $r = r_c$ and $p = p_c$.

Invasion percolation in a network can be simulated using a sorted list of threshold capillary pressures corresponding to throat radii. Initially, the list only contains throats adjacent to the inlet, and are ranked in order of entry pressure, with the lowest pressure ranked top. This top throat is filled first together with the adjoining pore, and then the entry pressures of bounding throats, not already on the list, are added and sorted. Then the top-ranked throat is filled again, new throats are added to the list and the process continues. We assume that the wetting phase is not trapped, and so every throat can be accessed. Computationally most time is spent ordering the list of capillary pressures. For a list with n items the most efficient algorithms take a time of order $\ln n$ to sort a new entry using a binary-tree search. Therefore to simulate the filling of n throats takes a time that scales as $n \ln n$ (Sheppard et al., 1999; Masson and Pride, 2014). This algorithm is computationally efficient, allowing the displacement through many millions of pores and throats to be computed (Sheppard et al., 1999). A sequence of filling and local capillary pressures is determined: if needed, the minimum local pressures corresponding to pore filling can be determined. Additional information can also be computed: from a representation of the geometry of the pores, the amount of wetting phase in the corners can also be found, allowing the saturation to be calculated.

The macroscopic capillary pressure is only unambiguously defined when the local capillary pressure reaches a new maximum value: that is a value larger than the threshold pressures for all the previously filled elements. If this is not the case, then the local capillary pressure will not be the same as the macroscopic value imposed externally, even in the limit of a low flow rate, as discussed previously. Moreover, the fluid configuration will not be correct except at these local maxima, as the invasion percolation algorithm does not account for local non-wetting phase retraction and fluid re-arrangement during a Haines jump: it does not allow the wetting phase to re-invade previously filled pores to supply fluid to fill larger pores in the vicinity. As we show below, this has a significant impact on the capillary pressure and estimated throat size distribution for throats with radii larger than r_c.

Invasion percolation allows a very powerful suite of statistical analyses for a porous medium that is macroscopically homogeneous. While the throat and pore sizes may be highly variable, at some scale there is no spatial correlation in sizes (that is, large pores aggregated together) with the same distribution of size and connectivity in different parts of the rock. This is a reasonable approximation for the quarry rocks so far presented in this book: this was tested, for instance, for Berea in Fig. 3.8, where the capillary pressures measured on two samples were similar. While the samples we have studied have a complex pore structure with a wide range of throat size (evident in, for instance, Fig. 3.12), at a larger, cm-scale, they are indeed homogeneous, in that one piece of Bentheimer or Berea sandstone resembles another piece in terms of pore structure. There is a reason for this: Berca and Bentheimer are used as building stones, while Ketton, for instance, is frequently used for making statues. Their utility is predicated on the fact that they appear at a macroscopic level to be homogeneous, and therefore give a uniform appearance at the cm to metre scales. Unfortunately, reservoir rocks, chosen for study because they contain hydrocarbon, are less likely to be so macroscopically homogeneous, which somewhat limits the applicability of the statistical analysis presented below: indeed, generally in subsurface systems we see correlated spatial structure over all scales from the micron level upwards to the field.

In a statistically homogeneous sample, as the fluid advances, up to and at the point that the non-wetting phase spans the system, it has a self-similar or fractal structure, as in ordinary percolation. To progress through the medium, the local capillary pressure has to be close to P_c^* and therefore the arrangement of fluid resembles the connected pathway in ordinary percolation at the threshold. This immediately allows us to appreciate that during this initial advance the Haines jumps may be of arbitrary size, as moving through just one restrictive throat near the inlet of radius r_c allows the entire system to be spanned without a further increase in pressure.

In terms of fluid patterns, the arrangement of fluid displays similar structures over all scales from the pore to the overall system size, with loops within loops and multiply nested branches. This was first seen experimentally in the micro-model experiments of Lenormand and Zarcone (1985), Fig. 3.2. This displacement pattern is illustrated for a two-dimensional system in Fig. 3.16 where air (the non-wetting phase) is injected into a quasi two-dimensional porous medium comprised of a monolayer of glass beads initially saturated with a water/glycerin mixture (the wetting phase).

3.4.1 Scaling Relations in Invasion Percolation

One significant feature of percolation is its universality. While the percolation threshold is dependent on the precise nature of the network, the scaling properties – addressed below – are the same for any network. There are differences dependent on the physical nature of the displacement and the dimensionality of the system (which in reality is always three-dimensional), but not on the connectivity or throat size distribution itself.

We can define a correlation length ξ: from this length down to the pore scale, l, we see self-similar structures. As the fluid advances through the pore space until it connects across the system, $\xi = L$, the distance that the invading pattern has moved. l is a typical size of a pore, or the distance between pores. Usually in a

Primary Drainage

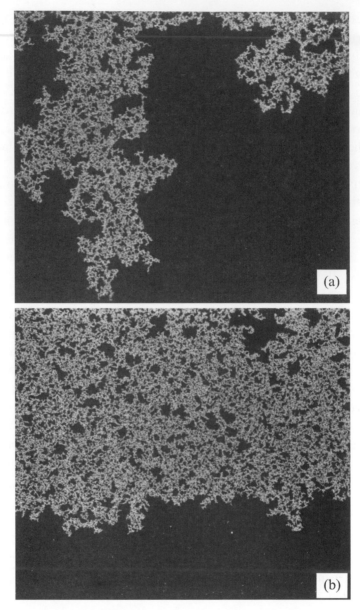

Figure 3.16 Two-dimensional invasion percolation, when the non-wetting phase (white) displaces wetting phase (dark). Here air displaces a water/glycerin mixture (black) through a mono-layer of glass beads at a low flow rate. The top figure shows a displacement in a horizontal cell at the percolation threshold, when the non-wetting phase first spans the system. The lower figure shows a gravity-stable displacement: in this case the pattern is percolation-like below a correlation length, but more filled-in when viewed at larger scales. Picture from Meakin et al. (2000).

packing of a granular medium, it is taken as the average grain diameter. Then the number of invaded throats n_t scales as

$$n_t \sim \left(\frac{L}{l}\right)^D, \tag{3.28}$$

where D is the fractal dimension which is approximately 2.52–2.53 for a network embedded in three-dimensional space, while it is $91/48 \approx 1.896$ in two dimensions (Stauffer and Aharony, 1994).

If n_{tot} is the total number of throats in the system then $n_{tot} \sim (L/l)^d$, where d is the dimension of space ($d = 3$ obviously, but it is also instructive to consider two-dimensional, $d = 2$ systems as well). Then using Eq. (3.28),

$$S_{nw} \sim \left(\frac{L}{l}\right)^{D-d} \int_{r_c}^{r_{max}} \frac{G(r)}{r} dr \sim \left(\frac{L}{l}\right)^{D-d}, \tag{3.29}$$

which, when L is sufficiently large, tends to zero, since $D < d$.

While we use a $\sim$ symbol to imply scaling, Eq. (3.29) can be used to estimate the saturation at which the non-wetting phase first connects across the system, even if we ignore the term in $G(r)$. Consider, for example, our Bentheimer sandstone. From an image of size 3 mm (3,000 μm) we found 54,741 throats (Table 2.1), implying that the sample is approximately $58,741^{1/3} \approx 38$ throats across. We can identify $l \approx 3,000/38 \approx 79$ μm as the distance between throats, or a typical pore length, so that in this example $L/l = 38$. Then Eq. (3.29) with $D - d \approx -0.48$ gives a value of around 17% for the non-wetting phase saturation or a wetting phase saturation of 83%. This is the saturation when the mercury is first connected across the rock sample: in our model predictions the computed wetting phase saturation at this point is 85%. The sample used for the mercury injection capillary pressure test is approximately 1 cm across. If we use the same value of l now $L/l = 127$. Then Eq. (3.29) gives a value of around 10% for the non-wetting phase saturation, or a wetting phase saturation of 90%. If we considered a larger system, for instance, a block 1 m across, this spanning saturation would be only 1%.

We can define an apparent entry pressure when we first see a finite or measurable shift in wetting phase saturation, rather than invasion into a single large pore at the inlet. This is, for a sufficiently large system, similar to the capillary pressure when the non-wetting phase spans the system, since both cases are governed by the critical radius r_c: $P_c^{entry} \approx P_c^*$.

This analysis has some implications for the estimated throat size distribution, $G(r)$. For a sufficiently large system, we cannot detect any throat radii $r > r_c$; any apparent discernible displacement for larger radii is essentially a finite-size effect and not a reflection of the true distribution of throat size: this is evident in Fig. 3.14.

As the displacement continues further, the percolation probability exceeds the threshold value and the pattern of non-wetting phase begins to fill in. Now we only see a self-similar structure over a more restricted range of scales from the correlation length ξ to l, where ξ decreases as p increases. We can still use percolation theory to predict the capillary pressure behaviour for $\Delta p = p - p_c$ when $\Delta p \ll 1$.

The correlation length ξ can be related to the departure of the percolation probability from its critical value by a power-law relation:

$$\frac{\xi}{l} \sim \Delta p^{-\nu}, \tag{3.30}$$

where ν is equal to 4/3 in two dimensions and 0.876 in three dimensions (Wang et al., 2013).

The fraction of filled throats on the spanning cluster is defined as:

$$\frac{n_t}{n_{tot}} \sim \left(\frac{\xi}{l}\right)^\beta, \tag{3.31}$$

where in this context β describes a scaling exponent.

We now replace L – the correlation length in a finite-sized system at the threshold – with ξ in Eq. (3.29) to obtain from Eq. (3.30):

$$S_{nw} \sim \frac{n_t}{n_{tot}} \sim \Delta p^{\nu(d-D)} \equiv \Delta p^\beta, \tag{3.32}$$

using Eq. (3.31), which defines $\beta = \nu(d - D)$. $\beta \approx 0.42$ in three dimensions.

Note that the saturation now has a finite value, even in an infinite system, whose value is controlled by the departure from the exact percolation threshold. Let's consider a concrete example. For a random lattice with an average coordination number of 4 the bond percolation threshold is approximately 3/8 using Eq. (3.25) (Stauffer and Aharony, 1994): this is representative of the threshold for a well-connected sandstone or a sand pack. With some degree of consolidation, the coordination number decreases: for instance, for Bentheimer the coordination number is 3.8 (see Table 2.1), giving an estimated percolation threshold of 39% from Eq. (3.25). This means that the capillary pressure does not give any indication of the largest 39% of the throats. We also see a very large shift in saturation for a small change in capillary pressure for throat sizes just above r_c. Imagine that we now fill throats up to a radius where $p = 0.41$ and $\Delta p = 0.02$. From Eq. (3.32) we find $S_{nw} \approx 0.19$. Just allowing the pressure to rise to allow a further 2% of available throats to fill allows the saturation to increase to 19%. Even in a finite size system, this is significant: for the sample used for the mercury injection measurements, this is a change in saturation from 10 to 19% through allowing the filling of just a further 2% of the accessible throats.

The result of rapid filling near the percolation threshold is to make the throat size distribution $G(r)$ inferred from the capillary pressure peak at r_c: this is valuable as it gives the critical radius necessary to span the system, but does not accurately capture the actual distribution of sizes. For Bentheimer, for instance, we see a peak in $G(r)$ at around $r = 24$ μm (see Fig. 3.14) from which we can say $r_c \approx 24$ μm but little else about the real distribution for large radii unless we do an explicit network analysis, as in Fig. 3.14.

In a system with distinct regions of micro-porosity, such as Ketton, the second peak in the distribution $G(r)$ derived from the capillary pressure represents the critical radius to access this micro-porosity and may again misrepresent the sizes of some of the larger micro-pores. Indeed, there is some evidence of this where the largest visible intra-granular pores apparent in Fig. 2.7 do seem larger than 0.062 μm in radius, implied by the capillary pressure, Fig. 3.8.

This style of analysis can be continued to derive scaling relationships for various properties at and near this critical percolation threshold, when the incipient cluster of non-wetting phase first advances across the rock, when it first spans the system and shortly afterwards. As an example, let us now consider flow of the non-wetting phase. We can distinguish between the backbone of non-wetting phase, through which there is flow, and the stagnant dangling ends, or branches, Fig. 3.15. For a large system, only an infinitesimal fraction of the non-wetting phase occupies the backbone, which has a smaller fractal dimension, D_b, defined such that the number of throats on the backbone scales as $n_b \sim (\xi/l)^{D_b}$, where $\xi \equiv L$. We can then write the following for the fraction of filled throats on the backbone

$$\frac{n_b}{n_t} \sim \left(\frac{\xi}{l}\right)^{D_b - D}, \qquad (3.33)$$

which tends to zero as $\xi \to \infty$.

For transport, we may also wish to consider the shortest path through the backbone from inlet to outlet. This avoids long excursions in loops, and again occupies only an infinitesimal fraction of the backbone with its own lower dimension D_{min}. However, $D_{min} > 1$, which means that flow cannot take a straight-line path through the network, but takes a tortuous route whose tortuosity τ (defined as the length of the flow path to that of a straight line) diverges as the system size increases:

$$\tau \sim \left(\frac{\xi}{l}\right)^{D_{min} - 1}. \qquad (3.34)$$

There is a set of so-called red bonds on the backbone, which, if broken, will disconnect the spanning cluster. This will be of significance when we consider imbibition in the following chapter, when the wetting phase displaces the non-wetting phase: filling any single red bond (throat) will disconnect the non-wetting phase. The number of these bonds scales as

$$n_{red} \sim \left(\frac{\xi}{l}\right)^{D_{red}}, \tag{3.35}$$

where $D_{red} = 1/\nu$ or approximately 1.141 in three dimensions (Ben-Avraham and Havlin, 2000).

Finally, we return to the size of the Haines jumps close to the percolation threshold. As mentioned previously, the number of pores filled in a single jump – before the local capillary pressure reaches a new maximum – can be of any size, since invading a throat close to the critical radius r_c may allow a burst of filling from a single pore to the size of the system. It should now come as no surprise to note that there is a power-law distribution of bursts: the probability of a Haines jump occurring which fills s pores is given by

$$n_H \sim s^{-\tau'}, \tag{3.36}$$

where the exponent τ' can be related to other percolation exponents through (Martys et al., 1991b)

$$\tau' = 1 + \frac{D_H}{D} - \frac{1}{D\nu}. \tag{3.37}$$

D_H is the fractal dimension of the interface or hull of the percolating cluster: this excludes regions that cannot move, since the neighbouring throats are already filled. For invasion percolation in two dimensions, $D_H < D$, and we find $\tau' = 1.30 \pm 0.05$ in two dimensions (Martys et al., 1991b): note that a different expression was derived by Roux and Guyon (1989), which gives slightly different values for τ'. The distribution of bursts is truncated at a maximum size, related to the correlation length: $n_H(s) \to 0$ for $s > \xi^D$. Eq. (3.37) has been confirmed by micro-model experiments of drainage, combined with sensitive pressure measurements for very low flow rate injection, and confirmed by simulations of the process, which account for the accessibility of fluid during Haines jumps (Furuberg et al., 1996). Aker et al. (2000) also observed this scaling by simulating drainage through two-dimensional network model, combined with sensitive pressure measurements in experiments on a monolayer of glass beads.

In three dimensions, the structure is more open and $D_H = D$ (Strenski et al., 1991) giving $\tau' = 1.55$ from Eq. (3.37), or 1.63 using Roux and Guyon (1989). An approximate power-law relationship of jump sizes has been seen experimentally using from rapid pore-scale imaging of drainage in Bentheimer sandstone (Bultreys

Table 3.1 *Fractal dimensions computed from simulations of percolation processes. D is the fractal dimension of the displacement pattern, while D_{min} is the dimension of the minimum path and D_b is the dimension of the backbone. IP stands for invasion percolation. Values from Sheppard et al. (1999).*

Displacement process	D	D_{min}	D_b
Two dimensions			
IP without trapping	1.8949 ± 0.0009	1.129 ± 0.001	1.642 ± 0.004
Site IP with trapping	1.825 ± 0.004	1.214 ± 0.001	1.217 ± 0.020
Bond IP with trapping	1.825 ± 0.004	1.2170 ± 0.0007	1.217 ± 0.001
Three dimensions			
IP without trapping	2.528 ± 0.002	1.3697 ± 0.0005	1.868 ± 0.010
Site IP with trapping	2.528 ± 0.002	1.3697 ± 0.0005	1.861 ± 0.005
Bond IP with trapping	2.528 ± 0.002	1.458 ± 0.008	1.458 ± 0.008

et al., 2015b). Acoustic measurements have also seen a power-law distribution of events with an exponent 1.70 ± 0.15 (DiCarlo et al., 2003).

A listing of some of these fractal dimensions is provided in Table 3.1, calculated from simulations of percolation (Sheppard et al., 1999). For reference, and for the discussion in Chapter 4, we include the exponents for different processes. If invasion percolation describes primary drainage into a wetting phase that is everywhere interconnected through layers, then there is no trapping. However, if the displaced phase can only escape if it has a connected path through the centres of pores or throats to the inlet, we have a distinct process called invasion percolation with trapping (Dias and Wilkinson, 1986). In this case we can consider two cases, where the flow is either controlled by the filling of throats (bond percolation) or pores (site percolation).

The properties of a network of one phase as it just connects across a system are important in two types of process. The first is primary drainage during the initial invasion of a non-wetting phase through a porous medium: examples include primary oil migration when the first filament of oil progresses from the source rock to the cap rock through which it cannot pass, and the rise of a CO_2 plume in a storage aquifer. The second is when we consider the disconnection of a displaced phase, which will be presented in detail in Chapter 4.6.

3.4.2 Displacement under Gravity and Gradient Percolation

The applications of percolation theory in primary drainage mentioned above involve a non-wetting phase that moves under the influence of gravity. This means that the local capillary pressure is not fixed, but varies with depth. If we have a continuous phase of density ρ in the pore space (however tortuous the connected path) with a negligible viscous pressure gradient caused by flow, the fluid pressure obeys

$$\frac{dP}{dz} = \rho g \tag{3.38}$$

for both phases, where z is depth, which increases as we go deeper. Hence the capillary pressure, or pressure difference between the phases, is:

$$P_c = \Delta \rho g (z_0 - z) = \Delta \rho g h, \tag{3.39}$$

where $\Delta \rho = \rho_w - \rho_{nw}$ is the density difference between the phases; see Eq. (1.18). We have defined a datum depth, $z = z_0$ in the subsurface where the pressure of non-wetting and wetting phases are equal, meaning that the capillary pressure is zero. This is a somewhat hypothetical level, in that no non-wetting phase (oil) is necessarily present at this depth: until a finite entry capillary pressure is applied, the wetting phase saturation will be 1. The height, h, above z_0 is $z_0 - z$.

We can relate the percolation probability to the capillary pressure using Eqs. (3.27) and (3.1) to find:

$$\frac{\partial p}{\partial P_c} = \frac{\partial p}{\partial r}\frac{\partial r}{\partial P_c} = \left(-\frac{G(r)}{r}\right)\left(-\frac{r}{P_c}\right) = \frac{G(r)}{P_c}, \tag{3.40}$$

where the overall positive sign of the expression indicates that the fraction of accessible throats p increases with capillary pressure.

Then, using Eq.(3.39) we can relate p to height h:

$$\frac{\partial p}{\partial h} = \frac{\Delta \rho g G(r)}{P_c}. \tag{3.41}$$

Normally we consider displacement under gravity in terms of a Bond number, or the ratio of the pressure change due to buoyancy at the pore scale to a typical capillary pressure. Over a characteristic length, l, density differences lead to a pressure difference (or change in local capillary pressure) of $\Delta \rho g l$. The capillary pressure is given by Eq. (3.1), which is a function of throat radius r. Traditionally, the Bond number simply uses $r = l$ (see, for instance, Meakin et al. (2000)), which tends to underestimate the effect of capillary forces, since the throat radius will be smaller than a typical distance between pores. Here we define, following Blunt et al. (1992), a Bond number

$$B = \frac{\Delta \rho g l r_t}{\sigma \cos \theta} \tag{3.42}$$

using r_t as the characteristic throat radius, since this controls the advancing front in drainage, as discussed above. For drainage we use the receding contact angle in Eq. (3.42).

Now we can write Eq. (3.41) in terms of B as

$$\frac{\partial p}{\partial h} = B \left(\frac{G(r)}{r} \right) \left(\frac{r^2}{r_t l} \right). \tag{3.43}$$

Close to the percolation threshold, $r \approx r_t$ and Eq. (3.43) simplifies to

$$\frac{\partial p}{\partial h} = B \left(\frac{G(r_t)}{l} \right). \tag{3.44}$$

In Figs. 3.11, 3.12 and 3.14, $G(r)$ is of order 1 close to $r = r_t$, so that in Eq. (3.44), $\partial p / \partial h \approx B/l$.

Typically, gravitational forces are small at the pore scale compared to the capillary pressure. Consider our Bentheimer example, with $l = 79$ μm and $r_t = r_c = 24$ μm. For an oil/water system at reservoir conditions we can take $\cos \theta_R \approx 1$ and $\sigma = 20$ mN/m. A typical density difference between a reservoir brine and oil is around 300 kg/m^3 and $g = 9.81$ m/s^2, which from Eq. (3.42) gives $B \approx 2.8 \times 10^{-4}$. This implies that at the pore scale, the capillary pressure imposed by buoyancy forces changes slowly, with a corresponding gradual change in the percolation probability p with height.

Our example displacements involved a non-wetting phase that is less dense rising upwards to displace a denser wetting phase (alternatively we can consider the sinking of a denser non-wetting phase). The non-wetting phase will progress upwards as soon as it is connected: hence the non-wetting phase everywhere is at the percolation threshold. However, the gradient in capillary pressure imposes a correlation length to the percolation pattern: the non-wetting phase progresses in a series of fingers of finite width with statistically the same percolation pattern repeated as the non-wetting phase rises. The non-wetting phase cannot progress through the porous medium at a lower value of p than the percolation threshold, nor will it acquire a higher value, since – for slow flow – the non-wetting phase would rather continue upwards than displace more wetting phase lower down. It only accumulates and fills the porous medium at a higher saturation (larger p) when it collects under a barrier to upwards movement.

We can use percolation theory to predict the correlation length, which is equivalent to the finger width. This is invasion percolation with a gradient, which has been extensively studied numerically and experimentally (Wilkinson, 1984, 1986; Birovljev et al., 1991; Blunt et al., 1992; Meakin et al., 1992). Over a distance ξ we use Eq. (3.44) to find $\Delta p \approx B\xi/l$, then using Eq. (3.30) we find the relation

$$\xi \sim l \left(B \frac{\xi}{l} \right)^{-\nu}, \tag{3.45}$$

from which we find

$$\xi \sim l B^{-\nu/(1+\nu)}, \tag{3.46}$$

where $\nu/(1 + \nu) \approx 0.47$ (Wilkinson, 1984, 1986). For our Bentheimer sandstone example, Eq. (3.46) yields $\xi/l = 136$, or a correlation length of around 11 mm, with a minimum saturation, from Eq. (3.29) with $L = \xi$, of approximately 9%.

An illustration of this gravity fingering is shown in Fig. 3.17, where a gravitationally unstable drainage experiment is performed in a transparent porous medium (Frette et al., 1992; Meakin et al., 2000). The non-wetting phase rises upwards in a finger whose width can be predicted using percolation theory. In addition, statistically similar patterns can be simulated using an invasion percolation algorithm where the ranking of threshold pressure is based on a local threshold $p - Bh/l$, where p is found from Eq. (3.27) and h is the height of the throat above some datum, favouring the filling of available throats where h is largest. Physically, this process is also seen during the infiltration of water into a water-repellent soil, where the water is the non-wetting phase. This problem has been studied both experimentally and theoretically (Bauters et al., 1998, 2000): we will consider this problem in more detail in Chapter 4.3, where we present wetting phase invasion.

For a gravity-stable displacement, where the denser phase rises upwards, as shown in Fig. 3.16, the correlation length is also given by Eq. (3.46) and indicates the scale of likely fluctuations in the advancing front. We see a self-similar

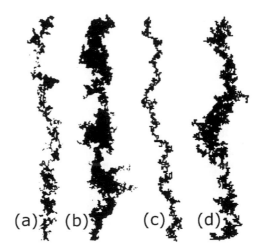

(a)   (b)   (c)   (d)

Figure 3.17 Parts (a) and (b) show displacement patterns generated by the buoyancy-driven penetration of dyed non-wetting phase through a transparent three-dimensional porous medium at Bond numbers of 0.40 and 0.046, respectively. Parts (c) and (d) show projections of the displacement patterns generated using the three-dimensional gradient destabilized invasion percolation model at similar values of the Bond number. From Meakin et al. (2000).

advance at the top of the upwards-advancing non-wetting phase to a depth ξ: as we go lower, the local percolation fraction increases, the correlation length decreases, and the pattern fills in with a progressively increasing non-wetting phase saturation with depth.

Regardless of the displacement process, as the non-wetting phase saturation increases, we go beyond the percolation threshold, filling throats that have radii smaller than the critical value, the network begins to fill in and the correlation length of loops and branches decreases. Here, I will not elaborate further, as it has been amply and ably covered in other work, where using scaling arguments close to the percolation threshold forms a powerful springboard for a range of fascinating analysis and generic predictions of geometric, flow and transport properties: see Larson et al. (1981); Sahimi (1993); Sahini and Sahimi (1994); Stauffer and Aharony (1994); Hunt (2001); Hunt and Ewing (2009); Sahimi (2011). For primary drainage, the discussion here will not continue, since in most of the situations we are interested in, such as the complete filling of an oil or gas reservoir or significant quantities of CO_2 collecting under caprock, or indeed most laboratory experiments, displacement continues to higher non-wetting phase saturations, where the non-wetting phase is much better connected and percolation theory has more limited applicability, since for $\Delta p \sim O(1), \xi \approx l$.

3.4.3 Invasion Percolation, Normal Percolation and Flow

In an invasion percolation process, the injected phase connects across the rock at a low saturation, Eq. (3.29): even if we account for the finite correlation length introduced by gravitational forces, the non-wetting phase will span the system at saturations around 10% or less; see Eq. (3.46), page 108; we will find similar minimum saturations when we account for finite correlation lengths due to viscous forces in Chapter 6.4.4. When there is an invasion percolation advance of an injected phase, it rapidly hogs the fast flow pathways at low saturation, leaving the displaced phase moving slowly through the narrower regions. This is bad for recovery, if we inject water or gas to displace oil.

In contrast, we can have ordinary percolation displacements when layer flow in corners allows displacement throughout the pore space – this will be described in the next chapter when we consider imbibition. Now we only see significant connectivity, with the injected phase occupying a spanning path of filled pores and throats, at the percolation threshold, when there is a substantial saturation of the injected phase. This means that the invading phase has a very poor flow conductance – limited by layers – until the percolation threshold is reached. This holds back the advance, allows the displaced phase to escape, and is much more favourable for recovery.

This important concept is only mentioned briefly here, but will be a major theme of the discussion throughout this book. Oil recovery is controlled principally by whether we have an invasion percolation or percolation-like advance, since there is a stark difference in the saturation at which the injected phase connects across the rock, as shown in Fig. 3.15.

3.5 Final Saturation and Maximum Capillary Pressure

Primary drainage continues until some maximum capillary pressure, or initial wetting phase saturation, is reached. In an experiment with mercury, since the vacuum can always be displaced, the wetting phase saturation – with a sufficiently large capillary pressure – can be driven down to arbitrarily low values, as shown in Figs 3.8 and 3.9.

In a laboratory experiment where two liquids, such as oil and water, are used, there is always a finite minimum wetting phase saturation. This is for three reasons. The first is that the wetting phase may be trapped in the pore space. While we generally assume that the wetting phase collects in roughness and that this roughness is all mutually connected, there can be cases where water resides next to a smooth surface and is not able to be displaced. This has been observed in bead packs: with smooth beads there is an irreducible saturation of around 9% with the water trapped in pendular rings; these have been imaged directly (Turner et al., 2004; Armstrong et al., 2012). However, when the beads are made rough by etching in acid, final wetting phase saturations of close to 1% have been achieved after a long period of drainage, with displacement continuing, albeit very slowly (Dullien et al., 1989). The second reason is that only a finite capillary pressure can be reached. The third, and generally the most significant reason, is that insufficient time is allowed for the wetting phase to escape.

Near the end of a primary drainage experiment, where the wetting phase saturation is low, it may take a very long time to allow capillary equilibrium to be attained. This is another situation where the macroscopic and local capillary pressures can be different. A large external capillary pressure may be imposed. At a local level, water is still retained in the corners and the local capillary pressure is lower. The non-wetting phase is at high saturation, occupying most of the larger pores and throats, is well connected, and hence maintains a constant pressure throughout the pore space. However, the wetting phase is poorly connected and has a tiny flow conductance. If the local capillary pressure is lower than the external value and the non-wetting phase pressure is constant, then the wetting phase pressure locally is higher than externally. This provides a driving force for movement of the wetting phase out of the system; however, if the flow conductance is small, it may take a very long time to see an appreciable change in saturation. In experiments, this may be interpreted as a trapped saturation and the measurements prematurely terminated. When we introduce flow in Chapter 6, we will show that equilibrium times of days or longer are typical. It is often not feasible to wait this long for the experiment to equilibrate in most laboratory settings.

In the field, for instance, when we consider primary oil migration, there is sufficient time to reach equilibrium, since the process takes place over geological time scales, or millions of years. In this case, the limitation is the imposed capillary pressure. As discussed previously, at rest, the fluid pressure increases with depth z (or decreases with height) according Eq. (3.38), with the capillary pressure given by Eq. (3.39).

Now, let us consider typical values of the capillary pressure. While there is a huge variability, the top of the oil column in hydrocarbon fields is typically 10s m above the free water level (where the capillary pressure is zero): here we consider a range for $z_0 - z$ between 10 and 100 m, using a density difference between a reservoir brine and oil of around 300 kg/m^3, giving capillary pressures of around 0.03–0.3 MPa. We cannot though use these values to compare with the measured capillary pressures in Fig. 3.8, 3.9 and 3.13 since the interfacial tensions and contact angles for mercury are not representative of field settings.

For most subsurface primary drainage processes, bearing in mind invasion over a rough yet water-wet surface, the receding contact angle $\theta_R \approx 0$. We retain our value of 20 mN/m for the oil/brine interfacial tension. Then a capillary pressure of 1 MPa represents a throat radius, using Eq. (3.1), of 0.04 μm, or 20 nm. Alternatively, a mercury injection capillary pressure of 1 MPa represents an oil/water pressure of only 0.061 MPa.

Maximum capillary pressures of 0.03–0.3 MPa in the field represent the invasion of throats with radii approximately 1.3–0.13 μm. This corresponds to a mercury injection capillary pressure in the range 0.49–4.9 MPa. If the reservoir were composed of Bentheimer sandstone, then Fig. 3.13 would imply a minimum saturation of less than 5%: the rock would be almost completely saturated with oil. In contrast, for Ketton, Fig. 3.8, the minimum water saturation would be in the range of 45–55%, with the micro-porosity still containing brine. In complex rocks, with many small pores and throats, it is misleading to contend that all of these elements will ever be invaded by oil (or CO_2) in sub-surface settings: the required capillary pressure is infeasibly large.

Eq. (3.39) implies that the saturation distribution as a function of height above the free water level, $z_0 - z$, is the re-scaled capillary pressure: the shape of the saturation-height function is the same as the capillary pressure. This is presented in more detail in Chapter 6.2.3 where we introduce the Leverett J function.

Once the initial water saturation is established, the wettability can change caused by the sorption of surface-active components of the oil to the solid surface, as presented in Chapter 2.3.3, page 62. The wettability will be a function of height and initial saturation, with the rock becoming more oil-wet and less water-wet with height above the free water level, for two reasons. Firstly, the amount of surface contacted by oil, and hence available for wettability alteration, will increase with height, making a larger proportion of the surface oil-wet (or at least of altered wettability). Secondly, the capillary pressure itself influences the change in wettability, through forcing the collapse of protective molecular films of water at a sufficiently high value. Hence, water films are more likely to be present, protecting the surface from wettability alteration, closer to the free water level where the water saturation is high; they are more likely to be absent, allowing a wettability change, near the top of the oil column. This trend in saturation and wettability is shown in Fig. 3.18. The free water level, as mentioned above, is defined where the capillary pressure is zero. However, we only have a finite oil saturation once we have exceeded P_c^*, which traditionally represents the location of the oil/water contact, or the depth where there is first appreciable quantities of oil in the pore space. The water saturation decreases and the oil saturation increases with height and may reach some minimum (trapped) value, S_{wc}, where c stands for connate, or the irreducible water saturation in the reservoir. In our example, this may be reached for a reservoir comprised of Bentheimer sandstone, but not for a micro-porous carbonate (Ketton limestone, for instance). The region where the oil saturation is changing with height is called the transition zone and may extend through the entire oil column in many – if not most – reservoirs. The wettability trend is from water-wet conditions near the oil/water contact, mixed-wettability in the transition zone and more oil-wet conditions near the top of the reservoir.

Fig. 3.18 gives a textbook style indication of the variation in water saturation with height in an oil reservoir, but can be misleading when used to interpret real data. Fluid pressures can be measured down-hole in a reservoir, from which the free water level may be determined. It is also possible to infer, rather than directly measure, the water saturation through the measurement of electrical resistivity. A sharp decrease in resistivity with depth is interpreted as an indication of a transition between largely oil-filled to largely water-saturated rock. Normally we locate the oil/water contact where we first detect oil: that is the depth (or height) corresponding to P_c^*: however, we know neither the capillary pressure nor the saturation in the reservoir rock precisely and so this is a somewhat overly theoretical definition: in reality, the operational oil/water contact is an indication of where the water

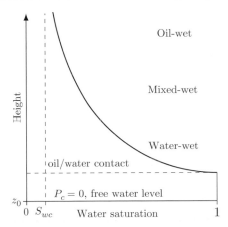

Figure 3.18 The variation of saturation and wettability encountered in an oil reservoir. The saturation change with height is a re-scaled capillary pressure curve, Eq. (3.39). z_0 is the depth of the free water level where the capillary pressure is zero. The oil/water contact is traditionally located where we first see oil in the pore space. The transition zone, which typically extends across the whole oil column, is the region where there remains some mobile water: theoretically the water saturation may reach a minimum value S_{wc}, although this is rarely evident. The wettability of the reservoir often varies from water-wet near the oil/water contact to mixed-wet in the transition zone, with more oil-wet conditions near the top of the oil column.

saturation decreases rapidly in the pore space, close to, but often slightly above, the threshold pressure P_c^*.

Furthermore, the reservoir rock is heterogeneous with a spatial variation in capillary pressures. Hence, the saturation does not smoothly decrease with height, but will experience significant fluctuations. For instance, layers of rock with tight pore spaces will tend to remain water saturated even if they lie far above the free water level. Attempting to define a transition zone as some limited region close to the oil/water contact where the water saturation is high, as done traditionally, ignores the changes in saturation that are typically observed.

The other hazy concept is the irreducible water saturation S_{wc}. As the capillary pressure increases, more and more water can be squeezed out: there is always some tight pore space, or a narrow crack or roughness that retains water and which could be displaced at a finite, but very large, capillary pressure. Hence it is misleading to assert that some fixed value is ever reached at some height above the free water level. This, combined with the inevitable heterogeneity of the reservoir, means that a constant S_{wc} is a fiction. All fields have a variation in water saturation throughout the oil column.

The trend in wettability shown in Fig. 3.18 has been observed in several oil reservoirs, including Prudhoe Bay, off the north coast of Alaska, which is the largest producing oil field in the United States (Jerauld and Rathmell, 1997). This will have significant implications for fluid displacement and oil recovery during subsequent water injection (Jackson et al., 2003), presented in Chapters 5 and 7. Even if there is no significant wettability change, such as in gas fields, or where there is CO_2 storage, the variation in initial water saturation still has an impact on the trapping of the non-wetting phase, as discussed in the following chapter.

4

Imbibition and Trapping

Rainfall soaking into soil, plant roots taking up water, a paper tissue mopping up a spill, or even filling a baby's nappy, are all commonly encountered examples of imbibition. This is the reverse process to drainage, where now it is the wetting phase that invades the non-wetting phase. Primary imbibition is the invasion of a wetting phase into a porous medium initially completely saturated with the non-wetting phase. It describes, for instance, water advancing into dry sand. This process is complicated by the formation of wetting films and layers in advance of water movement through the centres of the pore space (Dussan, 1979; de Gennes, 1985; Cazabat et al., 1997; de Gennes et al., 2003), although in many cases, as discussed below, wetting layer flow can be ignored. It is more normal to encounter secondary imbibition, where the wetting phase enters a porous medium in which some wetting phase is already present in layers and occupying the smaller regions of the pore space established after primary drainage. This is always the case for waterflooding an oil reservoir, the migration of CO_2 in an aquifer or the rise of the water table through damp soil.

We will describe two distinct processes, whose competition controls the nature of the displacement. The first is the swelling of wetting layers, leading to snap-off. The non-wetting phase can then become trapped in the larger pores, surrounded by the wetting phase in narrower regions. This has a major impact on oil recovery, the security of CO_2 storage and the amount of trapped air near the water table. The second process is piston-like advance, or the direct filling of pores and throats from adjacent pores filled with the wetting phase, which is the reverse of the displacement in drainage. However, there is a subtlety associated with the filling of pores, discussed below. Piston-like advance alone leads to a rather flat advance of fluid and very little trapping.

We will assume that the wetting phase is water, while the non-wetting phase is oil: this helps focus attention on applications in oil recovery and renders the text less abstract. We will also discuss CO_2 storage, where CO_2 is the non-wetting phase, and the infiltration of water into soil. We will allow for wettability alteration and contact angle hysteresis, but will assume that all the surfaces are still water-wet with contact angles less than 90°. Cases where the surfaces are not water-wet will be considered later, in Chapter 5.

4.1 Layer Flow, Swelling and Snap-Off

In an oil reservoir, primary drainage, or primary oil migration, occurs over hundreds of thousands or millions of years. As mentioned in Chapter 3.5, it is therefore reasonable to assume that the fluids are in capillary equilibrium; throughout a small region of rock the interfacial curvature between oil and water will be constant. This will be true for both the arc menisci (AM) in the corners and roughness of the pore space, and the terminal menisci (TM) that cross a pore or throat. The macroscopic and microscopic capillary pressures will be the same: its value will be governed by the height above the free water level.

Primary drainage is followed by waterflooding, where water is injected through a well into the reservoir to displace oil that is produced from another well. While near the well the injection pressure may be high and the flow rates rapid, in the middle of the oil reservoir flow rates are generally much lower. The typical distance between wells is hundreds of metres to kilometres, while waterflooding may continue for decades until the water saturation has increased throughout the field. At some location away from the wells, the water saturation will increase rather slowly, over months or years. We can assume that there is a gradual increase in water pressure and saturation with a corresponding decrease in capillary pressure.

If the initial capillary pressure is high, oil, as the non-wetting phase, will occupy the centres of most of the pore space, confining the water to layers in the corners and roughness, and some of the tighter regions of the pore space. We assume that these layers are connected throughout the rock. As the water pressure increases, these layers will swell; the process is slow, but allows a uniform increase in saturation throughout the system. Let us first imagine an extreme case, where the water is only contained in layers. Then the water is unable to displace oil directly from an adjacent water-filled pore. Hence, the water layers continue to swell until it is no longer possible to establish a three-phase contact between the water, oil and the solid. At this point we have snap-off, where water spontaneously fills the centre of the pore space. This mechanism has a significant impact on the fluid distributions and the amount of oil (or other non-wetting phase) that can be recovered. Snap-off will occur in the narrowest region of the pore space – the throat – and then cause partial filling of the two adjacent pores.

Figs. 4.1 and 4.2 illustrate the process in a single idealized region of the pore space. The pore space is triangular in cross-section: as has been emphasized

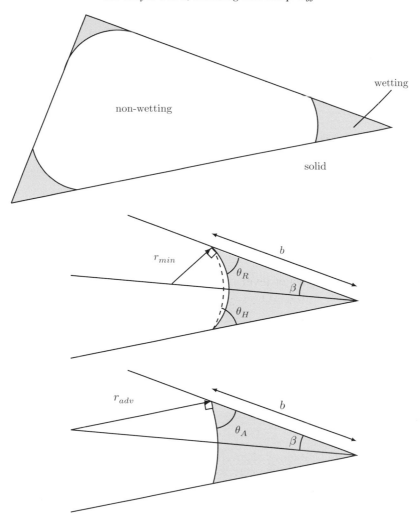

Figure 4.1 Wetting phase in the corner of the pore space. Top: a cross-section through a pore or throat is shown at the end of primary drainage with water, the wetting phase (shaded), in the corners, and oil, the non-wetting phase, in the centre. Middle: a detailed view of a corner, showing the corner half-angle β and the receding contact angle θ_R with a corresponding radius of curvature r_{min}. There is initially a length b of the surface contacted by water in the corner. As the capillary pressure decreases, initially the contact between water, oil and the solid is pinned – it cannot move. The radius of curvature increases, as does the contact angle. The dashed line shows the interface with a hinging contact angle θ_H. Bottom: the contact line only moves once the advancing contact angle θ_A is reached, with a radius of curvature r_{adv}, where $r_{adv} > r_{min}$.

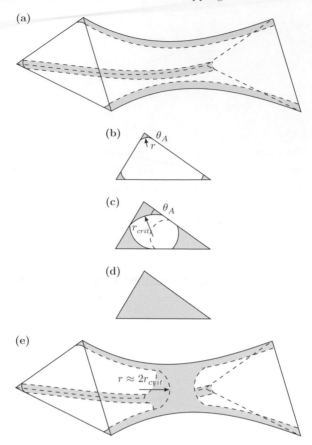

Figure 4.2 An illustration of snap-off in imbibition. Water, the wetting phase, swells in layers occupying the corners of the pore space, as shown in Fig. 4.1. (a) The initial distribution of the wetting phase, shaded, is shown. (b) The throat in cross-section at its narrowest point, showing the radius of curvature of the arc meniscus, AM and the advancing contact angle, θ_A, as the wetting layers swell. (c) There is a critical curvature, r_{crit}, when the three-phase contact between the two fluids and the solid is lost. This first occurs in the narrowest region of the pore space, defined as the throat. The AM fills the throat rapidly – a transient position is shown by the dashed line. (d) Snap-off fills the throat completely. (e) Once snap-off occurs two terminal menisci, TM, are formed, as shown: these TM span the throat and invade the narrower portions of the pore space rapidly, and, if possible, will establish a new position of capillary equilibrium at the capillary pressure given by Eq. (4.7): the new radius of curvature is approximately $2r_{crit}$ since the TM is curved in two directions.

already, there is no need for the throat to be this shape. The essential features are that the cross-section is not smooth, so that the wetting phase may reside in the corners or roughness when the centre is occupied by the non-wetting phase, and that there is some variation in the size (or inscribed radius) of the cross-section.

At the beginning of waterflooding, water resides in the corners of the pore space with a radius of curvature defined such that $P_c^{max} = \sigma/r_{min}$, where P_c^{max} is the maximum capillary pressure obtained after primary drainage and r_{min} is the corresponding average radius of curvature. Fig. 4.1 shows a cross-section of the pore space: if the curvature of the solid surface (and hence the fluid interface) in the plane perpendicular to the diagram is small, then r_{min}, as shown, is the radius of curvature of the AM in the throat in the plane of the diagram. The contact angle is θ_R, or the receding value, since the wetting phase distribution was established by drainage. When the water pressure increases, the three-phase contact line between water, oil and the solid cannot move until the contact angle becomes θ_A, the larger advancing value: this is simply because water cannot move across the surface at a lower contact angle. The capillary pressure decreases (the water pressure increases), with a corresponding increase in the radius of curvature, but the contact point is fixed.

The three-phase contact is pinned, and the contact angle which maintains the imposed capillary pressure is the hinging angle θ_H, where $\theta_A \geq \theta_H \geq \theta_R$. This hinging angle is seen for a capillary pressure $P_c^{max} \geq P_c \geq P_c^{adv}$,

$$\theta_H = \cos^{-1}\left[\frac{P_c}{P_c^{max}}\cos(\beta + \theta_R)\right] - \beta, \tag{4.1}$$

where b is the length of the solid in the corner contacted by water after primary drainage (see Fig. 2.29) which from Eq. (3.6) is,

$$r_{min} = \frac{b \sin \beta}{\cos(\beta + \theta_R)}, \tag{4.2}$$

or

$$b = r_{min}(\cos \theta_R \cot \beta - \sin \theta_R). \tag{4.3}$$

In terms of the maximum capillary pressure attained during primary drainage, $P_c^{max} = \sigma/r_{min}$, Eq. (4.3) is

$$b = \frac{\sigma}{P_c^{max}}(\cos \theta_R \cot \beta - \sin \theta_R). \tag{4.4}$$

The layers will start to swell by moving across the surface when a pressure $P_c^{adv} = \sigma/r_{adv}$ is reached, where r_{adv} is the radius of curvature when the hinging contact angle increases to reach the value θ_A. Now, when the water pressure increases further, the contact angle remains at θ_A while the AM moves further into the throat. The curvature at which the wetting layer begins to advance can be found using Eq. (4.2)

$$r_{adv} = r_{min}\frac{\cos(\beta + \theta_R)}{\cos(\beta + \theta_A)}, \tag{4.5}$$

or a capillary pressure

$$P_c^{adv} = P_c^{max} \frac{\cos(\beta + \theta_A)}{\cos(\beta + \theta_R)}. \qquad (4.6)$$

As the water pressure increases further, the wetting layers increase in volume with the water-oil-solid contact moving out from the corner as shown in Fig. 4.2: the radius of curvature r is found from $P_c = \sigma/r$ and will increase over time. There comes a point when the three-phase contacts from two corners meet, at some critical radius of curvature r_{crit}. This will occur in the narrowest region of the pore space, the throat, that is not already filled with water: everywhere else, the contacts will be separated.

What happens when the water pressure (and corner volume) is increased further? We now have an unstable situation, since adding more water will cause the meniscus from the two corners to lose contact with the surface: however, there is no way to do this without the radius of curvature *decreasing* corresponding to an increase in capillary pressure. If we consider that the oil (non-wetting phase) pressure remains constant, an increase in capillary pressure leads, locally, to a rapid decrease in water pressure. Water will flow from high to low pressure and rapidly moves into the throat and fills it completely. In so doing, two TM are formed, since now the water spans the throat: the water fills the throat and the narrow regions of the two adjoining pores bounded by TM. This leads to an instantaneous increase in local capillary pressure. If, later, we establish capillary equilibrium at the same local pressures as just before the snap-off event, then the average curvatures are equal to $1/r_{crit}$. However, since the TM is curved in two directions – it is roughly hemispherical in shape – the radius of curvature of the interface is approximately $2r_{crit}$ for circular throats, meaning that the TM rapidly advance into wider regions of the pore space. A more accurate calculation of the radius of curvature for pores of arbitrary shape can be performed considering an energy or force balance.

It is possible that for some time, the TM are confined to narrower regions where the curvature is larger than $1/r_{crit}$, resulting in an increase in local capillary pressure. This will be discussed later when we consider the dynamics of filling in imbibition.

This process is snap-off, which was first described by Pickell et al. (1966) to interpret and explain trapping in imbibition. It was observed directly at the pore scale in the pioneering micro-model experiments of Lenormand and co-workers (Lenormand et al., 1983; Lenormand and Zarcone, 1984). While the paper Lenormand et al. (1983) is in a prestigious publication – *The Journal of Fluid Mechanics* – I consider the conference proceedings paper, Lenormand and Zarcone (1984), more discursive and instructive as a teaching tool: I thoroughly recommend a close study. The micro-model was a square array of small rectangular glass ducts where displacement processes could be observed directly under a microscope. A numerical and theoretical interpretation of snap-off has been provided by several authors, starting from the work of Lenormand and Zarcone (1984), and Mohanty et al. (1987).

Geometrical arguments can be used to compute the critical radius of curvature and the corresponding capillary pressure for snap-off. Here we will assume that there is negligible curvature of the AM in the direction perpendicular to the plane of the throat: a more detailed analysis, considering this curvature and different non-circular cross-sectional shapes of the throats has been presented by Deng et al. (2014).

If we have a throat whose corner angles are all β and sides of length $2l$, the capillary pressure at which snap-off occurs can be derived from Eq. (4.3) by replacing b with l, θ_R with θ_A and identifying $P_c = \sigma/r_{crit}$ to find

$$P_c = \frac{\sigma}{l} \left(\cos \theta_A \cot \beta - \sin \theta_A \right). \tag{4.7}$$

The inscribed radius of the throat $r_t = l \tan \beta$ and so we can rewrite Eq. (4.7) as

$$P_c = \frac{\sigma \cos \theta_A}{r_t} \left(1 - \tan \theta_A \tan \beta \right). \tag{4.8}$$

Consider two simple cases. The first is a pore of rectangular cross-section. Here l represents the half-length of the shorter side: the inscribed radius of the throat $r_t = l$. $\beta = 45°$ and $\cot \beta = 1$, giving

$$P_c = \frac{\sigma \cos \theta_A}{r_t} \left(1 - \tan \theta_A \right), \tag{4.9}$$

while for an equilateral triangular pore $\beta = 30°$ and we find

$$P_c = \frac{\sigma \cos \theta_A}{r_t} \left(1 - \frac{\tan \theta_A}{\sqrt{3}} \right), \tag{4.10}$$

where $\tan(30°) = 1/\sqrt{3}$.

We will use these equations in the following section where we compare the capillary pressure for snap-off with piston-like advance. At this stage there are two features to note. Firstly, the pre-factor in these equations does not contain 2, since there is only one curvature in the AM, unlike expressions for direct advance with a TM, where there are two finite radii of curvature; see, for instance, Eq. (1.16). This makes the threshold capillary pressure lower, or less favoured, during water injection (a higher water pressure is needed). The second comment is that snap-off is only possible if we have a positive capillary pressure with a concave AM: if not, the layer is either completely absent or remains hinged. The AM can only move during imbibition at a positive capillary pressure, indicative of a water-wet medium. From Eq. (4.7) we can see that snap-off will only occur for $P_c > 0$ if $\tan \theta_A \tan \beta < 1$ or

$$\theta_A + \beta < \frac{\pi}{2}. \tag{4.11}$$

Snap-off is only possible for a combination of small contact angles and sharp corners.

Snap-off leads to the rapid filling of the throat and the adjoining narrow regions of a pore. It is confined to a small region of the pore space. This appears to contrast with displacement in drainage, where once a TM had passed a narrow throat, this could engender a cascade of further pore and throat filling. However, as we show below, while generally less dramatic, snap-off can also allow further displacement at the prevailing externally imposed capillary pressure through the direct (piston-like) filling of neighbouring pores and throats.

The most important impact of snap-off is to establish TM across throats that block the flow of oil. If all the throats bounding a pore are filled by snap-off, then the oil remaining in the pore is trapped and cannot be displaced by further water injection (assuming slow flow where viscous forces are neglected). This is the principal mechanism by which oil is trapped in the pore space of a hydrocarbon reservoir, or how the migration of CO_2 is impeded in a storage aquifer.

Note that snap-off has not been analysed through consideration of an energy balance: if a TM is present in an adjacent pore then the throat will fill with water at a lower water pressure (higher, more favourable, capillary pressure) by piston-like advance, as described later in this chapter. For snap-off, the displacement event represents a discrete change in interfacial energy.

4.1.1 Roof Snap-Off during Drainage

Roof (1970) observed snap-off directly in idealized pore geometries during drainage, associated with a Haines jump. Here the process is shown schematically in Fig. 4.3. Once the non-wetting phase has invaded a throat, it will rapidly pass into the adjoining pore, as we described previously in Chapter 3.1 (see Fig. 3.1). When this occurs, the local capillary pressure will drop. This allows the wetting layers in the corners of the pore space to swell. It is possible that the capillary pressure drops sufficiently far that the threshold capillary pressure for snap-off in the throat is met and hence the throat spontaneously fills with water (Ransohoff et al., 1987). This is, locally, an imbibition process but occurs during a drainage displacement, in that overall the non-wetting phase saturation increases. A new position of capillary equilibrium is reached with the three TM shown in Fig. 4.3 by the solid

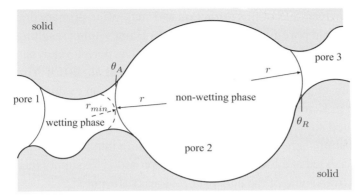

Figure 4.3 A representation of Roof snap-off in drainage. The non-wetting phase first enters a throat, as shown in Fig. 3.1. The critical radius of curvature of the terminal meniscus, TM, is r_{min}, shown by the dashed arc. Once the throat has been passed, the non-wetting phase can rapidly fill the adjacent pore. This leads to a drop in local capillary pressure; if this falls below the threshold for snap-off, the throat is re-invaded by wetting phase with a contact angle θ_A. A new configuration of fluid at local capillary equilibrium is established with a radius of curvature r in the TM, shown by the solid arcs. There remains a trapped ganglion or blob of non-wetting phase in the pore space. The local capillary pressure is now lower than before, $r > r_{min}$: as the externally imposed capillary pressure is increased, and more non-wetting phase is injected, it will once more invade the throat and reconnect the trapped blob.

arcs: those formed after snap-off will encounter the surface at the advancing contact angle θ_A while the TM furthest into the pore will have a contact angle θ_R. The local capillary pressure will be lower than before, with larger radii of curvature of the TM.

This process will occur whenever the local capillary pressure to pass through the widest portion of the pore is lower than the threshold for snap-off in the throat, although there is a natural time scale for the process, dependent on how quickly the fluids are able to move to reach new positions of equilibrium (Ransohoff et al., 1987; Kovscek and Radke, 1996). As we show below, snap-off requires movement into a pore that is at least twice the inscribed radius of the adjacent throat. If we consider long pores, or groups of several pores, we also need to account for changes in local capillary pressure as the result of rapid fluid flow: in this case the analysis becomes more involved, and requires an accurate computation of the fast-evolving flow field.

As a result, a ganglion, or blob, of non-wetting phase is trapped in the pore. This non-wetting phase is stranded and cannot move. The dynamics of this process has been studied in detail; it will occur even when there are no wetting layers in the corners of the pore space – the non-wetting phase will still snap-off through thin film flow in a throat of circular cross-section (Roof, 1970; Gauglitz and Radke, 1990).

In rocks, the dynamics of this retraction and snap-off in primary drainage is complex, with multiple pore-filling events during a Haines jump, resulting in retraction of the non-wetting phase throughout the system. The pore-level dynamics has been observed directly using rapid X-ray imaging (Berg et al., 2013; Armstrong et al., 2014b; Andrew et al., 2015; Bultreys et al., 2016a). As an example, Fig. 4.4 shows images taken in Ketton limestone during CO_2 injection into brine. The images have been segmented so that only the CO_2 phase, which is non-wetting, is shown. During a Haines jump there is a sequence of pore filling into a region of the pore space that can be accessed at a lower local capillary pressure. To supply this CO_2 rapidly, CO_2 retracts from some of the narrower regions of the pore space, which can result in snap-off and disconnection of the non-wetting phase, as shown. Note that unlike the cartoon shown in Fig 4.3, the non-wetting phase that becomes trapped is not necessarily in the pore that has just been partially filled, but can be in another region some distance away. This distinguishes two types of snap-off in a drainage process: local snap-off, where the non-wetting phase is disconnected in the pore that it has just entered, and distal snap-off, where the event occurs further away, shown in Fig. 4.4.

Fig. 4.5 shows a detail of Fig. 4.4 in the region where distal snap-off occurred stranding a region of the non-wetting phase. The curvature of the fluid interface is mapped. After snap-off, the curvature has decreased, indicating a drop in local capillary pressure and the attainment of a new position of equilibrium. Later, as the displacement proceeds, the local capillary pressure will rise again, allowing the throat to be re-accessed by non-wetting phase, reconnecting the stranded ganglion. However, this reconnection is not instant: there is significant further displacement throughout the sample until the local capillary pressure rises again to a new maximum.

Micro-second resolution X-ray scanning in three dimensions indicates that the whole cascade of pore filling continues for around 1 s in rock samples (Armstrong et al., 2014b), although individual filling events are completed much faster. The pore-scale dynamics has also been studied in analogue micro-model experiments with high-speed photography. Again, a decrease in local capillary pressure is observed, with local flow rate of up to 1,000 m/day, compared to typical average flow rates of less than 0.1 m/day in reservoirs, with pore filling events completed in micro-seconds (Armstrong and Berg, 2013). The use of pore-image velocimetry in micro-models has also confirmed pore-scale flow rates, which are thousands of times faster than those imposed externally (Kazemifar et al., 2016). This demonstrates that even if the injected flow rate is infinitesimal, the perturbation of local equilibrium during a Haines jump leads to very rapid flows in the pore space: this will be discussed and quantified further in Chapter 6.

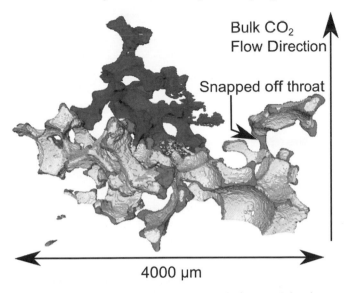

Figure 4.4 An illustration of snap-off in primary drainage. CO_2, the non-wetting phase, displaces brine in Ketton limestone. The experiment was performed at reservoir conditions (10 MPa pressure and 50°C temperature), so that CO_2 is a super-critical phase. The rock and fluids were imaged using a synchrotron X-ray beam: a sequence of images was taken with a time lag of approximately 45 s. Only the CO_2 phase is shown. The pore structure of the rock is illustrated in Fig. 2.6. The yellow (palest shade) represents the distribution of CO_2 before a Haines jump. Blue (darkest shade) indicates the CO_2 that invades the pore space after the jump, while red shows where the CO_2 has retracted, marking local imbibition events. Note that snap-off can isolate a region of the non-wetting phase that is not directly involved in the Haines jump. From Andrew et al. (2015). (A black and white version of this figure will appear in some formats. For the colour version, please refer to the plate section.)

It has been suggested that careful measurements of the minimum capillary pressure attained during a Haines jump could be used to infer information associated with the pore (as opposed to throat) size and volume (Yuan and Swanson, 1989). However, the subtle dynamics where the filling of a throat can impact the arrangement of fluid some distance away makes a simple interpretation difficult, despite some successful attempts to analyse pressure signals in low rate micro-model experiments (Furuberg et al., 1996). In a core-scale experiment though, only the externally imposed macroscopic capillary pressure can be measured, which is not always the same as the local capillary pressure, even in the limit of very slow flow, and which – as we have shown – fluctuates strongly during substantial re-arrangement of the fluid during a Haines jump.

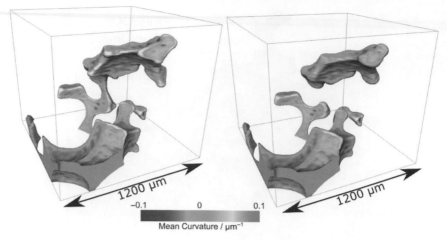

Figure 4.5 Detail of part of Fig. 4.4 illustrating the disconnection of the non-wetting phase, CO_2 after a cascade of pore filling. The curvature of the CO_2 interface is shown. The negative curvature represents CO_2 in contact with the Ketton grains which bulge out into the non-wetting phase. The regions of positive curvature indicate a fluid/fluid interface between CO_2 and brine. Notice that after the snap-off event, the curvatures have decreased, indicating a drop in local capillary pressure. As the local capillary pressure increases again during the displacement, the CO_2 will re-connect. From Andrew et al. (2015). (A black and white version of this figure will appear in some formats. For the colour version, please refer to the plate section.)

4.2 Piston-Like Advance and Pore Filling

4.2.1 Piston-Like Throat Filling

The second type of displacement process in imbibition is essentially the reverse of drainage, where a terminal meniscus moves through a throat that initially contains a non-wetting phase. An energy balance approach will be used, similar to that developed in Chapter 3.1.2, to find the threshold capillary pressures. If there is no contact angle hysteresis, so that $\theta_A = \theta_R$, or if the wetting layers are beginning to swell ($P_c \leq P_c^{adv}$ in Eq. (4.6)), the threshold capillary pressure can be obtained through analogy to Eq. (3.10):

$$P_c = \frac{\sigma \left(1 + 2\sqrt{\pi G}\right) \cos \theta_A F_d(\theta_A, G)}{r_t} \equiv \frac{C_{It} \sigma \cos \theta_A}{r_t}. \qquad (4.12)$$

defining a coefficient C_{It} where $2 \geq C_I \geq 1$.

However, if the wetting layers are still pinned in the corners, the calculation of the threshold capillary pressure is more complex. Considering a balance of free energy, from Eq. (3.5), for a throat with n corners of length l, we can write

$$\frac{\Delta A}{r} - \Delta L_{wnw} - 2n(l - b)\cos\theta_H = 0, \tag{4.13}$$

where ΔL_{nws} from Eq. (3.5), the change in length of the surface between the non-wetting phase and the solid as a result of the displacement, is $2(l - b)$ for each corner, with b is given by Eq. (4.3); it is not a function of the radius of curvature r.

Similarly to Eq. (3.7), the length of the AM before the displacement is

$$L_{AM} = 2r\left(\frac{\pi}{2} - \theta_H - \beta\right) \equiv S_{LI}r, \tag{4.14}$$

while from Eq. (3.8) the area of the wetting phase in the corner is

$$A_{AM} = r^2\left(\frac{\cos\theta_H\cos(\theta_H + \beta)}{\sin\beta} - \frac{\pi}{2} + \theta_H + \beta\right) \equiv S_{AI}r^2, \tag{4.15}$$

where S_{LI} and S_{AI} are geometrical coefficients and θ_H is found from Eq. (4.1), substituting σ/r for P_c:

$$\theta_H = \cos^{-1}\left[\frac{\sigma}{r P_c^{max}}\cos(\beta + \theta_R)\right] - \beta. \tag{4.16}$$

Employing the same approach as for drainage, we arrive at the following expression using Eqs. (4.13)–(4.15) to obtain:

$$A_t - nS_{AI}r^2 - nS_{AL}r^2 - 2n(l - b)\cos\theta_H = 0. \tag{4.17}$$

It is not possible to find a closed-form expression for r and hence P_c, since θ_H is itself a function of r through Eq. (4.16): solutions to the system of equations need to be acquired iteratively (Øren et al., 1998). Despite these complexities, as in drainage, for a largely circular pore at low values of θ_A, $C_{It} \approx 2$ in Eq. (4.12), whereas for more slit-like pores $C_{It} \to 1$.

One interesting observation is that C_{It} in Eq. (4.12) can be negative for $\theta_A > \pi/2$ ($\cos\theta_A < 0$), meaning that we can have spontaneous filling at a positive capillary pressure, even if the contact angle implies that the surface is non-wetting to water (Ma et al., 1996; Øren et al., 1998). The physical interpretation is that when the TM of water advances across the AM in the corner, this is a displacement of water over water with a contact angle of 0; this makes the throat appear more water-wet than one completely full of oil. The net result is that the throat, containing oil and water, only appears oil-wet for contact angles somewhat greater than 90°. The maximum contact angle for spontaneous displacement is (Øren et al., 1998)

$$\cos\theta_A^{max} = -\frac{4nG\cos(\theta_R + \beta)}{P_c^{max}r_t/\sigma - \cos\theta_R + 12G\sin\theta_R} \tag{4.18}$$

for a polygonal throat of shape factor G with n sides and corner half angle β. Note that as the maximum capillary pressure imposed during primary drainage, P_c^{max} tends to infinity, the throat becomes completely oil-saturated and $\theta_R^{max} = \pi/2$ or 90°; however, as more of the corners of the throat are filled with water, invasion is allowed at larger contact angles.

Fig. 4.6 shows the maximum contact angle for spontaneous advance for different shaped throats as the initial water saturation varies, defined as the ratio of the area of water initially in the corners after primary drainage to the total cross-sectional area: nA_{AM}/A_t; see Eq. (4.15). Note that displacement with a contact angle up to 130° is possible if sufficient water is initially present in the corners.

4.2.2 Cooperative Pore Filling

Since we decrease the capillary pressure during imbibition, the processes with the highest capillary pressures are most favoured. Hence, advance through a throat will always occur preferentially to the filling of an adjacent pore with the same contact angle and shape factor. However, an adjacent pore must be filled with the wetting phase first before this process is allowed; otherwise the only possible filling

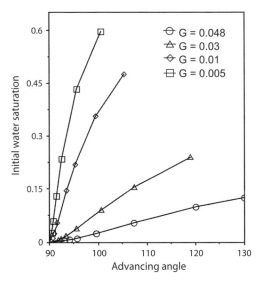

Figure 4.6 Graph of the maximum advancing contact angle for spontaneous throat filling (a positive threshold capillary pressure) for throats with different shape factor G as a function of the initial water saturation, defined as the ratio of the area of water in the corners after primary drainage to the total cross-sectional area. Note that spontaneous displacement is possible for contact angles substantially higher than 90° if there is initially a lot of water in the corners. From Øren et al. (1998).

mechanism is snap-off. Piston-like advance, when it can occur, is always more favoured than snap-off.

Therefore, the filling of the pore space in imbibition is impeded by the pores, not the throats. The calculation of the threshold capillary pressure necessary for a TM to pass through the centre of a pore (at its widest point) is once more an exercise in geometry employing energy balance: it is necessary to find the largest radius of curvature as the TM moves. The threshold pressure is dependent not only on the size and shape of the pore, but the arrangement of water – in particular, what bounding throats are also water filled. This is illustrated in Fig. 4.7, where the pore has either one or two bounding throats that remain filled with the non-wetting phase.

These pore filling mechanisms are called I_n after Lenormand et al. (1983) and Lenormand and Zarcone (1984), where n is the number of throats connected to the pore that contain non-wetting fluid. I_0 is impossible, since in this case the non-wetting phase in the pore is surrounded by water and trapped: it cannot be displaced since it has no connection to the outlet. I_1 is most favoured with a threshold capillary pressure that is, approximately, given by:

$$P_c = \frac{C_{Ip}\sigma\cos\theta_A}{r_p},\qquad(4.19)$$

where r_p is the inscribed radius of the pore. Similar to Eq. (4.12), C_{Ip} is a coefficient dependent on the pore geometry and contact angle, with a range from 2 for a circular pore to 1 for more slit-like shapes. In theory, if the pore geometry is

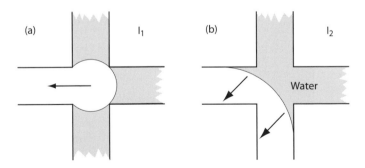

Figure 4.7 An illustration of pore filling in imbibition. The threshold capillary pressure is determined by the largest radius of curvature necessary for the terminal meniscus to move through the centre of the pore. This radius is smaller (making the process more favoured) for I_1, where only one adjacent throat is filled with non-wetting phase, than for I_2, where two adjacent throats are filled with non-wetting phase. Taken from Valvatne and Blunt (2004) based on the work of Lenormand et al. (1983).

known, C_{Ip} can be derived from energy balance in the same way as performed for throat filling, starting from Eq. (2.15) or (4.13) as appropriate.

I_2 is less favoured: the critical, or largest, radius of curvature for filling is bigger than in I_1 (see Fig. 4.7) as the TM spans the pore and moves close to the bounding throats. I_3 is less favoured still and is rarely seen, since in this case, at least for the geometry shown in Fig. 4.7, the wetting phase has to bulge out into the non-wetting phase to advance across the pore, resulting in a negative threshold capillary pressure, as shown in Fig. 4.8: the critical curvature is now when the wetting phase first emerges from the throat. Furthermore, the threshold pressures are also controlled by the location of the throats which are filled with water, giving different pressures for I_2 and other filling events even in the same pore.

The threshold capillary pressure for pore filling is sensitive to the geometry of the pore, which throats are filled with wetting phase and how they are arranged, making a precise determination of their value difficult. In network modelling, to date, rather simplified empirical expressions have been proposed to quantify these threshold pressures (Blunt, 1997b; Øren et al., 1998): in principle though, as stated above, their calculation is a geometric energy-balance exercise where the threshold pressure is determined by the largest radius of curvature needed for the TM to move across the pore.

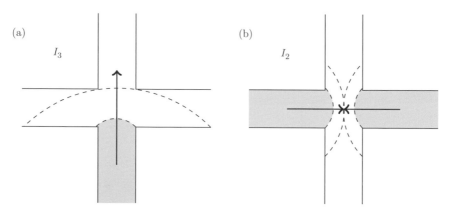

Figure 4.8 Pore filling with a negative threshold capillary pressure, even when the advancing contact angle is less than 90°: here $\theta_A = 45°$. The point of maximum water pressure, or most negative curvature, is shown shaded, as well as when the terminal menisci touch or reach the other side of the pore. (a) An I_3 event where the minimum capillary pressure needed to traverse the pore is negative: unlike the I_1 and I_2 processes illustrated in Fig. 4.7, the pressure in the advancing phase has to exceed that of the displaced phase (the interface bulges outwards). (b) The limiting curvatures for an I_2 displacement where the advancing phase approaches the pore from opposing throats. Again, the minimum capillary pressure is negative, and the same value as in (a). This event is therefore less favoured than I_2, with filling from adjacent throats, shown in Fig. 4.7.

For reference, approximate expressions for I_n filling processes in the literature are listed here. The first quantitative consideration of this problem for general three-dimensional systems was by Jerauld and Salter (1990), who proposed

$$P_c(I_n) = \frac{2\sigma}{nr_p}, \tag{4.20}$$

where n is the number of oil-filled throats and it is assumed that the contact angle $\theta_A \approx 0$. For I_1, a circular pore shape is assumed with $C_p = 2$ in Eq. (4.19).

To account for a full range of contact angle, Øren et al. (1998) suggested

$$P_c(I_n) = \frac{2\sigma \cos \theta_A}{r_p + \sum\limits_{i=1}^{n} b_i r_{ti} x_i}, \tag{4.21}$$

where r_{ti} is the inscribed radius of throat i filled with oil, b_i are dimensionless coefficients and x_i are random numbers between 0 and 1. If we assume that $C_p = 2$ in Eq. (4.19), then $b_1 = 0$. While Eq. (4.21) makes filling less favourable when there are more oil-filled throats, it does not allow the capillary pressure to become negative, as illustrated in Fig. 4.8.

Blunt (1997a) proposed a similar, if slightly simpler model

$$P_c(I_n) = \frac{2\sigma \cos \theta_A}{r_p + \sum\limits_{i=1}^{n} b_i x_i}, \tag{4.22}$$

where now b_i are coefficients with units of length and magnitude equal to a typical throat size, while as before x_i are random numbers between 0 and 1. This expression also suffers from the limitation that the threshold pressures are always positive.

Blunt (1998) refined these models to allow negative threshold pressures and suppress the spontaneous filling of pores except by I_1 and, rarely, by I_2:

$$P_c(I_n) = \frac{2\sigma \cos \theta_A}{r_p} - \sigma \sum\limits_{i=1}^{n} b_i x_i, \tag{4.23}$$

where now b_i are coefficients whose size are approximately the inverse of an average throat radius, except for b_1, which is assumed to be zero, to obey Eq. (4.19) for $C_{Ip} = 2$.

All of these models are rather approximate, and while they are able to reproduce the qualitative features of the displacement, they are likely to be inaccurate for the quantitative characterization of imbibition where these pore filling processes are important. Also note that they all assume that Eq. (4.19) is valid for I_1 filling, which assumes a largely circular pore cross-section.

When the contact angle is greater than some threshold value (normally assumed to be 90°, but strictly dependent on pore geometry), the capillary pressure for pore invasion is no longer dependent on the number of adjacent oil-filled throats, and we have a drainage process. In this case, the following simple expression is assumed,

$$P_c = \frac{2\sigma \cos\theta_A}{r_p}. \qquad (4.24)$$

Once a pore has been filled, the TM will pass through the bounding throats rapidly, since, by definition, this will require a larger (more favoured) capillary pressure, assuming that the contact angle in the throat is similar. This is called piston-like advance to contrast it with snap-off. This event, occurring at a capillary pressure given by Eq. (4.12), as discussed above, is always favoured over snap-off when it can occur. However, snap-off is a possibility in any throat containing a connected wetting phase in layers, whereas piston-like filling can only occur if an adjacent pore is filled with wetting phase.

Pore filling alone leads to very little trapping. Indeed, the I_n mechanisms favour the development of a flat frontal advance of the wetting phase. The growth of protuberances of the wetting phase requires a disfavoured I_{z-1} filling, where z is the coordination number of the pore. In contrast, any invaginations are filled by the more favoured I_1 process. It is still possible to trap non-wetting fluid through bypassing, where frontal advance provides a pincer movement of wetting phase surrounding some region of the pore space.

Cooperative pore filling is illustrated in Fig. 4.9, which shows how a combination of I_3 followed by a line of I_2 in a regular lattice leads to a macroscopically flat advance with no trapping: in a more heterogeneous system the advancing front is more ragged, as discussed below, but still there is little trapping.

4.2.3 *Competition between Snap-Off and Cooperative Pore Filling*

The nature of the displacement, and the amount of non-wetting phase that is trapped, is controlled by the competition between piston-like filling and snap-off. We will first consider the ratio of the threshold capillary pressures for snap-off and piston-like advance in the same throat using Eqs. (4.8) and (4.12):

$$\frac{P_c^{snap}}{P_c^{piston}} \equiv P_{cR}^t = \frac{1 - \tan\theta_A \tan\beta}{C_{It}}. \qquad (4.25)$$

$P_{cR}^t \leq 1$ since $C_{It} \geq 1$, meaning that piston-like advance is always favoured, when possible, over snap-off.

(a)

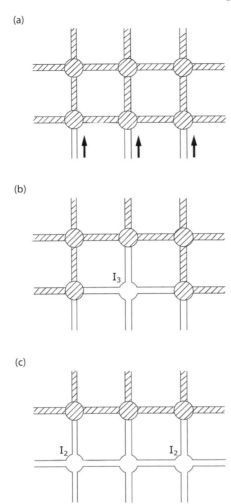

(b)

(c)

Figure 4.9 An illustration of cooperative pore filling in imbibition. (a) Water, the wetting phase (white), displaces non-wetting phase, as shown by the arrows. (b) A pore – the smallest pore – is filled by an I_3 event. (c) There then follows a sequence of more favoured I_2 filling, leading to a flat frontal advance with no trapping. Based on Blunt et al. (1992).

Now we find the ratio of the threshold pressure for snap-off in a throat, Eq. (4.8), to that for the most favoured, I_1 pore filling mechanism, Eq. (4.19):

$$P_{cR}^I = \frac{r_p}{C_{Ip} r_t} \left(1 - \tan \theta_A \tan \beta\right). \tag{4.26}$$

If this ratio is greater than 1, snap-off is favoured; otherwise, pore filling is preferred. This ratio can, in theory, be applied to any throat and any pore in the system; however, it is most instructive to consider this as a ratio between I_1

pore filling and snap-off in the largest throat bounding this pore, as this controls trapping.

Here we will discuss this ratio mainly in the context of imbibition displacements, but a similar criterion can be used to determine if snap-off occurs after a Haines jump in drainage. In this case, the competing piston-like advance occurs at the receding contact angle (the non-wetting phase is moving through a pore, while we consider snap-off in the throat behind it). Hence, the appropriate ratio is between Eq. (4.8) for snap-off in a throat and Eq. (3.13) applied for piston-like advance through the pore in drainage through replacing r_t by r_p:

$$P_{cR}^D = \frac{r_p}{r_t} \frac{\cos\theta_A}{\cos\theta_R} \frac{(1 - \tan\theta_A \tan\beta)}{C_D}, \tag{4.27}$$

where snap-off will occur for $P_{cR}^D > 1$. This is an approximate analysis, since we have neglected the considerable impact of viscous flow which perturbs the local capillary pressures. However, this does provide a good guide to the likelihood of Roof snap-off occurring.

Returning to imbibition, Eq. (4.26), more snap-off ($P_{cD}^I > 1$ for most pores) leads to more trapping of the non-wetting phase. There are four features that control this. Firstly, to favour snap-off, the throat radius r_t has to be less than half the pore radius r_p for $C_{Ip} = 2$. The ratio of pore to throat radii is commonly called the aspect ratio.

The aspect ratio has various definitions. For a given pore, the aspect ratio is the ratio of the pore radius to the mean of the bounding throat radii. Then we can define an average aspect ratio as the mean value for all the pores in a network.

The aspect ratio needs to be greater than two to allow snap-off. Strictly speaking, the radius of the largest bounding throat must be less than half the pore radius (assuming largely circular elements), since it is the filling of this last throat that will determine trapping – the non-wetting phase is only trapped in the pore once all the throats bounding it are filled; if instead I_1 is more favoured, the wetting phase will invade the pore once all but one of the throats are filled, with the non-wetting phase escaping through this last, largest throat.

Secondly, snap-off is more favoured if we have throats with sharp corners ($\tan\beta$ in Eq. (4.26) is small): this is intuitively obvious, since more wetting phase in layers encourages throat filling. The extreme limit here is to consider slit-like pores and throats where $C_{pI} \approx 1$ since the TM is only curved in one direction, across the slit, while we ignore any curvature in the other direction; see Eq. (1.17). Similarly, the threshold pressure for snap-off is given by Eq. (4.8) in the limit of $\beta \to 0$, giving

$$P_{cR}^I = \frac{r_p}{r_t}, \tag{4.28}$$

implying that piston-like advance and snap-off occur at the same capillary pressure for elements of the same size. In drainage displacements, with little contact angle hysteresis, this can allow repeated piston-like filling and snap-off events in the same throat, since the capillary pressures for both processes are similar. Here the ratio of capillary pressures is, from Eq. (4.27):

$$P_{cR}^D = \frac{r_p \cos \theta_A}{r_t \cos \theta_R},$$ (4.29)

for $C_D = 1$.

Thirdly, snap-off is more likely for more wetting conditions: that is small values of θ_A. The precise amount of trapping in a porous medium with large pores relative to throats is very sensitive to θ_A.

Lastly, although only implicit in Eq. (4.26), snap-off leading to trapping is less likely in a well-connected pore space, where every pore has many bounding throats, since the non-wetting phase has more opportunity to escape during imbibition. As mentioned above, it is the largest bounding throat that determines if the non-wetting phase will be trapped: with many connecting throats one is likely to be relatively large, making I_1 filling more favoured. Hence we expect less trapping for large values of the coordination number, or Euler characteristic, χ from Eq. (2.12). χ quantifies the number of redundant loops in the structure, Fig. 2.24. $\chi_V = \chi / V$ is the Euler characteristic per unit volume. $-\chi_V$, which is a positive number for a well-connected pore space, can be interpreted, approximately, as the number of throats that can be filled by snap-off per unit volume without disconnecting the non-wetting phase: the larger this is, the more snap-off that can occur before the non-wetting phase is trapped.

4.2.4 Frequency of Different Filling Events

We can quantify the frequency of different displacement processes through pore-scale modelling. This helps to illustrate the competition between filling mechanisms and how this changes for different pore structures and contact angles. We will compare displacement statistics for some of the network models whose properties are presented in Table 2.1, page 47. We use Eq. (4.23) for pore filling with weights b_n for I_n filling of 15,000 m^{-1} (1/67 μm^{-1}) for $n \geq 2$ in all cases: $b_1 = 0$.

We start with our standard exemplar, Bentheimer sandstone, in Table 4.1. The commonest displacement process, regardless of contact angle, is piston-like filling of throats: whenever a pore is filled, the neighbouring throats are likely to fill as well. There is a competition between I_1 pore filling and snap-off, with fewer snap-off events in throats at larger contact angles, as expected. We do though see more

Table 4.1 *Displacement statistics for imbibition in Bentheimer sandstone. The number of events for different processes is shown for imbibition with different ranges of contact angle. The network statistics are shown in Table 2.1.*

Contact angle range (degrees)	0–30	30–60	60–90	90–120
Uninvaded pores	9,492	7,525	7,328	21,217
Pore snap-off	155	1,184	2,286	0
I_1 pore filling	9,993	8,154	6,786	474
I_2 pore filling	6,735	7,616	7,005	975
I_3 pore filling	1,834	3,110	3,658	1,200
I_4+ pore filling	392	1,012	1,538	4,735
Uninvaded throats	6,436	8,444	9,896	45,259
Throat snap-off	18,687	16,957	12,211	0
Piston-like throat filling	29,618	29,340	32,634	9,482
Residual oil saturation	0.52	0.46	0.36	0.27

snap-off in pores: when the contact angle is close to 90° it is possible for the capillary entry pressure for pore snap-off to be more favourable than for the adjoining throats if they have larger contact angles and are less slit-like. That there are snap-off events when the contact angle is close to 90° shows that some elements have very sharp corners. However, we do not observe an increase in I_1 events as the contact angle increases; instead we see more I_n events with $n > 1$ as these become more favoured relative to snap-off.

The residual, or trapped, oil saturation decreases from 52% to 36% when the contact angle range changes from 0–30° to 60–90°. This is, however, not evident from the number of uninvaded elements: while there is less trapping in pores, there are more uninvaded throats as the contact angle increases. At the lowest contact angles, the largest elements are trapped, with filling governed by pore size. With a range of contact angles, particularly when these values are close to 90°, with a variety of pore shapes, the order of entry pressure is not the same as the order of size: we tend instead to find trapping in more rounded pores and throats with larger contact angles. Hence, as we progress to more neutrally wet conditions, trapping no longer preferentially selects the larger elements and the volume trapped decreases. This is demonstrated in Fig. 4.10, where the volume-weighted fraction of all elements (pores and throats) trapped (oil filled) at the end of waterflooding is plotted as a function of the inscribed radius of curvature. We see the same behaviour in all our other examples: for lower contact angles, filling is controlled by inscribed radius, trapping is confined to the largest elements, and the residual oil saturation is largest. At larger contact angles some trapping occurs in smaller elements, and the residual saturation decreases.

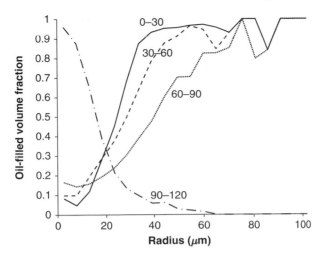

Figure 4.10 The volume-weighted fraction of oil-filled (trapped) pores and throats at the end of waterflooding in a Bentheimer network model, as a function of the inscribed radius of the element. The contact angle ranges in degrees are indicated on the graph. Note that as the contact angles increase we see a less strict size-segregation leading to a lower residual saturation; see Table 4.1. When the contact angle exceeds 90° trapping occurs preferentially in the smallest elements.

For reference, the final column in Table 4.1 is where all the contact angles are just above 90°: this is now a drainage process, and so we have pore filling events for all I_n indifferently and no snap-off. We also see significant trapping with a large number of uninvaded pores and throats: the oil phase, when surrounded by water, is unable to escape in wetting layers: these are only seen when the system is more strongly oil-wet, discussed in Chapter 5. However, this trapping is now in the smaller elements, Fig. 4.10, and the residual saturation decreases slightly to 27%.

In Berea, Table 4.2, the trends are similar. As the average contact angle approaches 90° we see a very sharp decrease in residual oil saturation, with a rapid transition away from lots of trapping and frequent snap-off. This will allow a very sensitive interpretation (and matching) of trapping and flow properties in Chapters 4.6.2 and 7.1.3.

For Mount Gambier, Table 4.3, a limestone sample, with the most open structure but, on average, the lowest coordination number of 2.86, we see the same behaviour. There are well-connected larger pores, but also many smaller pores with only one or two connections in which there are frequent pore snap-off events for larger contact angles. If the contact angle exceeds 90° the residual saturation increases, as we have a large number of trapped elements.

Similar behaviour is also seen for Estaillades and Ketton limestones, shown in Tables 4.4 and 4.5, respectively. Again, there is less snap-off and a transition towards pore filling independent of the number of already-invaded throats as the

Table 4.2 ***Displacement statistics for imbibition in Berea sandstone.*** *The number of events for different processes is shown for imbibition with different ranges of contact angle. The network statistics are shown in Table 2.1.*

Contact angle range (degrees)	0–30	30–60	60–90	90–120
Uninvaded pores	5,107	4,233	3,477	635
Pore snap-off	14	39	140	0
I_1 pore filling	2,274	1,646	1,437	650
I_2 pore filling	2,889	3,319	3,674	2,252
I_3 pore filling	1,810	2,682	3,183	5,645
I_4+ pore filling	255	430	438	3,167
Uninvaded throats	6,880	9,239	7,926	3,074
Throat snap-off	7,679	7,447	6,711	0
Piston-like throat filling	11,587	9,460	11,509	23,072
Residual oil saturation	0.49	0.47	0.30	0.02

Table 4.3 ***Displacement statistics for imbibition in Mount Gambier limestone.*** *The number of events for different processes is shown for waterflooding with various ranges of advancing contact angle. The network statistics are shown in Table 2.1.*

Contact angle range (degrees)	0–30	30–60	60–90	90–120
Uninvaded pores	35,760	31,978	30,900	53,666
Pore snap-off	227	1,688	3,317	0
I_1 pore filling	18,125	16,182	14,500	2,292
I_2 pore filling	9,352	11,022	10,707	2,086
I_3 pore filling	2,411	4,284	5,070	1,841
I_4+ pore filling	404	1,125	1,785	6,394
Uninvaded throats	22,761	24,842	26,813	77,720
Throat snap-off	28,285	25,454	18,363	0
Piston-like throat filling	44,382	44,432	49,502	16,958
Residual oil saturation	0.45	0.40	0.28	0.32

contact angle increases: the residual saturation also declines as the trapped elements are no longer necessarily the largest pores and throats. Estaillades has a similar coordination number to Mount Gambier, but a less open pore space with throat sizes that are around an order of magnitude smaller, Fig. 3.12, page 91. Estaillades has fewer well-connected pores and a broader distribution of throat radius. We see more trapping and very little pore filling: the invasion of a few throats strands oil in most of the pores. Ketton has larger pore sizes, so the number of elements in the network, based on an image of the same physical size, is

Table 4.4 **Displacement statistics for imbibition in Estaillades limestone.** *The number of events for different processes is shown for waterflooding with various ranges of advancing contact angle. The network statistics are shown in Table 2.1.*

Contact angle range (degrees)	0–30	30–60	60–90	90–120
Uninvaded pores	65,700	63,280	64,576	74,414
Pore snap-off	550	606	1,558	1
I_1 pore filling	8,736	7,526	5,337	992
I_2 pore filling	5,956	6,736	5,512	1,407
I_3 pore filling	2,056	3,396	3,522	1,476
I_4+ pore filling	601	1,528	2,567	4,782
Uninvaded throats	74,791	77,523	84,052	108,221
Throat snap-off	18,380	14,268	8,362	1
Piston-like throat filling	27,696	29,075	28,453	12,645
Residual oil saturation	0.59	0.57	0.56	0.30

Table 4.5 **Displacement statistics for imbibition in Ketton limestone.** *The number of events for different processes is shown for waterflooding with various ranges of advancing contact angle. The network statistics are shown in Table 2.1.*

Contact angle range (degrees)	0–30	30–60	60–90	90–120
Uninvaded pores	664	538	466	1,281
Pore snap-off	23	95	137	0
I_1 pore filling	701	635	648	40
I_2 pore filling	425	473	497	81
I_3 pore filling	95	156	150	141
I_4+ pore filling	8	19	18	373
Uninvaded throats	580	644	521	2,711
Throat snap-off	1,062	938	812	0
Piston-like throat filling	1,861	1,921	2,170	792
Residual oil saturation	0.42	0.35	0.20	0.37

smaller. It also has a larger coordination number and a narrower distribution of larger throats (see Fig. 3.11), resulting in less trapping.

This exercise reveals that an analysis of single events does not give a complete picture of what happens in a network. We see two surprising emergent phenomena when we have a distribution of contact angles. Firstly, we observe pore snap-off for weakly water-wet systems, when the adjoining throats have less favourable entry pressures. Secondly, the decrease in residual oil saturation as the average contact angle is increased is caused not primarily by a suppression of snap-off, but more by a shift away from trapping in the larger elements.

Computationally, the network model follows an algorithm similar to that for invasion percolation, in that the threshold capillary pressures are ranked in a sorted

(a)   (b)

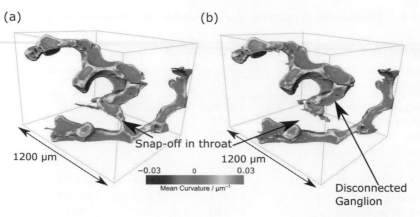

Figure 4.11 An image of snap-off in imbibition. The non-wetting phase is CO_2 and the experimental conditions are similar to those described in Fig. 4.4. Images are shown before and after snap-off, which has disconnected a ganglion of the non-wetting phase. In contrast to drainage, the snap-off event does not lead to a significant change in curvature in either the connected or disconnected phases, indicating a constant local capillary pressure. The presence of wetting layers allows a smooth progression of positions of local capillary equilibrium as displacement proceeds. From Andrew (2014). (A black and white version of this figure will appear in some formats. For the colour version, please refer to the plate section.)

list; the most favourable (highest capillary pressure) event is chosen and the list is updated to reflect the new phase occupancy of the network and to update the pressures for I_n filling events. One new feature though is to disallow the invasion of trapped oil through the identification of clusters of the non-wetting phase. The standard algorithm to identify clusters in percolation with a single sweep through the network was developed by Hoshen and Kopelman (1976), extended to irregular lattices by Al-Futaisi and Patzek (2003a). However, it is not necessary to identify clusters across the whole system after every displacement event; instead, a burning algorithm is performed to test for trapping. The element to be filled is labelled. Then all its adjacent neighbours also containing oil are labelled (these will be bounding throats for a pore or adjacent pores for a throat). Then the neighbours of these neighbours are found and so on. Either this process reaches an external boundary of the network, and the oil can be displaced, or no new neighbours are found. In the latter case we have a trapped cluster and all the identified elements are labelled as such. If the whole network needs to be scanned every time an element is filled (or considered for filling) then n calculations are needed, where n is the number of elements, making the whole calculation take of order n^2 operations, as opposed to $n \ln n$ for invasion percolation without trapping; see page 97. However, the burning algorithm is very quick and the simulation of the displacement of many millions of elements is still possible in just a few seconds of computer time.

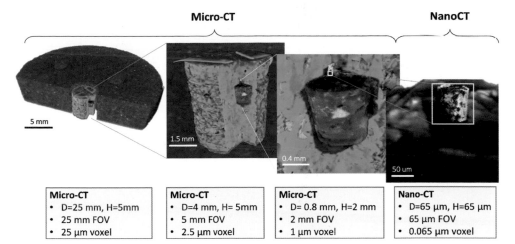

Micro-CT	Micro-CT	Micro-CT	Nano-CT
• D=25 mm, H=5mm	• D=4 mm, H= 5mm	• D= 0.8 mm, H=2 mm	• D=65 µm, H=65 µm
• 25 mm FOV	• 5 mm FOV	• 2 mm FOV	• 65 µm FOV
• 25 µm voxel	• 2.5 µm voxel	• 1 µm voxel	• 0.065 µm voxel

Figure 2.8 An illustration of imaging across multiple scales. Here a tight sandstone with clay cement is imaged at both the core (cm) scale and at sub-micron resolution to resolve the pore space between kaolinite particles. FOV stands for field of view, while D and H refer to the diameter and height of the samples, respectively. Adapted from Roth et al. (2016).

Figure 2.15 A two-dimensional cross-section of a three-dimensional image of Bentheimer sandstone. In the pore space (shown dark) we have superimposed the pores (balls) and throats (cylinders) of the equivalent network. This illustrates how a network model captures the connectivity of the pore space. From Bultreys et al. (2016b).

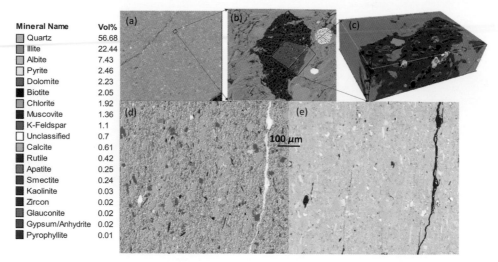

Mineral Name	Vol%
Quartz	56.68
Illite	22.44
Albite	7.43
Pyrite	2.46
Dolomite	2.23
Biotite	2.05
Chlorite	1.92
Muscovite	1.36
K-Feldspar	1.1
Unclassified	0.7
Calcite	0.61
Rutile	0.42
Apatite	0.25
Smectite	0.24
Kaolinite	0.03
Zircon	0.02
Glauconite	0.02
Gypsum/Anhydrite	0.02
Pyrophyllite	0.01

Figure 2.9 Images of a shale gas sample from the Sichuan Basin China showing the state of the art in nm-scale imaging. (a) A back-scattered electron microscopy (BSE) image approximately 1 mm^2 in area with a pixel size of 10 nm: there are 10^{10} pixels in the image. (b) A magnified portion of the image of size 34.2×29.5 μm showing organic material (dark). (c) A three-dimensional FIB-SEM image showing nano-pores within the organic matter. Each voxel is approximately 6 nm cubed; the physical size is $8.8 \times 5.6 \times 2.5$ μm. (d) A QEM-SCAN (Quantitative Evaluation of Minerals by SCANning electron microscopy) image showing details of the mineralogy using the key on the left. The image has an area of 699×636.9 μm with a pixel size of 1 μm. (e) The corresponding BSE image. Courtesy of iRock Technologies, Beijing.

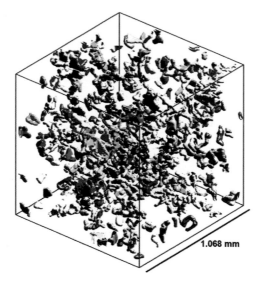

Figure 2.20 Throats identified in the pore space of Berea sandstone using a watershed algorithm. The throats are surfaces dividing different pores; see Fig. 2.19. In a digital image, these throats are adjoining voxels cutting across the pore space with different pores on either side. From Rabbani et al. (2014).

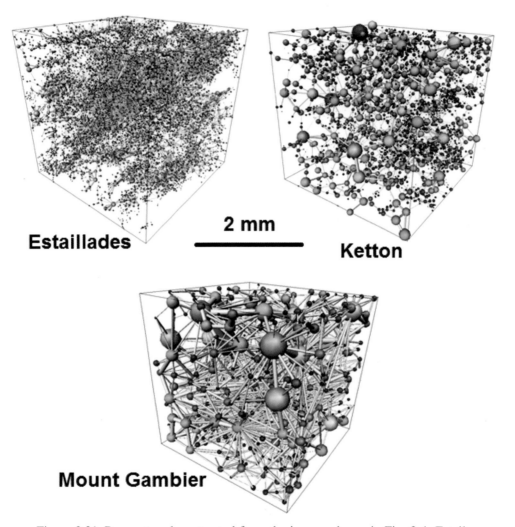

Estaillades

2 mm

Ketton

Mount Gambier

Figure 2.21 Pore networks extracted from the images shown in Fig. 2.4: Estaillades, Ketton and Mount Gambier. The pore space is represented as a lattice of wide pores (shown as spheres) separated by narrower throats (shown as cylinders, even though strictly they are surfaces separating pores). The size of the pore or throat indicates the inscribed radius. Adapted from Blunt et al. (2013).

Figure 2.23 A visualization of a dual-porosity network model for Estaillades limestone. Here, in contrast to the network shown in Fig. 2.21, we also accommodate micro-porosity. On the left is the macro-network derived from a pore-space image similar to that shown in Fig. 2.4: the image size is approximately 3.1 mm. The middle image shows the micro-porous structure with different colours for different clusters: in blue is the largest connected cluster. On the right is an illustration of the two-scale network superimposed on the underlying image, showing pores and throats at both the macro and micro scales. From Bultreys et al. (2015a).

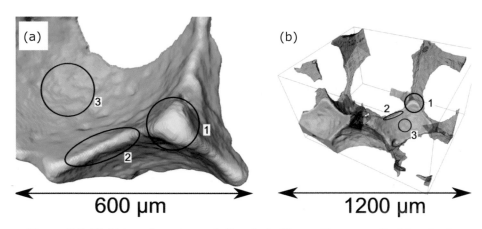

Figure 3.5 Fluid interfaces imaged directly in Ketton limestone. In (a) only the non-wetting phase, CO_2, is shown. There are three types of interface. The first, labelled 1, is a terminal meniscus, TM, where the two principal radii of curvature are approximately equal, straddling a pore or throat. The second is an arc meniscus, AM, where one of the principal radii of curvature is much smaller than the other, indicating a wetting layer in the corner or roughness of the pore space. The third type has a negative curvature and is the direct contact of CO_2 with the solid grains which bulge out into the pore space. (b) shows in pink the location of the wetting phase, allowing a visual confirmation of the types of interface present. The colours on the contour of the CO_2 phase indicates the curvature, which is approximately the same for both the TM and AM, suggesting local capillary equilibrium. From Andrew et al. (2015).

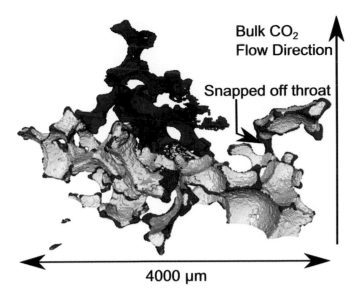

Bulk CO₂ Flow Direction

Snapped off throat

4000 µm

Figure 4.4 An illustration of snap-off in primary drainage. CO_2, the non-wetting phase, displaces brine in Ketton limestone. The experiment was performed at reservoir conditions (10 MPa pressure and 50°C temperature), so that CO_2 is a super-critical phase. The rock and fluids were imaged using a synchrotron X-ray beam: a sequence of images was taken with a time lag of approximately 45 s. Only the CO_2 phase is shown. The pore structure of the rock is illustrated in Fig. 2.6. The yellow (palest shade) represents the distribution of CO_2 before a Haines jump. Blue (darkest shade) indicates the CO_2 that invades the pore space after the jump, while red shows where the CO_2 has retracted, marking local imbibition events. Note that snap-off can isolate a region of the non-wetting phase that is not directly involved in the Haines jump. From Andrew et al. (2015).

(a)   (b)

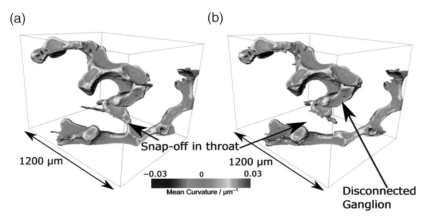

1200 µm

Snap-off in throat

1200 µm

−0.03   0   0.03

Mean Curvature / µm⁻¹

Disconnected Ganglion

Figure 4.11 An image of snap-off in imbibition. The non-wetting phase is CO_2 and the experimental conditions are similar to those described in Fig. 4.4. Images are shown before and after snap-off, which has disconnected a ganglion of the non-wetting phase. In contrast to drainage, the snap-off event does not lead to a significant change in curvature in either the connected or disconnected phases, indicating a constant local capillary pressure. The presence of wetting layers allows a smooth progression of positions of local capillary equilibrium as displacement proceeds. From Andrew (2014).

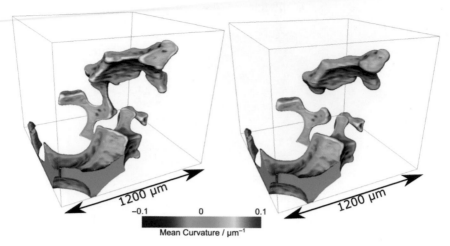

Figure 4.5 Detail of part of Fig. 4.4 illustrating the disconnection of the non-wetting phase, CO_2 after a cascade of pore filling. The curvature of the CO_2 interface is shown. The negative curvature represents CO_2 in contact with the Ketton grains which bulge out into the non-wetting phase. The regions of positive curvature indicate a fluid/fluid interface between CO_2 and brine. Notice that after the snap-off event, the curvatures have decreased, indicating a drop in local capillary pressure. As the local capillary pressure increases again during the displacement, the CO_2 will re-connect. From Andrew et al. (2015).

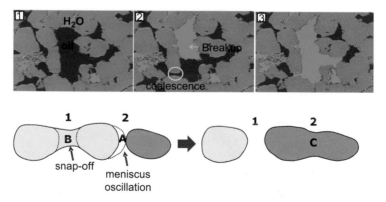

Figure 4.12 The upper figure shows sequence of images of oil, the non-wetting phase (red) and the wetting phase, water (blue), during an imbibition experiment in a sandstone. The voxel size is 2.2 μm and the images were acquired every 40 s. Water invades the pore space (pictures 1 and 2), but this leads to a local drainage event where the non-wetting phase re-connects (2), before disconnecting again as the displacement proceeds (3). The lower figure shows a schematic of this process, with snap-off at location 2 (event A), then at 1 (B), followed by reconnection of a throat at location 2 (C). Adapted from Rücker et al. (2015).

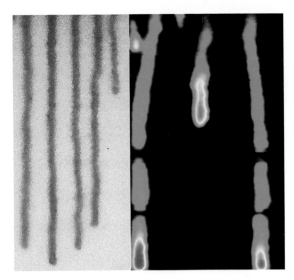

Figure 4.22 Infiltration fingers showing the progression of an unstable wetting front through an initially dry porous medium. On the left is a photograph of water, dyed blue, progressing though a bead pack. The beds are have a diameter between 1 and 1.3 mm and the average finger width is approximately 0.8 cm. Picture courtesy of Abigail Trice. The right figure shows imbibition visualized in a slab chamber using light transmission. The coloured regions are the flow paths: the hotter colours correspond to higher water saturation. Notice that the saturation at the tip of the finger is higher than further up. From DiCarlo (2013).

Figure 6.8 Images of the pore space of a bead pack (a), Bentheimer sandstone (b) and Portland carbonate (c). The second row shows normalized pressure fields with a unit pressure difference across the model for the bead pack (d), Bentheimer sandstone (e) and Portland carbonate (f). The bottom row shows the normalized flow fields, where the ratios of the magnitude of the velocity at the voxel centres divided by the average flow speed are shown using a logarithmic scale from 5 to 500 for the bead pack (g), Bentheimer sandstone (h) and Portland carbonate (i). Green and red indicate high values, while blue indicates low values. From Bijeljic et al. (2013b).

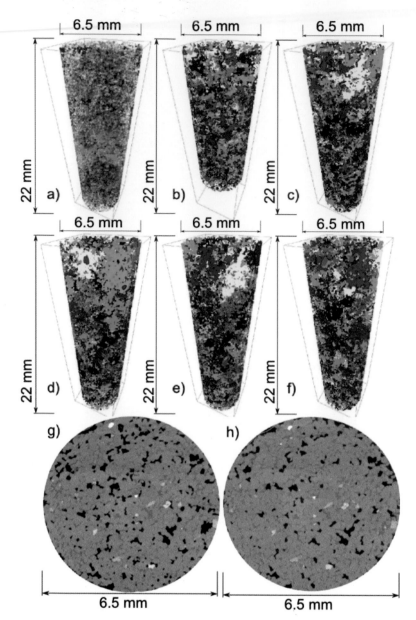

Figure 4.28 Micro-CT X-ray images of CO_2 as a non-wetting super-critical phase in Bentheimer sandstone with a voxel size of around 6 μm. The top left figure shows the distribution of CO_2 after drainage: the blue cluster is connected, while snap-off at the end of the process allows some disconnection shown by the clusters of other colours. The five other pictures show the distribution of trapped CO_2 after brine flooding: the colours indicate individual clusters. These are replicate experiments on the same rock: while the exact location of trapping is different in each experiment, the statistical properties, such as the overall amount of trapping and the cluster size distribution, are similar. Approximately one-third of the pore space contains residual non-wetting phase. The bottom row of figures shows raw images before processing after drainage (left) and water injection (right). Each image contains just over 3 billion voxels. From Andrew et al. (2014c).

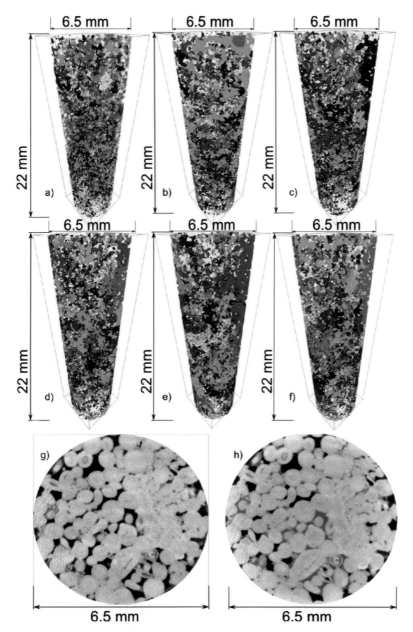

Figure 4.29 A similar figure to Fig. 4.28, except here trapping in Ketton limestone is shown. The voxel size is approximately 4 μm. Again, we see a significant amount of trapping with clusters of all size. From Andrew et al. (2014c).

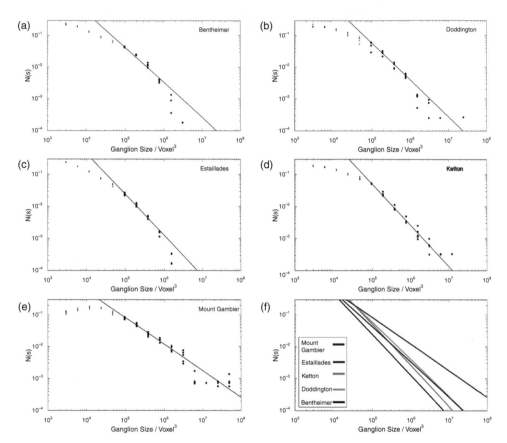

Figure 4.30 Trapped cluster size distributions measured on five rock samples: two cases are illustrated in Figs. 4.28 and 4.29. $N(s)$ is the relative frequency of clusters of size s voxels (not pores); however, the scaling relationship Eq. (4.34) should still hold with the same exponent. In all cases we see a truncated power-law distribution with an exponent τ between 2.1 and 2.3, broadly consistent with percolation theory (the exponent is predicted to be 2.19). The one outlier is Mount Gambier with an exponent of 1.8; this rock has a very porous and open pore structure (see Figs. 2.3 and 3.12). From Andrew et al. (2014c).

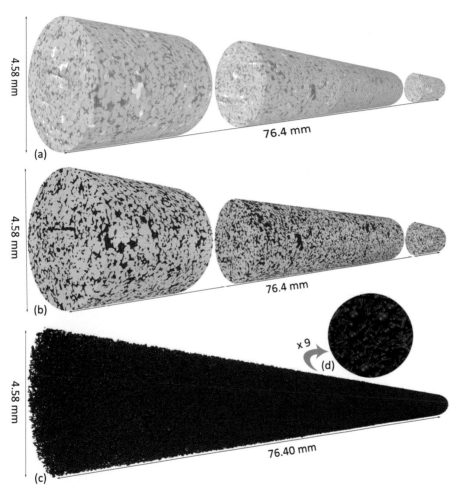

Figure 6.22 Pore network representation of a Berea sandstone core sample. (a) Greyscale images with a voxel size of 2.49 μm. (b) A segmented image where red and grey represent the pore and grain voxels, respectively. (c) The pore network generated from image (b). Red and blue represent the pores and throats in the network, respectively. For illustrative purposes only pores and throats are shown with circular cross-section. (d) A magnified image of a small section of the network. From Aghaei and Piri (2015); the network was generated by iRock Technologies, Beijing.

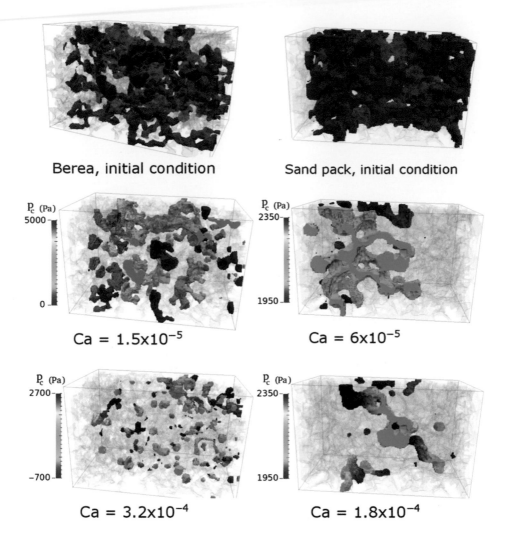

Berea, initial condition

Sand pack, initial condition

P_c (Pa)
5000 —

0 —

Ca = 1.5x10^{-5}

P_c (Pa)
2350 —

1950 —

Ca = 6x10^{-5}

P_c (Pa)
2700 —

−700 —

Ca = 3.2x10^{-4}

P_c (Pa)
2350 —

1950 —

Ca = 1.8x10^{-4}

Figure 6.23 Visualizations of the non-wetting phase in images of Berea sandstone (left) and a sand pack (right) at the start (top) and at the end of water injection (middle and lower figures). Simulations are performed at the different capillary numbers shown. The colours indicate the local capillary pressure. Through capturing flow and the geometry of the pore space accurately, the residual saturation, particularly for the sand pack, can be predicted more accurately than with network modelling. From Raeini et al. (2015).

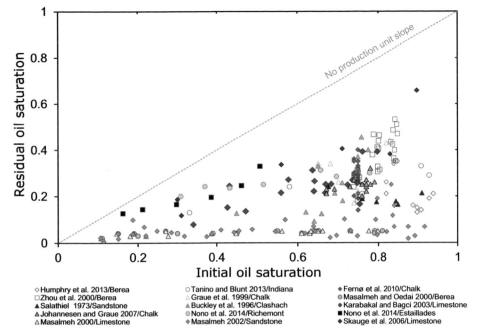

◇ Humphry et al. 2013/Berea   ○ Tanino and Blunt 2013/Indiana   ◆ Fernø et al. 2010/Chalk
□ Zhou et al. 2000/Berea   △ Graue et al. 1999/Chalk   ◉ Masalmeh and Oedai 2000/Berea
▲ Salathiel 1973/Sandstone   ▲ Buckley et al. 1996/Clashach   ◆ Karabakal and Bagci 2003/Limestone
△ Johannesen and Graue 2007/Chalk   ◉ Nono et al. 2014/Richemont   ■ Nono et al. 2014/Estaillades
△ Masalmeh 2000/Limestone   ◆ Masalmeh 2002/Sandstone   ◆ Skauge et al. 2006/Limestone

Figure 5.14 A compilation of trapping data for rocks of altered wettability, showing the residual oil saturation as a function of initial saturation. Details of the different experimental studies are provided in Table 5.7. There is considerable scatter because of the different rock samples studied and variations in wettability. However, in general we see less trapping than in water-wet systems; compare this graph with Fig. 4.35. From Alyafei and Blunt (2016).

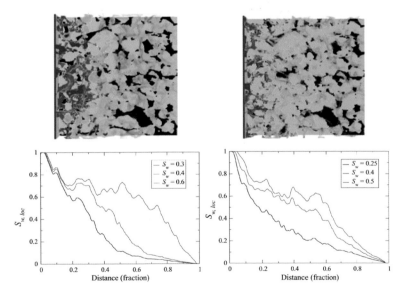

Figure 6.24 Fluid configurations for imbibition at an average saturation $S_w = 0.4$ (top) with associated local saturation profiles (bottom) for two flow rates in a Bentheimer image with 256^3 voxels using lattice Boltzmann simulation. The wetting fluid saturation is displayed with red the largest value towards blue as zero. The pore space is black. The capillary number, $Ca \approx 5 \times 10^{-5}$ in the left picture and 5×10^{-6} in the right. From Ramstad et al. (2012).

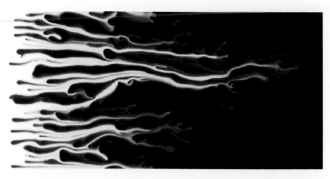

Figure 6.26 The concentration field during the unstable displacement of a more viscous fluid (dark) by a fully miscible, less viscous fluid (light). $M \approx 33$. Unlike DLA, the pattern does fill in from the inlet, but we still see a fractal interface between the fluids with a cascade of unstable fingering. From Jha et al. (2011).

Figure 8.3 A photograph of water and oil on the surface of leaves. The leaves are waxy, or oil-wet. On the left water forms a non-wetting drop. Some olive oil has been placed on the leaf to the right; this has spread over the surface and soaked into the dry leaf underneath (the darker shading). In the presence of air, water is the non-wetting phase. This allows gas exchange through stomata (small holes in the leaf). Imbibition of water is prevented; were it to occur, water would saturate the pores inside the leaf, restricting the ingress of carbon dioxide for photosynthesis and the escape of water vapour for transpiration. The same phenomenon is seen with ducks: the feathers are oily and repel water, keeping the space within the feathers full of air and insulating. In a mixed-wet or oil-wet rock, this implies that gas is not necessarily the non-wetting phase.

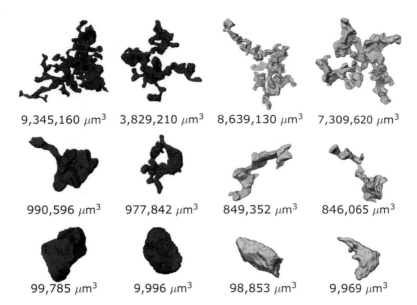

9,345,160 μm^3 3,829,210 μm^3 8,639,130 μm^3 7,309,620 μm^3

990,596 μm^3 977,842 μm^3 849,352 μm^3 846,065 μm^3

99,785 μm^3 9,996 μm^3 98,853 μm^3 9,969 μm^3

Figure 8.28 Ganglia of oil (two left-hand columns) and gas (right-hand columns) after carbonated water injection in a water-wet Berea sandstone. The volume of the ganglia are shown: the pictures are not all to the same scale. From Alizadeh et al. (2014).

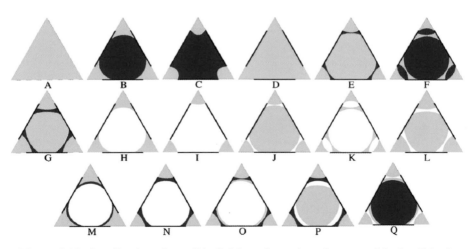

Figure 8.12 A collection of possible fluid configurations for water (blue), oil (red) and gas (yellow) in a triangular pore or throat. The bold line indicates a surface of altered wettability, Fig. 2.29, but this is not necessarily oil-wet. From Helland and Skjæveland (2006b).

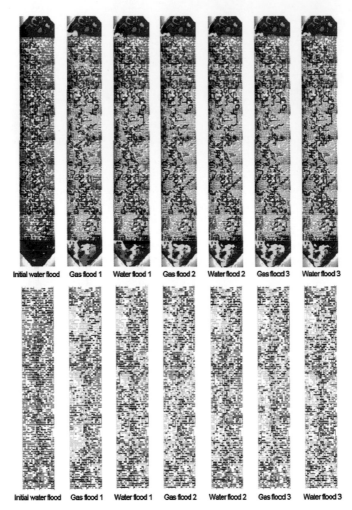

Figure 8.15 The distribution of three phases (water in blue, oil in red and gas in yellow) in an oil-wet micro-model for waterflooding followed by repeated cycles of gas and water injection. Injection occurs from the bottom of the system with recovery from the top. Experimental results from Sohrabi et al. (2004) (top) are compared with the network model of Al-Dhahli et al. (2013) (bottom). Similar displacement patterns are observed: these are the result of multiple displacement processes at the pore scale with a continual decline in remaining oil saturation, Fig. 8.14. Water is the most non-wetting phase, then gas and then oil.

4.2.5 Dynamics of Filling

Like a Haines jump, but for imbibition, the filling of a single pore could precipitate a cascade of filling of neighbouring smaller pores and throats with higher threshold capillary pressures until a new position of capillary equilibrium is reached. Pore filling also leads to the displacement of oil out of all the adjacent throats, unless the oil is trapped. Similarly, filling a throat by snap-off may allow a more favoured pore-filling event to occur spontaneously. The filling of a throat changes the threshold capillary pressure in its two adjacent pores. If the pore has n oil-filled throats initially, then the threshold for displacement after snap-off changes from that for an I_n process to I_{n-1}, which is more favoured, and could therefore lead to further filling at the prevailing macroscopic capillary pressure.

Any local increase in capillary pressure could then lead to the retraction of water from regions previously filled, allowing drainage and new positions of equilibrium with a higher local capillary pressure (a lower water pressure) to be reached. The portions of the pore space in which retraction occurred would be re-filled later, once more water is injected.

This cascade of filling is, in general, less dramatic than encountered in drainage, since the wetting phase is already present in the pore space, fills some of the narrower throats already, and fills others by snap-off in approximate order of size during the displacement. As a result, there is less free, unfilled, pore space for the wetting phase to explore, and any changes in local capillary pressure are modest. This shows the importance of wetting layers, which provide the ability for displacement to occur throughout the pore space, in contrast to drainage, where filling is limited to regions adjacent to the connected, invading, non-wetting phase cluster.

The dynamics of filling is illustrated at the pore scale in Fig. 4.11, where the non-wetting fluid configuration in Ketton limestone is shown before and after a snap-off event. Snap-off leads to the disconnection of a ganglion of the non-wetting phase, but there is no noticeable change in local capillary pressure, either in the ganglion or in the connected phase, and no evidence of retraction of the wetting phase. In measurements that recorded the sonic signature of individual displacements, fewer large bursts were observed compared to drainage, which confirms that Haines-like jumps are impeded in imbibition (DiCarlo et al., 2003).

This discussion though appears to conflict with the schematic filling sequence illustrated in Fig. 4.9, where a single I_3 event leads to a burst of I_2 filling in a regular lattice. This is indeed possible if there is little or no flow in wetting layers, and will still occur in irregular media, as shown in the next section.

Moreover, changes in local capillary pressure have been observed in Bentheimer sandstone (Rücker et al., 2015). Fluctuations in local meniscus curvatures can lead

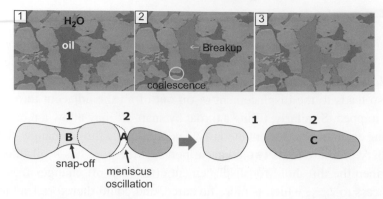

Figure 4.12 The upper figure shows sequence of images of oil, the non-wetting phase (red) and the wetting phase, water (blue), during an imbibition experiment in a sandstone. The voxel size is 2.2 μm and the images were acquired every 40 s. Water invades the pore space (pictures 1 and 2), but this leads to a local drainage event where the non-wetting phase re-connects (2), before disconnecting again as the displacement proceeds (3). The lower figure shows a schematic of this process, with snap-off at location 2 (event A), then at 1 (B), followed by reconnection of a throat at location 2 (C). Adapted from Rücker et al. (2015). (A black and white version of this figure will appear in some formats. For the colour version, please refer to the plate section.)

to successions of snap-off (imbibition) and reconnection (drainage) events. This can be quantified through a study of the ratio of the threshold pressures for snap-off, Eq. (4.8), to piston-like advance in the same throat in drainage, Eq. (3.13)

$$ P_{cR}^{ID} = \frac{\cos\theta_A}{\cos\theta_R} \frac{1 - \tan\theta_A \tan\beta}{C_D}. \tag{4.30} $$

$P_{cR}^{ID} \leq 1$ since $C_D \geq 1$ and $\theta_A \geq \theta_R$; however, this ratio becomes closer to one for slit-like throats and contact angles close to zero with little hysteresis.

Rather than a sequence of capillary controlled steady-states with some constant capillary pressure, we can see a more dynamic displacement with the exchange of a non-wetting phase between temporarily disconnected ganglia (Rücker et al., 2015). This process occurs where P_{cR}^{ID} in Eq. (4.30) is close to 1.

An example sequence of displacement is illustrated in Fig. 4.12 (Rücker et al., 2015). The non-wetting phase can disconnect and reconnect through sequential snap-off and piston-like drainage events. This is explained from a consideration of the local capillary pressure changes shown schematically in Fig. 4.13, which represent the events shown in the lower, conceptual, diagram in Fig. 4.12. Initially, oil occupies the centres of the pore space, including the throats labelled 1 and 2 in Fig. 4.12. As the capillary pressure drops, the most favoured event is snap-off at location 2 (event A in Fig. 4.13). Let us assume, for the sake of argument, that this is a throat with a slit-like cross-section. The local capillary pressure will increase,

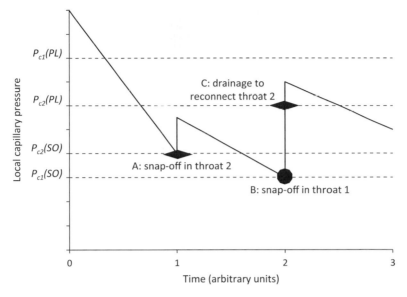

Figure 4.13 Putative local capillary pressure changes for the displacement sequence in Fig. 4.12. The local capillary pressure declines as more water enters the system. The first event is snap-off in throat 2, event A. This leads to an instantaneous increase in pressure as two terminal menisci are formed. We assume that the jump in capillary pressure is modest. The next event is B, snap-off in throat 1. In this case the pressure increases sufficiently to allow drainage in throat 2, event C, reconnecting the non-wetting phase. We see disconnection and reconnection of the non-wetting phase, even in a displacement at infinitesimal flow rate and controlled macroscopically by capillary forces.

as initially the TM have curvature in two directions. To a good approximation, since the TM first form in the throat, the local capillary pressure instantly rises to the threshold value for piston-like advance after snap-off, an increase by a factor of $1/P_{cR}^t$ which is close to 1 in this case, where P_{cR}^t is given by Eq. (4.25). We use this ratio since we are comparing two imbibition displacements.

The local capillary pressure then falls again as more water is injected into the system and the TM move into wider regions of the pore space. Note that this rise in capillary pressure does not necessarily represent the $P_{c2}(PL)$ value marked in Fig. 4.13, since this is the threshold for piston-like advance in *drainage* with $\theta_R \leq \theta_A$, which may occur at a higher pressure than piston-like advance in imbibition: its value is a factor $1/PcR^{ID}$ higher than for the snap-off event, where P_{cR}^{ID} is given by Eq. (4.25) and $P_{cR}^{ID} \leq P_{cR}^t$.

The next event is snap-off at location 1 in Fig. 4.12, event B in Fig. 4.13. Again, an instantaneous increase in local capillary pressure is seen as the TM are formed. In this case we hypothesize that we have a more circular throat. This means that $1/P_{cR}^t$ from Eq. (4.25) is larger (greater than 2) for throat 1, resulting in a more dramatic increase in pressure when the TM are formed. We can easily have a case where the capillary pressure for snap-off in throat 1 is lower (less favoured) than for 2, $P_{c1}(SO) < P_{c2}(SO)$, while the threshold pressures for piston-like advance is higher for throat 1 than for throat 2, $P_{c1}(PL) > P_{c2}(PL)$. This means that the local capillary pressure increases sufficiently to allow piston-like filling by *drainage* in throat 2, event C. This reconnects the throat and leads to the transport of non-wetting phase rightwards in Fig. 4.12 through a sequence of snap-off and reconnection. After this event, again the capillary pressure decreases as more water enters the system, and snap-off re-occurs in throat 2, finally disconnecting the non-wetting phase.

Note that this process will occur even for displacement at an infinitesimal flow rate; it is an inevitable consequence of instabilities in the local capillary pressure after a snap-off event. This phenomenon leads to changes in the connectivity of the non–wetting phase during the displacement; changes that both disconnect and reconnect ganglia. If we consider displacement involving more pores and throats, it is possible to extend this argument to conceive of situations where there are repeated cycles of snap-off and reconnection in a particular throat as the local capillary pressure fluctuates.

4.2.6 Displacement as a Series of Metastable States

The discussion above leads to an elegant conceptual picture of displacement processes, first advanced by Morrow (1970) and refined and placed on a more rigorous footing by Cueto-Felgueroso and Juanes (2016). In Fig. 4.13, we see two distinct types of change in capillary pressure. The first is a smooth variation in local capillary pressure over time, indicative of a correspondingly gradual alteration in saturation. These are reversible processes, in that we could change the flow direction and retract the menisci without altering the topology of the fluid pattern. If there is no contact angle hysteresis, this would occur without any jump in capillary pressure. Morrow (1970) termed such processes isons: where these reach local maximum in the injected phase pressure, they represent points that, for a low injection rate, the local and macroscopic capillary pressures are equal. Isons can occur in both imbibition and drainage and represent, for instance, the gradual ingress of a TM into a throat, or the swelling of AM in wetting layers. During an ison we have a state which is in local capillary equilibrium.

The second process is called a rheon and corresponds to a sudden change in local capillary pressure at fixed saturation. This is where the fluids rearrange themselves in the pore space as the result of a disequilibrium in local capillary pressure: the topology of the fluid arrangement may change and the process is irreversible. Fluid rearrangement after snap-off and Haines jumps are examples of rheons.

Displacement can then be considered as a transition between metastable states of capillary equilibrium, with a cascade of displacement processes when there is a jump, or rheon, from one state to another (Cueto-Felgueroso and Juanes, 2016). It is the fundamental reason for capillary pressure hysteresis or why the macroscopic capillary pressure as a function of saturation in imbibition and drainage are different. Locally, we see a succession of irreversible transitions, while the macroscopic pressure only records isons at a local maximum (for drainage) or minimum (for imbibition) in the local capillary pressure.

Before we illustrate capillary pressure hysteresis with large-scale measurements of capillary pressure, we first need to present the types of displacement pattern, since – in some cases – a smoothly varying capillary pressure as a function of saturation cannot be defined.

4.3 Displacement Patterns in Imbibition

Dependent on the competition between snap-off and pore filling we can observe different patterns during imbibition. There are four generic types of pattern that we will describe, based on the work of Chandler et al. (1982), Lenormand and Zarcone (1984), and Cieplak and Robbins (1988, 1990): percolation with trapping, invasion percolation, frontal advance and cluster growth. For reference two-dimensional simulations on a square lattice of pores and throats, illustrating these cases is shown in Fig. 4.14 (Blunt and Scher, 1995).

4.3.1 Percolation with Trapping

Consider a highly heterogeneous medium, with pores and throats of vastly different size. As we showed in Chapter 3, Fig. 3.12, the throat radius can vary by four orders of magnitude in quarry carbonates, while the variation is typically even larger in reservoir rocks. In these cases, to a first approximation, imbibition leads to the filling of the pore space simply in order of size, with the smallest regions filled first. The subtleties of corner geometry, or the exact arrangement of fluid for pore filling, is secondary to the dominant contribution of inscribed radius on the threshold pressures. In these cases, while a throat is, by definition, narrower than its adjacent pores, the pore and throat size distributions overlap, so that a small pore is smaller than many throats.

If we allow connected wetting layer flow throughout the sample, then wetting phase filling is initiated by snap-off. This will then lead to pore filling in the smallest pores, in preference to further snap-off in larger throats. Roughly speaking, the pores simply fill in order of size, with associated throat filling and snap-off to trigger the process. The non-wetting phase can be trapped, making

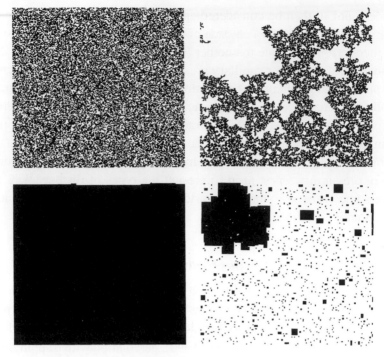

Figure 4.14 Different types of wetting invasion simulated on a 200×200 square network. Pores filled with wetting fluid are shown at breakthrough. Upper left: site percolation. Pores are filled throughout the network in order of size. Flow in crevices provides fluid for filling a pore or throat anywhere in the system. Upper right: site invasion percolation. This is similar to ordinary percolation, except that there is no crevice flow and hence we can only fill pores next to the connected front. Lower left: flat frontal advance. Cooperative pore filling mechanisms allow the fluid to advance as a uniform front. Lower right: cluster growth. This is similar to frontal advance, except that flow in crevices allows pores and throats to be filled anywhere. These act as nucleation sites for cluster growth. In this figure we are before breakthrough. The largest cluster will continue to grow and eventually will fill the system. From Blunt and Scher (1995).

displacement possible only if it has a connected pathway to the outlet. This is site percolation with trapping. It is site percolation since now it is the pores (the sites in percolation parlance) – not the throats (bonds) – that impede the wetting phase advance and which are, approximately, filled in order of size. It is possible to consider a variant on this process, where all the pores are larger than the throats and the only displacement process is snap-off. Here we have bond percolation with the throats filled in order of size, with the non-wetting phase trapped in all the pores. However, in most natural systems, there is some degree of pore filling; if not, then the trapped saturation is extremely high.

The important feature of this type of percolation displacement pattern is that filling can occur throughout the system, mediated by layer flow in a sequence controlled by pore size. It is the process most commonly encountered in reservoir settings, such as waterflooding and CO_2 migration through storage aquifers, as well as water invasion through wet soil. This happens when there is a wide range of pore size with a wetting phase saturation initially present in the pore space.

In terms of waterflood oil recovery, trapping is unfavourable, but the nature of the percolation-like advance means that the water remains poorly connected until the percolation threshold is reached, which holds back its advance and favours recovery of the oil, as discussed in Chapter 3.4.3. As we will show later, the efficiency of oil recovery is controlled by the trade-off between trapping and poor water connectivity.

4.3.2 Invasion Percolation with Trapping

If instead we do not allow layer flow, since we have a primary imbibition process into a completely dry rock or soil, or a medium comprised of smooth grains (such as a packing of spheres) where wetting layers cannot connect across the system, then we do not see snap-off. In this case, advance is only possible from a pore or throat already filled with wetting phase and connected through the centre of the void space to the inlet. We have an invasion percolation process, since we still fill the pore space in order of size, but from a connected wetting front.

While this is invasion percolation, this is distinct from the process described for drainage in the previous chapter for two reasons. Firstly, filling of narrow regions is favoured with advance limited by the wider pores, in contrast to drainage where advance is easy through the pores and limited by the throats. As such we have a site invasion percolation process, whereas in drainage we see bond (throat) invasion percolation.

Secondly, the non-wetting phase is trapped, whereas for invasion percolation in drainage, the wetting phase can escape through layers. Hence this process is called site invasion percolation with trapping (Dias and Wilkinson, 1986). As discussed in Chapter 3.4, invasion percolation develops a fractal pattern when the invading phase advances, with scaling laws for various geometric and flow properties. A listing of some of the fractal dimensions for this process was provided in Table 3.1. Trapping in imbibition leads to a significant suppression of filling in two dimensions, leading to lower fractal dimensions for the incipient cluster (the fractal dimension of the cluster at the percolation threshold) and other related properties, such as the backbone and minimum path dimensions. However, in three dimensions, the obvious focus of interest here, the structure is more open, and simulation results cannot detect any difference in the scaling properties of invasion percolation with or without trapping (Sheppard et al., 1999).

4.3.3 Frontal Advance

We now treat less heterogeneous cases, where cooperative pore filling plays a significant role. This means that fluctuations in pore size throughout the sample do not dominate over the differences in threshold pressure for different I_n processes. In Fig. 4.14 we consider a square network where every I_2 displacement is favoured over snap-off and I_3; the behaviour in disordered lattices will be described later.

If we do not allow layer flow, preventing snap-off, then a connected frontal advance is observed. An example is when sea water advances through dry sand, or the soaking of water into a paper towel, illustrated in Fig. 4.15. As in our second pattern, site invasion percolation, the wetting phase has to advance through the porous medium from the inlet by piston-like filling alone. The difference here is that the local arrangement of wetting phase controls the advance, rather than the variation of pore size, favouring in-filling of the front (I_1 processes) and suppressing ramified advance (I_{z-1}). In a regular lattice, such as a square grid realized in micro-model experiments or simulations, a smooth faceted invasion is possible, with one I_3 process inducing a cascade of I_2 filling as the network fills line by line, Fig. 4.9. In this case there is no trapping; even in an irregular pore space, though, the amount of trapping is very small.

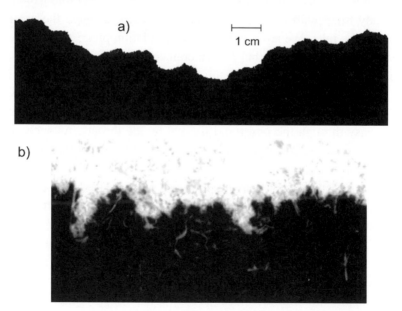

Figure 4.15 An example of frontal advance in a disordered porous medium. A wetting front of black ink soaks into a paper towel. (a) A digital photograph with approximately 1,200 pixels horizontal resolution, where the dark and light grey values were enhanced to black and white. (b) A high-resolution scan (1,000 dots per cm) of a small part (around 0.8 cm wide) in greyscale. From Alava et al. (2004).

4.3.4 Cluster Growth

The final pattern, cluster growth, is similar to frontal advance, except that the filling is nucleated within the porous medium itself, leading to the growth of patches of wetting phase, rather than advance from the inlet. For this to occur, we must allow wetting layer flow and have some throats that can fill by snap-off, or at least some throats initially filled with wetting phase at the end of primary drainage, to initiate the process. Here, while there is corner flow, throat filling by snap-off is strongly disfavoured in comparison to pore filling. This could occur, for instance, in a pore space where the pores are of similar size to the throats. Here there is little or no trapping in the wetting patches, although some non-wetting phase can be bypassed when these regions merge.

4.3.5 Phase Diagrams for Capillary-Controlled Displacement

These four generic imbibition patterns are fundamentally different, with either substantial or very little trapping, and disparate connectivity of the phases. Hence, it is important to know what regime a given displacement falls into. So far we have distinguished between them in a somewhat qualitative sense, dependent on pore space geometry and the availability of layer flow. We will now develop a more statistical characterization and discuss a putative phase diagram to assign flow to different regimes.

Robbins and co-workers have defined the second and third patterns – those involving connected advance with no layer flow – using statistical scaling arguments based on two-dimensional simulations and micro-model experiments (Cieplak and Robbins, 1988, 1990; Martys et al., 1991a, b; Koiller et al., 1992). Fig. 4.16 shows simulations of fluid advance through an array of discs of different size arranged on a regular square lattice. The upper figure illustrates a drainage invasion percolation pattern (the contact angle measured through the advancing fluid is almost 180°): here displacement occurs through the widest pore spaces and proceeds in a ramified or fingered pattern with a self-similar structure, as discussed in Chapter 3.4. What happens as we decrease the contact angle and progress towards more wetting conditions in the same porous medium? While we still have a drainage displacement, with contact angles above 90°, the behaviour is similar, since filling of the wider regions continues to be favoured. However, we do not observe some discontinuous transition in flow pattern when the contact angle first drops below 90° and we have, technically, an imbibition process. The complex nature of the pore space means that there is always a range of threshold capillary pressures for different local advances of the TM: in general we do not see a flip from positive to negative capillary pressures at $\theta = 90°$, but a smooth change. Instead, for lower contact angles we tend now to favour invasion into smaller regions of the pore space, and cooperative filling – that is, invasion controlled by the local arrangement of fluid – becomes more significant.

The consequence of local cooperative filling is that the fingers of fluid fatten, as the advancing phase is more likely to advance where, locally, small dips in the front are filled in, while disfavouring the formation of narrow tongues penetrating through the pore space. This is shown in Fig. 4.16 where for a lower contact angle, 58° in this case, the advance becomes more blocky with very wide fingers and very little trapping. At scales larger than the finger width, W, but smaller than the system size we still have, statistically, an invasion percolation pattern with a self-similar structure. This finger width is larger, and often significantly larger, than a typical pore size. A small change in contact angle can result in a significant thickening of the fingers and a huge reduction in the degree of trapping.

There is a critical contact angle where the finger width diverges and we change from an invasion percolation pattern to connected frontal advance. This, as discussed above, is a generically different pattern with distinct scaling properties and a marked reduction in the amount of non-wetting phase that is trapped. The critical contact angle depends on the geometry of the porous medium: while this has not been directly studied, it is likely that we see a similar transition from invasion percolation to frontal advance as the disorder or heterogeneity in the pore space decreases for a fixed contact angle.

In frontal advance, the spanning cluster at breakthrough does not have a fractal structure: it is compact filling a finite fraction (indeed, the vast majority) of the pore space. However, in an irregular porous medium, the wetting phase front is not smooth, see Fig. 4.14, but has a more ragged structure, as shown in Fig. 4.15.

The invasion pattern is self-affine, in that the connected front has scaling properties, although the structure overall is not self-similar or fractal. We can define $h(\mathbf{x})$ as the height (or minimum distance) of the front from the inlet at a location $\mathbf{x}$, which defines a surface for a three-dimensional displacement, or a line in two dimensions, as shown in Fig. 4.17. Then denoting $< f >$ as the mean value of a function f,

$$\langle |h(\mathbf{x} + \mathbf{r}) - h(\mathbf{x})| \rangle \sim |\mathbf{r}|^H, \tag{4.31}$$

where $\mathbf{r}$ is some displacement distance and H is the Hurst or roughness exponent. For self-affine surfaces we have $1 \geq H \geq 0$; in this specific case the displacement has $H = 0.81$ (Martys et al., 1991b). This means that while the advance is compact with very little trapping, the front is rough and has structure at all scales.

(a)

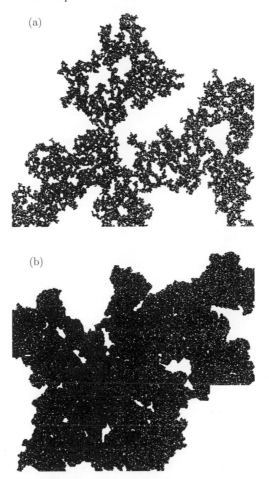

(b)

Figure 4.16 Simulations of meniscus advance in a random array of two-dimensional discs. (a) An invasion percolation pattern with a contact angle of 179°. (b) Wetting phase advance with a much larger finger width where the contact angle is 58°. From Cieplak and Robbins (1988).

Such self-affine structures have been seen experimentally, as shown in Fig. 4.15. As another example, Fig. 4.18 shows an imbibition front advancing in a fracture made by two roughened glass plates. In this case we have a disordered two-dimensional porous medium. The advancing front is ragged with the same roughness exponent, $H = 0.81$, as found in simulations of the process (Geromichalos et al., 2002).

As indicated in Fig. 4.19, just as in drainage, displacement proceeds in a succession of bursts, with one unfavourable filling event followed by a subsequent series of filling at a higher capillary pressure (lower water pressure). We also see a power-law distribution of burst size, as in Eq. (3.36) but with a lower exponent than for invasion percolation: in this case $\tau' = 1.125 \pm 0.025$ (Martys et al., 1991b).

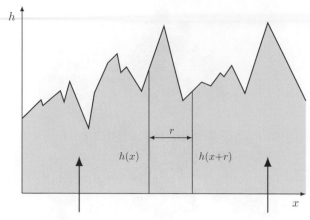

Figure 4.17 A schematic of a self-affine surface. The wetting phase, shaded, is advancing as indicated by the arrows. Unlike invasion percolation, cooperative pore filling leads to a connected frontal advance with little or no trapping. However, the interface between the wetting and non-wetting phases is not smooth in a disordered medium. $h(x)$ is the distance of the interface from the inlet as a function of distance, x, as shown: in three dimensions $h(\mathbf{x})$ describes a rough surface. If we look at the difference in h for a displacement r, then the average value of $|h(x + r) - h(x)|$ scales as a power of r, Eq. (4.31), with a roughness exponent H.

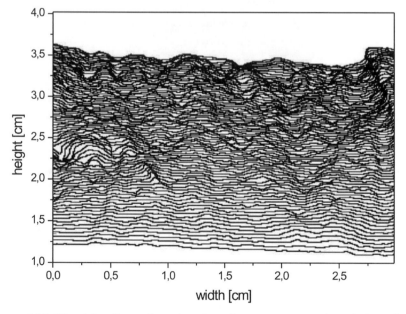

Figure 4.18 The rising front of wetting phase between two roughened glass plates with an average separation of 10 μm at a series of equidistant times. The time elapsed between two snapshots is 10 s. From Geromichalos et al. (2002).

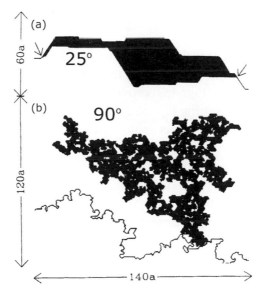

Figure 4.19 The area invaded when a single terminal meniscus becomes unstable is shown in black. The contact angles are shown on the diagram. In both frontal advance (a) and invasion percolation (b), a power-law distribution of burst sizes is seen, Eq. (3.36), but with different exponents. *a* is the distance between pores. From Martys et al. (1991a).

Connected advance in imbibition is just one of a number of growth patterns that generate an advancing rough surface. Further details providing a statistical description of the dynamics and structure of such processes can be found in Vicsek (1992) and Meakin (1993). Alava et al. (2004) provides a review of the dynamics of imbibition, with an emphasis on scaling and mathematical models of interface motion.

In most natural settings, where we have water-wet conditions and a wide variation in pore and throat sizes, we see displacement controlled largely by snap-off and I_1 pore filling. This is, therefore, percolation with trapping. However, locally, cooperative pore filling may control the exact sequence of filling. Using the analogy with connected advance, we might expect to see a percolation process but with a finite finger width, meaning that the displacement pattern is filled in from the pore scale to the finger width, with a ramified structure apparent only at larger scales; the pattern resembles a percolation cluster, with self-similarity at the percolation threshold, for length scales beyond the finger width. This implies that while there is little trapping at the scale of a single pore, there will be significant trapping in larger clusters. In reality, the displacement pattern is more complex than this, with structure at all scales down to the level of single pores with the details governed by the exact nature of the local geometry of the void space; however, when we describe trapping in more detail later this conceptualization of the displacement will help explain the behaviour.

We will now synthesize these ideas by constructing phase or regime diagrams to indicate which type of pattern is expected for different contact angles and degrees of pore-scale heterogeneity. Fig. 4.20 provides a possible phase diagram for cases where there is no flow in wetting layers. There is a transition from frontal advance to invasion percolation as the contact angle increases. Furthermore, the critical contact angle at which this occurs is a function of the structure of the porous medium, with a larger angle required for more regular systems. As the critical angle is approached in the invasion percolation regime, the finger width, W increases and diverges at the transition between patterns. In theory, if we have a completely regular structure, with no variation in threshold capillary pressure for any displacement, then a single advance leads to complete filling: this is frontal advance, regardless of contact angle. For more heterogeneous rocks, with a wide range of pore size, the advance becomes invasion percolation regardless of contact angle, since invasion is controlled purely by pore size rather than cooperative filling. While the regime diagram is instructive, we do not have a precise quantification of what is meant by heterogeneity: strictly, in this context we are comparing the variation in threshold capillary pressure for different pores and throats with the difference in I_n pressures for an individual pore. When the former dominates, we have invasion percolation; when the latter is more important, frontal advance is seen. Furthermore, in any real system there is a range of local contact angle, yet the invasion has to fall into one distinct pattern or another when observed at a sufficiently large length scale: the contact angle here is some appropriate average that captures the essential features of the displacement.

We can extend this treatment to consider cases where layer flow and snap-off are possible: Fig. 4.21. As before, we consider the full range of advancing contact angle. As a consequence, layer flow and snap-off are only allowed for contact angles below some threshold value. For a completely homogeneous structure, we call this threshold θ^*: this will depend on what regular structure we consider. For instance, if we have pores and throats with a square cross-section, from Eq. (4.11), $\theta^* = \pi/4$ or 45°; for triangles $\theta^* = 60°$. As the medium becomes more heterogeneous, we may assume that we see grooves, corners and crevices over a wide range of angle β, allowing some layer flow for larger angles. In the limit of an extremely heterogeneous rock, we could suppose that there will always be some very sharp valleys in the pore structure with β close to zero, allowing layer flow and snap-off for contact angles up to $\pi/2$ (90°). Therefore, for any degree of heterogeneity there is a division of the phase diagram between low contact angles where layer flow is possible, and larger angles where layer flow is not allowed. Where there is no layer flow, the phase diagram is similar to that presented in Fig. 4.20 with a transition

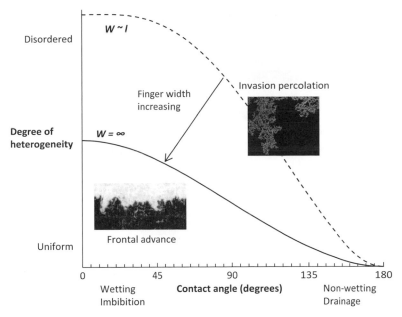

Figure 4.20 A phase diagram showing different displacement regimes when there is no layer flow and snap-off. We see either invasion percolation or frontal advance, dependent on the contact angle and the heterogeneity of the porous medium. Invasion percolation is favoured as the invading fluid becomes more non-wetting (larger contact angles), and the medium becomes more heterogeneous, such that the advance of fluid is controlled primarily by the size of the pores and throats. Frontal advance, controlled by cooperative pore filling, is observed for more regular or ordered systems at a low contact angle. The finger width W in invasion percolation diverges at the transition to frontal advance, as shown. A possible contour showing where the finger width is approximately equal to the pore size is also indicated.

from frontal advance to invasion percolation for larger contact angles and/or more heterogeneous media.

Where layer flow is allowed, we assume that a regime or pattern change, with a divergent finger width, occurs at the same contact angle and degree of heterogeneity as without layer flow, but now we see a transition from percolation with trapping to cluster growth. This then divides the full range of heterogeneity and contact angle into four possible patterns, as indicated in Fig. 4.21.

We can estimate the likely trend in finger width for the two percolation-like patterns. We know that $W \to \infty$ at the transition to cluster growth or frontal advance. However, the finger width will remain finite for the change from percolation to invasion percolation. Furthermore, while without layer flow, a decrease in contact angle encourages cooperative pore filling and leads to an increase in finger width, with layer flow, low contact angles also favour snap-off, leading to thinner fingers

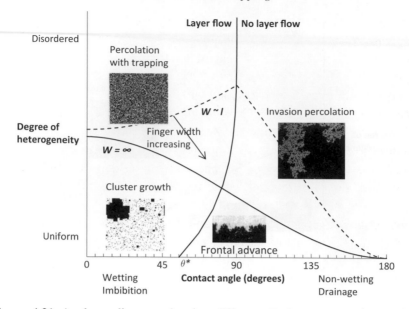

Figure 4.21 A phase diagram showing different displacement regimes when layer flow and snap-off are allowed. Four possible displacement patterns are possible – percolation with trapping, invasion percolation, cluster growth and frontal advance – dependent on the contact angle and the heterogeneity of the porous medium. There is a division between cases where layer flow is and is not allowed that runs approximately vertically in the diagram, since this is largely controlled by contact angle: for a uniform network this transition occurs at θ^* and the critical angle increases towards $\pi/2$ as the system becomes more disordered. This threshold is superimposed over the phase diagram for no layer flow, Fig. 4.20. Where layer flow is observed, the flow pattern is either cluster growth or percolation with trapping. The two percolation-like patterns have a finite finger width that diverges at the transition to cluster growth or frontal advance. The dashed lines indicate possible contours of constant finger width W. While the finger width increases with decreasing contact angle for invasion percolation, if layer flow is allowed, low contact angles favour snap-off and trapping and will tend to decrease the finger width.

and more trapping. Hence, in the percolation regime we must allow the finger width to increase with increasing contact angle in some cases. A possible trend of W is shown in Fig. 4.21.

These phase diagrams are different from those proposed by other authors, such as Lenormand and Zarcone (1984), since here we only consider capillary-controlled displacement. The flow pattern is controlled solely by the pore structure and contact angle and makes no account of flow rate or viscous forces; these are considered later in Chapter 6.6.

While these phase or regime diagrams help clarify the conditions under which different flow patterns evolve, to date there has been no direct experimental confirmation of their structure. This is a possible subject for future work, particularly with reference to pore-scale imaging in three dimensions which could elucidate the different types of flow pattern and the conditions under which each is observed.

For most of our examples, concerned with oil recovery, CO_2 storage and flow near the water table, we always have an initial water saturation before wetting phase invasion and the porous media are, in almost all realistic cases, highly heterogeneous (or at least much more disordered than a micro-model or a bead pack, for instance). As a result, for wetting phase invasion (contact angles below 90°) we see percolation with trapping, while for non-wetting phase advance (strictly drainage) we see invasion percolation.

A truly connected advance with no snap-off and little trapping is normally only observed for displacement into initially a completely dry sample; this is commonly encountered when rain infiltrates through soil after a drought, an important flow pattern discussed next.

4.3.6 Infiltration or Unstable Imbibition under Gravity

The movement of rainfall through soil towards the water table is called infiltration. This is capillary-controlled imbibition under gravity, similar to the drainage process described in Chapter 3.4.2. The nature of the displacement pattern depends on the presence of wetting layers. If the soil is damp, when more water is introduced, the wetting layers swell, snap-off can occur and a uniform, stable increase in wetting phase saturation results in a percolation-like displacement pattern. However, if the soil is completely or almost completely dry, wetting layer flow is prevented, meaning that the displacement pattern is either invasion percolation or frontal advance.

Invasion percolation under gravity was discussed in Chapter 3.4.2 and is observed for infiltration into a water-repellent soil, where water becomes the non-wetting phase (Bauters et al., 1998, 2000): fingers of the advancing phase move downwards through the system. The correlation length, which is the width of these downwards-moving fingers (note that this is distinct from, and larger than, the finger width controlled by cooperative pore filling discussed previously) can be predicted using percolation theory, Eq. (3.46): the displacement resembles invasion percolation with a self-similar structure from the pore scale l to the correlation length. In theory the same pattern could be observed even if the water wets the soil in a heterogeneous system if we have an invasion percolation flow pattern. There would be two intermediate length scales: that of the width of the downwards moving finger, ξ, predicted from the correlation length in percolation theory, and the width of the fingers within the structure itself, W, which is a function of contact angle and the nature of the pore structure. Over a length x where $W \geq x \geq l$ we would see a compact pattern, for $\xi \geq x \geq W$ we see self-similar invasion percolation, while for $x > W$ we observe downwards propagating fingers.

The correlation length ξ can be found from a modification of Eq. (3.30)

$$\xi \sim W \Delta p^{-\nu}. \tag{4.32}$$

Then using $\Delta p \approx B\xi/l$ (see page 108), with B defined by Eq. (3.42) we find

$$\xi \sim W \left(\frac{BW}{l}\right)^{-\nu/(1+\nu)}. \tag{4.33}$$

Experiments and simulations of infiltration in water-wet media tend to be performed on relatively homogeneous systems – usually bead or sand packs – and so an invasion percolation pattern is not seen. Instead we observe the unstable version of frontal advance, with a filled-in connected finger of wetting phase moving downwards. This process is shown in Fig. 4.22. The wetting phase begins a connected frontal advance, but now filling is favoured for pores that are lower down (the water pressure increases with depth for slow flow) and so an unstable displacement is seen with a saturated tip of the wetting phase propagating downwards.

This process has alarming implications for agriculture: watering a dry soil does not uniformly wet the subsurface. So the gardeners' tip is never to let the soil dry out in the first place! This type of gravitationally driven instability is also seen in many other situations: the first analysis by Hill (1952) was related to sugar refining. In the hydrology literature, starting from the work of Parlange and Hill (1976), fingering has been explained in detail on the basis of laboratory experiments and field studies (Glass et al., 1989c; Glass and Nicholl, 1996; DiCarlo et al., 1999; DiCarlo, 2004) with related theoretical analysis (Glass et al., 1989a; Selker et al., 1992; DiCarlo et al., 1999); see the review by DiCarlo (2013).

Two important features of the fingers will be mentioned. Firstly, while the advancing tip is almost completely saturated with water (the wetting phase saturation is close to 1), behind the tip the saturation is lower. That we have complete saturation and little trapping during a type of frontal advance, albeit one that is unstable, is consistent with our previous discussion. For the saturation to decrease as the finger moves downwards requires a drainage displacement. This can be explained from the differences in threshold capillary pressures in imbibition and drainage (capillary pressure hysteresis) (Glass et al., 1989a). The water pressure decreases with height up from the tip of the finger, since it is denser than air, meaning that the capillary pressure, or the difference between the air and water pressures, increases. Now consider the change in capillary pressure at a

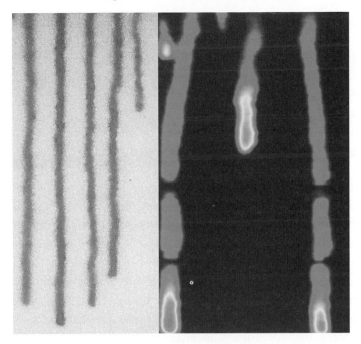

Figure 4.22 Infiltration fingers showing the progression of an unstable wetting front through an initially dry porous medium. On the left is a photograph
of water, dyed blue, progressing though a bead pack. The beds are have a
diameter between 1 and 1.3 mm and the average finger width is approximately 0.8 cm. Picture courtesy of Abigail Trice. The right figure shows
imbibition visualized in a slab chamber using light transmission. The coloured
regions are the flow paths: the hotter colours correspond to higher water saturation. Notice that the saturation at the tip of the finger is higher than
further up. From DiCarlo (2013). (A black and white version of this figure
will appear in some formats. For the colour version, please refer to the plate
section.)

fixed location as the wetting finger moves past. Initially the capillary pressure
is very high, in that the wetting phase saturation is low (remember that we
do not have layer flow). As the water first moves past, the capillary pressure
decreases as the water pressure increases and we have, obviously, an imbibition
process.

Let us say that the capillary pressure necessary for frontal advance is P_c^a. For
a horizontal displacement we would almost completely saturate the soil at this
pressure. However, for vertical displacement, this must represent the pressure difference between air and water at the bottom of the finger. As the water propagates
further downwards, the capillary pressure at a fixed location will increase: without flow it would be $P_c^a + \Delta\rho g h$ where $\Delta\rho$ is the density difference and h is the
distance to the bottom of the finger. h increases with time. This then is a *drainage*
process, associated with the invasion of air into soil saturated with water. Hence

air invades the pore space, leading to a lower water saturation behind the front than at the tip. The air cannot completely disconnect the water, as it needs to support the rate of infiltration at the surface. Furthermore, the water does not imbibe sideways to spread out the finger in the absence of layer flow: this requires a capillary pressure P_c^a, which is much lower than the prevailing value, meaning that the water pressure is too low to allow any further imbibition. This is capillary pressure hysteresis, which will be discussed in more detail in the next section: the capillary pressure for the imbibition process is lower, generally much lower, than the pressure for drainage.

The second, and related, feature of the fingers is their persistence (Glass et al., 1989b). If the rainfall continues, the water will flow in the channels established by the infiltration fingers. This will happen even if water eventually wets the whole system (sufficient time is given for wetting layers to become established and allow partial saturation of the soil between the fingers) or if the soil is allowed to dry partially: further infiltration results in the water moving through the same fingers as before. Again, this is the result of capillary pressure hysteresis. In equilibrium, we expect to see a capillary pressure that increases with height above the water table; see, for instance, Eq. (3.39). However, this does not mean that the water saturation is a constant. In the soil between the fingers, this saturation is established by imbibition, whereas in the fingers themselves, because of the desaturation behind the front, the saturation is reached by drainage. For a given capillary pressure, the saturation for a drainage displacement is much higher than for imbibition (or, alternatively, to fill a given region of the pore space, the threshold pressure for drainage is higher than for imbibition). This means that the water saturation in the fingers remains higher than between the fingers, which then provides a preferential high-conductivity flow path for further rainfall.

This phenomenon implies that in a dry soil, rainfall will not lead to a uniformly wetted system. Instead, preferential and persistent flow channels are established, which substantially reduces the volume of soil in which plant roots can access moisture.

We can also observe the reverse process, or a stable advance of a denser wetting phase upwards: two examples of this case were presented in Figs. 4.15 and 4.18. Here we see imbibition only, where we can use scaling arguments to define a maximum fluctuation in the width of the front due to the stabilizing influence of gravity (Geromichalos et al., 2002).

A fuller discussion of infiltration at the macro-scale requires the definition of a macroscopic capillary pressure for imbibition and its comparison to drainage, presented in the next section. We will then revisit this phenomenon in the context of its averaged description in Chapter 6.5.1.

4.4 Macroscopic Capillary Pressure

As for primary drainage, it is possible to define a macroscopic capillary pressure for imbibition, $P_{cm}(S_w)$. Even in the limit of a low flow rate, or an infinitesimally slow decrease in capillary pressure, this only corresponds to the local capillary pressure when the system is at rest in equilibrium when a new minimum in local capillary pressure has been reached. Unlike drainage, however, the existence of wetting layers where present, providing the invading phase throughout the system, suppresses violent leaps in filling and consequent retraction of the displacing phase.

It is important though to emphasize that we can only observe a macroscopic capillary pressure with a smooth variation in pressure with saturation if the flow pattern is either percolation or invasion percolation. In these cases, we can define some region of the rock where there is a variation in saturation with imposed capillary pressure. However, for frontal advance and cluster growth there is essentially just a single macroscopic value of capillary pressure: at P_c^a the wetting phase may advance and, bar some minor fluctuations, this is the pressure as the saturation increases from zero (or close to zero) to close to 1 (since there is little trapping). As a consequence a macroscopic description of this type of displacement based on some continuum characterization of properties as smooth functions of saturation is flawed: this conundrum will be addressed again when we consider fluid flow in Chapter 6.

It is possible to measure the imbibition capillary pressure using mercury after primary drainage by retracting the mercury from the pore space. However, since the invading phase is a vacuum, it is impossible to apply a positive pressure (a negative capillary pressure) even locally, which makes the initiation of pore filling by, for instance, an I_2 or I_3 process unlikely. Furthermore, a vacuum cannot be imposed in the mercury itself and so the displacement stops at a positive capillary pressure equal to atmospheric pressure (around 10^5 Pa). As a consequence, displacement is entirely controlled by snap-off with a very high remaining non-wetting phase saturation (typically much greater than 50%), which is not representative of fluid displacements at subsurface conditions. Since the wetting phase is a vacuum, snap-off does not require layer flow, but is a cavitation process when the pressure in the mercury drops below the threshold capillary pressure for the displacement.

Example mercury retraction (imbibition) capillary pressures are shown in Fig. 4.23; the primary drainage capillary pressures for the Ketton and Guiting samples were shown previously in Figs. 3.8 and Fig. 3.9, respectively. As mentioned above, the displacement stops artificially at a finite capillary pressure, leaving a large fraction of the non-wetting phase, mercury, in the pore space.

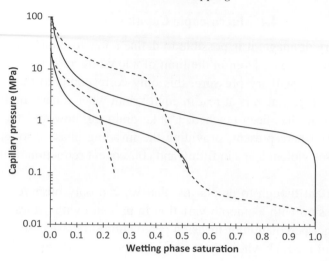

Figure 4.23 The mercury injection and retraction capillary pressures measured on Guiting (solid lines) and Ketton (dotted lines) limestones. This graph is equivalent to Figs. 3.8 and Fig. 3.9, but the lower curve in both cases shows the capillary pressure during retraction of mercury, which is an imbibition process. Note that the displacement stops at a finite capillary pressure when a significant fraction of the pore space is still full of the non-wetting phase: in this case the process is controlled by snap-off, trapping the non-wetting phase in most of the pores. Data from El-Maghraby (2012).

While a mercury retraction experiment is straightforward to perform, it does not necessarily capture the capillary pressure during imbibition for fluid pairs, such as oil and water. In this case a displacement experiment is necessary, which is time consuming since positions of capillary equilibrium have to be established with the average saturation measured during a drainage and imbibition (waterflooding) experiment. An example curve for a Voges sandstone (another exemplar quarry rock with a well-connected pore space) is shown in Fig. 4.24 measured for oil and water with a centrifuge method (Fleury et al., 2001). Notice that here, at the maximum imposed capillary pressure at the end of primary drainage, there remains a wetting phase saturation of approximately 20%. Even when the capillary pressure can drop to zero, almost half the pore space still contains oil. Further displacement may be possible at a negative capillary pressure (some pore-filling processes, for instance) but this is not recorded.

Fig. 4.25 shows another example for a CO_2/water system measured at different conditions of temperature and pressure for a coarse sand (Plug and Bruining, 2007). The curves overlie each other, implying that the temperature and pressure do not significantly affect the wetting properties of the fluid system: this will be confirmed in the next section where we consider the capillary trapping of CO_2. These curves

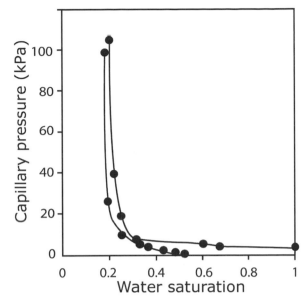

Figure 4.24 Primary drainage (upper curve) and imbibition (lower curve) capillary pressures measured for oil and water in Voges sandstone using a centrifuge technique. The residual saturation is approximately 0.45. Data from Fleury et al. (2001).

were measured using a porous plate technique: this is a water-wet ceramic disc with small pore spaces (a high capillary pressure) that allows only water to flow through: the capillary entry pressure prevents the passage of the non-wetting phase. Hence, a controlled amount of non-wetting phase (CO_2 in this example) can be injected into a core sample.

The capillary pressure in imbibition is always lower – and often substantially lower – than the capillary pressure at the same saturation in primary drainage. This is capillary pressure hysteresis and occurs for three reasons.

(1) **Contact angle hysteresis.** Even for the same type of displacement in the same region of the pore space, the threshold capillary pressure is lower because the advancing contact angle in imbibition θ_A is larger than the receding angle θ_R in drainage. Regardless of the exact expression for threshold capillary pressure, in all cases this pressure decreases as contact angle increases.
(2) **Different filling processes.** In imbibition, we have displacement that is principally controlled by snap-off and cooperative pore filling. These two processes do not occur in drainage and for the same pore or throat have a lower threshold pressure, even if there is no contact angle hysteresis. Filling a throat by snap-off generally requires less than half the threshold pressure in imbibition than piston-like advance in drainage; see Eq. (4.26). This is evident in Fig. 4.23,

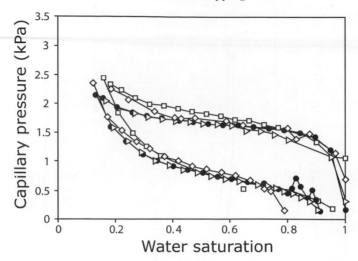

Figure 4.25 Primary drainage (upper curves) and imbibition (lower curves) capillary pressure measured for CO_2/water and N_2/water systems at different temperatures and pressures for a coarse sand. There is a minimum wetting phase saturation of approximately 15%, but very little trapping: the residual saturation is only around 10%. This is typical for a sand pack and indicates a displacement dominated by cooperative pore filling. The smooth change in capillary pressure with saturation implies that the invasion pattern in imbibition is percolation-like, rather than nucleated frontal advance, which would give a single capillary pressure value over most of the saturation range. From Plug and Bruining (2007).

where the capillary pressures during imbibition are around one-third to almost one-tenth the pressures in drainage, characteristic of a wetting phase displacement dominated by snap-off. In contrast, the shift in capillary pressure is more modest in Fig. 4.24, which is indicative of a more homogeneous medium with pores and throats of similar size where waterflooding is controlled largely by I_1 pore filling. The different filling sequence in drainage and imbibition is the main reason behind the thermodynamic interpretation of hysteresis based on a succession of metastable states of capillary equilibrium, discussed above in Chapter 4.2.6.

(3) **Trapping.** The wetting phase cannot displace all the non-wetting phase to return to a fully saturated system, because the non-wetting phase becomes trapped, or stranded, in the pore space when it is surrounded by the wetting phase. This does not directly reduce the capillary pressure, but shifts the imbibition curve to the left compared to drainage: the same pore or throat in imbibition will be filled at a lower wetting phase saturation than in drainage, with the difference in saturation representing the trapped non-wetting phase. Trapping will be discussed in more detail later in Chapter 4.6.

4.5 Interfacial Area

In the previous section we introduced the macroscopic capillary pressure. It is a function not only of saturation, but of the displacement path, being different in drainage and imbibition. As we will discuss later in Chapter 6.5.3, it has been suggested that a more complete description of capillary pressure should include its dependence on both saturation and the interfacial area between the fluids: this characterization then may reduce or even eliminate the apparent hysteresis seen when P_{cm} is described as a function of saturation only. Furthermore, the rate of mass transfer in dissolution or fluid/fluid reaction is also governed by the interfacial area.

Interfacial areas between two fluids can be measured by adding surfactants that preferentially reside on the interface and recording the amount absorbed (Kim et al., 1997; Saripalli et al., 1997; Schaefer et al., 2000; Jain et al., 2003). However, the introduction of surfactant alters the interfacial tensions and, potentially, the contact angle. At present, the method has had a relatively limited range of application confined to unconsolidated systems. Another core-scale technique is the use of nuclear magnetic resonance to determine the surface area to volume ratio applied to trapped ganglia in a glass bead pack (Johns and Gladden, 2001).

The advent of direct pore-scale imaging potentially offers a flexible and powerful approach to the measurement of interfacial area, since this can be recorded from a segmented image of the pore space and the fluid phases within it. Results for drainage and imbibition were first reported by Culligan et al. (2004) with further studies on water-wet granular packings (Culligan et al., 2006; Brusseau et al., 2007; Costanza-Robinson et al., 2008; Porter et al., 2010).

Fig. 4.26 shows the capillary pressure measured during a sequence of drainage and imbibition cycles on a bead pack with grain diameters between 0.6 and 1.4 mm. Two fluid systems were considered: oil/water and air/water; the difference in the magnitude of the capillary pressure is attributed to the difference in interfacial tension. A typical capillary pressure for a throat radius of 0.3 mm (half the smallest bead diameter) is around 470 Pa, or a water height of around 4.8 cm using Eq. (3.1) for the air/water system with a contact angle of zero, consistent with the values shown in Fig. 4.26. While there is significant hysteresis between imbibition and drainage, different cycles for the same displacement process have similar capillary pressures. The trapped non-wetting phase saturation is around 10% or less, similar to results for unconsolidated media, such as those shown in Fig. 4.25. The apparent irreducible water saturation is much larger for the oil/water experiment, implying the presence of pendular rings of trapped wetting phase with less wetting conditions (a larger contact angle) than the air/water system, preventing layer flow and complete displacement of the wetting phase.

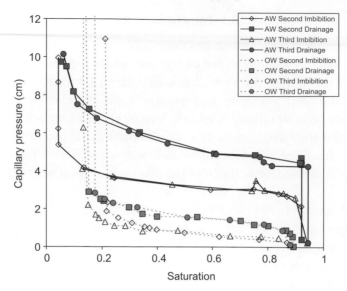

Figure 4.26 Capillary pressure curves for air/water (AW, solid lines) and oil/water (OW, dashed lines) experiments on a bead pack. Open and closed symbols designate imbibition and drainage curves, respectively. The pressure scale is in cm of water: 1 cm is approximately 98 Pa. The results here are qualitatively similar to those seen in Fig. 4.25. The minimum wetting phase saturation is higher for the oil/water experiments, implying less water-wet conditions in this case. From Culligan et al. (2006).

The corresponding interfacial areas are shown in Fig. 4.27. The specific surface area is shown: this is the area between the two fluid phases per unit volume of the porous medium. The units are 1/length, and a typical magnitude, if we see menisci throughout the pore space, should be around $1/l$, where l is some representative pore length. In this example, with bead diameters of around 1 mm, we anticipate specific surface areas of order 1 mm^{-1}, which is what is observed.

There is a linear rise in area with the non-wetting phase saturation: this is caused by the increasing number of terminal menisci in the pore space as more oil invades. A maximum is reached at a water saturation of around 0.3 (a non-wetting phase saturation of 0.7): this is when we have many terminal menisci between the non-wetting phase in the larger regions of the pore space and the wetting phase in smaller pores and throats. As the water saturation declines further, the area decreases sharply: the non-wetting phase now invades even the smaller throats, while layer flow allows the escape of the wetting phase, removing terminal menisci. For the oil/water system drainage and imbibition give similar areas for the same saturation: in both displacements we have a percolation-like pattern with a finger width that is close to the pore size. However, for the air/water experiments the area is lower in imbibition. The reason for this is that in imbibition, for

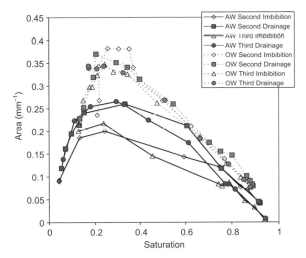

Figure 4.27 Interfacial area per unit volume as a function of water saturation for air/water (AW, solid lines) and oil/water (dashed lines) experiments. Open and closed symbols designate imbibition and drainage curves, respectively. The lower interfacial areas and the evident hysteresis for the air/water system are consistent with a larger finger width and a more blocky invasion pattern in imbibition. From Culligan et al. (2006).

a strongly water-wet homogeneous packing, there is likely to be significant cooperative pore filling and a large finger width; see Fig. 4.21: more blocky fingers will have a lower interfacial area than the wispier invasion percolation structures developed during drainage. The areas are also lower for the air/water system compared to oil/water, again consistent with an overall more compact pore-scale configuration of fluid. In the experiments, as expected, the air phase was trapped in fewer, larger clusters, while the oil was confined to many more, but smaller, ganglia (Culligan et al., 2006).

These experiments are, however, limited by the resolution of the image and cannot capture most of the arc menisci or the interfaces between the phases in the corners and roughness of the pore space. In comparisons with surfactant adsorption measurements, Brusseau et al. (2007) showed that imaging indeed tended to underestimate the area since layers were not accounted for. If they are included, the interfacial area continues to rise for low water saturations as the arc menisci cover more of the pore-scale roughness with increasingly fine detail (Schaefer et al., 2000; Brusseau et al., 2007).

Or and Tuller (1999) developed a model of interfacial area and capillary pressure for liquid/vapour systems which accounted for wetting layers and the adsorption of films. They showed that the contribution to the total area of these films and wetting layers in the crevices of the pore space may be two orders of magnitude larger than the area in terminal menisci, straddling the pore space.

There is a distinction between terminal menisci, that block flow through spanning across pores and throats, and arc menisci confined close to the solid surface. At low wetting phase saturations, the arc menisci will dominate the fluid/fluid interfacial area, but this is difficult to measure accurately. Direct imaging is a promising technique, but at present appears to be largely limited to capturing terminal menisci, although at the finest resolutions, some arc menisci can be seen, as shown in Fig. 3.5.

As mentioned at the beginning of this section, interfacial area is important as it controls the rate of mass transfer in dissolution and reaction processes, particularly for trapped phases which cannot flow; also significant, of course, is how much is trapped in the first place. This is addressed next.

4.6 Capillary Trapping and Residual Saturation

4.6.1 *Direct Imaging of Trapped Clusters and Percolation Theory*

In the literature, the pore-scale distribution of the trapped non-wetting phase has received extensive attention. Early work employed two-dimensional micro-model experiments which allowed fluid distributions to be analysed directly (Chatzis et al., 1983; Lenormand and Zarcone, 1984; Mayer and Miller, 1993). To extend the work to more realistic systems, an ingenious experiment was designed using styrene monomer as the non-wetting phase in core flooding experiments. After waterflooding, the trapped styrene polymerized to a solid, the rock grains were leached in acid and the remaining residual phase could be visualized under a microscope (Chatzis et al., 1983).

More recently, direct three-dimensional micron-resolution X-ray imaging has allowed trapped phases to be observed *in situ*. This research has studied a wide variety of porous media from bead packs to sandstones and carbonates, using different fluids and under various conditions of temperature and pressure, with application to oil recovery, CO_2 storage, freeze-thaw cycles in soil and gas entrapment in the capillary fringe (Coles et al., 1998; Kumar et al., 2009; Karpyn et al., 2010; Suekane et al., 2010; Iglauer et al., 2010, 2011; Singh et al., 2011; Wildenschild et al., 2011; Chaudhary et al., 2013; Georgiadis et al., 2013; Herring et al., 2013; Andrew et al., 2013, 2014c; Geistlinger and Mohammadian, 2015; Geistlinger et al., 2015; Herring et al., 2015; Kimbrel et al., 2015; Mohammadian et al., 2015; Pak et al., 2015). Other imaging methods have also been employed to study trapped phases, such as confocal microscopy (Krummel et al., 2013) or nanometre-resolution X-ray imaging in shale (Akbarabadi and Piri, 2014). In all cases, the experiments have confirmed that a substantial fraction of the pore space contains a trapped non-wetting phase.

We show examples of a trapped non-wetting phase (super-critical CO_2) for two of our standard rocks: Bentheimer and Ketton in Figs. 4.28 and 4.29, respectively. In these experiments, approximately one-third of the pore space is occupied by residual non-wetting phase with clusters of all size, from the scale of a single pore to multi-pore blobs that almost span the system. In these experiments we observe a secondary imbibition process, where a substantial volume of wetting phase is initially present in the pore space before water injection. This means that wetting layer flow will occur, confining the likely flow patterns to either percolation-like or cluster growth.

Percolation theory can be used to predict the distribution of ganglia, or clusters of non-wetting phase, trapped during imbibition and the overall trapped saturation, which can then be tested against the results of direct imaging experiments (Iglauer et al., 2010, 2011). So far we have considered percolation of the injected phase. However, we can also consider the defending or displaced phase – the non-wetting phase in this case. The percolation probability $p_{nw} = 1 - p$, where p is the percolation probability of the wetting phase. Again, there is a threshold value of p_{nw}, p_c, when the non-wetting phase is first connected, or since we start from a high value of p_{nw} and decrease its value as displacement proceeds, a critical value when the non-wetting phase first completely disconnects. The original application of this concept was by Larson et al. (1981), who suggested that Eq. (3.25) could be used to estimate the percolation threshold, which would, for relatively homogeneous media with similar pore sizes, be similar to the residual saturation: $S_{nwr} \approx p_c$. Subsequent work tended to focus on relating the residual saturation to the properties of the pore space, such as its connectedness and the aspect ratio (size of pores relative to throats) (Chatzis et al., 1983; Jerauld et al., 1984; Jerauld and Salter, 1990; Mayer and Miller, 1993; Coles et al., 1998; Wildenschild et al., 2011); this can be extended to relate the topology of the non-wetting phase itself to how much is trapped using the Minkowski functionals introduced in Chapter 2.2.4 (Herring et al., 2013, 2015).

From a pore-scale perspective, the amount of trapping is controlled by snap-off and how it is favoured over piston-like advance and cooperative pore filling. As discussed previously (see Eq. (4.26)), this in turn is a function of contact angle (more trapping for lower contact angles, assuming that there is wetting layer flow), coordination number and Euler characteristic (less trapping for well-connected pore spaces, consistent with Eq. (3.25)), aspect ratio (more trapping if the pores are much larger than the throats) and the amount of wetting layer flow, governed by the half-angle of the corners. In principle, if we know the contact angle and the pore structure, we should be able to predict the amount of trapping through the simulation of a sequence of displacements using the threshold capillary pressures presented in this chapter. This will be shown later for some of

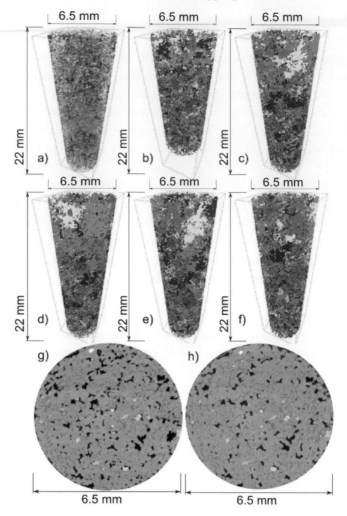

Figure 4.28 Micro-CT X-ray images of CO_2 as a non-wetting super-critical phase in Bentheimer sandstone with a voxel size of around 6 μm. The top left figure shows the distribution of CO_2 after drainage: the blue cluster is connected, while snap-off at the end of the process allows some disconnection shown by the clusters of other colours. The five other pictures show the distribution of trapped CO_2 after brine flooding: the colours indicate individual clusters. These are replicate experiments on the same rock: while the exact location of trapping is different in each experiment, the statistical properties, such as the overall amount of trapping and the cluster size distribution, are similar. Approximately one-third of the pore space contains residual non-wetting phase. The bottom row of figures shows raw images before processing after drainage (left) and water injection (right). Each image contains just over 3 billion voxels. From Andrew et al. (2014c). (A black and white version of this figure will appear in some formats. For the colour version, please refer to the plate section.)

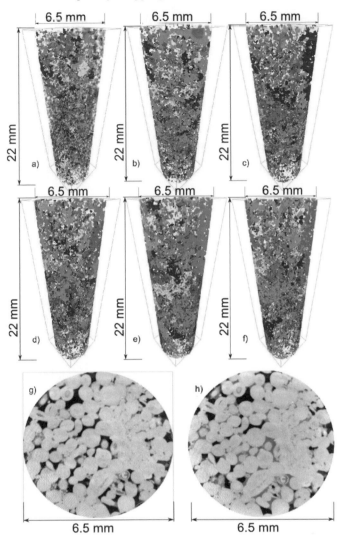

Figure 4.29 A similar figure to Fig. 4.28, except here trapping in Ketton lime-stone is shown. The voxel size is approximately 4 μm. Again, we see a significant amount of trapping with clusters of all size. From Andrew et al. (2014c). (A black and white version of this figure will appear in some formats. For the colour version, please refer to the plate section.)

our exemplar systems. However, there is no straightforward way to quantify these relationships, since they depend on the subtle balance of snap-off and cooperative filling. The primary control is the type of imbibition pattern, with percolation-type displacements allowing much more trapping than frontal advance. In this section the emphasis will be on assessing the likely flow regime and using percolation theory where appropriate. It is driven by the wealth of information on cluster size distribution made available from three-dimensional imaging.

In a percolation process, wetting phase invasion with cooperative pore filling has a finite finger width, W. While, at a large scale, we may see percolation-like behaviour, at the pore scale there tends to be a more compact arrangement of fluid, with a pore filled with wetting phase likely to be adjacent to other filled pores and throats. This inevitably introduces a constraint in the distribution of trapped clusters, with fewer blobs smaller than W seen than would be predicted for $W = l$, where l is the pore size. For this reason we cannot apply rather simple arguments, such as the use of Eq. (3.25), to estimate the residual saturation; instead we need to first quantify the nature of the displacement and the finger width and then consider the entire distribution of trapped clusters.

Imagine the ultimate displacement that finally disconnects all the non-wetting phase. Before this event, we must have a single critical spanning cluster of the non-wetting phase: this is a self-similar fractal with the same statistical properties as the invasion percolation patterns discussed in Chapter 3.4. However, there will also be some non-wetting phase outside this cluster which is already trapped. Then we allow this final imbibition event and all the non-wetting phase is trapped: it is comprised of clusters (ganglia) with a power-law distribution of size

$$n(s) \sim s^{-\tau}, \tag{4.34}$$

where $n(s)$ is the number of clusters containing s pores and here τ is the Fisher exponent with a value 2.189 ± 0.002 in three dimensions (Lorenz and Ziff, 1998).

This power-law is only observed over a limited range of length (Blunt and Scher, 1995). There is a lower cut-off: this is the width of the individual fingers of wetting phase. As explained in Chapter 4.3.5, cooperative pore filling leads locally to filled-in regions of the pore space, with percolation-like patterns only evident beyond a width W. In a pure percolation process $W = l$ and we follow Eq. (4.34) with a cut-off $n(s) = 0$ for $s < 1$. With a larger finger width, we still trap some small clusters: as the direct images show, Figs. 4.28 and 4.29, single pore blobs are possible, but $n(s)$ will be lower than predicted by Eq. (4.34) for small s. From a purely percolation perspective though, the finger width W now becomes the apparent minimum pore length; the smallest cluster is W/l pores across of size $s \approx (W/l)^d$: this defines an approximate lower limit to $n(s)$. Secondly, for the upper limit, no cluster can span the system size, L, else it will be connected. If we have a correlation length ξ where either $\xi = L$, or it is imposed by some external force such as gravity (see Eq. (4.33)), then $n(s) \to 0$ for $s > (W/l)^d (\xi/W)^D$ where D is the fractal dimension (2.52 in three spatial dimensions, d; see Table 3.1).

The lower-scale cut-off, or truncation, of the power-law is significant as it controls the overall amount of trapping. We can use Eq. (4.34) to compute the residual non-wetting phase saturation

$$S_{nwr} \approx \left(\frac{l}{L}\right)^d \int_0^\infty sn(s)ds, \tag{4.35}$$

since the total number of pores is $(L/l)^d$. Now, this is only approximate, as it assumes that all the pores contain the same volume. Then, using the scaling law, Eq. (4.34), and imposing, for the sake of argument, strict upper and lower bounds on $n(s)$

$$S_{nwr} \sim \int_{(W/l)^d}^{(W/l)^d (\xi/W)^D} s^{1-\tau}ds = \frac{1}{\tau-2} \left[s^{2-\tau}\right]_{(W/l)^d}^{(W/l)^d (\xi/W)^D} \tag{4.36}$$

and evaluating the limits

$$S_{nwr} \sim \frac{1}{\tau-2} \left(\frac{W}{l}\right)^{-d(\tau-2)} \left[1 - \left(\frac{\xi}{W}\right)^{-D(\tau-2)}\right]. \tag{4.37}$$

Then, in the limit of a large system so that $\xi \to \infty$ and where $\tau > 2$,

$$S_{nwr} \sim \left(\frac{W}{l}\right)^{-d(\tau-2)}, \tag{4.38}$$

which is independent of the overall size of the system, but controlled by the local finger width, with smaller residual saturations for larger values of W. The exponent $d(\tau-2)$ is approximately 0.57. This scaling of residual saturation with finger width has been seen in three-dimensional simulations of wetting with cooperative pore filling (Blunt and Scher, 1995).

Fig. 4.30 shows the measured distribution of trapped clusters for the two examples illustrated in Figs. 4.28 and 4.29 as well as for a further three samples, illustrated in Fig. 2.3 and whose drainage capillary pressures were shown in Figs. 3.8 and 3.9. These are the results of five replicate experiments on five rocks samples, where each experiment involves images with around 3 billion voxels. Such an analysis is now more feasible experimentally than through direct simulation: while we can model trapping at the pore scale, current computer technology precludes such an extensive analysis; see Chapter 6.4.9. We observe scaling over approximately two orders of magnitude in ganglion size with a clear small-scale cut-off.

Consider, as an example, the results for Bentheimer sandstone. The good fit to a percolation-like power-law starts at a ganglion size of around 10^5 voxels. Since the voxel size is 6 μm this represents a volume of around 2×10^7 μm^3. We have already determined an average distance between pores $l = 79$ μm (see Chapter 3), which, on average, is a void volume of $\phi l^3 \approx 10^5$ μm^3: note the porosity factor ($\phi \approx 0.2$ in this case). This gives an onset of percolation scaling at a ganglion size of around 200 pores, representing a finger width of approximately 6 pores across:

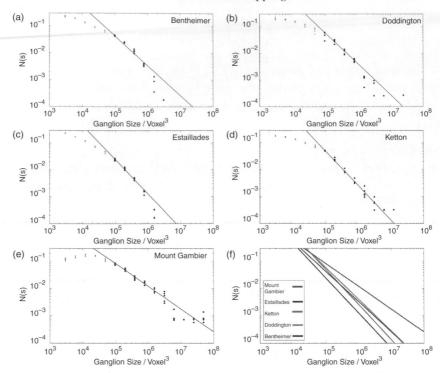

Figure 4.30 Trapped cluster size distributions measured on five rock samples: two cases are illustrated in Figs. 4.28 and 4.29. $N(s)$ is the relative frequency of clusters of size s voxels (not pores); however, the scaling relationship Eq. (4.34) should still hold with the same exponent. In all cases we see a truncated power-law distribution with an exponent τ between 2.1 and 2.3, broadly consistent with percolation theory (the exponent is predicted to be 2.19). The one outlier is Mount Gambier with an exponent of 1.8; this rock has a very porous and open pore structure (see Figs. 2.3 and 3.12). From Andrew et al. (2014c). (A black and white version of this figure will appear in some formats. For the colour version, please refer to the plate section.)

$W \approx 0.5$ mm. There *are* smaller trapped clusters, indeed more small blobs than larger ones, but fewer than would be predicted by percolation theory, evident in Fig. 4.30, since the local disconnection of the non-wetting phase is suppressed by cooperative pore filling. If we input our estimated value of W/l into Eq. (4.38), we find a residual saturation of 35%, which is close to the measured value of 32% (Andrew et al., 2014c); bearing in mind the approximate nature of the calculation this agreement is somewhat co-incidental, but does serve to emphasize that such scaling arguments can be used to judge trends and the likely magnitude of trapping. We can perform a similar analysis for the other cases: we see, at the largest scales, a percolation-like behaviour, but with a significant under-representation of the smallest clusters which then controls the overall residual saturation.

While not directly tested, the simulation results presented earlier to illustrate flow patterns imply that the amount of trapping is likely to be sensitive to the contact angle. While for connected advance, lowering the contact angle increases the finger width, leading to less trapping, when we have layer flow, low contact angles favour snap-off and we can see more trapping, with smaller W, as illustrated schematically in Fig. 4.21.

While this type of argument to determine the residual saturation may be physically appealing, from a practical perspective it only produces an approximate expression dependent on finger width. This explains why most studies instead focus on a direct link between pore geometry and the amount of trapping. Future work to provide more precision to the somewhat speculative regime diagram shown in Fig. 4.21 could help provide a systematic characterization of flow pattern, finger width and trapping related to contact angle and pore structure.

Not all experimental measurements of the distribution of trapped clusters have shown percolation behaviour. The Mount Gambier example in Fig. 4.30 has a best-fit exponent of 1.8, which is inconsistent with the expected value of 2.2 and which, for sufficiently large systems, would have the total residual saturation dominated not by the small-scale cut-off, but by the largest almost-sample-spanning cluster since the apparent $\tau < 2$; see Eq. (4.37): in the experiment the largest single cluster contributed 26% to the total volume, as opposed to 6% or lower for the other rocks. Similarly, experiments on bead packs have also found a very different cluster size distribution, with most of the residual saturation contained in one large, dominant ganglion (Georgiadis et al., 2013) or in a few of the largest ganglia (Krummel et al., 2013). These cases have either a rather homogeneous porous medium (bead packs) or a very open pore space (Mount Gambier). Imbibition is then controlled almost entirely by cooperative pore filling. In the experiments, since the wetting phase is initially present throughout the system at the beginning of waterflooding, the displacement pattern cannot be frontal advance. However, bursts of cooperative pore filling can be initiated throughout the porous medium, from regions already filled with water, giving a nucleated cluster growth. The non-wetting phase can be trapped when these clusters meet and merge, but overall we do not see a percolation process, with the typical ganglion size now determined by the size of the clusters, which is likely to be much greater than the pore size. Even in the bead pack experiments of Georgiadis et al. (2013), a power-law distribution of trapped cluster size was seen, consistent with percolation theory; however, the overall residual saturation was dominated by the largest ganglion. Other bead pack experiments have instead reported that the vast majority of the trapped saturation is in smaller clusters, although these were less ramified than would be expected from percolation theory with a surface area A that scaled with the volume V of the cluster as $A \sim V^{0.84}$ (Karpyn et al., 2010), as opposed to a linear scaling expected in three-dimensional percolation (Stauffer and Aharony, 1994).

The nature of trapping is not so easily assessed from the pore structure alone: Geistlinger and Mohammadian (2015) performed an extensive series of trapping experiments on bead packs and found results entirely consistent with percolation theory and with other experiments in sand packs and two-dimensional micro-models (Geistlinger et al., 2015). They had an air/water system with carefully cleaned glass beads which were likely strongly water-wet, allowing layer flow and favouring snap-off. They also measured the surface area between the fluid phases and showed that it scaled linearly with its volume, indicative of an open, fractal structure, and in contrast to the experiments of Karpyn et al. (2010). The discrepancy here is likely to be related to how strongly water-wet the system is, the ability to establish wetting layer flow and the amount of snap-off, with some experiments on bead packs in the cluster growth regime, while others are percolation-like.

This is an example of how the generic type of displacement pattern manifests itself in qualitatively different types of trapping with significant consequences for recovery and displacement of the non-wetting phase. In the cluster growth regime, the exact nature of trapping is controlled by the amount and location of the wetting phase present after primary drainage, from which pore filling starts. If the sample is initially dry we would expect to see frontal advance and very little trapping at all. This implies that the amount of trapping is not just controlled by wettability (the contact angle) and pore structure, which in turns governs the imbibition pattern and finger width, as illustrated in Figs. 4.20 and 4.21, but by the initial wetting phase saturation. This is addressed in the next section, where a dependence on trapping with initial saturation is also seen for percolation-like displacement.

4.6.2 Effect of Initial Saturation

In a macroscopic (cm-scale) experiment the residual non-wetting phase saturation can be measured at the end of imbibition. This saturation will be a function of the wetting phase saturation initially present in the rock, established at the end of primary drainage. We can define a trapping curve $S_{nwr}(S_{nwi})$ where S_{nwr} is the residual saturation and S_{nwi} is the initial saturation (at the start of the imbibition process). This relationship is important in several contexts. For instance, as presented in Chapter 3.5, initially there is a distribution of water saturation in a reservoir with a generally decreasing trend above the free water level. During subsequent waterflooding to displace oil, the amount that will be trapped (and hence not recovered) will depend on the initial saturation: taking the residual saturation from an experiment with a low initial water saturation will tend to overestimate the degree of trapping and hence underestimate the oil recovery. We see similar behaviour in a gas field: here water is not injected, but may ingress into the reservoir from the aquifer as the gas pressure drops. Gas trapping leads to reduced recovery and its magnitude is a function of how much water was initially present.

The second important application is in CO_2 storage: the migration of CO_2 after injection is a combination of a drainage and an imbibition process. At the leading edge of the plume, the CO_2 displaces brine in the aquifer as a primary drainage process. However, at the trailing edge, brine displaces CO_2, which is imbibition, leaving behind a trail of trapped non-wetting phase. This is good for long-term storage, since this CO_2 may dissolve or react with the host rock, but cannot now flow and escape from the aquifer.

The third application is to gas trapping near the water table: the rise and fall of the water table represents a repeated cycle of imbibition and drainage; the amount of air trapping is a significant control on chemical and biological processes in the capillary fringe, which itself is governed by the non-wetting phase saturation reached during the previous drainage displacement (Mohammadian et al., 2015).

Fig. 4.31 shows the sequence of capillary pressure changes required to measure the trapping curve. Traditionally, a primary drainage experiment continues to some high imposed pressure to drive the wetting phase saturation down to a low value – often assumed to be the irreducible or connate value S_{wc} – marked as point A in Fig. 4.31. The water, taken here to be the wetting phase, is injected in an imbibition process. The imbibition capillary pressure is lower than for drainage, for the reasons described above, and ends at point B when the non-wetting phase is trapped. To measure the trapping curve, however, we need to adjust this sequence and also perform a series of experiments where primary drainage proceeds to some intermediate saturation, followed again by waterflooding. The non-wetting phase saturation obtained in these experiments will be different: we plot these values as a function of initial saturation, as shown in Fig. 4.32. In general, for a percolation-like imbibition pattern where trapping is controlled by snap-off, the amount of trapping increases with the initial saturation (the more non-wetting phase you put in, the more you trap). Of course, you cannot trap more than was present initially, and so the curve has to lie below the line $S_{nwr} = S_{nwi}$. However, the proportion of the non-wetting phase trapped, S_{nmr}/S_{nwi}, may be close to 1 at low initial saturation for the reasons discussed below.

Some example capillary trapping curves are shown for oil/water systems using the porous plate method on three quarry sandstones: Berea, whose primary drainage capillary pressure is shown in Fig. 3.8, and two other well-connected samples, Stainton and Clashach (Pentland et al., 2010). Also shown are predictions made using pore-scale modelling with networks extracted from micro-CT images of the three rocks combined with simulations of primary drainage and imbibition performed assuming the pore-scale displacements described in this chapter

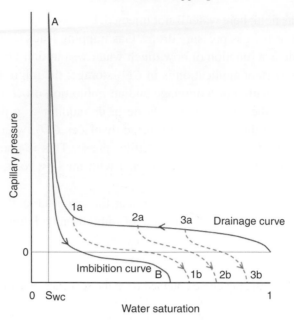

Figure 4.31 An illustration of capillary pressure scanning curves. Normally, an experimental primary drainage experiment continues to a high value of capillary pressure, point A, driving the wetting phase saturation close to the connate or irreducible water saturation S_{wc}. Then waterflooding is performed: the imbibition capillary pressure is lower and we reach point B with a large residual non-wetting phase saturation. This sequence defines the bounding capillary pressure curves. To determine the trapping curve, we need to perform experiments where we stop primary drainage at intermediate saturations; subsequent waterflooding, dashed lines, results in less trapping. From Pentland (2011).

(Valvatne and Blunt, 2004). As emphasized previously, the amount of trapping is controlled principally by the competition between snap-off and cooperative pore filling, which is sensitive to contact angle. We do not know, directly, the contact angle in the rock samples, and so this has to be assumed, or tuned to match the data. In this case, assuming a very strongly water-wet case, with contact angles randomly assigned to pores and throats with a uniform distribution between 0 and 30°, overestimates the amount of trapping through allowing too much snap-off to occur. A more realistic assignment of angles is to allow larger effective values, accounting for roughness and the complexities of the pore geometry: in this case a range between 35° and 65° provides a reasonable fit to the experimental measurements. This shows that it is possible to reproduce experiments, but does not immediately explain why this trend is seen.

We will now interpret the results in the context of the pore-scale analysis provided earlier in this chapter. Percolation theory has been employed to determine the trapping curve using an extension of Eq. (3.25): the initial non-wetting phase

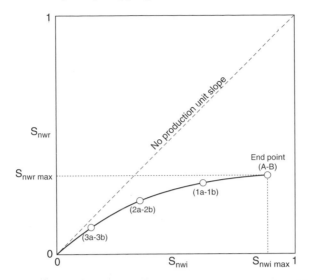

Figure 4.32 The trapping curve obtained from the sequence of capillary pressure changes shown in Fig. 4.31. We see an increase in residual saturation with initial saturation: the curve must lie below the line of unit slope representing no displacement of the non-wetting phase. From Pentland (2011).

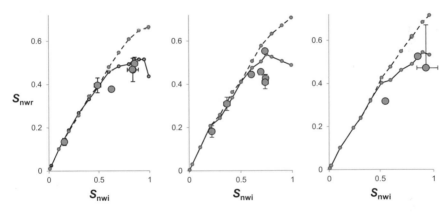

Figure 4.33 Comparison of measured (circle) and predicted trapping curves $S_{nwr}(S_{nwi})$ in (left) Berea, (middle) Clashach and (right) Stainton cores. Predictions using the pore network simulator of Valvatne and Blunt (2004) are shown, assuming intrinsic contact angles distributed in the range 0 to 30° (dashed line) and 35° to 65° (solid line) are included. For Berea, the data correspond, approximately, to the results shown in Table 4.2 for $S_{nwi} = 0.76$. From Pentland et al. (2010).

creates a network with some coordination number z_{nw}, which is lower than that of the whole pore space. This smaller value is then used to estimate p_c and the residual saturation, as mentioned above, using $S_{nwr} \approx p_c$ (Larson et al., 1981). However, this approach is of limited utility, since the nature of the displacement, as discussed above, is rather more complex than simple percolation. However, we can use percolation and topological arguments to explain the behaviour, although they do not necessarily yield a quantitative prediction of the amount of trapping.

To marshal the arguments, we will first consider the topology and order of filling in imbibition and how it relates to the preceding displacement sequence in drainage. Fig. 4.34 is a schematic of a displacement pattern at the end of primary drainage on a 4×4 square lattice shown simply for illustrative purposes. If the non-wetting phase is above – but not far above – the percolation threshold, the structure is composed of loops and dangling ends, as well as isolated clusters which connect to the inlet, but not to the outlet. If we had a snap-shot of the displacement during a Haines jump, there could also be isolated clusters of non-wetting phase present in the middle of the pore space: however, since we normally stop drainage when a constant, maximum external capillary pressure is imposed, we will assume that any such clusters have been reconnected.

Now we consider imbibition, starting from the fluid arrangements shown in Fig. 4.34: this normally proceeds by a combination of snap-off and I_1 pore filling. In Fig. 4.34, the dashed lines indicate pores and throats that are dangling ends. These can all be filled by a sequence of I_1 pore filling: the pore at the end of the chain is displaced by the wetting phase, followed by the last throat. This allows the next pore along to fill and so on. Even if there are branches in the structure, once

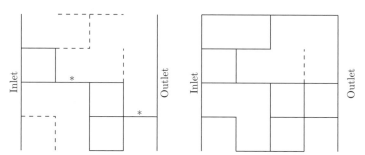

Figure 4.34 A schematic of fluid invasion in a two-dimensional square lattice to illustrate the topology of imbibition displacements and their impact on trapping. The lines are throats filled with a non-wetting phase at the end of primary drainage. The pores are located at the junctions: all pores next to a filled throat are also filled with a non-wetting phase. The solid lines indicate the backbone of the cluster, while the dashed lines are dangling ends: these can all be displaced by I_1 pore filling in imbibition without any trapping. In contrast, those throats indicated by a * are so-called red bonds. If a single one of these is displaced by snap-off, all the non-wetting phase becomes disconnected. On the left is a pattern at a low initial non-wetting phase saturation with many dangling ends and some red bonds; most of the non-wetting phase is rapidly trapped. The figure on the right has a higher saturation with no red bonds and fewer dangling ends. Here a smaller fraction of the initial saturation is trapped, as initially imbibition cannot disconnect the non-wetting phase.

both branches have been displaced then the pore at the junction can also be filled by an I_1 process. The remaining non-wetting phase, which comprises the backbone, running from inlet to outlet and with loops in the structure, cannot be displaced this way. Hence, if we have I_1 filling alone and no snap-off, or at least if this process is highly favoured, then we displace all the dangling ends with no trapping. This can be seen, for instance, in rather uniform media, such as sand packs.

If we consider a low initial non-wetting phase saturation, then we can use percolation concepts: the initial distribution of non-wetting phase has a self–similar or fractal structure below some correlation length ξ; see Eq. (3.30). Then, referring to Table 3.1, we note that the fractal dimension of the entire structure is larger than that of the backbone, implying that the vast majority of the non-wetting phase resides in dangling ends. The relative infrequency of loops in the structure is also evident from a study of the Euler characteristic. This has indeed been observed directly on images of non-wetting phase established after drainage: in Bentheimer sandstone, for instance, the Euler characteristic is positive for a non-wetting phase saturation of less than 40%, implying that there are more individual clusters of non-wetting phase than loops; see Eq. (2.10) (Herring et al., 2013).

This argument may imply that we see very little trapping for low initial saturations, since all the dangling ends are nibbled away by I_1 filling. However, we have ignored snap-off. In addition to dangling ends, we also have so-called red bonds (indicated by the * in Fig. 4.34): if any of these are filled by snap-off, the non-wetting phase becomes trapped, in that it is no longer connected from inlet to outlet. As evident from Eq. (3.35), there is a finite number of such critical throats at the percolation threshold: even above p_c if we cut one red bond in each volume of size the correlation length ξ we should expect to disconnect the non-wetting phase completely. Hence, just a few snap-off events are sufficient to disconnect the non-wetting phase at low initial saturation, which is also evident from the low value of the Euler characteristic.

As a consequence, for low initial non-wetting phase saturation, we trap most of the initial saturation during imbibition: the ratio S_{nwr}/S_{nwi} is close to one.

We now consider the other extreme, which is a high initial non-wetting phase saturation where most of the throats have been filled. The last throat to be filled in drainage will be one of the smallest ones obviously. Now consider imbibition. With few throats still filled with wetting phase, displacement must be initiated by snap-off. What throat snaps-off first? Well, clearly the smallest one – indeed, the one that was last filled during drainage. The threshold capillary pressure is different – indeed, considerably lower – but the location of the displacement is the same. We can continue this argument to consider a sequence of snap-off events in the smallest throats which replicates the order, in reverse, of the filling during drainage. What about pore filling? In drainage, once a throat is filled, so is the adjoining pore. To a good approximation, in a well-connected network, the first throat to be filled is the largest: thus the largest throat is filled, followed by the pore, with displacement later in the other throats, in order of size. Now we consider imbibition and we see the reverse order of displacement: around each pore the smallest throats are filled by snap-off. This then allows I_1 pore filling and the advance of water through the largest throat. In drainage there is no trapping overall, since the non-wetting phase has to be connected to the inlet at some time to progress into the pore space. If imbibition follows precisely opposite displacement sequence, then nothing is trapped either.

This argument again implies no trapping in imbibition. The complexity is the presence of Haines jumps: during these events in drainage, the filling of a relatively narrow throat is followed by a cascade of filling of wider pores and throats. In imbibition, this process does not occur in reverse. The relatively narrow throat will be filled, followed perhaps by I_1 filling of the adjoining pore, but then, as we have already shown, Fig. 4.11, layer flow inhibits the dramatic changes in local capillary pressure experienced during drainage, and will instead allow the next displacement to occur somewhere else in the system, in a pore or throat with a similar threshold capillary pressure. This breaks the symmetry: during imbibition, with a more strict sequence of filling in reverse order of size, the larger regions of the pore space become surrounded and trapped with the power-law distribution of ganglion size predicted by percolation theory, Fig. 4.30.

Trapping occurs mainly in regions of the pore space filled, in drainage, by Haines jumps: these events are most significant in the early part of drainage, for low non-wetting phase saturation. This is consistent with our previous discussion: we trap a greater fraction of the initial saturation when this saturation is low. At higher saturation, well above the percolation threshold, the non-wetting phase is well connected throughout the pore space (with a large and negative Euler characteristic) and drainage proceeds principally as a series of throat-filling events in order of size with few small Haines jumps, if any. During imbibition, the wetting phase displaces these throats (and some pores) in a reverse sequence with little trapping. The ratio S_{nwr}/S_{nwi} decreases with increasing S_{nwi}. For homogeneous structures with a small aspect ratio (the size of pores relative to throats), cooperative pore filling is important, with I_1 always occurring where possible, displacing any dangling ends of the non-wetting phase; in these cases, we may expect S_{nwr} to be approximately constant above some threshold saturation, since imbibition strictly follows the reverse order of drainage and there is no trapping. This is indeed what we will see below for sand packs.

The final twist in the argument is to note that the pore-network simulation results in Fig. 4.33, if not the experimental data, show a drop in residual saturation when the initial saturation is very high. This appears counter-intuitive, as the more non-wetting phase initially in the system, the more that can be trapped. The explanation here is associated with the effective contact angles used to match the data: for the largest (intrinsic) angles, the threshold capillary pressures for pore filling may be negative. This then swaps the filling sequence from favouring small pores, to displacing first the larger pores, where the threshold pressure, although negative, is not so large. Hence, for high initial non-wetting phase saturations, we favour the filling of some large pores (which in a more strongly water-wet system would be trapped). This explains why less is trapped than for lower contact angles: there are fewer snap-off events, while trapping can occur in some smaller, rather than larger, pores; see Chapter 4.2.4. The decrease at large initial saturations is caused by a transition from significant trapping controlled largely by the poor connectivity of the initial distribution of non-wetting phase, as explained previously, to a different filling sequence, determined by the distribution of contact angles established after primary drainage, with trapping now in those pores that have the lowest capillary pressures for filling. This effect of wettability will be discussed further in the next chapter.

A compilation of literature data on trapping in water-wet systems is shown in Fig. 4.35 (Al Mansoori et al., 2010). The smallest amount of trapping, with a maximum residual saturation of only around 10–15% is seen for unconsolidated media, such as sand packs. Here displacement, as mentioned above, is controlled largely by cooperative pore filling with relatively little snap-off. Furthermore, we see little increase in residual saturation above some threshold initial value, consistent with the argument presented above: the first phase of imbibition simply follows the reverse filling sequence to drainage with no trapping. The largest residual saturations – with maximum values of 50% or more and a continuing increasing trend with initial saturation – are observed for more highly consolidated rocks or other porous media of low connectivity.

The topological analysis of porous media, and the fluids within them, introduced in Chapter 2.2.4, suggest an approach to predicting the residual saturation, or at least finding a correlation between residual saturation and pore structure. Indeed, network modelling has demonstrated that the amount of trapping is sensitive to the details of the pore-space connectivity (Sok et al., 2002), which in turn is controlled by the diagenetic history of the rock (Prodanović et al., 2013). However, to date, only the porosity is routinely measured on rock samples for which residual saturations are also determined. As shown in Fig. 4.36, lower porosity media tend to be more poorly connected with larger maximum residual saturations than high-porosity samples, although there is a considerable scatter related to the details of the pore structure and the (unknown) contact angle.

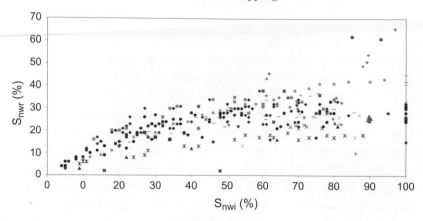

Figure 4.35 A compilation of literature data showing the relationship between residual and initial non-wetting phase saturation for different water-wet porous media. Each symbol represents a different experimental dataset. The wide variation in behaviour is a consequence of the different systems studied, from sand packs, with least trapped, to more consolidated rocks, with higher residual saturations. From Al Mansoori et al. (2010).

While we can use qualitative arguments to explain the amount of trapping, it is not possible to make quantitative predictions without performing an explicit simulation of displacement in the porous medium of interest. Even in these cases, as we have shown, we need some independent assessment of contact angle, which is not generally available. Instead of direct prediction, for modelling purposes, the trapping curve is frequently fit to some empirical form for convenience, although the functional form has no particular physical basis. The most commonly applied model is due to Land (1968), which was originally developed to quantify the trapping of gas during water influx:

$$S_{nwr} = \frac{S_{nwi}}{1 + C \frac{S_{nwi}}{1 - S_{wc}}} \qquad (4.39)$$

with a single fitting parameter C that is called the Land constant. This expression does require a determination of the connate water saturation S_{wc}; normally this is simply assumed to be the smallest water saturation reached during primary drainage.

A second empirical form was developed based on predictions of network modelling and applied to both CO_2 storage and mixed-wet reservoirs, see Chapter 5 (Juanes et al., 2006; Spiteri et al., 2008):

$$S_{nwr} = \alpha S_{nwi} - \beta S_{nwi}^2, \qquad (4.40)$$

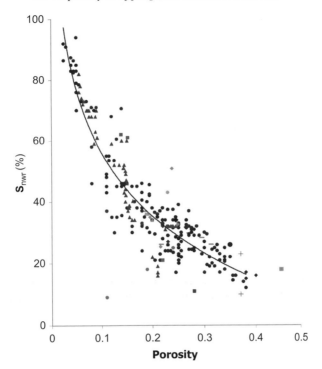

Figure 4.36 Literature data showing the maximum measured residual saturation as a function of porosity. As expected, lower porosity samples, with poorer connectivity, have higher residual saturations. There is, though, considerable scatter in the relationship, associated with differences in pore structure and contact angle for the media studied. Each symbol represents a different experimental dataset. The trendline is a logarithmic dependence proposed by Jerauld (1997). From Al Mansoori et al. (2010).

where the two parameters α and β are matched to the data. This is a two-parameter fit, which will be needed when we consider mixed-wet systems in the next chapter, where the residual saturation does not monotonically increase with initial saturation, as also seen in Fig. 4.33.

The third empirical form is appropriate for unconsolidated media, where we see an approximately linear S_{nwr} increase with S_{nwi} until some threshold value, beyond which the residual is constant (Aissaoui, 1983):

$$S_{nwr} = a S_{nwi} \qquad S_{nwi} \leq S_{nwr}^{max}/a \qquad (4.41)$$

$$S_{nwr} = S_{nwr}^{max} \qquad S_{nwi} \geq S_{nwr}^{max}/a. \qquad (4.42)$$

Eq. (4.41) has two adjustable parameters: a and the maximum trapped saturation S_{nwr}^{max}. This provides a good match, as indicated previously, for trapping in unconsolidated media (Al Mansoori et al., 2010; Pentland, 2011).

Much of the recent literature has focussed on the trapping of CO_2 for storage applications. If a significant fraction of the initial saturation (established after a period of CO_2 injection) is trapped, then this limits the migration of mobile CO_2 post-injection, making leakage unlikely. Some core-scale experiments have suggested, however, that in a saline aquifer, CO_2 is not the non-wetting phase which

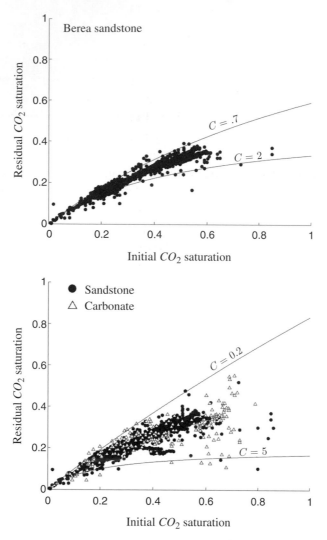

Figure 4.37 A compilation of measured data on trapping curves for CO_2 in (top) Berea sandstone and (bottom) different sandstones and carbonates. The lines show fits using the Land model, Eq. (4.39), with different values of C indicated. For Berea we see a consistent trend for experiments performed at different conditions of temperature and pressure, and with different brine compositions. There is a larger variation in the lower curve, since we are considering trapping in a wide range of rock types. From Krevor et al. (2015).

means, as we will show in the next chapter, that very little may be trapped (Plug and Bruining, 2007; Kneafsey et al., 2013). Fig. 4.37 shows a compilation of experimental data on CO_2 trapping in the literature (Krevor et al., 2015); see also Pentland et al. (2011), Akbarabadi and Piri (2013) and Niu et al. (2015). The results have been fitted to the Land model, Eq. (4.39): overall the results for Berea sandstone show a remarkable consistency between different experiments at different conditions of temperature and pressure and various brine compositions, and give residual saturations close to those, if possibly slightly lower, than those measured on oil/brine systems, Fig. 4.33.

When experiments on different sandstones and carbonates are compiled, there is a wide scatter in the data, Fig. 4.37. This is a result of the range of different pore structures considered. However, there is no fundamentally, or generically, different trend in trapping seen for carbonate rocks compared to sandstone: in both cases the residual saturation is dependent on the same physics and topological constraints. In all cases a significant fraction of the CO_2 can be trapped in the pore space, indicative of a water-wet system in a percolation regime, consistent with direct contact angle measurements (Iglauer et al., 2015).

While the application to CO_2 storage is informative, providing some assurance that at the pore scale, CO_2 can be safely trapped, its application to oil recovery is less useful. This is because with a change in initial water saturation, we see a trend in wettability, with lower water saturations associated with more oil-wet conditions, as discussed in Chapter 3.5. To treat this problem we need to analyse pore-scale configurations and displacement for mixed-wet systems and their impact on capillary pressure and residual saturation, which is the topic of the next chapter.

5

Wettability and Displacement Paths

5.1 Definitions and Capillary Pressure Cycles

We will now consider displacement after primary drainage when the invading phase is no longer, or at least not necessarily, wetting. This is the more likely case in oil-field applications, where, as described in Chapter 2.3.3, most reservoir rocks have a mix of water-wet and oil-wet pores during waterflooding. The distribution of contact angles for different displacement processes is determined by the mineralogy and roughness of the rock surface, the oil and brine compositions, as well as the capillary pressure and water saturation at the end of primary drainage; it will vary on a pore-by-pore basis.

Hitherto, we have defined imbibition to refer to a displacement where the advancing phase has a contact angle of less than 90°. While this is a convenient definition for simple cases, it is neither precise nor practical for displacement in porous media for two reasons. Firstly, the contact angle is not constant; many rocks have contact angles both less than and greater than 90°. Secondly, as seen in Chapter 4.2.2 for cooperative pore filling processes, even if the contact angle is less than 90°, the threshold capillary pressure for displacement may be negative, implying that the apparently wetting phase needs a higher pressure than the non-wetting phase to advance. In macroscopic experiments the externally imposed capillary pressure is measured: this cannot distinguish between a negative capillary pressure for an event with a contact angle less than 90° from one where the advancing phase is non-wetting with a contact angle greater than 90°.

From now on, a more pragmatic definition of imbibition and drainage will be employed, appropriate to describe macroscopic displacement processes in porous media, based on the sign and direction of change of the capillary pressure. The oil/water capillary pressure is defined as the externally imposed pressure difference between oil and water in a displacement at a very slow flow rate. $P_{cm} = P_l - P_d$, where d refers to the denser phase and l to the less dense phase, following the same convention used to define contact angle in Chapter 1. Then a drainage displacement is one where P_{cm} is positive and increasing (the saturation of the denser phase – typically water – is decreasing). Imbibition refers to $P_{cm} > 0$ and decreasing, representing an increase in the saturation of the denser phase (water). How then do we refer to situations where the capillary pressure is negative? It is drainage if $P_{cm} < 0$ and the saturation of the denser phase is increasing, while it is imbibition if $P_{cm} < 0$ and the saturation of the denser phase is decreasing.

To explain this more explicitly, as discussed in Chapter 2.3.3, most reservoir rocks undergo some degree of wettability alteration after primary drainage, rendering some surfaces oil-wet. Then the capillary pressure for displacement – for instance, piston-like advance through a throat, Eq. (4.12), for a contact angle θ_A greater than 90° – can be negative. In this case, if we are injecting water, it has a higher pressure than oil, and we have technically a drainage (a secondary drainage) displacement. Similarly, if we now re-introduce oil into the system and we have advance through oil-wet regions of the pore space with $P_{cm} < 0$, the oil pressure is lower than the water pressure, and this is imbibition. In other words, drainage is a displacement where the pressure of the advancing phase is greater than the receding phase; it is imbibition if the advancing phase has a lower pressure.

The almost universal, and incorrect, terminology in the oil industry is to refer to water injection, regardless of the sign of the capillary pressure, as imbibition. Drainage then simply refers to oil invasion. This is redundant (as they refer to the same thing) and misleading. However, to try to avoid confusion, water displacement at a positive capillary pressure (the water has a lower pressure than oil, so is not forced into the pore space) will be called spontaneous imbibition. When the capillary pressure becomes negative, rather than call this drainage (which it is) and invite a muddle with oil injection processes, we will call this forced water injection (the water is at a higher pressure than the oil). For oil advance, it is drainage for a positive capillary pressure and will be called spontaneous oil invasion when $P_{cm} < 0$. When oil is re-injected after waterflooding, this is a secondary displacement process, called secondary drainage where $P_{cm} > 0$.

An example displacement sequence is illustrated in Fig. 5.1 showing primary drainage, spontaneous imbibition, forced water injection, spontaneous oil invasion and secondary drainage. In this chapter we will use a pore-scale analysis of fluid configurations and displacement sequence to explain and interpret the macroscopic behaviour shown schematically in Fig. 5.1. For simplicity from now on, when we refer to the capillary pressure we mean the macroscopic value and will drop the subscript m: we assume that we can measure P_c as a function only of saturation and displacement history.

In detail, the displacement sequence we consider has five steps. Step 1 is primary drainage (Chapter 3) where we assume that the pore space is initially water-wet.

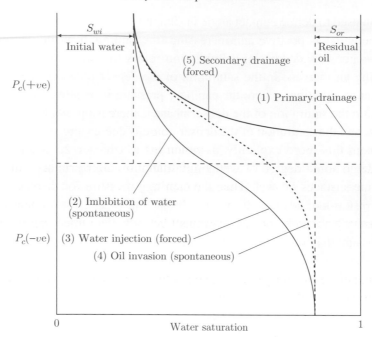

Figure 5.1 Capillary pressures in a mixed-wet system. Initially the porous medium is fully saturated with water (the saturation is 1). There then proceeds a primary drainage displacement, bold solid line (step 1), where we assume that the system is water-wet. At the end of primary drainage, the wettability of those surfaces contacted by oil may be altered. During subsequent water invasion the capillary pressure can be both positive (spontaneous imbibition, step 2) indicating displacement of oil from water-wet pores, or negative (step 3), as water is forced into oil-wet regions of the pore space in a drainage process, which we will call forced water injection. At the end of waterflooding we reach the residual oil saturation. We can also consider secondary oil invasion, the dotted line, where oil is introduced again into the system. We can see both negative (spontaneous oil invasion, step 4) and positive (secondary drainage, step 5) capillary pressures.

At the end of primary drainage, we reach some minimum water saturation and those portions of the pore space in contact with oil may alter their wettability, as described in Chapter 2.3. This then is followed by water invasion. This first occurs at a positive pressure, step 2, filling most of the water-wet pores and throats. This is spontaneous imbibition, Chapter 4, which is followed by forced displacement, step 3, where the water pressure exceeds that of the oil. This process ends when we reach the residual oil saturation. The final displacement sequence is the re-invasion of oil. Initially, the oil-wet pores and throats are filled in an imbibition process, step 4, which we will refer to as spontaneous oil invasion, followed by secondary drainage, step 5, when the oil pressure exceeds the water pressure. In most cases, if a sufficiently high positive capillary pressure is imposed, the saturation returns to its value at the end of primary drainage; however, as we discuss below, it is possible to trap water as the non-wetting phase.

The invasion sequence shown in Fig. 5.1 is not the only possible history of saturation change. As illustrated in the previous chapter, primary drainage may end at some intermediate saturation, and the subsequent water injection may then yield a different residual oil saturation, and hence a different initial condition for oil re-invasion. Furthermore, it is possible to consider additional cycles of water and oil invasion. While this is less common in oilfields, the fluctuation in height of the water table during periods of rainfall and drought gives a continual oscillation between wetting and drainage, which may begin and end at any arbitrary intermediate saturation.

The capillary pressure for a displacement between the minimum and maximum saturation of the advancing phase is called a bounding curve: these are the bold lines in Fig. 5.2 representing waterflooding and secondary oil injection. The waterflooding and oil re-injection cycles span a saturation range between the irreducible water saturation and the residual oil saturation. If there is no change in wettability

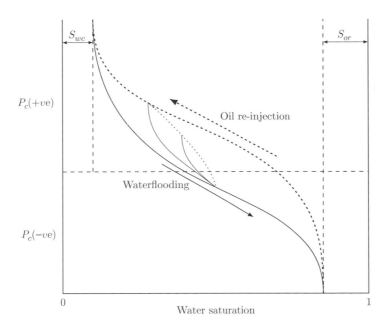

Figure 5.2 Scanning curves for a displacement sequence similar to that shown in Fig. 5.1. The thick lines are bounding curves, when the displacement proceeds all the way between residual and irreducible saturations. The solid lines are for waterflooding – the water saturation increases – while the dotted lines are for oil injection where the water saturation decreases. Scanning curves are transitions between bounding capillary pressures, starting and ending at intermediate saturations, shown as the fainter lines. In all cases the capillary pressure for oil invasion lies above that for water injection.

during this injection sequence, then other displacement cycles, shown by the fainter lines, will have capillary pressures between these curves. A scanning curve transitions between the bounding curves, representing water injection, or oil invasion, that starts at some intermediate saturation: the oil invasion capillary pressure will always be above the waterflood capillary pressure, as indicated in Fig. 5.2. More complex displacement sequences, beginning on a scanning curve, are also possible. Again, because of capillary pressure hysteresis, discussed in Chapter 4, regardless of the wettability of the sample, the capillary pressure for a waterflood will always lie below that for oil re-invasion.

We now show two datasets to illustrate these concepts. The first, Fig. 5.3, is measured on a water-wet sand pack. Here the minimum wetting phase saturation is around 30%: this is not the true irreducible saturation but the value at the highest imposed capillary pressure. The flooding sequence is primary drainage, spontaneous imbibition and then secondary drainage. In this case there is no forced invasion of water or spontaneous imbibition of oil. The trapped saturation is just under 10%.

The second example, Fig. 5.4, shows the capillary pressure for Berea sandstone. Note firstly that there is a large remaining water saturation of over 30% after primary drainage, even when a high capillary pressure is imposed. This is due to the presence of clays in the rock which remain water-saturated. This sample is

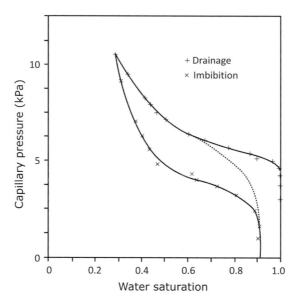

Figure 5.3 Capillary pressures showing primary drainage, spontaneous imbibition of water and secondary drainage (dotted line) as a function of water saturation. This is for a water-wet sand with little trapping and no recovery by forced displacement of water. From McWhorter (1971).

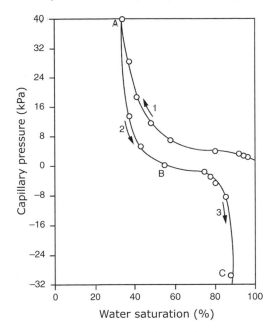

Figure 5.4 Measured primary drainage and waterflood capillary pressure curves measured on Berea sandstone. The displacement sequence, similar to that shown in Fig. 5.1, is primary drainage (step 1 to point A), spontaneous imbibition of water (step 2 to point B) and forced water injection (step 3 to point C). To reach the residual oil saturation a negative capillary pressure is needed: the water pressure is higher than the oil pressure during forced injection. Data from Anderson (1987a).

considered to be water-wet, in that it has not been subjected to ageing in crude oil: we will show a more oil-wet case later in this chapter. However, there is a significant amount of displacement for a negative capillary pressure during water-flooding, indicating that water needs to be injected into the sample at a higher pressure than the oil to displace all the mobile oil. The negative capillary pressures are likely to be necessary for some pore-filling displacements and for pores and throats where the effective advancing contact angle over a rough surface is greater than 90°.

There are two other complexities associated with these capillary pressures. Firstly, in strongly oil-wet media, water may be trapped as the non-wetting phase during secondary oil injection, and this residual water saturation may be higher than the irreducible water saturation reached after primary drainage, when water was the wetting phase. Secondly, scanning curves originating from a primary drainage displacement can cross each other and lie outside the bounding curves. The explanation lies in a careful consideration of trapping, layers and wettability alteration, which are the topics of the next section.

5.2 Oil and Water Layers

Implicit in the capillary pressure curves shown in Figs. 5.1 and 5.4, waterflooding proceeds in two steps. Firstly, the water-wet regions of the pore space are displaced at a positive capillary pressure, in imbibition, as described in the previous chapter. The difference though is that not all the pores and throats may be available for filling until the capillary pressure is negative. Secondly, and subsequently, the more oil-wet portions are filled in order of increasing water pressure necessary for displacement.

5.2.1 Pinned Water Layers and Forced Snap-Off

During imbibition, while the displacement processes are the same as described in Chapter 4, there are some differences if oil-wet pores and throats are also present in the pore space. The first is that water remains pinned in the corners of the oil-wet elements over a wide range of capillary pressure. We can use Eq. (4.6) to find the capillary pressure at which the wetting layer can begin to move out of the corner: if we have $\beta + \theta_A > \pi/2$ this is a *negative* capillary pressure – generally a large negative value.

The contact between oil and water at the solid surface remains pinned, with a hinging contact angle, as illustrated in Fig. 5.5. This hinging angle is seen for a capillary pressure $P_c^{max} > P_c > P_c^{adv}$ with $\theta_A > \theta_H > \theta_R$, similar to Eq. (4.1) presented for imbibition, but now where we can have $P_c < 0$:

$$\theta_H = \cos^{-1} \left[\frac{P_c}{P_c^{max}} \cos(\beta + \theta_R) \right] - \beta, \tag{5.1}$$

using Eq. (4.4).

$\theta_H = \theta_A$ when we attain a capillary pressure,

$$P_c^{adv} = P_c^{max} \frac{\cos(\beta + \theta_A)}{\cos(\beta + \theta_R)}, \tag{5.2}$$

at which point the wetting layer will start to advance across the surface (Mason and Morrow, 1991): this will happen first in the sharpest corner (with the smallest half-angle, β). This is the limiting capillary pressure for $\theta_A \leq \pi - \beta$. For $\theta_A > \pi - \beta$ the displacement happens sooner when the smallest possible (negative) radius of curvature is reached:

$$P_c^{adv} = -\frac{P_c^{max}}{\cos(\beta + \theta_R)}. \tag{5.3}$$

In both cases, the radius of curvature $r = -\sigma/P_c^{adv}$, ignoring curvature out of the plane of the diagram in Fig. 5.5.

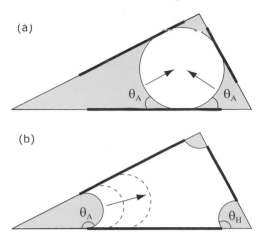

Figure 5.5 Water layer advance and snap-off in a throat when the solid contacted by oil (thick lines) is not necessarily water-wet. (a) As described in Chapter 4.1, if the surface is water-wet, the wetting layer can advance across the surface at a pressure given by with Eq. (4.6) at θ_A, the advancing contact angle. When the menisci meet we have snap-off and the throat fills with water. (b) For oil-wet surfaces the water layer is pinned at an angle θ_H until a large negative capillary pressure is reached. Once the corner layer can advance across the surface, as shown in one corner, the displacement is unstable and the throat fills with water. This is forced snap-off. From Valvatne and Blunt (2004).

Unlike a water-wet system, or at least one for which P_c^{adv} is positive, once the wetting layer starts to swell by moving across the solid surface, the local capillary pressure increases: the radius of curvature increases, representing, locally, a lower water pressure. Water flows from high to low pressure, rapidly filling the throat. This is a snap-off event, but occurs at a negative capillary pressure and does not require the intersection of arc menisci; instead, once an arc meniscus can move it will rapidly cause the filling of the throat.

We call this process forced snap-off (Valvatne and Blunt, 2004). However, as apparent in Eqs. (5.2) and (5.3), this only occurs at a very negative capillary pressure, near the end of forced water injection. Normally, by then, most of the throats are already filled by piston-like advance, which as we show below, will occur at a higher (less negative) capillary pressure. As a consequence, this displacement mechanism is uncommon.

The more significant observation is that for a very wide range of capillary pressure, the wetting layers remained pinned in the corners of the pore space: while the water may bulge out into the oil, the water volume remains low. This contrasts with water-wet systems, where the water layers may swell throughout spontaneous imbibition. The pinned water layers allow very little flow in comparison to the fatter ones in water-wet pores and throats. Exactly how much flow is allowed, and the threshold pressure at which the throat will fill during forced snap-off, is in turn controlled by the maximum pressure P_c^{max} reached during primary drainage. It transpires that this has a rather important impact on the flow behaviour, described in Chapters 6 and 7.

5.2.2 Forced Water Injection and Oil Layer Formation

As mentioned previously, the first and most favoured displacements during water injection will be those with the highest capillary pressure, and – if there are water-wet pores and throats – these will be snap-off, piston-like advance and pore filling as described in Chapter 4. Snap-off can occur spontaneously in any water-wet throat for which $\beta + \theta_A < \pi/2$. However, the opportunities for cooperative pore filling are more limited, because the most favourable process, I_1, is only possible if all but one of the surrounding throats is water filled. If any of these are oil-wet, this will not happen until the capillary pressure is negative. Furthermore, once a pore is filled, only the water-wet throats will be invaded, or more precisely those whose threshold capillary pressure is equal to or greater than that for pore fill-ing. The result is that for a mixed-wet system spontaneous imbibition is largely confined to snap-off with a rather limited amount of pore filling. The pores and throats fill largely in order of size, with the smallest pores filled first and the largest last.

When the capillary pressure becomes negative, the order of filling reverses, with displacement more favourable for the larger pores and throats, with progressively smaller elements filled as the capillary pressure becomes more negative. The larger pores are filled for low capillary pressures (the water-wet ones when P_c is positive and the oil-wet ones when P_c is negative). This results in a large change in satu-ration with a small change in capillary pressure, as opposed to the smaller change when smaller, lower volume, pores are filled near the end-points at S_{wc} and $1 - S_{or}$. This explains the shape of the waterflood capillary pressure curves, where dP_c/dS_w is small, close to $P_c = 0$, and larger (more negative) at the beginning and end of the displacement, respectively, as evident in Fig. 5.4.

Forced water injection is a drainage process, proceeding as described in Chap-ter 3, as a succession of piston-like displacements, limited by advance through the throats with filling of the adjacent pores occurring (for the same contact angle) at lower (that is, less negative) capillary pressures.

There are, however, two significant differences between forced water inva-sion and primary drainage. Firstly, the water can be connected through layers throughout the pore space and will occupy some of the smaller pores and throats after primary drainage. Moreover, in a mixed-wet system, spontaneous imbi-bition will have allowed some water-wet pores and throats to become water

filled. As a consequence, drainage can initiate from any adjacent water-filled
element. This is distinct from primary drainage, where only advance from the
invading cluster is allowed: here water need not be connected through the centres of pores and throats to the inlet; instead, connectivity is preserved through
water layers. The result is a percolation-like displacement with filling throughout the pore network controlled by size and contact angle, as opposed to invasion
percolation.

The second difference is the formation of oil layers in the pore space after the
centre of a pore or throat has been invaded by water. This is illustrated in Fig. 5.6:
during waterflooding the invasion of a pore or throat may result in complete filling of the element (as occurs during spontaneous displacement). However, if the
element is oil-wet, water is the non-wetting phase and may fill the centre, leaving
a wetting layer of oil attached to the oil-wet portions of the solid, sandwiched
between water in the corners and the water in the centre. The local capillary
pressure is negative: the arc menisci bulge out into the oil.

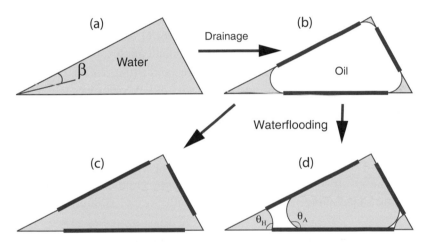

Figure 5.6 Pore and throat filling processes during waterflooding. (a) Initially, at
the start of primary drainage, the pore or throat is filled with water. The element
is shown in cross-section with angular corners. β is the half-angle of a corner. (b)
During primary drainage, step 1 in Fig. 5.1, oil has entered the element, confining water to a layer in the corners of the pore space. Where the oil is in direct
contact with the surface, shown by the bold line, the contact angle may change;
see Fig. 2.29. (c) During waterflooding, steps 3 and 4 in Fig. 5.1, water may fill
the element completely, as shown. (d) If the pore or throat is now oil-wet, filling
will only occur during forced water injection, step 4. Water may enter the centre
of the element as the non-wetting phase, leaving a layer of oil sandwiched in the
pore space between water in the corners and water in the centre. The capillary
pressure is negative and hence water bulges out into the oil. From Valvatne and
Blunt (2004).

It is possible to use geometric arguments to constrain when oil layers can form. For the arrangement (d) of fluid shown in Fig. 5.6 to be possible,

$$\theta_A - \beta > \frac{\pi}{2} \tag{5.4}$$

similar to the criterion for wetting layer formation, Eq. (3.3).

Note that the arc meniscus, AM, bounding the water in the corner contacts the solid at the advancing contact angle, θ_A. The AM bounding water in the corner is pinned with a hinging angle given by Eq. (5.1). As the capillary pressure becomes more negative, as displacement proceeds, the AM in the corner remains pinned and the contact angle changes continually, while the AM nearer the centre retains an angle θ_A and moves further towards the corner. At some point, the two AM on either side of the layer will touch: at this stage the layer is no longer stable and will collapse.

While Eq. (5.4) is a necessary condition for an oil layer to form, it is not sufficient. Some pore-scale models have used this geometric criterion to assess whether or not layers were present, using Eq. (5.1) and computing the pressures at which the two AM touched to determine when the layer collapsed – if you could draw a layer at the prevailing capillary pressure and contact angles, then it existed (Blunt, 1997b; Øren et al., 1998; Patzek, 2001; Valvatne and Blunt, 2004). However, layers are only present for a more restricted range of capillary pressures when their presence is thermodynamically, or energetically, favourable; that is, fluid configuration (d) in Fig. 5.6 has a lower free energy than arrangement (c). A complete discussion of this is somewhat algebraically involved, but conceptually simple, using energy balance, as in previous chapters, starting from Eq. (3.5) to find the fluid configurations for a given local capillary pressure, pore geometry and contact angles (Helland and Skjæveland, 2006a; van Dijke and Sorbie, 2006b; Ryazanov et al., 2009).

There are three types of displacement, or configuration change, that are possible. We start from a cross-section containing oil in the centre and water in the corners: arrangement (b) in Fig. 5.6. Then displacement with water leads to either complete filling of the cross-section with water, arrangement (c), or filling of the centre with water with an oil layer, (d). The third displacement is the collapse of the oil layer, or a transition from arrangement (d) to (c). These displacements can occur for every corner in the pore space. In each case, the threshold pressures for the displacement can be computed using conservation of energy (van Dijke and Sorbie, 2006b). In pore-scale models, as the system becomes more strongly oil-wet, we see more oil layer formation followed by collapse as the capillary pressure decreases further; for contact angles close to 90° oil layers can only form in the very sharpest corners (Ryazanov et al., 2009).

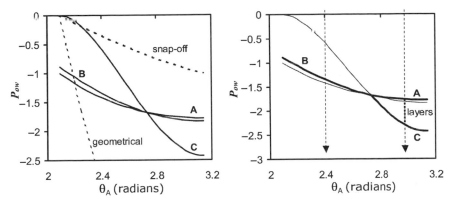

Figure 5.7 Normalized capillary pressures for filling a throat of constant cross-section as a function of advancing contact angle (shown in radians) for different displacement processes during forced water injection. On the left are shown as solid lines the threshold pressures for different displacements. The line A represents water invasion leading to the formation of oil layers. Line B represents complete filling of the cross-section with water with no layers. Line C represents layer collapse. The dotted lines show the geometric criterion for layer existence: this is always more permissive than the correct thermodynamic calculation (line C). For capillary pressures below the dotted line labelled snap-off, arc menisci may exist in the pore space: this is a necessary condition for oil layers to be present. The right-hand figure shows two example displacement sequences as the capillary pressure decreases. The first, at the lower contact angle, is when displacement proceeds without the formation of an oil layer when line B is crossed; at a larger angle layers are formed when the displacement crosses line A, which collapse later when the capillary pressure crosses line C. From van Dijke and Sorbie (2006b).

As an example, Fig. 5.7 shows the normalized threshold capillary pressures for different displacements as a function of contact angle (in radians). For a given geometry, contact angle and capillary pressure there is a unique configuration that is favoured energetically. The example shown is for a corner half angle of 30° and where the final capillary pressure, at the end of primary drainage, is ten times larger than the entry pressure, while the receding contact angle is zero. For comparison, the geometric criterion is also shown: in all cases, this is always more permissive than the correct thermodynamic calculation, and so is not useful for quantifying the range of capillary pressure over which layers are present in the pore space. The figure can be used to calculate possible displacement sequences. For instance, for the two cases shown, at the lower contact angle the first allowable displacement is when the declining capillary pressure crosses line B, representing water invasion without the formation of oil layers. When the contact angle is larger, under more oil-wet conditions, the first displacement occurs on crossing line A, indicating water invasion leaving behind oil layers. These layers will collapse when line C is crossed.

The principal impact of oil layers is to provide continuity of the oil phase: even in pores and throats invaded by water, the oil can still flow, albeit slowly, through these layers. This prevents trapping leading, as we discuss below, to low residual oil saturations (Salathiel, 1973). This is different from a water-wet system, where water completely fills pores and throat when it displaces oil, disconnecting the oil from adjacent elements.

5.2.3 *Recap of Displacement Processes*

We have now presented displacement processes and fluid configurations for a complete displacement sequence: primary drainage (Chapter 3), spontaneous imbibition (Chapter 4) and now forced water injection. If oil is re-injected, we again use the same displacement mechanisms and processes as before, using the appropriate contact angle. This conceptualization of displacement on a pore-by-pore basis has been used successfully by several researchers to calculate the sequence of invasion and capillary pressure for water-wet, oil-wet and mixed-wet systems (McDougall and Sorbie, 1995; Blunt, 1997b; Dixit et al., 1998b; Øren et al., 1998; Patzek, 2001; Al-Futaisi and Patzek, 2003b; Øren and Bakke, 2003; Valvatne and Blunt, 2004). To recap, the sequence is as follows, using the steps defined in Fig. 5.1.

(1) **Primary drainage.** The porous medium is initially completely water saturated, $S_w = 1$. A non-wetting phase invades the pore space (oil, which we will consider as the default fluid here, gas, or CO_2) with a contact angle θ_R. We assume that $\theta_R < 90°$: typically, if we have an initially water-wet surface, it is reasonable to estimate $\theta_R \approx 0$. Invasion is limited by ingress through throats with a local capillary pressure given by Eq. (3.13). Displacement proceeds by piston-like advance in an invasion percolation process, see Fig. 4.21. Direct contact of oil with solid surfaces can lead to a wettability alteration, giving different contact angles for subsequent water and oil injection: the contact angles for these processes will, typically, vary from the pore-to-field scales. At the end of the process, water is retained in the smaller regions of the pore space and as layers in corners and roughness.

(2) **Spontaneous imbibition**. If we have some water-wet regions of the pore space with $\theta_A < 90°$, then water injection begins with a combination of snap-off, Eq. (4.8), and cooperative pore filling events, Eqs. (4.19) to (4.23). After a pore is filled, it is generally more favourable to fill any adjoining throats by piston-like advance, Eq. (4.12). Oil can be trapped in the pore space surrounded by water.

(3) **Forced water injection.** Here the capillary pressure is negative. The transition from positive to negative capillary pressure is not necessarily associated

with the contact angle changing from less than 90° to more than 90°, but is instead determined by the threshold pressure for different displacement processes. In particular, a negative pressure may be necessary to fill a pore, even for $\theta_A < 90°$. Displacement proceeds as a series of piston-like throat and pore-filling events (see Eq. (4.12)), with the same equations as for spontaneous imbibition, but where $\cos\theta_A$ may be negative; see Eqs. (4.19) to (4.23) or, for a sufficiently large contact angle, Eq. (4.24). There can be forced snap-off, but this is unusual. When water invades the centre of a pore or throat, a layer of oil may remain. Layers maintain connectivity of the oil phase and prevent trapping. At a sufficiently large and negative capillary pressure these layers will collapse.

(4) **Spontaneous oil invasion.** Oil re-enters the porous medium. The contact angle for this process will be smaller than θ_A because of contact angle hysteresis, but will be larger than θ_R for primary drainage, if there has been some degree of wettability alteration. We call the contact angle for this secondary process θ_{R2}, where $\theta_A \geq \theta_{R2} \geq \theta_R$. If we have $\theta_{R2} > 90°$ then oil is the wetting phase and may spontaneously enter the pore space at a negative capillary pressure (that is, the oil pressure is lower than the water pressure and hence oil is not forced into the system). We see the same processes as for water imbibition, namely snap-off and cooperative pore filling, with displacement proceeding as the oil pressure rises, representing an increase in capillary pressure. Here snap-off occurs through the formation and swelling of oil layers in the pore space. The equations for threshold capillary pressure are the same as for step 2, but with θ_{R2} substituted for θ_A where events with the most negative capillary pressure are favoured.

(5) **Secondary drainage of oil.** This process is similar to forced water injection, with now oil as the non-wetting phase. As with the previous step, the displacement sequence can be quantified using the same mechanisms and equations for threshold capillary pressure as in step 3, but with θ_{R2} replacing θ_A. Water can be trapped in the centres of the pore space, surrounded by oil in layers and occupying the centres of adjacent elements. This central water is then isolated from other water in the centres of the pore space and from water layers in the corners. This trapping is often ignored – see, for instance, Fig 5.1 – but can give an irreducible water saturation that is larger than that seen after primary drainage.

5.3 Capillary Pressures and Wettability Indices

The wettability of a rock is determined by the distribution of contact angles. However, despite the recent development of methods to determine these angles directly from pore-scale imaging, see Figs. 2.32 and 2.33, this is not a routine measurement and, in any event, requires a detailed level of characterization, since we are primarily interested in effective angles for displacement, which depend on the flow direction, pore geometry and mineralogy on a pore-by-pore or even sub-pore basis.

It is more common to identify wettability through a macroscopic measurement of capillary pressure (Anderson, 1986). Since there is no definitive relationship between capillary pressure and contact angle, there is some ambiguity in how this is quantified. There are two widely used approaches: the USBM index (Donaldson et al., 1969) and the Amott indices (Amott, 1959). These methods define wettability based on displacement during waterflooding and oil re-injection cycles, Fig. 5.8. For an Amott test we start with a core sample at the waterflood residual oil saturation. We then measure the change in saturation $\Delta S_{os} = S_o^* - S_{or}$ during spontaneous invasion of oil, and the saturation change during forced injection, $\Delta S_{of} = 1 - S_{wi} - S_o^*$. This is then followed by a (tertiary) waterflood, with $\Delta S_{ws} = S_w^* - S_{wi}$ representing the saturation change during spontaneous imbibition of water and $\Delta S_{wf} = 1 - S_{or} - S_w^*$ the change during forced water injection: $\Delta S_{os} + \Delta S_{of} = \Delta S_{ws} + \Delta S_{wf} = \Delta S_{wt}$. The * indicates the saturation where the capillary pressure is zero.

The Amott water index is defined as

$$I_w = \frac{\Delta S_{ws}}{\Delta S_{wt}}, \tag{5.5}$$

with an oil index

$$I_o = \frac{\Delta S_{os}}{\Delta S_{wt}}. \tag{5.6}$$

$1 \geq I_w \geq 0$ and $1 \geq I_o \geq 0$; water-wet systems have $I_w \approx 1$ and $I_o = 0$, neutral wettability occurs for $I_w \approx I_o \approx 0$, while mixed-wet rocks have $I_w > 0$ and $I_o > 0$. An oil-wet medium has $I_0 > 0$ but $I_w = 0$.

Amott is one of the few authors who use the terminology for forced and spontaneous displacement correctly; he does not confuse imbibition with waterflooding and drainage with oil injection. It is a pity so few subsequent researchers have followed his example. And if that weren't enough, for the embarrassingly large number of people who seem unable to cope with two numbers, a single value to characterize wettability is often employed, called the Amott-Harvey index (see, for instance, Morrow (1990))

$$I_{AH} = I_w - I_o, \tag{5.7}$$

where $1 \geq I_{AH} \geq -1$. While this is convenient, it should never be quoted without reference to the individual Amott indices, as alone it results in an unnecessary loss

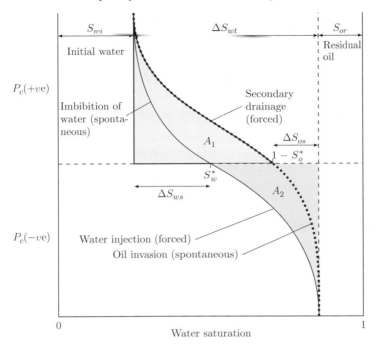

Figure 5.8 Capillary pressure curves, see Fig. 5.1, for waterflooding and oil re-injection with the definition of Amott and USBM wettability indices. The Amott test starts with waterflood residual oil. The oil index is the saturation change for spontaneous oil displacement, ΔS_{os} divided by the saturation change for both spontaneous and forced invasion, $\Delta S_{wt} = 1 - S_{or} - S_{wi}$. During a subsequent waterflood, the water index is the ratio of the saturation change during spontaneous water imbibition, ΔS_{ws} to the total saturation change. In the USBM test the area under the capillary pressure curve (assumed to be a positive value) is found during forced oil injection (A_1) and during forced water injection (A_2). The wettability index is $\log(A_1/A_2)$.

of information and cannot distinguish, for instance, between a neutrally wet system where $I_w \approx I_o \approx 0$ and a mixed-wet one where $I_w \approx I_o > 0$. It seems to have been introduced when it was assumed, axiomatically, that rocks had a single value of contact angle and hence it was impossible to imbibe both oil and water, so that one Amott index had to be zero.

The advantage of the Amott test is that it is possible to characterize the wettability without measuring capillary pressure: instead the saturation change during spontaneous displacement is found from surrounding a core with either water or oil, but without injection, and recording, for instance, its change in weight; then the saturation after subsequent forced injection is measured.

The USBM (US Bureau of Mines) index, in contrast, requires a measurement of capillary pressure during forced injection of both water and oil, making it of little value when used in isolation, since it too results in a loss of information if it is quoted without showing the full capillary pressure curves. It is determined from the ratio of the two areas indicated in Fig. 5.8 defined as:

$$A_1 = \int_{S_{wi}}^{1-S_o^*} P_{c2o} dS_w, \tag{5.8}$$

where P_{c2o} is the capillary pressure during secondary oil invasion, while

$$A_2 = -\int_{S_w^*}^{1-S_{or}} P_{cw} dS_w, \tag{5.9}$$

where P_{cw} is the waterflood capillary pressure. The USBM index is then defined as

$$I_{USBM} = \log(A_1/A_2): \tag{5.10}$$

$\infty \geq I_{USBM} \geq -\infty$. Water-wet rocks have $A_1 > A_2$ and a positive index, while oil-wet rocks have $A_2 > A_1$ and a negative index. However, this characterization makes no account of spontaneous displacement of either water or oil and so is unable to distinguish between neutrally wet and mixed-wet systems: $I_{USBM} \approx 0$ in both cases. This problem combined with the necessity to measure capillary pressures makes this index generally of limited value compared to the Amott test.

An example capillary pressure for an oil-wet rock is shown in Fig. 5.9 (Hammervold et al., 1998): there is no imbibition of water ($I_w = 0$) but a significant amount of spontaneous oil invasion ($I_o \approx 0.7$).

The effect of wettability on capillary pressure has been studied by several authors (Killins et al., 1953; Anderson, 1987a; Morrow, 1990); the difficulty with a quantitative discussion is that at present we cannot relate measured pore-scale contact angles to the macroscopic capillary pressure. However, this relationship can be explored using pore-scale network modelling (McDougall and Sorbie, 1995; Øren et al., 1998; Dixit et al., 2000; Øren and Bakke, 2003; Valvatne and Blunt, 2004; Zhao et al., 2010), which is touched upon in the following two sections.

5.3.1 Wettability Trends and Relationships between Indices

The Amott and USBM wettability indices probe different characteristics of the rock and hence in combination may reveal information about the pore-scale distribution of contact angle, which, as we have said, is difficult to find directly. Fig. 5.10 shows a compilation of literature measurements of the Amott-Harvey and USBM wettability indices (Donaldson et al., 1969; Crocker, 1986; Sharma and Wunderlich, 1987; Torsæter, 1988; Hirasaki et al., 1990; Yan et al., 1993; Longeron et al., 1994) compiled by Dixit et al. (1998a, 2000). Also shown are predictions using pore-scale network modelling that help to interpret the results. If the larger pores are oil-wet, the USBM index tends to indicate more water-wet conditions than the

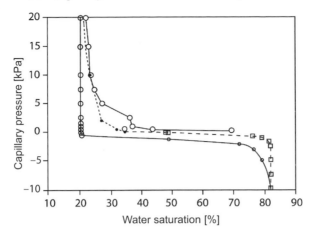

Figure 5.9 Example capillary pressure curves for an oil-wet sandstone. Primary drainage (top curve) is followed by water injection. There is no displacement at a positive capillary pressure: water has to be forced into the pore space to displace oil (lower curve). This followed by spontaneous imbibition of oil (dashed line, squares) and secondary drainage (dashed line, solid dots). $I_w = 0$ and $I_o \approx 0.7$ indicating oil-wet conditions. From Hammervold et al. (1998).

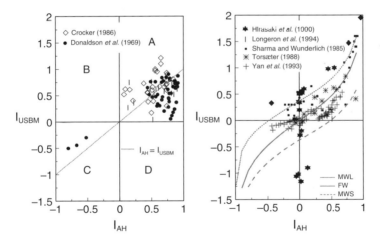

Figure 5.10 A compilation of literature data comparing the Amott-Harvey and USBM wettability indices. The left figures show experiments performed on different cores of the same rock type, while the right figure shows tests on the same core samples. The lines on the right are predicted relationships between the two indices from network modelling. MWL refers to a mixed-wet system where the larger pores become oil-wet; MWS is when the smaller pores become oil-wet, while FW refers to fractional wettability, where the contact angle is assigned at random, regardless of size. From Dixit et al. (1998a).

Amott-Harvey index, since forced water invasion of these pores occurs at a relatively low negative capillary pressure, making A_2 lower than were smaller pores
oil-wet. This observation could be used to determine if wettability alteration has
favoured larger or smaller pores. The experimental evidence though is inconclu-
sive, with results consistent with both larger and smaller pores being preferentially
oil-wet.

As discussed in Chapter 3.5, page 113, there may be a wettability trend in an oil
reservoir as a function of height above the free water level, as more oil contacts
the solid surface. As an example, Fig. 5.11 shows the measured Amott wettability
indices as a function of subsea depth for Prudhoe Bay, located off the northern
coast of Alaska (Jerauld and Rathmell, 1997); this is one of the largest sandstone reservoirs in the world. The reservoir is mainly mixed-wet but becomes more
water-wet deeper down, where more of the rock surface remains water-saturated
with a modest wettability alteration.

The largest oilfield in the world is a carbonate, Ghawar, in Saudi Arabia. Again,
the same trend in wettability is seen, with more water-wet conditions near the
oil/water contact and more oil-wet conditions, with less spontaneous imbibition
of water, as we move further up the oil column (Okasha et al., 2007). This trend
for the Arab-D reservoir (which comprises a major part of the field) is shown in

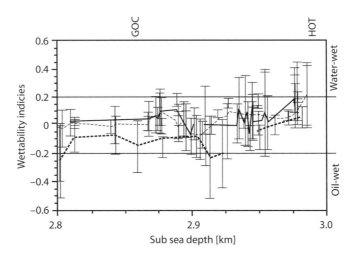

Figure 5.11 Wettability indices as a function of subsea depth for the Prudhoe Bay
oilfield. The points are the Amott-Harvey index, Eq. (5.7), while the apparent
error bars indicate the individual Amott indices: upper bar for water and lower
bar for oil (with a change of sign). GOC is the gas/oil contact where a gas cap is
seen, while HOT refers to a layer of heavy tar at the base of the reservoir. Most
of the samples are mixed-wet with spontaneous imbibition of both water and oil
where $I_w > 0$ and $I_o > 0$. From Jerauld and Rathmell (1997).

Table 5.1 *Definitions of wettability based on measurements of the Amott wettability index.*

Wettability state	Amott water index, I_w	Amott oil index, I_o
Completely water-wet	1	0
Water-wet	> 0	0
Neutrally wet	0	0
Oil-wet	0	> 0
Completely oil-wet	0	1
Mixed-wet	> 0	> 0

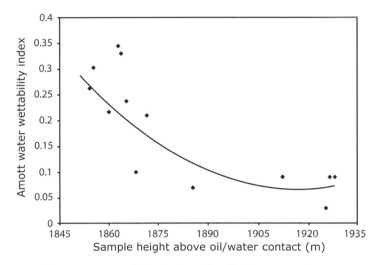

Figure 5.12 The water wettability index, Eq. (5.5), measured on core samples, plotted as a function of height above the oil/water contact for the Arab-D reservoir of the giant Ghawar field. Notice how high above the contact oil is found. There is a tendency to encounter less water-wet conditions with height: the points are measurements with a trend line through the data. From Okasha et al. (2007).

Fig. 5.12. We will revisit this behaviour for both Prudhoe and Ghawar in the context of fluid flow in Chapter 7.2.3.

The presentation of wettability indices now allows us to provide a macroscopic description of different wettability states, as presented in Table 5.1. This contrasts with the definitions based on a single value of contact angle in Table 2.3: all natural rocks contain a variation in contact angle which is, in general, unknown: it is instead more useful to define wettability based on core-scale measurements. A water-wet rock imbibes only water, an oil-wet rock imbibes only oil, a mixed-wet rock imbibes both oil and water, while one that is neutrally wet imbibes neither phase.

Table 5.2 **Displacement statistics for waterflooding in Mount Gambier**
limestone. *The number of events for different processes is shown for various*
fractions of oil-wet elements. The network properties are shown in Table 2.1,
while Table 4.3 shows the displacement statistics for water-wet cases.

Fraction of oil-wet elements	0	0.25	0.5	0.75	1
Uninvaded pores	31,978	35,102	31,182	29,059	28,668
Pore snap-off	1,688	1,618	1,490	1,216	0
I_1 pore filling	16,182	15,182	17,721	18,432	15,224
I_2 pore filling	11,022	9,547	9,033	8,293	8,044
I_3 pore filling	4,284	3,623	3,640	3,732	5,063
I_4+ pore filling	1,125	1,207	3,213	5,547	9,280
Uninvaded throats	24,842	29,991	25,304	24,385	29,257
Throat snap-off	25,454	24,758	21,956	16,622	5
Piston-like throat filling	44,382	39,929	53,671	65,416	58,154
Remaining oil saturation	0.40	0.29	0.14	0.08	0.06
Amott water index	1	0.80	0.44	0.20	0
Amott oil index	0	0.16	0.55	0.80	1
USBM index	∞	0.95	0.059	-0.50	-5.4

5.3.2 Displacement Statistics in Mixed-Wet Systems

We now present the displacement statistics for different pore and throat filling events for mixed-wet and oil-wet networks during waterflooding. This complements the discussion in Chapter 4.2.4 for water-wet systems and is used to illustrate the competition between different pore-scale processes in a network. We use the same models as before, but now consider cases where a fraction, f, of oil-filled pores and throats becomes oil-wet after primary drainage. The initial water saturation is only 2–3%. The oil-wet regions have a contact angle range 120–150°, while the water-wet regions have a range 30–60°. The oil-wet patches are clustered with a correlation length of seven pores. This better represents the connectivity of the phases seen experimentally, which we discuss further in Chapter 7.2 (Valvatne and Blunt, 2004). Waterflooding proceeds until a minimum capillary pressure of -4×10^5 Pa is reached. We also record the wettability indices and remaining oil saturations. Algorithmically, the simulation proceeds as for imbibition (see page 210), with the successive filling of a sorted list of capillary pressure. Trapping is also accounted for with a clustering algorithm that accounts for the presence or absence of oil layers.

Table 5.2 shows the number of displacement events in the Mount Gambier network. Compared to the statistics shown in Table 4.3 there are fewer snap-off events and less trapping as the oil-wet fraction increases. There is snap-off in pores where the surrounding throats have less favourable capillary entry pressures (they are oil-wet or have larger contact angles). Oil layers can form in the oil-wet pores and

Table 5.3 *Displacement statistics for waterflooding in Estaillades limestone.*
The number of events for different processes is shown for various fractions of
oil-wet elements. The network properties are shown in Table 2.1, while Table 4.4
shows the displacement statistics for water-wet cases.

Fraction of oil-wet elements	0	0.25	0.5	0.75	1
Uninvaded pores	63,280	65,113	67,144	62,108	57,495
Pore snap-off	606	725	761	639	0
I_1 pore filling	7,526	6,823	6,146	8,160	7,883
I_2 pore filling	6,736	6,028	5,231	5,622	5,567
I_3 pore filling	3,396	3,021	2,558	2,939	4,039
I_4+ pore filling	1,528	1,362	1,232	3,604	8,088
Uninvaded throats	77,523	80,789	84,453	79,369	79,629
Throat snap-off	14,269	14,185	13,966	11,732	7
Piston-like throat filling	29,075	25,893	22,448	29,766	41,231
Remaining oil saturation	0.57	0.56	0.53	0.35	0.05
Amott water index	1	0.91	0.80	0.28	0
Amott oil index	0	0.09	0.20	0.72	1
USBM index	∞	1.5	1.0	-0.22	-5.3

throats. When the oil-wet regions span the network, residual saturations below 10% are reached. When $f = 1$, we see the least trapping: there do remain many unfilled elements, but these are the smallest pores and throats. Since there is a spatial correlation in the assignment of contact angle, water-wet and oil-wet patches easily connect across the network. This allows the water-wet regions to be filled by spontaneous imbibition and the oil-wet regions by oil imbibition. The Amott wettability indices reflect, approximately, the fractions of water-wet and oil-wet elements: $I_w \approx 1 - f$ and $I_o \approx f$.

During forced water injection, the larger pores are the first to fill: these pores may have only a single adjoining water-saturated throat. There is a large increase in water saturation (the big pores have a large volume) but with a negligible boost to connectivity and conductance until a cluster of water-filled pores and throats spans the system. This is a percolation-like advance. The exceptions are when the rock is completely or mainly oil-wet: there are few water-filled elements initially to nucleate filling, and advance is therefore limited to invasion percolation through a path of larger pores and throats from the inlet. This allows a more rapid connection of the water and better flow conductance, which has a major impact on flow properties, discussed in Chapter 7.2.4.

For Estaillades, Table 5.3, only when $f > 0.5$ do the oil-wet elements span the system and connected oil layers allow flow down to low saturations. This is seen in the Amott oil index that only registers a significant amount of displacement by spontaneous oil invasion for $f = 0.75$. For the completely oil-wet case, the

Table 5.4 **Displacement statistics for waterflooding in Ketton limestone.** *The number of events for different processes is shown for various fractions of oil-wet elements. The network properties are shown in Table 2.1, while Table 4.5 shows the displacement statistics for water-wet cases.*

Fraction of oil-wet elements	0	0.25	0.5	0.75	1
Uninvaded pores	538	599	441	376	309
Pore snap-off	95	86	74	51	0
I_1 pore filling	635	608	633	605	508
I_2 pore filling	473	433	408	357	268
I_3 pore filling	156	145	180	205	310
I_4+ pore filling	19	45	180	322	521
Uninvaded throats	644	743	562	542	486
Throat snap-off	938	951	850	664	0
Piston-like throat filling	1,921	1,809	2,091	2,297	3,017
Remaining oil saturation	0.35	0.21	0.07	0.03	0.02
Amott water index	1	0.84	0.47	0.23	0
Amott oil index	0	0.15	0.51	0.74	1
USBM index	∞	0.86	0.028	−0.49	−5.9

Table 5.5 **Displacement statistics for waterflooding in Bentheimer sandstone.** *The number of events for different processes is shown for various fractions of oil-wet elements. The network properties are shown in Table 2.1, while Table 4.1 shows the displacement statistics for water-wet cases.*

Fraction of oil-wet elements	0	0.25	0.5	0.75	1
Uninvaded pores	7,525	10,101	6,307	5,300	4,657
Pore snap-off	1,184	1,084	940	696	0
I_1 pore filling	8,154	7,287	9,473	9,709	7,926
I_2 pore filling	7,616	6,564	6,489	5,755	4,899
I_3 pore filling	3,110	2,639	2,716	2,799	3,470
I_4+ pore filling	1,012	926	2,672	4,342	7,649
Uninvaded throats	8,444	12,322	7,707	7,119	9,554
Throat snap-off	16,957	16,627	14,947	11,691	0
Piston-like throat filling	29,340	25,792	32,087	35,931	45,187
Remaining oil saturation	0.46	0.40	0.07	0.03	0.01
Amott water index	1	0.97	0.48	0.23	0
Amott oil index	0	0.02	0.50	0.76	1
USBM index	∞	1.8	0.08	−0.44	−5.5

remaining saturation falls below 10%, but there are still many trapped elements: these, however, are the smallest pores and throats. While their number is large, their volume is small; see Fig. 3.12.

The statistics for Ketton, Bentheimer and Berea are similar (Tables 5.4, 5.5 and 5.6), indicative of well-connected rocks. As in Mount Gambier, we see little

Table 5.6 **Displacement statistics for waterflooding in Berea sandstone.** *The number of events for different processes is shown for various fractions of oil-wet elements. The network properties are shown in Table 2.1, while Table 4.2 shows the displacement statistics for water-wet cases.*

Fraction of oil-wet elements	0	0.25	0.5	0.75	1
Uninvaded pores	4,233	4,960	1,323	593	242
Pore snap-off	39	40	44	43	0
I_1 pore filling	1,646	1,468	3,142	3,059	783
I_2 pore filling	3,319	3,082	3,529	3,224	2,246
I_3 pore filling	2,682	2,409	3,112	3,587	5,839
I_{4+} pore filling	430	390	1,199	1,843	3,239
Uninvaded throats	9,239	9,512	3,234	1,659	1,335
Throat snap-off	7,447	8,072	8,395	6,899	0
Piston-like throat filling	9,460	8,562	14,517	17,588	24,811
Remaining oil saturation	0.47	0.46	0.08	0.02	0.01
Amott water index	1	1	0.46	0.23	0
Amott oil index	0	0	0.54	0.76	1
USBM index	∞	2.8	0.06	−0.44	−5.1

trapping once we have a spanning cluster of oil-wet elements and a drainage-type displacement dominated by piston-like throat filling, with significant I_{3+} filling of pores when completely oil-wet.

5.4 Trapping in Mixed-Wet and Oil-Wet Media

The amount of trapping in oil-wet regions of the pore space is controlled by the presence and stability of oil layers. While layers are present, these retain the connectivity of the oil phase, allowing displacement to low saturations; when they collapse, oil may be trapped.

Fig. 5.13 from Salathiel (1973) shows the remaining oil saturation in a mixed-wet sandstone sample from an oilfield in Texas as a function of the number of pore volumes of water injected through the rock (if the rock volume is V and the porosity is ϕ then one pore volume is ϕV). Note that the oil saturation can be driven down to values of less that 10%, consistent with the pore-network model results in the previous section, lower than would be observed in a water-wet sample (see Fig. 4.36), but only after the injection of an enormous amount of water.

Fig. 5.14 shows a compilation of literature data showing the residual (or remaining) oil saturation as a function of initial saturation for different samples whose wettability has been altered: details of the rocks studied and the ageing fluids are provided in Table 5.7. There is considerable scatter in the data, since we have different rocks with variable wetting states. It is also not clear in many cases if

Table 5.7 *Rocks and fluids used for the trapping data on altered-wettability rocks shown in Fig. 5.14.*

Reference	Rock	Ageing fluid
Salathiel (1973)	Sandstone	Crude oil
Buckley et al. (1996)	Clashach sandstone	Crude oil
Graue et al. (1999)	Chalk	Crude oil
Masalmeh and Oedai (2000)	Berea sandstone	Crude oil
Masalmeh and Oedai (2000)	Limestone	Crude oil
Zhou et al. (2000)	Berea sandstone	Crude oil
Masalmeh (2002)	Sandstone	Crude oil
Karabakal and Bagci (2004)	Limestone	Mineral oil
Skauge et al. (2006)	Limestone	Crude oil
Johannesen and Graue (2007)	Chalk	Crude oil
Fernø et al. (2010)	Chalk	Crude oil
Humphry et al. (2013)	Berea sandstone	Crude oil
Tanino and Blunt (2013)	Indiana limestone	Organic acid
Nono et al. (2014)	Richemont limestone	Crude oil
Nono et al. (2014)	Estaillades limestone	Crude oil

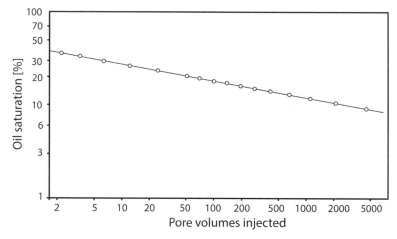

Figure 5.13 The oil saturation remaining in an oil-wet sandstone core as a function of the number of pore volumes of water injected. When a sufficiently large amount of water is injected, the remaining oil saturation drops below 10%. From Salathiel (1973).

the true residual is reached or there is a remaining saturation after some finite amount of water injection. However, overall, we tend to see less trapping, lower residual saturations, than for water-wet systems, with no wettability alteration – see Fig. 4.35. The saturation values are consistent with the network model results shown previously, although only if many of the samples are mixed-wet rather than

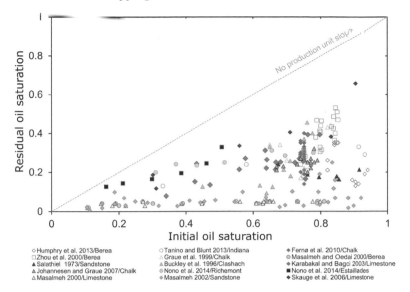

Figure 5.14 A compilation of trapping data for rocks of altered wettability, showing the residual oil saturation as a function of initial saturation. Details of the different experimental studies are provided in Table 5.7. There is considerable scatter because of the different rock samples studied and variations in wettability. However, in general we see less trapping than in water-wet systems; compare this graph with Fig. 4.35. From Alyafei and Blunt (2016). (A black and white version of this figure will appear in some formats. For the colour version, please refer to the plate section.)

completely oil-wet. In CO_2 systems as well, much less trapping is observed in CO_2-wet bead packs compared to water-wet ones (Chaudhary et al., 2013).

The final collection of data, Fig. 5.15, is from Tanino and Blunt (2013), which shows the remaining oil saturation as a function of initial saturation for systems aged in organic acid: it is assumed that most surfaces contacted by oil after primary drainage become oil-wet, while regions of the pore space still containing water remain water-wet. There are two features to note. The first is that the oil saturation continues to decrease, albeit slowly, as more and more water is injected. Secondly, there is a non-monotonic trend in residual (or remaining) oil saturation with initial saturation: the amount of trapping first increases, then decreases and then increases again.

5.4.1 Layer Connectivity as a Function of Initial Water Saturation

The pore-scale explanation for the behaviour presented in Figs. 5.13 and 5.14 is the existence of oil layers in the pore space, as illustrated in Fig. 5.6, which maintain connectivity of the oil during water injection. This situation is analogous to the irreducible water saturation encountered in primary drainage, since in both cases we

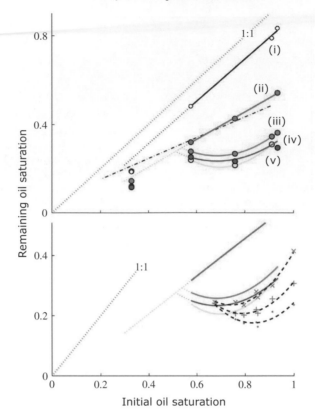

Figure 5.15 The remaining oil saturation as a function of the initial saturation under mixed-wet conditions. The upper graph shows the remaining oil saturation after: (i) spontaneous imbibition of water; (ii) the injection of a further 0.7 pore volumes of water during forced injection; (iii) 23–25 pore volumes of water injected; (iv) 99–101 pore volumes injected; and (v) 193–217 pore volumes injected. The solid lines are fits to the data. The dashed line depicts the residual saturation under water-wet conditions. The lower graph shows similar data recompiled from the results of Salathiel (1973). The points are measured at 1, 5 and 20 pore volumes of water injected. The dashed lines are fits to the data, while the solid lines are the fits from the upper graph for comparison purposes. From Tanino and Blunt (2013).

consider the displacement of a wetting phase: this only represents a truly trapped value if an infinite capillary pressure is applied and an infinite amount of time have been allowed for flow, neither of which pertain. In field situations, only a finite amount of water can be feasibly injected and so the oil saturation after waterflood represents a remaining value which may be significantly different from a true residual. This concept will be discussed in more detail in Chapter 9, but it is important to emphasize here that the residual saturation in a mixed- or oil-wet system is not a good nor even a particularly relevant indicator of recovery, since it will not be reached in most field-scale displacements.

While the explanation above illustrates why we see lower residual saturations in general than for water-wet systems and why this saturation is only reached after the injection of large amounts of water, it does not immediately provide an explanation for the non-monotonic trapping curves shown in Fig. 5.15. This is explained from the stability of wetting layers for different initial water saturations (Spiteri et al., 2008), as shown in Fig. 5.16. We assume that after primary drainage, surfaces of the rock in direct contact with the solid become oil-wet, while those regions of the pore space where water is retained in the corners remain water-wet. If the final capillary pressure in primary drainage is large, this forces water far into the corners, the water saturation is small and the initial oil saturation is close to 1. When water is now injected to displace oil, it occupies the centre of the pore space as the non-wetting phase, leaving a layer of oil sandwiched between water in the corner and water in the centre, as described previously. For a high initial oil saturation, a thick oil layer is established that will remain stable over a wide range of capillary pressure. This allows the oil to maintain connectivity and be displaced to a low residual saturation. In contrast, if the final capillary pressure after primary drainage is lower, giving a higher initial water saturation and a lower initial oil saturation, the

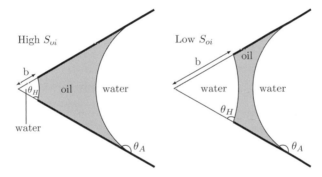

Figure 5.16 An illustration of oil layers with different capillary pressures at the end of primary drainage. We consider a system where surfaces directly contacted by oil at the end of primary drainage become oil-wet (shown bold). During subsequent waterflooding, water enters the centre of the pore space, leaving an oil layer sandwiched between water in the centre and oil in the corner, as shown in Fig. 5.6. On the left an oil layer in the corner of the pore space is shown where the maximum capillary pressure is large, forcing oil far into the corner; *b* is small, giving a low initial water saturation or a high initial oil saturation. The oil layer is relatively thick and remains stable over a wide range of capillary pressure during forced water injection. On the right is the situation if the maximum capillary pressure is lower, corresponding to a lower initial oil saturation. The oil enters less of the corner, leaving more of it water-wet and leading to a thinner layer, as shown. This layer will collapse early during forced water injection, allowing oil to become trapped and leading to a higher residual saturation than for the situation on the left. This leads to the seemingly counter-intuitive observation that a higher initial oil saturation leads to a *lower* residual oil saturation.

oil layers are thinner. This is because water in the corners now occupies a greater volume and is closer to the water in the centre of the pore space. Oil layers will collapse sooner during forced displacement, allowing oil to be stranded in the pore space and leading to a higher residual saturation.

This non-monotonic variation of residual saturation as a function of initial saturation was first observed in pore-scale modelling studies (Blunt, 1997b, 1998; Spiteri et al., 2008) and is captured in the empirical trapping model, Eq. (4.40). It is seemingly counter-intuitive, since more oil initially leads to less oil trapped: the reason is simply that a higher initial oil saturation forces oil to contact more of the solid surface, which becomes oil-wet and leads to more stable oil layers.

When the highest initial oil saturations are reached, the residual (or at least the remaining) oil saturation increases again. This is simply the result of a competition between layer stability (lower residual for higher initial saturation) and the obvious result that with more oil initially present, there is more oil to be trapped. The oil is forced into smaller pores and throats; for oil-wet elements, this is where trapping is more likely to occur at the end of waterflooding. Similar behaviour has been predicted using pore-scale network modelling as a function of the fraction of oil-wet pores (Blunt, 1997b).

One consequence of this non-monotonic trapping behaviour is that scanning capillary pressure curves, see Fig. 5.2, can cross: for a given saturation and capillary pressure there is not a unique saturation pathway dependent simply on whether the water saturation is increasing or decreasing, but different paths dependent on the whole history of displacement. A saturation path for waterflooding from a high initial oil saturation will cross one from a lower initial saturation, as shown schematically in Fig. 5.17. Although this behaviour is not captured in empirical models of capillary pressure hysteresis (Killough, 1976; Carlson, 1981), it is a physical manifestation of the different pore-scale configurations of fluid and contact angle as a result of wettability changes. As we will discuss later in Chapter 9, this has significant implications for waterflood recovery in transition zone reservoirs where there is a variation in initial oil saturation above the free water level (Jackson et al., 2003); see Fig. 3.18.

5.4.2 Pore-Scale Observation of Trapping in Mixed-Wet Systems

Fluid distributions in porous media of altered wettability have been studied using pore-scale imaging (Prodanović et al., 2006; Al-Raoush, 2009; Iglauer et al., 2012; Murison et al., 2014; Herring et al., 2016; Rahman et al., 2016; Singh et al., 2016). This work confirms that the morphology of trapped phases is different in water-wet and oil-wet (or mixed-wet) systems. As discussed in Chapter 4.6, in a water-wet rock, the oil (the non-wetting phase) is trapped in the centres of the larger pore

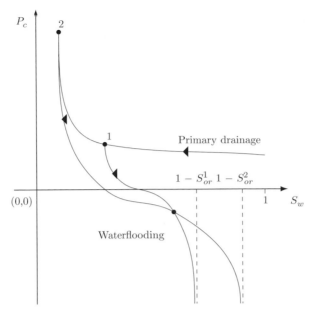

Figure 5.17 A representation of how a non-monotonic dependence of residual saturation on initial saturation must lead to capillary pressure scanning curves which cross at the point shown. Primary drainage to points 1 and 2 is followed by waterflooding. With the lower initial water saturation, 2, we have stabler oil layers during forced injection and less trapping: $S_{or}^1 > S_{or}^2$, Fig. 5.16. While this phenomenon is not captured in traditional empirical models, it is a manifestation of different wettability states at the end of primary drainage. This behaviour is expected for transition zone oil reservoirs where there the initial saturation varies with height above the free water level, Fig. 3.18.

spaces with ganglia that vary in size from a single pore to the extent of the experimental system, with a power-law distribution of size consistent with percolation theory, Eq. (4.34). In contrast, in an oil-wet system, oil tends to reside in layers clinging to oil-wet surfaces of the pore space. These layers can collapse, but they do not all do so at the same capillary pressure, since the threshold depends on the pore geometry and initial arrangement of water in the corners. Hence, oil can be trapped in layers, or the smaller pores and throats, after some layers surrounding these regions have disconnected. The result will be a sheet-like arrangement of trapped oil and overall less trapping than in a water-wet rock. This is difficult to image directly, since the layers are, by definition, smaller than the pores themselves. However, a distinct difference in shape is observed in comparison to water-wet systems, with thin structures present exposing a large surface area to both the surface and the water phase.

As an example, Fig. 5.18 shows residual clusters of oil in an oil-wet sandstone compared to those in an equivalent water-wet system (Iglauer et al., 2012). These

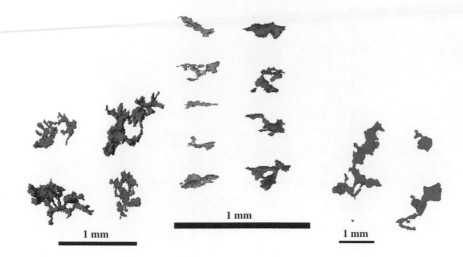

Figure 5.18 Illustration of trapped clusters in an oil-wet sandstone. Left: the four largest residual oil clusters (size 10,917–15,653 voxels). Middle: medium-sized residual oil clusters (clusters on the right contain 1,000–1,100 voxels while those on the left have 500–600 voxels). Right: residual oil clusters in the same sandstone under water-wet conditions: the largest cluster has a volume of 25,193 voxels. The residual clusters in the oil-wet rock have a flatter sheet-like structure, in layers and filling small pores, while the clusters in water-wet rock occupy the centres of the wider pore spaces. The voxel size is 9 μm. From Iglauer et al. (2012).

clusters display a wide range of size, but the biggest ones are not as large as those seen in a water-wet rocks, while they tend to have a flatter and more ramified structure. Overall the total trapped saturation, 19% is also lower than observed for a water-wet system, 35%, consistent with data on larger core samples; see Fig. 5.14.

Oil layers have been observed directly in the pore space of mixed-wet rocks, sandwiched between initial water in the corners and water in the centre of the pore-space which has invaded as the non-wetting phase (Singh et al., 2016). While the three-dimensional morphology of these layers is rather more complex than the described for illustrative purposes earlier (Fig. 5.16), the conceptual picture is similar: they lie between water-filled portions of the pore space and provide connectivity to the oil phase.

We have now finished the first half of the book, based on the concept of capillary equilibrium. However, any displacement requires flow, which introduces viscous forces and additional subtleties in its macroscopic description and pore-scale interpretation. Momentum and mass balance will be invoked in the following chapters to quantify how multiple fluid phases move through the pore space.

6

Navier-Stokes Equations, Darcy's Law and Multiphase Flow

6.1 Navier-Stokes Equations and Conservation of Mass

So far we have described the configuration of fluids at rest in capillary equilibrium in a porous medium; while we have treated displacement from one equilibrium state to the next, we have not addressed how the fluids flow. In this chapter we will first present the flow of a single fluid phase, both at the pore scale and its macroscopic, averaged, counterpart, before quantifying the flow of multiple phases.

Flow is governed by the Navier-Stokes equations, given here for an incompressible Newtonian fluid of fixed viscosity μ (Batchelor, 1967):

$$\mu\nabla^2\mathbf{v} = \rho\left(\frac{\partial\mathbf{v}}{\partial t} + \mathbf{v}\cdot\nabla\mathbf{v}\right) + \nabla P - \rho\mathbf{g}, \qquad (6.1)$$

where $\mathbf{v}$ is the vector velocity field, P is the fluid pressure and $\mathbf{g}$ is the acceleration due to gravity. The Navier-Stokes equation, which, like the Young-Laplace equation, is named after the French and English duo who first developed the ideas, is the expression of Newton's second law of motion ($\mathbf{F} = m\mathbf{a}$) applied to a continuous fluid: the driving force for flow is the pressure gradient and gravity (the force per unit fluid volume is $-\nabla P + \rho\mathbf{g}$ in Eq. (6.1)), while the total acceleration $\mathbf{a} \equiv d\mathbf{v}/dt$ for a moving fluid is given by $\partial\mathbf{v}/\partial t + \mathbf{v}\cdot\nabla\mathbf{v}$. The term involving viscosity is a representation of the viscous force per unit volume, resisting flow. In solid mechanics we use a generalization of Hooke's law to find a linear relationship between stress and strain: for a fluid, by analogy, the principal physical insight is to assume that the stress is proportional to the *rate* of change of strain (the strain or displacement itself can be arbitrarily large for a fluid). Strictly, in the Navier-Stokes equation, the viscous term is the divergence of the stress tensor.

Claude Navier first derived the equations of motion for a fluid and introduced a slip condition at the solid surface (Navier, 1823); Siméon Poisson, another French scientist, introduced the concept of viscous fluids (Poisson, 1831); while George

Stokes, later to hold the Lucasian Professorship of Mathematics at Cambridge University for 54 years (a post also held by Isaac Newton, Paul Dirac, Charles Babbage and Stephen Hawking; the current professor is Michael Cates), re-derived Navier's equations in a form similar to what we use today, and introduced the no-slip condition that we discuss below (Stokes, 1845). Conventionally, we recognize the contributions of Navier and Stokes in the name of Eq. (6.1) (but not Poisson, and others; however, Poisson gets his own equation, $\nabla^2 \phi = \rho$).

Eq. (6.1) is a vectorial equation for pressure and three coordinates of velocity. For a complete solution we need another equation invoking conservation of mass. If we consider some arbitrary volume of fluid V bounded by a surface $\mathbf{S}$ then the rate at which mass crosses the surface is the integral of the normal component of the fluid velocity times the density: this must be equal to the rate of change of mass within the volume. In mathematical form:

$$\int \frac{\partial \rho}{\partial t} dV = \int \rho \mathbf{v} \cdot d\mathbf{S}. \tag{6.2}$$

Using Gauss' theorem, we convert the surface integral into one over the same volume V,

$$\int \frac{\partial \rho}{\partial t} dV = \int \nabla \cdot (\rho \mathbf{v}) dV. \tag{6.3}$$

Then, since this relation holds for any arbitrary volume of space, the integrands must be equal and we have

$$\frac{\partial \rho}{\partial t} = \nabla \cdot (\rho \mathbf{v}). \tag{6.4}$$

The pressure drop across a pore is typically of order a few Pa (see later in Chapter 6.4.2) compared to absolute values of 10s of MPa at reservoir conditions. In these circumstances, the variation in density at the pore scale is small compared to the density itself, even for gases. If we assume that the density ρ is constant then

$$\nabla \cdot \mathbf{v} = 0. \tag{6.5}$$

The velocity field is divergence free for an incompressible fluid. This provides an additional equation, allowing both the fluid pressure and velocity to be calculated.

We now treat fluid flow in a porous medium. For the purposes of this work, we will consider a rigid porous medium where both the normal and tangential components of the fluid velocity are zero at the solid surface. This is a simplification to the real situation; however, a fuller discussion lies outside the scope of this book (Neto et al., 2005). If we consider low-density gases or flow in very small channels, whose inscribed radius is comparable or smaller than the molecular mean free path, we need to consider slip effects. More fundamentally, if the fluid velocity is zero, then it should be impossible to have multiphase displacement since this inevitably

involves the movement of a fluid/fluid contact across the solid. This is accommodated by some degree of slip at the molecular scale close to the surface, which
we will ignore here. The other boundary condition is a specification of pressures
and/or flow rates over the surface of the system of interest.

The Navier-Stokes equation, Eq. (6.1), combined with conservation of mass,
Eq. (6.4) or (6.5), as appropriate, and surface energy conservation, normally encapsulated in the Young-Laplace equation, Eq. (1.6), complete a system of expressions
to describe fluid configurations and movement in a porous medium once we apply
the (albeit complex) boundary conditions: no flow and defined contact angles at
solid surfaces. However, as should be apparent by now, a careful treatment of this
problem uncovers myriad complexities.

6.1.1 Flow in a Pipe

Before presenting some of the subtleties of multiphase flow, we will illustrate the
use of the Navier-Stokes equation though the derivation of one analytical solution,
which will prove useful later. Consider a cylinder with a circular cross-section
of radius R. For simplicity, we assume that the cylinder is aligned horizontal,
so there is no effect of gravity on the flow. A pressure P_{in} is applied at the
inlet of the tube and a lower pressure P_{out} is maintained at the outlet. We will
treat steady-state flow, $d\mathbf{v}/dt = 0$ (which implies laminar flow with no turbulence) and so the solution must have a constant pressure gradient along the tube,
$\nabla P = -(P_{in} - P_{out})/L$, where L is the tube length. The pressure gradient is negative, and flow proceeds from high to low pressure. If we define the x direction
as the coordinate along the length of the tube, then from symmetry, the pressure
gradient and flow must be aligned along x only. However, the velocity v_x is a function of radial coordinate, where $r = 0$ is the centre of the cylinder and $r = R$
defines the circumference. The no-slip boundary condition imposes $v_x(r) = 0$ at
$r = R$.

For an incompressible fluid, conservation of volume, Eq. (6.5), requires
$dv_x/dx = 0$, so that v_x is a function of r only. The Navier-Stokes equation, (6.1),
simplifies to:

$$\mu \frac{1}{r} \frac{d}{dr} \left(r \frac{dv}{dr} \right) = \nabla P, \tag{6.6}$$

where the term $\mathbf{v} \cdot \nabla \mathbf{v} = 0$ and we have written $v \equiv v_x$. Integrating Eq. (6.6) twice
we obtain:

$$v = \frac{\nabla P}{4\mu} r^2 + a \ln r + b \tag{6.7}$$

for integration constants a and b. We must have a finite v at $r = 0$, hence $a = 0$, while forcing $v_x = 0$ for $r = R$ yields

$$v = -\frac{\nabla P}{4\mu}\left(R^2 - r^2\right), \qquad (6.8)$$

noting that v is always positive, since ∇P is negative. Flow goes from high to low pressure, and so is positive for a negative pressure gradient. We always see a positive flow for a negative pressure gradient, so it is important that this concept is understood clearly.

It is useful to consider the total flow through the cylinder, Q,

$$Q = \int_0^R 2\pi r v(r)dr = -\frac{\pi R^4}{8\mu}\nabla P. \qquad (6.9)$$

This is called Poiseuille flow after the French scientist who first published this expression and demonstrated it experimentally (Poiseuille, 1844); it is sometimes termed Hagen-Poiseuille flow to credit the German scientist Hagen who independently found the same relationship.

The average velocity is defined as $\bar{v} = Q/A$, where A is the cross-sectional area, πR^2 in this case,

$$\bar{v} = -\frac{R^2}{8\mu}\nabla P = -\frac{A}{8\pi \mu}\nabla P. \qquad (6.10)$$

The main feature to note is that the total flow scales as the fourth power of radius, or as the square of the cross-sectional area, Eq. (6.9).

If we had considered electrical current in a cylindrical wire, doubling the radius would have increased the area, and hence the current and conductance, by a factor of four; in contrast, a doubling of radius in Eq. (6.9) leads to a 16-fold increase in flow. The electrons making the current travel at a given speed, proportional to the imposed potential gradient. If we double the area, we allow twice the current; increasing the area by a factor of four increases the flux in the same proportion. But fluid flow is different. We impose a zero flow rate at the solid boundaries; the velocity increases away from the solid, as described in Eq. (6.8). If we increase the radius, we allow higher velocities to be reached in the tube. Thus, we have a larger area *and* a larger velocity, giving a double boost to the flow.

The easiest way to conceptualize fluid flow harks back to the transport analogy developed for networks in Chapter 2.2.1. The maximum traffic flow on a four-lane motorway is not four times larger than on a single-lane road: it is considerably higher. Why? Because on the motorway, the cars can travel faster. We see exactly the same for fluid flow: larger pores allow faster flow, making the flow rate through the porous medium extremely sensitive to size.

There is a range of up to four orders of magnitude in throat radius in a single rock sample, Fig. 3.12: if each of these throats experience the same pressure gradient, this equates to eight orders of magnitude in average flow speed, Eq. (6.10), and sixteen orders of magnitude variation in total flow, Eq. (6.9). This is indeed what we will encounter when simulations of flow through pore-space images are presented later: the real range of flow speed is even more stark though, since it is also influenced by the connectivity of the network of pores and throats. Furthermore, when we come to multiphase flow, with individual phases confined to sub-networks occupying only portions of the pore space and clinging near the solid walls, we will see even larger variations, with important consequences for the relative motion of oil and water, or CO_2 and water.

Before proceeding, we can follow the same approach as employed in Chapter 3.1.2 for capillary pressure, to generalize the analysis to cases where we have a throat of constant, but non-circular cross-section. For steady-state flow where y and z are the coordinates in the plane of the throat, the extension to Eq. (6.6) is

$$\mu \left(\frac{\partial^2 v}{\partial x^2} + \frac{\partial^2 v}{\partial y^2} \right) = \nabla P, \tag{6.11}$$

for which analytical expressions can be derived for a square and an equilateral triangle (Patzek, 2001). However, in general, a numerical solution of Eq. (6.11) is required; then the velocity is integrated over the cross-section to find the total flow.

We can define a dimensionless flow conductance, g as follows:

$$Q = -\frac{g A^2}{\mu} \nabla P, \tag{6.12}$$

$$\bar{v} = -\frac{g A}{\mu} \nabla P, \tag{6.13}$$

where A is the cross-sectional area. For scalene triangles, g is almost exactly proportional to the shape factor, the ratio of cross-sectional area to perimeter squared, Eq. (2.2); see Fig. 6.1 (Øren et al., 1998). The constant of proportionality is 3/5, the value known analytically for an equilateral triangle: this contrasts with the value 0.5623 for a square and 0.5 ($g = 1/8\pi$) for a circle.

The final result in this section is for a tube where the cross-sectional area varies slowly with distance. We assume that the shape remains constant (g is fixed) and that flow is aligned only along an axis through the centre of the tube, x. The flow rate Q must be constant from conservation of volume. Then from Eq. (6.12) we write:

$$\frac{\partial P}{\partial x} = -\frac{Q\mu}{g A^2(x)} \tag{6.14}$$

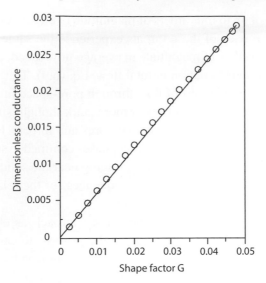

Figure 6.1 The dimensionless flow conductance, g, Eq. (6.12), for flow through a throat of constant scalene triangular cross-section plotted as a function of shape factor, G, the ratio of cross-sectional area to perimeter squared, Eq. (2.2). The results are found from a numerical solution of the steady-state Navier-Stokes equation, (6.11). From Øren et al. (1998).

from which the total pressure drop along a tube of length L is,

$$\Delta P = -\frac{Q\mu}{g} \int_0^L \frac{1}{A^2(x)} dx. \tag{6.15}$$

Then we write an expression similar to Eq. (6.12) but with some effective average area $\bar{A}$,

$$Q = \frac{g\bar{A}^2}{\mu} \frac{\Delta P}{L} \tag{6.16}$$

where we define

$$\bar{A}^2 = L \Big/ \int_0^L \frac{1}{A^2(x)} dx . \tag{6.17}$$

Using an electrical analogy, this is conductance in series, where the conductance of each element is proportional to the square of the cross-sectional area. This is only an approximation to the flow, but is a reasonable representation when the variation in inscribed radius occurs over a longer length-scale than the radius itself: $dr/dx \ll 1$ (Zimmerman et al., 1991).

Note that the total flow is controlled by the narrowest parts of the tube, where the integral in Eq. (6.17) is largest. Since $A \sim r^2$, where r is the inscribed radius of the tube, we see that the overall flow resistance is governed by the average of radius to the negative fourth power. To a reasonable approximation, it is often assumed that the minimum restriction defines the flow: this, from our network analysis, is the location of the throat. The throat area $A_t = \min(A(x))$ and we estimate:

$$Q \approx \frac{g A_t^2}{\mu} \frac{\Delta P}{L}. \tag{6.18}$$

6.1.2 The Washburn Equation

We now introduce two phases in a cylindrical tube to study the dynamics of imbibition. Consider a tube filled with the wetting phase at the inlet ($x = 0$) and non-wetting phase at the outlet ($x = L$): the pressure drop across the system is ΔP where, without loss of generality, we set $P(x = L) = 0$. There is a terminal meniscus, TM, across the tube with a pressure difference $P_c = P_{nw} - P_w$. The TM is at some location x at time t as shown in Fig. 6.2. Then we can write the following equations for flow, using Eq. (6.13):

$$\bar{v} = \frac{dx}{dt} = \frac{gA}{\mu_w}\left(\frac{\Delta P - P_w(x)}{x}\right) = \frac{gA}{\mu_{nw}}\left(\frac{P_w(x) + P_c}{L - x}\right), \tag{6.19}$$

from which we can find $P_w(x)$ that we substitute back into Eq. (6.19) to find (after some algebra),

$$\frac{dx}{dt} = gA\left[\frac{P_c + \Delta P}{x(\mu_w - \mu_{nw}) + L\mu_{nw}}\right], \tag{6.20}$$

which can be integrated, noting that $x(t = 0) = 0$ to find:

$$gA(P_c + \Delta P)t = \mu_{nw}Lx + \frac{(\mu_w - \mu_{nw})}{2}x^2. \tag{6.21}$$

This is the Washburn equation, describing multiphase flow in a single capillary tube. Strictly speaking, it is a variant on the original derivation by Washburn (1921); we have ignored slip at the solid walls, but allowed the non-wetting phase to have a finite viscosity. Eq. (6.21) is a quadratic equation for distance as a function of time. Rather than wade through more algebra to present an equation for $x(t)$, we will make four observations.

Firstly, in the limit where the non-wetting phase viscosity is negligible (representing, for instance, imbibition into a tube initially filled with air), the second term on the right-hand-side of Eq. (6.21) dominates and we find

$$x = \sqrt{Dt}, \tag{6.22}$$

where

$$D = \frac{2gA(P_c + \Delta P)}{\mu_w}. \tag{6.23}$$

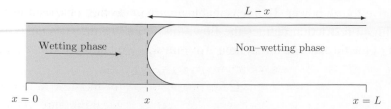

Figure 6.2 A schematic showing imbibition in a single capillary tube of length L. The terminal meniscus has reached a location x. The Washburn equation, (6.21), describes the dynamics of the process.

The distance imbibes scales as the square-root of time: all the pressure drop occurs through the wetting phase, with a fixed pressure difference. Hence the pressure gradient is inversely proportional to the distance moved; the wetting phase slows down as it advances.

Even in the general situation where $\mu_w \neq \mu_{nw}$, the time for the wetting phase to fill the system ($x = L$) scales as L^2: from Eq. (6.19), we find

$$t = L^2/D, \tag{6.24}$$

where now, instead of Eq. (6.23) we have

$$D = \frac{2gA(P_c + \Delta P)}{\mu_w + \mu_{nw}}. \tag{6.25}$$

The time to fill a pore scales with the square of the pore length, and is proportional to the average of the two fluid viscosities. We will revisit this scaling behaviour for spontaneous imbibition in a porous medium in Chapter 9.3.

The second observation comes from a consideration of the imbibition rate for different sized tubes, when we have spontaneous imbibition only: $\Delta P = 0$. From Eq. (6.21) the time-scale for displacement is inversely proportional to gAP_c. If the inscribed radius of the throat is r_t, then from Eq. (4.12), $P_c \sim 1/r_t$, while from simple geometry, $A \sim r_t^2$ and hence $t \sim 1/r_t$: larger throats imbibe faster. This appears to conflict with the whole thesis of Chapter 4, where we assumed that water-wet elements filled in order of size, with the smallest filled first. The resolution of this is to distinguish between flow in a single tube and flow in a porous medium. It is indeed correct that were a series of parallel tubes of different radii lined up (see Fig. 3.10) and water at zero capillary pressure introduced at the inlet, the tubes will fill in order of size, with the largest first – the opposite sequence from that observed for capillary-controlled displacement.

The reason for this is two-fold. Firstly, we described imbibition as a process where the capillary pressure is slowly decreased; this means that we start with a large and negative value of ΔP which then slowly increases to zero: flow (at any rate) is only allowed once $P_c < -\Delta P$, which restores the filling in order of size, with the smallest first. Secondly, though, we still fill the smallest elements first in a porous medium, even when the macroscopic boundary condition is a sudden change to a zero capillary pressure (this is, for instance, used during the Amott test described in the previous chapter, page 202). This is because the driving force for the flow, P_c, is largest in the smallest regions of the pore space, while the flow conductance which, for a single tube of constant cross-section is proportional to its area, is instead, for a rock, governed by all the pores and throats from the inlet to the particular element filling at that instant: there is no way that this can magically be only dependent on the local radius. To a good approximation, this conductance is much the same for every element being filled (with the possible exception of a few close by the inlet), meaning now that the smaller pores fill first, since they have larger capillary pressures. This faster filling of a smaller tube in preference to a larger one in a branched network has been observed directly in micro-fluidics experiments (Sadjadi et al., 2015).

The third observation is that the time-scale for pore filling is rapid: of an order of microseconds to milliseconds. If we consider a spontaneous event, such as a Haines jump, or filling after snap-off, the driving force for the fluid movement is of the same order as the capillary pressure to initiate the event: we see large fluctuations in local capillary pressure, as illustrated, for instance, in Fig. 4.5.

To obtain an order of magnitude estimate, consider the time taken to move a length $x = L$ through a cylindrical tube of radius r_t driven by a pressure $\sigma/r_t = P_c + \Delta P$. We further assume that the viscosities of the fluids are equal: $\mu_w = \mu_{nw} = \mu$. Then from Eq. (6.21) with $A = \pi r_t^2$ and $g = 1/8\pi$, we find:

$$t = \frac{8\mu L^2}{\sigma r_t}. \tag{6.26}$$

We can now estimate a time-scale for pore filling, based on our previous analysis of Bentheimer sandstone. We have determined an average distance between throats of 79 μm, see Chapter 3, page 101. A typical throat radius (at least the mode value from an analysis of capillary pressure, see Fig. 3.9) is around 24 μm. We will consider an oil/water displacement, where both fluids have a viscosity of 1 mPa.s. Then using $L = 79$ μm, $r_t = 24$ μm and a reservoir-condition $\sigma = 20$ mN/m we find $t \approx 100$ μs, from Eq. (6.26). While the analysis is somewhat approximate and, clearly, we can encounter a range of time scales dependent on the local pore geometry and size, as well as the disequilibrium of capillary pressure, any sensible calculation will reveal that individual pore and throat filling events are sub-second and generally sub-millisecond phenomena. This is significant: while, at the field scale, overall saturation changes in some small region of the pore space may occur over weeks to years, each individual pore-filling event occurs much, much more rapidly. This is implicit in the X-ray images presented so far: they cannot capture the dynamics of a filling event; even with fast tomography they can only record positions of local equilibrium between displacements. Acoustic measurements can detect displacement more rapidly and events do indeed occur within a millisecond (DiCarlo et al., 2003). In our example, the flow speed during filling is $L/t \approx$ 1.3 m/s or many 1,000s m/day, which is even faster than seen in micro-model experiments, see page 126.

Our final observation is to note that Eq. (6.21) is only strictly correct for tubes of circular cross-section, or for primary imbibition, since we have ignored the flow in wetting layers that may be present in the pore space of angular throats. This flow is not a major contributor to the behaviour in this example, but is significant when it provides the only connectivity for a phase, and is treated next.

6.1.3 Flow in Wetting Layers

We will now quantify flow rates in the presence of wetting layers. Firstly, it is common in pore-scale models to approximate the conductance of a phase in the centre of the pore space, even if it does not occupy the whole cross-section, by Eq. (6.12) with g given by the same expression as used in single-phase flow, but where A is now the area of the cross-section occupied by the phase. If water occupies the centre of the pore space and the corners, Eq. (6.12) only accounts for the water in the centres: wetting layers are considered separately as an additional flow element in parallel.

For layers themselves, the expressions for flow conductance are based on numerical solutions of the steady-state Navier-Stokes equation for a fixed and predetermined pore-space geometry and multiphase fluid configuration. For throats of constant cross-section, Eq. (6.11) can be used in each phase with appropriate boundary conditions at the interface between the fluids. We can write, in analogy to Eq. (6.12), the following expression for the total flow in a single layer (Ransohoff and Radke, 1988),

$$Q_l = -\frac{A_{AM}r^2}{\mu\beta_R}\nabla P, \tag{6.27}$$

where A_{AM} is the cross-sectional area of the layer (bounded by an arc meniscus, AM, see Eqs. (3.8) and (4.15)) while r is the radius of the curvature of the fluid interface (as we assume a throat of constant cross-section, $P_c = \sigma/r$). Eq. (6.27) defines a dimensionless layer resistance factor β_R. This can be applied to both wetting layers in corners and oil layers in oil-wet pores, albeit with different values of β_R. One complexity that arises in this description is the assignment of the boundary condition between the two fluids at their interface.

The viscous stress tensor $\mathcal{T}$ is, for an incompressible fluid of constant viscosity, using Einstein notation (Batchelor, 1967):

$$\mathcal{T}_{ij} = \mu \left(\frac{\partial v_i}{\partial x_j} + \frac{\partial v_j}{\partial x_i} \right). \tag{6.28}$$

The tangential component of the viscous stress needs to be continuous across the fluid interface (Batchelor, 1967; Li et al., 2005): continuity of the normal component leads to the Young-Laplace equation, (1.6). For steady-state flow, this analysis can be simplified through the consideration of flow normal and tangential to the interface. The normal velocity is zero, as the interface location remains fixed. Equating the non-zero tangential component of the stress in Eq. (6.28) across the interface we find:

$$\mu_1 \frac{\partial v_{t1}}{\partial n} = \mu_2 \frac{\partial v_{t2}}{\partial n}, \tag{6.29}$$

where n and t label the normal and tangential directions, respectively, while 1 and 2 label the two fluid phases. In addition, for continuity, we require $v_{t1} = v_{t2}$.

There are two limiting situations. The first is when the bulk fluid, 2, (the phase occupying the centre of the pore or throat) is much more viscous than phase 1 in the layer. In this case, the more viscous phase is essentially stationary which imposes $v_{t1} = v_{t2} \approx 0$. A no-slip, or no-flow, boundary condition is also seen if surface-active components of the oil reside at the fluid interface, creating a large interfacial viscosity.

The second limit is a so-called free boundary or no-stress condition, seen at the interface between a gas and a liquid with negligible surface viscosity. If phase 2 (the gas) has a negligible viscosity compared to the liquid, then $\mu_2 \approx 0$ in Eq. (6.29) and we find

$$\frac{\partial v_{t1}}{\partial n} = 0. \tag{6.30}$$

There is no viscous stress at the fluid interface.

In most applications a no-flow boundary condition is assumed at the fluid interfaces, making the presence of another fluid phase equivalent to the solid and limiting layer flow; however, this is not necessarily a good approximation for gas flows, or, for instance, CO_2/water displacements. Furthermore, if there is viscous coupling between the phases, the movement of one phase will affect the other, which will be discussed with regard to macroscopic properties later in Chapter 6.3.

Tables 6.1 and 6.2 provide calculated values of β_R in Eq. (6.27) as a function of corner half angle and contact angle (Ransohoff and Radke, 1988). Other work has found empirical expressions for layer conductance, including for oil-wet systems, based on solutions of Eq. (6.11), or approximate analytical formulae, whose accuracy has been confirmed experimentally (Zhou et al., 1997; Firincioglu et al.,

Table 6.1 *Calculated resistance factors β_R in Eq. (6.27) for flow in an angular throat with a contact angle of zero as a function of corner half angle. From Ransohoff and Radke (1988).*

Corner half angle (radians)	β_R, no stress boundary	β_R, no-flow boundary
$\pi/18$	9.981	13.60
$\pi/12$	12.95	19.91
$\pi/6$	31.07	65.02
$\pi/5$	46.67	108.1
$\pi/4$	93.93	248.8
$\pi/3$	443.0	1,411
$\pi/2.5$	3,185	11,390

Table 6.2 *Calculated resistance factors β_R in Eq. (6.27) as a function of contact angle. The first value is for throats of equilateral triangular cross-section and the second for throats of rectangular cross-section. From Ransohoff and Radke (1988).*

Contact angle (radians)	β_R, no stress boundary	β_R, no-flow boundary
0	31.07	65.02
	93.93	248.8
$\pi/36$	30.89	65.64
	93.99	253.2
$\pi/18$	31.94	68.24
	100.2	270.8
$\pi/9$	38.10	81.58
	139.0	374.9
$\pi/6$	54.09	115.9
	290.7	782.8
$\pi/5$	75.20	161.3
	698.2	1,881
$\pi/4$	165.2	355.5
	—	—

1999); these expressions have then been used to compute flow conductances in pore-scale network models (Øren et al., 1998; Patzek, 2001; Valvatne and Blunt, 2004). Experimental measurements of flow in capillary tubes of angular cross-section have suggested that a free or no-stress boundary condition is appropriate for an air/liquid interface, while flow in layers between oil and water is accurately captured using a no-flow boundary condition (Zhou et al., 1997; Firincioglu et al., 1999).

The same approach can be used to compute the conductance of oil layers in mixed-wet and oil-wet media; see Chapter 5.2. In this case, an expression similar to Eq. (6.27) is used, but with a resistance coefficient, β_R, which accounts for the presence of both curved boundaries between water in the pore centre and in the corners (Ransohoff and Radke, 1988). Alternatively, closed-form empirical expressions may be used that allow the convenient computation of conductance for any combination of contact angles, corner geometry and capillary pressure (Zhou et al., 1997).

We will now provide two calculations using these conductances to estimate the time necessary to fill a pore or throat in imbibition, to compare with flow through the centre of the throat, Eq. (6.26). We find the time needed to fill a pore completely with wetting phase, but where the fluid is provided by layer flow: it is the situation as shown in Fig. 6.2 but where we have a dead-end pore and wetting phase has to be provided through layers. The flow rate is given by Eq. (6.27), with a pore-scale pressure gradient of $\sigma/(r_t(L-x))$ driving the flow as before: we ignore any pressure gradient in the region of the pore space filled with wetting phase and any resistance to flow in the non-wetting phase. Then the TM moves with a rate

$$\frac{\partial x}{\partial t} = \frac{A_{AM}}{\beta_R A_t} \frac{r^2}{r_t(L-x)} \frac{\sigma}{\mu_w}. \tag{6.31}$$

As before, the throat length is L with area A_t. Integrating Eq. (6.31) with $x(0) = 0$ and finding the time for $x = L$ yields:

$$t = \frac{\beta_R A_t}{2A_{AM}} \frac{r_t L^2}{r^2} \frac{\mu_w}{\sigma}. \tag{6.32}$$

As an example, take a throat of equilateral triangular cross-section, $\mu_w = 10^{-3}$ Pa.s and an interfacial tension $\sigma = 20$ mN/m. Firstly, we consider a strongly water-wet case with $\theta_A = \theta_R = 0$, a throat with an inscribed radius of 24 μm (see page 227 and Fig. 3.9) and a length $L = 79$ μm (the average distance between throats, representing the pore length). Let us take a local capillary pressure in the layers of 20 kPa, or $r = 10$ μm. We use Eq. (3.8) with $\theta_R = 0$ to find $A_{AM} = r^2(3\sqrt{3} - \pi)$ (this is the area of the three wetting layers in the corners) and $A_t = 3\sqrt{3}r_t^2$. If we assume no flow at the fluid/fluid interface, $\beta_R = 65.02$, Table 6.1. Then Eq. (6.32) gives a time of around 25 ms. Note that this is almost two orders of magnitude slower than the time to fill by flow through the centre of the throat, but still takes much less than 1 second.

Our second example is where we have a pinned wetting layer. A typical maximum capillary pressure for an oil column of around 30 m and a density difference between oil and water of 300 kg/m^3 is of order 10^5 Pa, or a radius of curvature at the end of primary drainage of $r_{min} = 0.2$ μm for $\sigma = 20$ mN/m, see Chapter 3.5.

For our equilateral triangle with $\theta_R = 0$, b the length of wetting phase in the corner is around 0.34 μm, Eq. (3.6). We then use Eq. (4.16) with $r = 10$ μm as before to find $\cos(\beta + \theta_R) = \sqrt{3}/100$, which is close to zero: $\theta_H \approx \pi/3$ or 60°. While we can use Eq. (4.15) to find the area of the layer, to a good approximation the layer occupies an equilateral triangular area of side length b. We find, for three such layers, $A_{AM} \approx (9\sqrt{3}/4)r_{min}^2$. Then we use the expression for the conductance of a triangular throat of side length b, Eq. (6.12) with $g = 0.6G = \sqrt{3}/60$. The expression for the filling time is from Eq. (6.32) using Eq. (6.12) instead of Eq. (6.27):

$$t = \frac{A_t L^2 r_t}{2g A_{AM}^2} \frac{\mu_w}{\sigma} = \frac{160}{27} \frac{r_t^3 L^2}{r_{min}^4} \frac{\mu_w}{\sigma} \tag{6.33}$$

for this case. Note the three length scales: L is the pore length (how far we fill), r_t is the throat radius which controls both the volume filled and the driving capillary pressure, while r_{min} controls the flow rate in the pinned layer.

We obtain a time scale of around 16,000 s, or over four hours. Note that this is now many orders of magnitude larger than the equivalent time for filling a pore through its centre. Since conductance scales as the square of the area, or to the fourth power of radius, the ratio of the time to fill through a pinned layer, t_{AM}, to that through a pore, t_{pore}, scales as follows from Eq. (6.26) and Eq. (6.33):

$$\frac{t_{AM}}{t_{pore}} \sim \left(\frac{r}{r_{min}}\right)^4, \tag{6.34}$$

where r_{min} is the smallest radius of curvature during primary drainage, while r represents the typical pore or throat radius that governs the prevailing capillary pressure during waterflooding. In our example, $r = r_t \approx 100 r_{min}$ giving a 100-million-fold difference in the time to provide sufficient flow to fill a pore through a pinned layer, as opposed to through the pore centre. If the layer is not pinned, the layer swells and the difference in time scales, Eq. (6.32), although significant, is less dramatic.

This calculation helps explain the complexities of the dynamics of drainage and imbibition, and in particular the nature of Roof and distal snap-off; see page 125. When there are, locally, large gradients in capillary pressure, the rate at which fluid can move is controlled by the fluid configuration: flow through pore centres may be many orders of magnitude faster than flow in wetting layers. As a consequence, where fluid retraction and rearrangement occurs may not necessarily be in directly adjacent elements if the flow pathway to them has low connectivity, such as through layers, but in pores and throats that are more distant, but which can provide fluid more quickly.

This range of flow behaviour will also have major consequences for macroscopic multiphase flow, since the connectivity and conductance of layers will control how easily the wetting phase can move through the pore space. Furthermore, it presents a challenge to direct simulation of multiphase flow since thin layers need to be resolved in the computation. However, before developing these ideas further we return to single-phase flow and simplifications of the Navier-Stokes equations.

6.1.4 Reynolds Number and the Stokes Equation

We can make two simplifications to Eq. (6.1) appropriate for slow steady-state flows in porous media. The Reynolds number is

$$R_e = \frac{\rho v l}{\mu}, \tag{6.35}$$

which represents the ratio of inertial to viscous forces in the flow. v is a typical flow speed, while l is a characteristic length scale: for our applications this is a pore scale – in this context a representative pore length, or distance between throats – rather than a measure of the total size of the system.

In Eq. (6.1), the Reynolds number is a representative ratio of the non-linear term $\rho \mathbf{v} \cdot \nabla \mathbf{v}$ to the viscous dissipation $\mu \nabla^2 \mathbf{v}$. I will step through the derivation, since it is a helpful way to find the order-of magnitude of quantities in porous media flow. We estimate a derivative, $\nabla \mathbf{v}$, as v/l where l is the length scale over which we expect to see a significant change in flow speed: this length should be related to the pore size, since we already know from our analytical solutions that we see changes in speed from zero at the solid walls, to a maximum in the centre of the pore space. This is not a precise calculation based on a solution of the flow equations, but does facilitate a rapid assessment of the size of the different terms. Using this idea $\rho \mathbf{v} \cdot \nabla \mathbf{v}$ has a likely magnitude $\rho v^2 / l$ while $\mu \nabla^2 \mathbf{v}$ has a typical value $\mu v / l^2$: the Reynolds number is then the ratio of the first to the second of these terms.

There is no standard way to define a pore-scale l since we are dealing with a complex void structure with pore radii that may span several orders of magnitude (see, for instance, Fig. 3.12). Traditionally, for a granular system, $l = d$ the mean grain diameter; this is convenient to describe, for instance, bead and sand packs. However, for consolidated media and particularly for carbonates with a complex history of diagenesis, such a definition is problematic at best since grains cannot be identified in many cases: as an example, in Fig. 2.3 the granular structure is evident for Ketton limestone (ignoring the micro-porosity), the sand pack and Bentheimer sandstone, but it not so readily defined for the other five examples.

In our discussion of percolation theory in Chapter 3.4 (see page 95), we estimated the distance between throats, l, as the cube root of the number of throats per unit volume calculated from a network extracted from a pore-space image. This is a reasonable approach, but requires a network analysis of the pore space, which may not be available or necessary. Here, we will introduce an image-based definition of characteristic length, due to Mostaghimi et al. (2012), which starts from the observation that, for a uniform cubic packing of equally sized spheres, $d = \pi V / A$ where V is the volume of the system (pore and grain) while A is the surface area of the spheres. We then apply the same equation more generally, by computing the surface area between grain and void in a segmented image of a porous medium,

$$l = \pi \frac{V}{A}. \tag{6.36}$$

This definition is dependent on the voxel size of the image, since at a higher resolution smaller features of the pore-grain interface will be captured, increasing the computed surface area. However, it is a convenient method to define a length scale for the pores that have been imaged.

As an example, the use of Eq. (6.36) for Bentheimer based on the image shown in Fig. 2.3 gives $l = 140$ μm; for comparison, using the number of throats per unit volume yields $l = 79$ μm (page 101). Bearing in mind the simplifications associated with both methods, and that l itself is not precisely defined, either value can be used as a representative measure of a typical pore length scale.

To specify the Reynolds number, Eq. (6.35), we also need to find an average velocity in the pore space. As shown below, locally the velocity varies by many orders of magnitude. However, we can make an estimate based on average flow speeds. For instance, in oilfields, water is injected to displace oil: there is flow between wells typically a few 100s m apart that continues for 20 or 30 years before most of the mobile oil is displaced. This is equivalent to an average, or at least typical, flow speed of around 10 m/year, 3×10^{-6} m/s. Now, the flow speed near the wells, or in fractures, for instance, may be much larger, but the average in the middle of the field is of this order of magnitude. Natural groundwater flows are at most 1 m/day in coarse-grained shallow sediments, and may be as low as 1 m/year or less in deep aquifers used for CO_2 storage, for instance; it will be this flow that traps the CO_2 after injection through an imbibition process. These numbers give a range of flow rates from around $10^{-7} - 10^{-5}$ m/s.

We can now use this information to estimate the Reynolds number. Sticking with our Bentheimer example, with $l = 79$ μm, $v = 10^{-6}$ m/s, and the flow of water with $\rho = 1,000$ kg/m^3 and $\mu = 10^{-3}$ Pa.s, $Re \approx 10^{-4}$. If we consider faster flows in a sand with $v = 10^{-5}$ m/s and $l \approx 1$ mm, $Re = 10^{-2}$, while for the slowest flows in a rock with smaller pores than Bentheimer (say $l = 10$ μm), $Re = 10^{-6}$. In all cases we see that $Re \ll 1$. This means that inertial effects are small compared to the viscous dissipation: unlike other, familiar, applications of fluid dynamics, such as waves at sea, wind passing by buildings and airflow around an aeroplane, we have in porous media much slower speeds and smaller characteristic lengths, and as a consequence we do not see turbulent flow, which is only encountered for $Re \gg 1$.

For $Re \ll 1$ the inertial term can be ignored in Eq. (6.1) and,

$$\mu \nabla^2 \mathbf{v} = \rho \frac{\partial \mathbf{v}}{\partial t} + \nabla P - \rho \mathbf{g}, \qquad (6.37)$$

which is the Stokes equation. In the examples presented hitherto, we have assumed Stokes flow, since the inertial term was zero anyway because of the symmetry of the problem.

The second approximation is to assume steady-state flow, where the time-dependent term in Eq. (6.37) is neglected, meaning that $\mathbf{v}$ is a function of space only. This is a result of there being very different time scales associated with flow through a small region of a porous medium and the time over which there are significant saturation changes. Now, during a rapid displacement process, discussed previously, we see a time-evolving flow field over the millisecond movement during a fast jump from one position of capillary equilibrium to another. Flow mediated by thin layers is slower, but the layers themselves remain largely static, and is likely to be trumped, in most cases, by flow in pore centres to achieve local capillary equilibrium. Between displacement events, the fluid configuration remains constant. Hence, to a good approximation, at any moment, we can consider that we have steady flow through a fixed arrangement of fluid. Imagine, for instance, water injection into an oil reservoir. At the field scale, the water will displace oil in a series of pore-scale displacement events (not all the water, by the way, as some will simply flow through the reservoir and exit through a production well). The water will displace most of the mobile oil near the well, while the water saturation further from the well will be lower. However, we can define a length scale over which the saturation is, locally, approximately constant and the location of the fluid phases is unchanged over the time that it takes for flow across this length. For instance, the scale of most flow experiments in the laboratory is a few cm. With flow at a typical velocity of 10^{-6} m/s, fluid can traverse the system in a few hours: in a steady-state displacement experiment the saturation will, in contrast, be varied over a period of days or even months. Now consider a cm^3 piece of rock in the middle of an oilfield. It is not unreasonable to assume that over the time it takes for water to flow through this region, the pore-scale fluid configuration remains largely unchanged, since the displacement process continues over years. The water flowing through this region will displace oil somewhere else in the reservoir: only a tiny fraction of it will partake in displacements in this section of the rock; it has travelled to this location without displacement from the injection well. This concept is reasonable, but will be challenged below when we consider pore-scale dynamics further and mentioned again when we present solutions to the flow equations in Chapter 9.

We then arrive at the steady-state Stokes equation:

$$\mu \nabla^2 \mathbf{v} = \nabla P - \rho \mathbf{g}. \tag{6.38}$$

This can apply to both single-phase and multiphase flow: for the latter we also have to account for the pressure change across the fluid interfaces. We will now consider the averaged behaviour of flow in a porous medium governed by Eq. (6.38).

6.2 Darcy's Law and Permeability

In this section we will assume that we only have a single phase flowing in a porous medium. Steady-state Stokes flow, governed by Eq. (6.38), leads to a linear relationship between the pressure gradient and flow speed. To see this, imagine that we have a solution $\mathbf{v} = \mathbf{v_0}$ for the velocity. This solution is obtained with no flow at the solid walls and some imposed pressure gradient across the system. Now consider that we increase the force term $\nabla P - \rho \mathbf{g}$ by some arbitrary factor a. Therefore, the imposed pressure gradient across the system is increased by a factor a. Then, rather simply, we can see that $\mathbf{v} = a\mathbf{v_0}$ is the solution to Eq. (6.38): the velocity is proportional to the pressure gradient. The same argument can be applied to show that the velocity is inversely proportional to the viscosity. Hence we can write

$$\mathbf{v} = -\frac{f(\mathbf{x})}{\mu}(\nabla P - \rho \mathbf{g}), \tag{6.39}$$

where f is some positive function of position, $\mathbf{x}$, which depends only on the geometry of the porous medium and not on the fluid properties or the forces driving the flow. The minus sign in Eq. (6.39) indicates that fluid moves from high to low pressure.

On its own, Eq. (6.39) is not particularly useful, as the determination of f requires a solution to the original equation (6.38). We now return to the concept of volume averaging and the REV; see page 30. We will find the macroscopic flow integrated over some representative volume of space, containing many pores. We can define an average velocity $\mathbf{q}$ in one of two ways. Either as the average in some representative element of volume, V, see Eq. (2.1),

$$\mathbf{q} = \frac{1}{V}\int \mathbf{v} dV, \tag{6.40}$$

or as an average flux over an area $\mathbf{A} = \int d\mathbf{A}$:

$$\mathbf{q} \cdot \mathbf{n} = \frac{1}{A}\int \mathbf{v} \cdot d\mathbf{A}, \tag{6.41}$$

where **n** is a unit vector normal to the surface **A**. In both integrals we include both void space where the velocity may be finite, and the solid, where it is zero. q is the volume of fluid flowing per unit area per unit time.

We can also average the other terms in Eq. (6.39); however, the averaging of pressure P is not straightforward. In Eq. (6.1) the pressure is assigned for any point in the pore space and is undefined in the solid: we could consider again a volume average, as in Eq. (2.1), and correct for porosity. However, we need to define an average pressure gradient: for this, it is not appropriate to find the spatial average of the local gradient; instead we traditionally use a definition based on how pressure gradients are measured experimentally. Transducers record fluid pressure in a small region of space and are, effectively, point measurements, and the pressure gradient is calculated from the difference recorded over a macroscopic distance. We will not continue the discussion here, but indeed we may define a pressure gradient as the difference in two point values over a macroscopic length (Whitaker, 1986). For more details, a rigorous framework for volume averaging has been described by Gray and O'Neill (1976): a review of this methodology is provided in Gray et al. (2013).

Lastly, we need to average the function $f(\mathbf{x})$ in Eq. (6.39): this leads to a quantity that is dependent on the pore structure. However, to find a relationship for q, we need to find the average of the whole right-hand-side of Eq. (6.39) and then decompose it into terms related to the averaged pressure gradient and some rock property.

The description here simply suggests an approach to the construction of an equation relating macroscopic, or averaged, flow properties, but is not intended to present a rigorous derivation: this has been provided by several authors (Bear, 1972; Gray and O'Neill, 1976; Whitaker, 1986, 1999; Gray et al., 2013).

A proper averaging of Eq. (6.38) leads to Darcy's law, introduced as an empirical relationship to describe flow in sand filters for fountains by the French engineer Henry Darcy (Darcy, 1856), written here in its general form, first proposed by Nutting (1930) and Wyckoff et al. (1933):

$$\mathbf{q} = -\frac{K}{\mu}(\nabla P - \rho \mathbf{g}), \qquad (6.42)$$

where q is the Darcy velocity. Note that while q can be defined as a volume-average of the flow field, it is *not* a real velocity, but the volume of fluid flowing per unit area of the porous medium (and this area includes both solid and void). ∇P is no longer a microscopic pore-level quantity, but is defined over a distance that encompasses an REV in the porous medium.

K is the permeability and is an intrinsic property of the pore structure of the system. It has the units of length squared, or m^2 in SI units. Conventionally, though, permeability is reported in units of the Darcy, or D. Annoyingly, the definition of a Darcy is not based on a conversion to SI units, but was introduced by Wyckoff et al. (1933) in a paper describing some of the first measurements of permeability on small rock samples: a cubic block of rock 1 cm on all sides has a permeability of 1 D if it allows a flow of 1 cm^3/s for a potential difference of 1 atm. Since 1 atm $\approx$ 10^5 Pa, 1 D $\approx 10^{-12}$ m^2. A more precise conversion is 1 D = 9.869233×10^{-13} m^2. As we show later, permeability typically varies by orders of magnitude for different rock samples, and spatially over scales from cm to km: the quick conversion of 1 D = 10^{-12} m^2 is usually sufficiently accurate for engineering purposes.

K relates two vector quantities $\mathbf{q}$ and $(\nabla P - \rho\mathbf{g})$ and hence is a tensor: in component notation, Darcy's law (6.42) can be written

$$q_i = -\frac{K_{ij}}{\mu}\left(\frac{\partial P}{\partial x_j} - \rho g_j\right). \tag{6.43}$$

This treatment is somewhat brief, since Darcy's law and permeability are amply and ably treated in other classic texts, such as Bear (1972). However, we will provide some physical interpretation of the permeability, and the scaling of capillary pressure, before discussing direct computations of flow in pore space images, and then multiphase flow.

6.2.1 Permeability of a Bundle of Capillary Tubes

In Chapter 3.3 we introduced the idea of displacement in a bundle of tubes of different size; see page 89 and Fig. 3.10. If we have $f(r)dr$ tubes of size between r and $r + dr$ and again assume that they are circular in cross-section, Fig. 6.3, then we can use Eq. (6.9) to find, for a collection of tubes of length L

$$Q = -\frac{\pi}{8\mu}\nabla P \int_{r_{min}}^{r_{max}} r^4 f(r)dr. \tag{6.44}$$

The total flow is governed by the fourth moment of the radius distribution and hence the large tubes dominate. This is flow in parallel and contrasts with the radius distribution along a tube, considered previously, where the narrowest restriction controls the flow, Eqs. (6.16) and (6.17).

Imagine that the centres of the tubes are arranged on a square grid a distance d apart, Fig. 6.3. The cross-sectional area of (pore and solid) for each tube is d^2 while the void occupies an area πr^2. The number of tubes is

$$n = \int_{r_{min}}^{r_{max}} f(r)dr \tag{6.45}$$

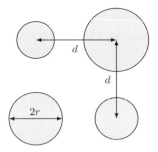

Figure 6.3 A cross-section through a bundle of tubes of different radius r: the tube centres are arranged on a square grid a distance d apart.

and the total area is $A = nd^2$. Then assuming that $d \geq r_{max}$ the porosity is

$$\phi = \frac{\int_{r_{min}}^{r_{max}} r^2 f(r) dr}{nd^2}, \qquad (6.46)$$

which is the ratio of the cross-sectional area of voids to the total area. Comparing Eq. (6.44) with Darcy's law, (6.42) where $Q = q/nd^2$, and using Eq. (6.46) we have:

$$K = \frac{\pi\phi}{8} \frac{\int_{r_{min}}^{r_{max}} r^4 f(r) dr}{\int_{r_{min}}^{r_{max}} r^2 f(r) dr}. \qquad (6.47)$$

The permeability is given by the ratio of the fourth to the second moments of the throat size distribution and its value is consequently dominated by the largest elements. If we have tubes of a single radius $f(r) = n\delta(r)$, then Eq. (6.47) becomes

$$K = \frac{\pi\phi r^2}{8}. \qquad (6.48)$$

This enables us to relate a typical pore or throat size r from two macroscopic quantities: ϕ and K,

$$r \sim \sqrt{\frac{K}{\phi}}. \qquad (6.49)$$

6.2.2 Typical Permeability Values

The permeability has the dimensions of an area, and this area, physically, is the cross-sectional area of a throat that controls the flow, as shown by Eq. (6.48). However, even if the permeability is proportional to the square of a typical throat radius, the constant of proportionality is much less than 1, to account for the tortuosity and limited connectivity of the pore space.

In the absence of direct images and a network analysis, it is easier to relate permeability, for granular systems, to the average grain diameter. For instance, Bourbie and Zinszner (1985) proposed the following empirical correlation for Fontainebleau sandstone for $\phi > 0.08$ (see Fig. 2.10):

$$K = 4.8 \times 10^{-3}\phi^{3.05}D^2, \tag{6.50}$$

where D is the grain diameter. Other correlations can be used, of which the most common is called Kozeny-Carman (Kozeny, 1927; Carman, 1956), which relates permeability to porosity and grain size with some measure of tortuosity: however, it is not really the tortuosity that limits the flow, but the need to pass through a few narrow throats. In its simplest form for a packing of equally sized spheres:

$$K = \frac{\phi^3}{180(1-\phi)^2}D^2. \tag{6.51}$$

For unconsolidated systems, with a porosity of around 35%, such as sand packs, $K \approx 6 \times 10^{-4}D^2$ (Bear, 1972), consistent with Eq. (6.51); this expression also holds for packings of grains of different size and shape, where D is the average diameter (Garcia et al., 2009).

For more consolidated systems, the relationship between permeability and a generally unknown grain size is less useful. In these cases, we need to relate pore size to permeability. However, since many media have pore or throat sizes spanning many orders of magnitude, there is no simple relationship, as it is controlled by the connectivity of the network and is sensitive to the conductance of the most restrictive element through which the flow has to pass, as shown in Eqs. (6.16) and (6.17). Again, a transport analogy is useful. The maximum flux of traffic into and out of a city, or the number of passengers that can be carried on a complex public transport network, is not readily calculated from a simplistic analysis of the number of cars, or passengers, that can be carried on a representative street or train line, but largely how the network connects and the maximum flux through those connections (or bottlenecks) that limit the flow: these bottlenecks may be all-to-evident to drivers and passengers, but are not obvious from an examination of a street map, otherwise, one would hope that they would be avoided!

However, it is still instructive to relate permeability to pore size. As mentioned above, this is the smallest throat that it is necessary to pass through to span the system: using a percolation analysis this would be the throat size associated with P_c^*; see Chapter 3.4. Alternatively, from a topological analysis, a representative size would be the radius at which the Euler number first became positive. In Fig. 2.25 and page 54, the connectivity of the pore space of the sands and Bentheimer sandstone studied was very good; however, the Euler characteristic is zero at a pore diameter of around 600 μm, or a radius of 300 μm, for the sands studied (Vogel et al., 2010); this is the radius that controls the permeability. The sand grains had an approximately uniform distribution between 0.6 and 2 mm, giving an average diameter of 1.3 mm. While the porosity and permeability of the sands was not reported, we can use the correlation mentioned above ($K = 6 \times 10^{-4}D^2$) to estimate a permeability of around 10^{-9} m^2 while the porosity for an unconsolidated system of similar-sized grains likely to be around 0.35.

For granular systems, we can find a relationship between average grain size, D, and pore size, l (Whitaker, 1999)

$$l = \frac{\phi D}{(1 - \phi)}. \tag{6.52}$$

If we assert that $l \propto r_t$ and use Eq. (6.51), we arrive at following empirical correlation,

$$K = a\phi r_t^2, \tag{6.53}$$

where a is some constant. This is also consistent with Eq. (6.49). $a \ll 1$ since we need to account for the tortuosity and restricted connectivity of the pore space. In the case of the sands mentioned above, we find $a = 3.2 \times 10^{-2}$ for $r_t = 300$ μm. A form similar to Eq. (6.53) was first proposed, without the ϕ term, by Hazen (1892) for sand packs.

In general, though, a topological (or network) analysis of the pore space is not available and for consolidated media we cannot easily find a grain size; however, the measurement of the primary drainage capillary pressure is routine; see Chapter 3.2. Here we take the mode throat radius inferred from the measurements, as this corresponds approximately to the minimum radius necessary to connect across the rock. Table 6.3 shows the measured porosity and permeability measured on the exemplar rocks presented in this book with the mode throat radius obtained from Figs. 3.11 and 3.12. In this case the constant a in Eq. (6.53) is smaller than for sand packs – in the range 0.9–2.3×10^{-2} – indicating the more consolidated nature of these systems. Note, however, that there is no precise way to determine permeability: our correlation simply gives some estimate to within a range of 2.

The general conclusion is that the permeability is proportional to some pore-scale length squared, or, more precisely, a throat radius squared. However, the constant of proportionality is generally much less than 1: if we use Eq. (6.53) using a throat radius as the characteristic length, $a \approx 10^{-2}$; if we use a mean grain diameter, the permeability is around 10^3–10^5 times smaller than D^2. For instance, in Table 6.3, where grain size is recorded, $K \approx 5 - 18 \times 10^{-5}\phi D^2$. If instead we used a pore length, l (or a distance between throats), the constant of proportionality is also small – around 1.5×10^{-3} – for instance, for Bentheimer using $l = 79$ μm (see page 101).

Table 6.3 *Permeabilities, pore sizes and grain sizes for different representative rock samples. D is the mean grain size (determined from a watershed segmentation of the solid from a pore-scale image). ϕ is the porosity measured using Helium adsorption. K is the measured permeability. r_t is a typical throat radius, defined here as the mode throat radius inferred from the primary drainage capillary pressure shown in Figs. 3.11 and 3.12. a is the constant in the empirical relationship $K = a\phi r_t^2$, Eq. (6.53). Data from El-Maghraby (2012); Tanino and Blunt (2012); Gharbi (2014); and Andrew (2014).*

Rock	D (μm)	ϕ	K (10^{-15} m^2)	r_t (μm)	a ($\times 10^{-2}$)
Doddington	250	0.192	1,040	18	1.7
Bentheimer	227	0.20	1,880	24	1.6
Berea	–	0.224	40	10	2.0
Ketton	500	0.234	2,810	23	2.3
Mt Gambier	–	0.552	6,680	32	1.2
Estaillades	–	0.295	149	7.5	0.90
Indiana	–	0.186	278	12	1.0
Guiting	–	0.287	3.85	0.9	1.7

The samples listed in Table 6.3 are quarry samples which generally have a higher porosity and permeability than many reservoir rocks containing oil and gas. Most reservoir rocks have permeabilities in the range 1–1,000 mD (10^{-15}–10^{-12} m^2), porosities around 0.1 to 0.3 and corresponding critical pore radii around 1 to 10 μm. In many carbonates, portions of the rock connected only by micro-porosity (which we will define as throat sizes less than 1 μm) will have lower permeabilities, often down to the μD range with pore sizes contributing significantly to flow as low as 0.1 μm.

Clays and shales may have pore radii that are 5 nm or even smaller with porosities for shales often only a few %. Using Eq. (6.53) with, for instance, $a = 0.01$ and $\phi = 0.04$, we predict permeabilities of around 10^{-20} m^2 or 10 nD: in general, we often encounter in the subsurface fine-grained or highly consolidated rocks with permeabilities as low as 10^{-24} m^2. At the other extreme, fractured rocks where the aperture size is a few mm, or unconsolidated media with grain sizes of around 1 mm, such as sands or well-sorted soils, will have permeabilities in the range of 10s to 1,000s D (10^{-11}–10^{-9} m^2).

Different rock types, from sands to shales, have a range of permeability than spans around ten orders of magnitude, controlled principally by the variation in throat size and the connectivity of the pore space. Within a single large body of rock, be it an oil reservoir or a storage aquifer, we will encounter this range of permeability, between clay lenses and sand channels, or from almost completely consolidated carbonate to areas where there are connected vugs (large pores of mm or cm size), for instance.

6.2.3 The Leverett J Function

The relationship between a representative throat radius and permeability can be used to scale macroscopic capillary pressure curves. The Leverett J function is defined as follows (Leverett, 1941),

$$J(S_w) = \frac{1}{\sigma \cos \theta} \sqrt{\frac{K}{\phi}} P_c(S_w) \tag{6.54}$$

such that the capillary pressure is

$$P_c(S_w) = \sigma \cos \theta \sqrt{\frac{\phi}{K}} J(S_w), \tag{6.55}$$

where $J(S_w)$ is a dimensionless function of saturation. The concept behind the relationship derives from Eq. (3.1), which provides a typical capillary pressure for a given throat radius, and then employing Eq. (6.53) to relate this to permeability. Usually this scaling is applied to the primary drainage capillary pressure, where $\theta \equiv \theta_R$, often assumed to be zero for a fluid/fluid displacement: the same approach can be applied to the waterflood capillary pressure, but here different wettabilities, cooperative pore filling and trapping add additional complexities.

For processes with an unknown wettability, or where there are significant variations in contact angle, we can use a simpler form of Eq. (6.55),

$$P_c(S_w) = \sigma \sqrt{\frac{\phi}{K}} J(S_w). \tag{6.56}$$

Of course, there is a range of throat radius in any real sample and so the computed values of J will vary significantly with saturation; however, the relationships above, approximately, remove the dependence on different throat sizes, and allows experiments performed with one set of fluids on one sample to be applied to other situations with different fluids and for similar rocks but with different porosities and permeabilities.

As an example, Fig. 6.4 shows the Leverett J function for Berea sandstone computed using Eq. (6.54): the results of a mercury injection test are compared to experiments where super-critical CO_2 displaced brine at different conditions of temperature and pressure, with correspondingly different interfacial tensions. To within experimental error all the points lie on one curve providing a quantification of the range of throat size for this sandstone. The value of this approach

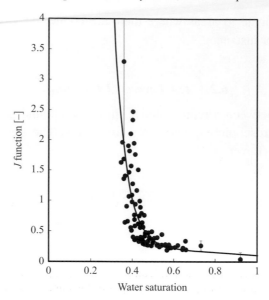

Figure 6.4 The Leverett J function, computed using Eq. (6.54), for Berea sandstone with different fluids. The solid line shows the mercury injection capillary pressure, while the points show experiments for CO_2 displacing brine at different conditions of temperature and pressure. The fine lines show estimated error bars. To within experimental error, all the points lie on the same curve. In contrast to the results shown in Fig. 3.8 there appears to be an irreducible water saturation of around 30%: this is due to the presence of clay in the samples. Replotted from Al-Menhali et al. (2015).

is implicit in the figure: mercury injection provides a quick and accurate measurement of capillary pressure, but is not performed on representative fluids at reservoir conditions. However, reservoir-condition fluid displacement experiments are much more difficult to perform, provide fewer data and are generally less accurate. Fig. 6.4 indicates that to determine a reservoir-condition capillary pressure, Eq. (6.54) could be used to find J from P_c, and then P_c for drainage at reservoir conditions can be found from Eq. (6.55) but now using values of K, ϕ and $\sigma \cos \theta$ ($\cos \theta = 1$ for the CO_2 experiments) for the situation of interest.

The second application of the J function is to compare different rock types. As examples, in Figs. 6.5 and 6.6 we show $J(S_w)$ for the exemplar rocks whose capillary pressures were given in Figs. 3.8 and 3.9, respectively. In these examples the J function represents the variation in capillary pressure compared to a typical value whose throat radius controls the permeability. In Fig. 6.5, the permeability is controlled by the larger throats of a size that first span the sample: in all cases the entry value of J^* (corresponding to the capillary pressure when the non-wetting phase first spans the system) is approximately 0.3. The reason for this is evident from

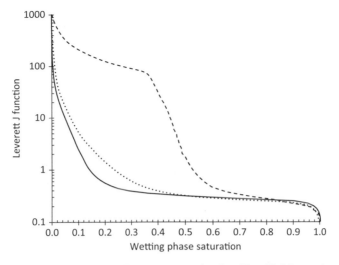

Figure 6.5 The Leverett J function, computed using Eq. (6.54) on the capillary pressure curves shown in Fig. 3.8. The shapes of the curves are preserved, but the magnitude is scaled by a typical throat size, related to the permeability through Eq. (6.49). The solid line is for Doddington sandstone, the dotted line for Berea sandstone and the dashed line for Ketton limestone. The dimensionless value of J when the non-wetting phase first enters the sample is similar for all samples, indicating that these throats control the permeability. The high values of J seen for Ketton are an indication of micro-porosity in the oolithic grains with radii of less than 1 μm, as shown in Fig. 2.7: this micro-porosity does not contribute significantly to the permeability.

Eq. (6.53) and Table 6.3. Using Eq. (3.1) with $r \equiv r_t$ in Eq. (6.55) and substituting K/ϕ from Eq. (6.53) we find:

$$J^* \approx 2\sqrt{a} \qquad (6.57)$$

which, from Table 6.3, is around 0.3 for Doddington, Ketton and Berea. In Fig. 6.4 the entry pressure is lower – around 0.2 – as the presence of clays gives a smaller permeability, 2.1×10^{-13} m^2, for this sample than recorded for Berea in Table 6.3: the inter-granular throat sizes are the same, but the calculated value of J is smaller from Eq. (6.54).

The two sandstone samples, Berea and Doddington, have a relatively narrow uni-model distribution of throat size (see Figs. 3.11 and 6.4): for $J = 1$ most of the pore space has been invaded by the non-wetting phase, with the wetting phase contained in the smaller regions of the pore space and roughness and corners. The limestone, Ketton, contains micro-porosity with intra-granular voids that are approximately one hundred times smaller than the large pores and throats between grains. These micro-pores do not contribute significantly to the permeability and are indicated by large values of J at low wetting phase saturations, of around 100 or larger.

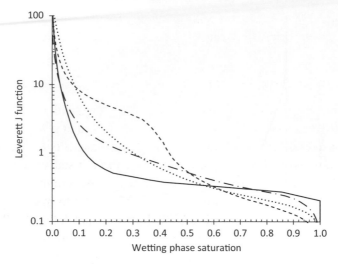

Figure 6.6 The Leverett J function, computed using Eq. (6.54) on the capillary pressure curves shown in Fig. 3.9. The solid line is for Bentheimer sandstone, the dashed line for Estaillades, the dotted line for Mount Gambier and the dot-dashed line for Guiting. The last three samples are limestones with a wide range of pore size: some of the largest throats with the lowest value of J are not necessarily well-connected through the sample; the permeability is controlled by the network of smaller elements with better connectivity.

In Fig. 6.6 we see a wide range of behaviour with one sandstone, Bentheimer, compared to three carbonates: Mount Gambier, Estaillades and Guiting. Bentheimer also has a relatively narrow throat size distribution, see Fig. 3.12, with an entry value of $J^* \approx 0.25$ similar to the other sandstone samples: again the permeability is governed by the largest throats that first connect across the sample, broadly consistent with Eq. (6.57). In contrast, the limestones have entry pressures given by $J^* < 0.1$ suggesting that while there are some relatively large throats, the overall permeability, indicating where the majority of the flow occurs, is governed by smaller throats that provide a better connected network. For the carbonates there is also micro-porosity that does not impact the permeability, represented by values of J at low wetting phase saturation which are larger than 1.

The Leverett J function scales the capillary pressure by the permeability to provide a dimensionless assessment of the range of throat size. Values of $J < 0.1$ indicate the presence of large throats, which while they might span the system do not provide a well-connected network. These are frequently seen in carbonates where large pores or vugs may be present, but where these elements do not control the permeability. We see the filling of those throats which do provide good connectivity at values of J around 0.3. By $J = 1$ the non-wetting phase has filled most of the elements that affect the overall flow: if there is micro-porosity, which does not affect permeability significantly, the wetting phase saturation need not be low at this point. This argument will be revisited when we look explicitly at the conductance of each phase as a function of saturation in Chapter 7.

6.2.4 Computing Flow Fields on Pore-Space Images

It is now standard to compute the single-phase flow field through pore-space images or statistical reconstructions, and from this to predict permeability. The first applications used numerical solutions of the steady-state Stokes equation, (6.38), on synthetic random and fractal porous media, and on a statistical representation of Fontainebleau sandstone (see Fig. 2.11). Here the voxels of the image were used to define a Cartesian mesh for a finite difference solution to find the pressure and velocity field in the pore space (Adler, 1990; Lemaitre and Adler, 1990). This work was later extended to compute permeability directly on a three-dimensional image of the same rock: the predicted permeability on the largest image studied (84 voxels cubed with a voxel size of 10 μm) was within 30% of the value measured on a larger core sample (Spanne et al., 1994). A similar approach has been employed to predict permeability for different sandstones using process-based representations of the pore space (Øren and Bakke, 2002) and for a variety of digitally reconstructed samples with successful comparison to experiment (Kainourgiakis et al., 2005; Piller et al., 2009).

Developments in both computational algorithms and increases in computer power, coupled with the availability of pore-scale images, have enabled the routine prediction of permeability, with calculations on samples containing up to a billion voxels now possible (see, for instance, Muljadi et al. (2016)), with – no doubt – continued improvements in the future. For a recent review see Adler (2013). The ability to solve directly for flow in billion-cell images means that usually this approach is used in favour of first extracting a network and finding the permeability of the network, using semi-analytic expressions of the flow conductance of each pore and throat. One caveat is micro-porous media where this micro-porosity contributes to flow and a single image cannot capture both micro- and macro-pores adequately; see Fig. 2.22, page 50. As we discuss later, the network approach is principally of value for computing multiphase flow to allow an accurate representation of capillary equilibrium and layer flow.

There are two principal approaches used to solve the flow equations. The first is a traditional finite difference or finite volume discretization of the governing Navier-Stokes equations (6.1), usually at steady-state: the incorporation of the non-linear term $\mathbf{v} \cdot \nabla \mathbf{v}$ does not introduce numerical problems for non-turbulent, low Reynolds number flows. The grid is typically the same as the underlying image, but may be an unstructured mesh that attempts to capture the pore-space geometry more accurately or which is less refined in the larger void spaces. Strict no flow at solid boundaries is usually imposed in the difference equations. The most common algorithm uses an initial estimate of the pressure (normally a constant gradient in the direction of the average flow) from which the velocity consistent with the Navier-Stokes equation is computed. The pressure is then updated to obey mass conservation, leading to a new computation of velocity (Patankar and Spalding, 1972; Spalding and Patankar, 1972; Issa, 1986). This is an iterative processes involving the manipulation of large, sparse matrices, which can now be performed efficiently using a number of approaches in the public domain (OpenFOAM, 2010; Bijeljic et al., 2011, 2013a, b; Mostaghimi et al., 2013).

The second approach uses particle-based methods whose statistically averaged properties are solutions of the Navier-Stokes equations (Rothman, 1988, 1990). The most popular technique is the lattice Boltzmann method that tracks the motion of an ensemble of particles on a lattice. The algorithm naturally incorporates complex pore-space geometries and is ideally suited to parallel computing. For a thorough review, see Chen and Doolen (1998). This approach has been employed by many researchers to compute flow fields and predict permeability (Pan et al., 2001; Keehm et al., 2004; Okabe and Blunt, 2004) with extensions to study transport and reaction (Kang et al., 2006).

As an example, Fig. 6.7 shows a comparison between measured and predicted permeability for Fontainebleau sandstone as a function of porosity; the simulations, with some scatter for different samples, fall within the trend of the measurements, showing that reliable predictions can be made on small samples, just a few mm across and resolved with around 3×10^7 voxels (Arns et al., 2004), consistent with the estimate of the size of the representative elementary volume presented in Fig. 2.12. The ability to perform many simulations was used to explore correlations between permeability and mercury injection capillary pressure, as well as pore-scale measures of the pore space, such as specific surface area (Arns et al., 2005).

The other commonly employed Lagrangian method is called smoothed particle hydrodynamics. Again, the approach is suited for parallel computation, but allows more flexibility in the assignment of fluid properties and inter-particle interaction compared to the lattice Boltzmann method. Smoothed particle hydrodynamics was first applied to porous media by Zhu et al. (1999) and then later used to simulate single-phase flow to predict permeability (Jiang et al., 2007a).

While often the methodology used to solve the flow equations seems to take on a quasi-religious significance, obscuring what the computations themselves reveal, there is no definitive reason to choose one computational tool over another: in comparisons of lattice Boltzmann and finite volume methods, the two algorithms produce similar results with similar computational times (Manwart et al., 2002; Mostaghimi et al., 2013). Yang et al. (2016) compared predictions of flow fields and permeability in a bead pack using network modelling, grid-based simulation,

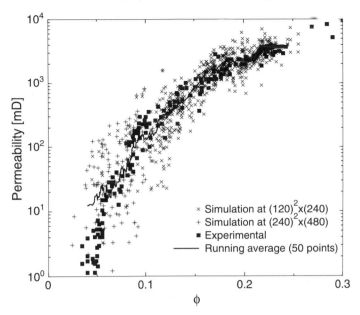

Figure 6.7 Numerical predictions of permeability on images of different voxel sizes, indicated, as a function of porosity, ϕ, for Fontainebleau sandstone, compared to experimental data. From Arns et al. (2004).

lattice Boltzmann simulation and smoothed particle hydrodynamics. The results of all four approaches were similar, with an agreement of predicted permeability $\pm 10\%$, although the network approach was much faster (see Chapter 6.4.9 for a fuller discussion), followed by the lattice Boltzmann model using massively parallel simulation. Andrä et al. (2013b) presented the results of an extensive comparative study and showed that both lattice Boltzmann simulation and a finite volume Stokes solver gave similar predictions of permeability, albeit with up to 50% difference in some cases, which were broadly consistent with experimental measurements for a range of porous media, from sphere packs to carbonates.

As a general guideline, the permeability computed on a good-quality image which accurately captures the main flow channels is likely to be within a factor of 2 of the results of measurements on larger samples of the same rock type, as seen in Fig. 6.7, although there is a sensitivity to image size and resolution (Alyafei et al., 2015; Shah et al., 2016). The reasons for any discrepancy may include inadequate image or numerical resolution, or measurement errors. However, the extreme sensitivity to pore size and connectivity means that even if the numerical computation were exact, the permeability of another sample of similar rock may indeed be significantly different.

Table 6.4 shows a comparison of computed permeability (using lattice Boltzmann, finite volume and network simulations) and porosity from mm-sized images compared to measurements for five samples. For the two more permeable and porous sandstone samples – Bentheimer and Clashach – the images capture the pore space accurately and the predictions are within 50% of the measurements. The results are less accurate for the Berea core, which has a significant amount of pore-blocking clay, Fig. 6.4, that is not easily captured in the image; indeed, the porosity is overestimated, which suggests that some clay is counted as void space in the computations. The large discrepancy in porosity for Ketton and Estaillades is because the image does not capture micro-porosity: however, in these cases it appears that this unresolved pore space does not have a significant impact on the overall connectivity of the rock, consistent with the J function; see Figs. 6.5 and 6.6. The network models significantly over-predict the permeability for Berea, Estaillades and Ketton through simplifying the pore-space geometry. The agreement can be improved, however, by adjusting the conductance of each element to match the results of a direct simulation, or experiment.

The purpose of this exercise is not to critique predictions of permeability or review in detail the methods employed to compute the flow field, but instead to provide an appreciation of current capabilities: given a sufficiently good-quality image it is possible to make reasonable predictions. For sandstones, with a relatively narrow distribution of pore size, images around 10 pores across (or at least 100^3 voxels) are sufficient (see, for instance, Mostaghimi et al. (2013)). For more complex carbonates, good predictions are only possible if the image captures the pore space which carries the majority of the flow, such as for Ketton and Estaillades in Table 6.4. The correlation structure of the flow field can be used to determine if we do indeed have a sufficiently large sample to be a representative elementary volume (Ovaysi et al., 2014).

The advantage of a numerical approach is that small samples of any shape, for which flow experiments are problematic, can be used for computations, as long as an image is available (Arns et al., 2004), and it is easy to study the whole permeability tensor, by allowing flow in all three directions (in our examples the samples are isotropic). They also serve as a useful benchmark for the trickier computation of multiphase flow, which is discussed later.

The principal value of the solution of the Navier-Stokes equations is, however, not the somewhat unrevealing agonizing over the accuracy of the predicted permeability, but to explore features that are not amenable to direct measurement, such as the flow field itself (Bijeljic et al., 2011, 2013b; Jin et al., 2016). Fig. 6.8 illustrates the pore space, pressure and flow field for three rock samples: a bead pack, our all-too-familiar Bentheimer sandstone and Portland limestone. For the bead pack we see rather uniform flow with approximately similar speeds throughout the pore space. For Bentheimer, we see evidence of channelling, but again most of the pore space experiences significant flow. However, for Portland, we see a qualitatively different behaviour, with most of the pore space effectively stagnant with the

Table 6.4 *Permeabilities predicted from a solution of the Navier-Stokes equations on* $1,000^3$ *pore-space images with a voxel size of* 4 μm *compared to measured values for different representative rock samples.* ϕ *and* K *are the porosity and permeability, respectively, measured on the same samples used for imaging. The results do not correspond to those in Table 6.3 since different cores are considered here.* ϕ_{image} *is the porosity determined from the segmented image, while* K_c *is the permeability computed on this image.* K_c^{NM} *is the permeability of the extracted network using the method of Dong and Blunt (2009). Permeability values are shown in D: 1 D = 9.87 $\times$ 10^{-13} m^2. For Ketton, the permeability was computed using a finite volume method; for all the other cases lattice Boltzmann simulation was employed. Data from Andrew et al. (2014a); Shah et al. (2016).*

Rock	ϕ	ϕ_{image}	K	K_c	K_c^{NM}
Bentheimer	0.190 ± 0.003	0.189	2.19 ± 0.14	2.80	3.05
Clashach	0.110 ± 0.002	0.113	0.365 ± 0.116	0.434	0.437
Berea	0.112 ± 0.004	0.129	0.0175 ± 0.007	0.209	0.421
Estaillades	0.256 ± 0.009	0.074	0.0607 ± 0.0026	0.042	0.381
Ketton	0.234	0.149	2.81	3.59	10.1

majority of the flow taken by a few fast paths. Similar behaviour is seen in other carbonates with a wide variation in pore size (Bijeljic et al., 2013a).

The distribution of velocity, computed at the centre each void voxel, is shown in Fig. 6.9. Since we impose no flow at the solid boundaries, any porous medium must show a distribution of speeds; for comparison, the analytically determined distribution for a circular cylindrical tube is shown to indicate the limit of a homogeneous system. The distribution of velocities for bead pack is close to this limit, but with some faster and slower regions. The Bentheimer and Portland samples display a much broader distribution with some eight orders of magnitude variation in flow speed with a significant fraction of the pore space that is effectively stagnant.

To appreciate the implications of this wide spread of flow speeds, the transport analogy, introduced in Chapter 2.2.1, page 33, is useful. The traditional quantification of flow and transport in porous media assumes that the velocity has some relatively narrow distribution around a mean value. Then species moving through this flow field experience the mean speed with some modest fluctuations leading to a dispersed plume of solute. The transport analogy here would be driving on an open freeway in the United States: let's assume that, on average, cars travel at 55 mph, but there will be, inevitably, some variation. If the distance to travel is, say, 110 miles, then it may be reasonable to assert that the journey time is around 2 hours, plus or minus 15 minutes. Of course, transport in porous media is not like that, nor indeed are most real journeys either. Try travelling around London! The

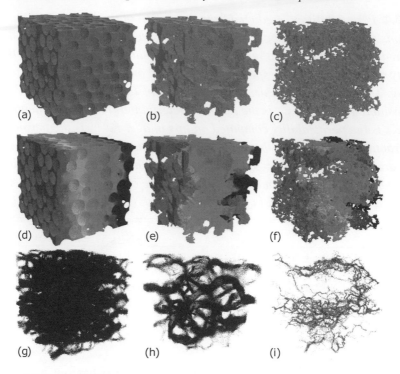

Figure 6.8 Images of the pore space of a bead pack (a), Bentheimer sandstone (b) and Portland carbonate (c). The second row shows normalized pressure fields with a unit pressure difference across the model for the bead pack (d), Bentheimer sandstone (e) and Portland carbonate (f). The bottom row shows the normalized flow fields, where the ratios of the magnitude of the velocity at the voxel centres divided by the average flow speed are shown using a logarithmic scale from 5 to 500 for the bead pack (g), Bentheimer sandstone (h) and Portland carbonate (i). Green and red indicate high values, while blue indicates low values. From Bijeljic et al. (2013b). (A black and white version of this figure will appear in some formats. For the colour version, please refer to the plate section.)

average speed of travel – and this has been much the same for more than 100 years – is around 11 mph (now I will give some sensible units: this is around 18 kmph or, for purists, approximately 5 m/s). This average, while it exists, is practically useless, as the typical speed of travel is very different: most people are walking, or stuck in traffic and moving much slower, while a few are speeding along on fast train lines. There is no sense of some average and a variation around this reliable average: the time to travel a fixed distance may vary by an order of magnitude or more through the availability of good roads and transport links, dependent, at any moment, on the traffic or reliability of public transport.

The true complexity of the flow field leads, inevitably, to a rich transport behaviour for species dissolved in the flowing phase. A discussion of this lies

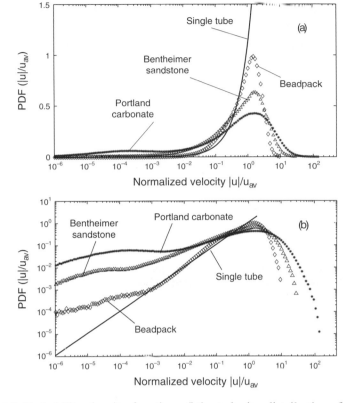

Figure 6.9 Probability density function of the velocity distributions for a bead pack, Bentheimer sandstone, and Portland carbonate presented on (a) semi-logarithmic axes and (b) doubly logarithmic axes. The solid line is the distribution for a single cylindrical tube, representing the homogeneous limit. From Bijeljic et al. (2013b).

outside the scope of this work, which is focussed on multiphase flow: the reader is referred to the excellent reviews and books on this fascinating and often controversial topic; see Berkowitz et al. (2006); Sahimi (2011); Adler (2013).

The reason for this tantalizing introduction to flow is to set the scene for multiphase displacement, where the fluid phases are confined to a subset of the pore space, and one, as we have already described, controlled by pore size. Returning to journeys, it is as though we compare travelling on foot, confined to pavements and paths, to flying. However, even this does not remotely capture the variability of the problem: the difference in speed between walking and an aeroplane is at most a factor of 200: we see differences that are a million times greater – this is even more extreme than comparing the speed of a rocket with, say, the trail of a snail. The consequence will be a huge range of local flow speeds, leading to an enormous variability in relative permeability (the permeability of each phase) as a function of saturation, a variation that is very difficult to capture accurately through direct measurement. We will also revisit some of the computational methods introduced here when we briefly review simulations of multiphase flow near the end of this chapter. First, however, we need to present the macroscopic flow equations.

6.3 The Multiphase Darcy Law and Relative Permeability

When multiple phases are present in the pore space, the extension of Darcy's law is as follows:

$$\mathbf{q}_p = -\frac{k_{rp}K}{\mu_p}(\nabla P_p - \rho_p \mathbf{g}) \qquad (6.58)$$

for phase p. k_{rp} is the relative permeability of phase p and represents the amount by which flow is restricted by the other phases present in the pore space: it is a function of saturation.

The multiphase Darcy law, Eq. (6.58), together with the capillary pressure P_c, comprise a macroscopic description of flow in porous media.

6.3.1 Historical Interlude: Muskat, Leverett and Buckingham

The development of Eq. (6.58) proceeded independently among petroleum engineers and soil scientists. The form of the equations for multiphase flow in porous media presented here was first proposed by Muskat and Meres (1936): Morris Muskat's book entitled *The Physical Principles of Oil Production* established reservoir engineering as a distinct discipline (Muskat, 1949). Wyckoff and Botset (1936) simultaneously presented an extensive set of measurements of gas/oil relative permeability in unconsolidated sands: we show these results later in Chapter 7.1. The concept of capillary pressure as a saturation-dependent function was first proposed in petroleum engineering by Leverett (1941). In addition, he introduced the dimensionless scaling of J function described previously, Eq. (6.54). Furthermore, he reported some of the first measurements of relative permeability and worked on three-phase flow, see Chapter 8, for his PhD thesis at MIT (Leverett, 1939). And if this were not sufficient, Miles Leverett also derived a conservation for multiphase flow, using Eq. (6.58) while working for Humble Oil, which later became part of Exxon (Buckley and Leverett, 1942): this is presented in Chapter 9.2. When war broke out, he joined the Manhattan project and spent the rest of his career working on nuclear power. He continued to be active into the 1980s, working mainly on nuclear safety; see Leverett (1987).

In the context of water/air configurations in soils and treating only the movement of water, similar ideas pertaining to multiphase flow had been introduced previously in hydrology. Edgar Buckingham provided a mathematical description of a moisture or capillary potential (what we call here the capillary pressure) (Buckingham, 1907), while Richards (1931) considered a saturation-dependent water conductivity (equivalent to the water relative permeability), appreciated the importance of hysteresis and derived the conservation equation for one-dimensional flow that is named after him. For an excellent historical review of the soil science literature in the first half of the twentieth century, see Philip (1974).

6.3.2 Assumptions Inherent in the Multiphase Darcy Law

The next chapter will present a detailed discussion of relative permeability focussed on its saturation dependence and hysteresis for different pore structures and wettability: here instead we will first examine and challenge the assumptions made in the presentation of Eq. (6.58). We have hypothesized that each phase flows in its own sub-network of the pore space, unaffected by the flow of the other phase. We also ignore any impact of displacement processes: we consider that in some representative elementary volume we have a static configuration of the phases. This fluid arrangement may change slowly, affecting the value of k_{rp}. Indeed, as should be evident from the preceding chapters, the pore-scale pattern of fluids is dependent on the whole displacement history, and so k_{rp} will not only be a function of saturation, but saturation history: it will be different in waterflooding than oil invasion, or primary drainage. However, this does not prevent the use of Eq. (6.58) and relative permeability: as we do for capillary pressure, we simply measure k_{rp} as a function of both saturation and displacement history.

The key assumption inherent in the multiphase Darcy law is that we have a percolation-like displacement pattern, such that there is a representative elementary volume within which we have an average saturation that evolves slowly in both space and time. Over the time-scale for flow across this elementary volume, the saturation is approximately constant (that is there is little displacement), the fluid configuration is fixed and the relative permeabilities are the flow conductances of each phase, which can, in principle, be computed using the expressions we have introduced previously. If instead we have frontal advance, the relative permeability is not defined as a smoothly differentiable function, in that it will vary from a value close to 1 to close to zero over a pore scale. We will return to this important concept later when we present flow patterns.

This traditional form of Eq. (6.58) also ignores any viscous coupling between the phases: at the pore scale it asserts that there are no-flow boundaries at the fluid/fluid interfaces; to a given phase, the other fluid phases behave as solid. If we relax this condition, we introduce cross-terms and can write

$$\mathbf{q}_p = -\lambda_{pq} K (\nabla P_q - \rho_q \mathbf{g}), \tag{6.59}$$

where λ is a mobility ($\lambda_{pp} \equiv k_{rp}/\mu_p$) while p and q label the phases. Here we have used component notation for the phases, so we sum over q in Eq. (6.59). This allows the flow of one phase to affect the movement of the other through the continuity of viscous stress at the interface, see Eq. (6.29). Normally these effects are ignored, if we assume that the interface is stationary, but as discussed previously, page 229, this may be a poor approximation for gas/liquid systems, or where there is a large viscosity contrast between the phases.

The impact of these viscous stresses on macroscopic flow behaviour has been studied through a comparison of the flow of two phases in the same direction (co-current flow) and in opposite directions (counter-current flow) theoretically, numerically and experimentally (Bourbiaux and Kalaydjian, 1990; Kalaydjian, 1990; Li et al., 2005). This work has shown that transfer across the fluid interfaces can be important, particularly when there is a large viscosity contrast between the fluids; for instance, relatively low viscosity water in wetting layers can force the flow of higher viscosity oil in the centres of the pore space, apparently boosting the oil flow: this is properly accounted for using the cross-terms λ_{ow} in Eq. (6.59). This topic will be revisited in Chapter 7.3.2.

If we have an anisotropic porous medium, with single-phase flow governed by Eq. (6.43), then we can treat K in Eq. (6.58) or (6.59) as a tensor as well. In general though, the orientation of the flow of multiple phases need not be the same as in single-phase flow. In theory this makes relative permeability a fourth-rank tensor as it relates two tensorial quantities: in its general form we can write:

$$q_{p,i} = -\Lambda_{pq,ij} \left(\frac{\partial P_q}{\partial x_j} - \rho_q g_j \right) \tag{6.60}$$

where the subscripts i, j and k label coordinate directions, and p and q phases with summation over j and q. Λ is a mobility, which can be related to the permeability tensor as follows:

$$\Lambda_{pq,ij} = \lambda_{pq,ijkl} K_{kl}, \tag{6.61}$$

with a summation over coordinate indices k and l which – as mentioned above – makes the mobility (or relative permeability) a fourth-rank tensor in terms of coordinate direction and a second-rank tensor in terms of phase.

This formulation of multiphase flow is now almost absurdly complex and is never used in practice, since each component is itself a function of saturation and displacement history; in experiments, as we show in the next chapter, it is difficult to obtain good measurements of k_{rp} for a displacement process of interest, without worrying about cross-terms or anisotropy. In what follows, we will only consider relative permeability defined through Eq. (6.58).

Furthermore, and most importantly, this mathematical tautology ignores the most important physical complexity associated with multiphase flow, which is the dependence on flow rate, treated next.

6.4 Capillary Number and Pore-Scale Dynamics

Viscous forces, or the impact of flow rate, can challenge the assumptions inherent in the multiphase extension to Darcy's law. In steady-state single-phase flow, while the fluid is moving, the velocity field does not change with time and we can define permeability, at least for low Reynolds number and Newtonian fluids, as a function only of the pore geometry.

The principal problem with the conventional description of multiphase flow using relative permeability is that we are describing a dynamic process, where one phase displaces another. As a result we do not have a genuine steady-state and the fluid pattern changes. There are two ways in which this can affect relative permeability. Firstly, the flow alters the microscopic configuration of fluid from that governed by capillary forces only, making the relative permeabilities dependent on q, leading to a non-linear relationship between Darcy velocity and pressure gradient. Secondly, it is possible that we never achieve a genuine steady-state even if we have, macroscopically, a constant saturation, since the fluid arrangement at the pore scale fluctuates between locations of local capillary equilibrium, as observed in high-speed pore-scale visualization experiments, see Chapter 4.2.5, page 142. We will consider the two potential problems in turn in the following sections. This is a subtle topic, with many details, so we first present a pedagogical description of flow patterns, before quantifying the behaviour.

6.4.1 Macroscopic Flow Patterns for Imbibition

We will now revisit some important, fundamental concepts which hitherto, since we ignored viscous effects, we have assumed implicitly, rather than discussed rigorously. Fig. 6.10 shows schematically the three length scales that we consider. Here our focus will be principally on water injection, since this is where the most significant impact of flow rate is seen, namely a reduction in residual saturation. As a concrete example, we show waterflooding an oil reservoir. The wells are usually several 100s m apart and over these lengths there is a distinct variation in water saturation, with the water at its maximum saturation, $1 - S_{or}$, near the injection well and at a lower value, initially S_{wi}, near the production well. This we define as the mega-scale, where there are gradients in macroscopic quantities and we see significant changes in pressure caused by viscous forces. If we were now, conceptually, to consider a small region of the reservoir, as shown, we can take a portion of the porous medium in which we do observe an approximately constant value of macroscopic, or averaged, quantities, such as saturation. This is the scale of a representative elementary volume and is reproduced, in the laboratory, by experiments on rock samples: this is the macro-scale, of order mm to cm, containing many pores. The third, smallest scale, typically of order μm, is that of the pores where we have analysed fluid configurations governed by positions of local capillary equilibrium.

Even for displacement at an infinitesimal flow rate we have made a distinction between the local capillary pressure and its macroscopic counterpart. We will revisit this concept here, before introducing the impact of flow rate. The local capillary pressure is the difference in pressure across a fluid/fluid interface at the pore scale. When this interface is stationary, the pressure difference is governed by the curvature and the interfacial tension as given by the Young-Laplace equation, (1.6). The macroscopic capillary pressure has a more ambiguous definition. It may be considered as some average of the local capillary pressures; however, here, in keeping with the definition of pressure gradient used in Darcy's law, we will consider it as the difference in pressure between the phases imposed externally on a sample of a porous medium. This, we assume, is a small portion of the rock across which there is no significant gradient in saturation, defined as the macro-scale in Fig. 6.10: it is at least as large as a representative elementary volume. The meaning of macroscopic capillary pressure is somewhat less clear in a field setting, since we do not impose pressures at this scale; we assume that we may use the experimentally measured capillary pressure and relative permeabilities measured at the macro-scale in the governing conservation equations that we will present in Chapter 9.

In an experiment, the macroscopic capillary pressure increases or decreases monotonically during a displacement, whereas locally new positions of pore-scale capillary equilibrium between the phases may be attained whose capillary pressures are different from that imposed externally. If these local pressures are found in isolated ganglia of fluid they may persist for a significant time during a displacement; see page 125. However, we have assumed that there are points during the displacement when both locally and externally the local and macroscopic capillary pressures, at least of the connected portion of each phase, are the same, when a new local maximum or minimum in the capillary pressure is reached. It is this concept that has allowed us to relate considerations of pore-scale capillary equilibrium to macro-scale measurements of capillary pressure.

However, when we introduce a flow rate, there is a gradient in pressure across the system and hence a gradient in capillary pressure and saturation. To illustrate this, Fig. 6.11 shows a seminal series of micro-model experiments where a wetting phase (shown dark) displaces a non-wetting phase (light) in a two-dimensional

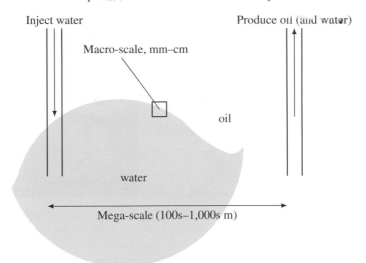

Figure 6.10 An illustration of length scales in porous media flow. The top figure shows a cartoon of a mega-scale displacement where water is injected into a reservoir to displace oil: the shading represents areas of high water saturation; in reality there will be a distribution of water throughout the reservoir. Over the distance between wells, 100s m, the water saturation and pressure will vary significantly. The small square shows one selected region that is shown enlarged in the lower figure. The macro-scale is defined as a section of the porous medium of the size of a representative elementary volume where we may define average properties, such as saturation. These properties will vary slowly over space and time at this scale as the megascopic displacement continues. The macroscopic scale is typically mm to cm and similar to the size of samples used in laboratory experiments. The pore scale, where capillary forces dominate, is of order μm. Specifically, in this case, a micro-CT image of Bentheimer sandstone is shown 3 mm across containing nitrogen (light) and brine. Lower figure from Reynolds (2016).

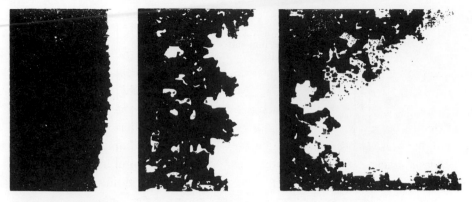

Figure 6.11 Imbibition in a micro-model. The wetting phase, injected from the left, is shown in black and the non-wetting phase is light. Flow patterns at capillary numbers, Eq. (6.62), of 3×10^{-4}, 1.4×10^{-5} and 6×10^{-7} are shown from left to right. The models are a square lattice of 42,000 ducts (throats) with 4 mm between junctions (pores). From Lenormand and Zarcone (1984).

square lattice (Lenormand and Zarcone, 1984). The pictures are shown at three flow rates.

A dimensionless capillary number Ca is traditionally defined as follows:

$$Ca = \frac{\mu q}{\sigma}, \tag{6.62}$$

where q is the Darcy velocity of the injected phase, while μ is its viscosity. In Fig. 6.11 Ca varies from 6×10^{-7} to 3×10^{-4}. While Ca is easy to determine, it is not the dimensionless ratio of viscous to capillary effects at the pore scale: as we show later, Chapter 6.4.2, it is necessary to multiply Ca by some geometry and saturation-dependent quantity whose value is much greater than 1 to find the true ratio. This explains how $Ca \ll 1$ in Fig. 6.11 even though viscous effects are evident at the pore scale.

We will first discuss the right-hand figure, at the lowest flow rate, where, at the pore scale, the configuration of fluid is controlled by capillary forces. The saturation profile is not uniform since there is a variation in capillary pressure across the system, with a higher saturation near the inlet. There is flow in wetting layers, and this flow penetrates through most of the model, allowing snap-off in the smaller throats. Nearer the inlet, we see more piston-like advance and trapping, with regions of non-wetting fluid stranded in the pore space, over a wide range of size, from single pores upwards. It is possible to imagine portions of the model, several pores across, in which the saturation is constant and filling proceeds largely in order of pore size, as described in Chapter 4, where we have a percolation-like displacement pattern: this would define the macro-scale in Fig. 6.10, while the pore scale is that of the individual ducts and the mega-scale, while it is not nearly the extent of an oilfield, is that of the whole model itself.

As the flow rate increases, not only is the extent of the region between filled and unfilled portions of the model compressed, as one might expect if a larger pressure gradient is imposed, but the pore-scale arrangement of fluid alters as well. There is less snap-off, with essentially none at the highest flow rate, and less trapping. Specifically, what trapping is observed is confined to smaller clusters, with only a few single pores stranded at the highest flow rate. It is now more difficult to define an obvious macro-scale and, in any event, the fluid configurations are different. This demonstrates that there is a flow-rate dependence on fluid patterns, and consequently their flow properties, which is not explicit in the multiphase Darcy law, Eq. (6.58).

Qualitatively, this can be explained since we know that there is a huge disparity between the conductance (and hence flow in response to a given pressure gradient) in layers and through the centres of the pore space; see Chapter 6.1.3. Since layer flow is slow, the amount of filling by snap-off, mediated by this flow, is limited in comparison to the rate at which the pore space can fill through piston-like advance supplied by fluid connected through the centre of the pore space to the inlet. If the imposed flow rate is sufficiently low, then there is enough time to fill throats in order of size by snap-off, by layer flow, in preference to piston-like advance. The slow flow means that the change in capillary pressure across the system is modest, and we can define macroscopic portions of the system, spanning many pores, where the filling sequence is capillary-controlled. On the other hand, if the flow is more rapid, the wetting phase pressure is higher near the inlet, favouring filling close to it, even if there is a smaller throat some distance away that would be filled by snap-off in a capillary-controlled displacement. There is insufficient time to allow restricted layer flow to fill elements by snap-off in comparison to the rush or burst of piston-like advance that can proceed much faster. The result is a suppression of snap-off, less trapping and a more connected advance as the flow rate is increased.

This generic behaviour is not confined to two-dimensional micro-model experiments, which, with touching modesty, were prefaced by the authors with *we don't try to draw a parallel between the etched network and a real porous medium* (Lenormand and Zarcone, 1984). We can, in fact, draw this analogy, by showing similar results in three dimensions. Fig. 6.12 illustrates secondary imbibition in a three-dimensional packing of sintered glass beads (Datta et al., 2014b); the results of the primary drainage experiment were shown in Fig. 3.3, page 78. The displacement pattern is similar to that in the low-rate experiment in Fig. 6.11, which was performed at the same capillary number: water invades the system in wetting layers initially, filling narrower regions of the pore space by snap-off. Closer to the inlet – with a higher water pressure (lower capillary pressure) – water also fills

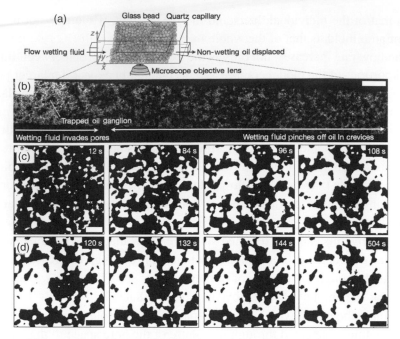

Figure 6.12 (a) Secondary imbibition through a packing of sintered glass beads imaged using confocal microscopy; similar experiments for drainage were shown in Fig. 3.3. (b) Optical section through part of the medium, taken as the wetting fluid displaces the oil at a capillary number of 6.4×10^{-7}. The bright areas show the wetting phase. The wetting fluid first snaps off oil in corners, as seen in the region spanned by the double-headed arrow, and then bursts into the pores of the medium, starting at the inlet, by piston-like advance, as seen in the region spanned by the single-headed arrow. Some oil ganglia remain trapped, as indicated. (c)–(d) Time sequence of zoomed confocal micrographs. These are binary images showing wetting fluid in white as it (c) initially snaps off oil, and then (d) invades the pores. The last frame shows an unchanging steady state; the arrow indicates a trapped oil ganglion. Scale bars in (b) and (c)–(d) are 500 μm and 200 μm, respectively. From Datta et al. (2014b).

the pore space by piston-like advance. The combination of these two displacement processes leads to trapping. The formation of trapped ganglia can be observed as the displacement proceeds.

This descriptive narrative has introduced four impacts of flow rate, that we present in detail below.

(1) Flow induces a gradient in local capillary pressure, favouring filling near the inlet and changing the sequence of pore and throat filling from that for a capillary-controlled displacement.
(2) There is also a competition in the time for filling through layer flow compared to piston-like advance. This tends to favour frontal advance at higher flow rates.

(3) The capillary pressure gradient leads to a finite correlation length in the system, most noticeable in a reduction in the largest trapped ganglia in imbibition. This is conceptually similar to the correlation length derived for displacement under the influence of gravity; see Chapter 3.4.2, page 107.
(4) There is also the possibility that viscous forces are sufficiently strong that ganglia, surrounded by wetting phase, can be pushed out of the rock.

6.4.2 Capillary Number and the Perturbative Effect of Flow Rate

Now we will quantify how the local capillary pressure varies to address the first point in the list above. Imagine that we have oil and water flow from left to right with some externally imposed capillary pressure $P_c^I = P_o^I - P_w^I$ at the inlet. Then we can find the pressures of the phases at the outlet, P^O, where L is the size of the sample, assuming one-dimensional flow and ignoring gravitational effects, using Eq. (6.58):

$$P_p^O = P_p^I - \frac{\mu_p q_p L}{K k_{rp}} \tag{6.63}$$

from which we can find the change in capillary pressure,

$$\Delta P_c = \frac{L}{K} \left(\frac{\mu_w q_w}{k_{rw}} - \frac{\mu_o q_o}{k_{ro}} \right), \tag{6.64}$$

where $\Delta P_c = P_c^O - P_c^I$. If viscous effects are to be insignificant at the pore-to-macro scales, we require that the flow rate does not perturb the fluid configurations, which, in turn, are controlled by local capillary equilibrium. Eq. (6.64) shows that the capillary pressure does vary with distance: this variation therefore must be small compared to a typical capillary pressure.

We define a ratio of viscous to capillary forces, $R_{vc} = \Delta P_c / P_c$. To relate the capillary pressure to rock and fluid properties, we use the Leverett J-function scaling, Eq. (6.56). Then, from Eq. (6.64) we find:

$$R_{vc} = \frac{L}{\sigma J \sqrt{\phi K}} \left(\frac{\mu_w q_w}{k_{rw}} - \frac{\mu_o q_o}{k_{ro}} \right). \tag{6.65}$$

We generalize Eq. (6.62) to define a dimensionless capillary number for each phase p,

$$Ca_p = \frac{\mu_p q_p}{\sigma}, \tag{6.66}$$

Then Eq. (6.65) becomes

$$R_{vc} = \frac{L}{\sqrt{\phi K} J} \left(\frac{Ca_w}{k_{rw}} - \frac{Ca_o}{k_{ro}} \right). \tag{6.67}$$

Notice that the ratio of viscous to capillary effects is the capillary number, Eq. (6.66), multiplied by porous-medium and saturation-dependent quantities (Hilfer et al., 2015a). We now estimate representative values for the capillary number, which for simplicity we will assume are similar for both phases. For waterflooding we may take a Darcy velocity of around 10^{-6} m/s (around 30 m/year), which is at the upper end of typical values. Then for a viscosity of 10^{-3} Pa.s and an interfacial tension of $20 - 25$ mN/m, we have $Ca = 4 - 5 \times 10^{-8}$: usual field-scale values are in the range 10^{-7} to 10^{-9}. Clearly these numbers are much less than 1 and hence we might naively propose that we can safely ignore viscous effects at the pore scale, in the same way that we used the low values of Reynolds number to dismiss turbulence in porous media flows. However, Ca does not properly represent the pore-scale balance of forces, leading to considerable complexities and subtleties in our analysis, as evident by the fact that Eq. (6.67) contains other parameters. This is also clear from Fig. 6.11, where a significant change in the displacement pattern and degree of trapping is observed for $Ca \ll 1$.

First, we consider a situation where indeed it is correct to assume that capillary forces dominate, or at least that there is no gradient in capillary pressure. In a steady-state flooding experiment, both oil and water are injected simultaneously with some pre-determined ratio of oil and water flow rates into a core sample; the saturation is monitored until a uniform distribution is obtained. In this case, by construction, the capillary pressure is constant across the system as the two terms in Eq. (6.67) cancel.

In an unsteady-state experiment, such as those shown in Figs. 6.11 and 6.12, or in field settings, one phase is injected to displace the other. R_{vc} is a function of saturation and so it is difficult to make a precise assessment without first having a solution for the evolution of the saturation profile in space and time: this will be presented in Chapter 9, but we can make some general comments here. Normally the total velocity, defined as $q_t = q_w + q_o$, is imposed and this does not vary with distance for a one-dimensional incompressible displacement, through conservation of volume. The two terms in Eq. (6.67) will tend to cancel, but not completely, as in a steady-state scenario, since there is some displacement.

If we have an intermediate saturation, where both the oil and water phases are well connected, then the relative permeabilities, while they will be less than 1, are not very small. J is also of order 1; see Fig 6.6 for instance. For the sake of argument, take $1/J(1/k_{rw} - 1/k_{ro}) \approx 10$ and $Ca_o = Ca_w = 4 \times 10^{-8}$. If the rock is our Bentheimer then, from Table 6.3, $\sqrt{\phi K} = 0.61$ μm. If we take L initially as a pore length, with a value 79 μm in this case, we find $R_{vc} = 5 \times 10^{-5}$. At the pore scale, capillary forces control the fluid configuration. However, to make a meaningful measurement or computation of averaged properties, we require that the local and macroscopic capillary pressures are similar over some length which encompasses several pores. Consider that we have a sample $L = 1$ cm across; putting this value into Eq. (6.67) we find $R_{vc} = 7 \times 10^{-3}$, which is still small: the variation in local capillary pressure is less than 1%. As a general guideline, it is considered that for $Ca < 10^{-6}$, the process is capillary-controlled at the pore scale, with local capillary equilibrium maintaining an approximately constant pressure over a length sufficiently large to define average properties, such as saturation and relative permeability, with reasonable accuracy. This is consistent with the flow patterns shown in Figs. 6.11 and 6.12 where, while at the mega-scale there was a gradient in saturation, on the scale of a few pores, the displacement was capillary-controlled.

6.4.3 Layer Conductance and Viscous Effects

However, there are cases where R_{vc} is of order 1, even over a pore length at typical flow rates. Consider displacement near the end of primary drainage. To make the exposition clear, imagine that the non-wetting phase is injected into a rock sample from one face; all the other faces are blocked to flow. The wetting phase therefore has to escape through the injection face. This is counter-current flow with $q_t = 0$ or $q_w = -q_o$, meaning that both terms in Eq. (6.67) have the same sign. Near the end of the displacement, the non-wetting phase saturation is high with a relative permeability k_{ro} close to 1, while the wetting phase, confined to small pores and layers, is less well connected and consequently has a much lower relative permeability. Therefore R_{vc} is controlled by the flow of the wetting phase and we may write

$$R_{vc} = \frac{LCa_w}{k_{rw}\sqrt{\phi K}J}. \tag{6.68}$$

Another way to achieve this situation experimentally is to place a low-permeability porous plate at the outlet. The capillary entry pressure in the porous plate for the non-wetting phase is larger than needed to saturate the rock sample: the wetting phase can pass through the plate, but not the non-wetting phase. This allows high capillary pressures to be imposed (Plug and Bruining, 2007).

We have already presented the flow conductances for throats and compared them with the much higher resistance in wetting layers, page 232. Eq. (6.34) presented a ratio of time scales of filling a pore through the centre of a throat to that through a layer. This is simply a ratio of the flow conductances,

$$\frac{g_{AM}}{g_{throat}} \sim \left(\frac{r_{AM}}{r_t}\right)^4, \tag{6.69}$$

where r_t is the throat radius and r_{AM} is the radius of curvature of the arc meniscus in a wetting layer. If we consider that at high non-wetting phase saturation, the wetting phase flow is confined largely to layers, then the layer conductance determines the relative permeability. If, in contrast, the wetting phase saturation is 1, the flow is entirely through filled pores and throats with a relative permeability of 1 by definition. Therefore, the ratio of conductances is a ratio of relative permeabilities

$$k_{rw} \sim \left(\frac{r_{AM}}{r_t}\right)^4, \qquad (6.70)$$

for flow controlled by wetting layers.

At the end of primary drainage, in a field setting, the capillary pressure, imposed by the gravitational pressure difference between the phases, is of order 10^5 Pa, with r_{AM} of around 1 μm or less: on page 232 we considered a minimum radius of 0.2 μm. Now, if we employ our Bentheimer example with $r_t = 24$ μm as a characteristic throat radius, Eq. (6.70) gives $k_{rw} \approx 4.8 \times 10^{-9}$. Here $J = 15$ for $P_c = 10^5$ Pa, see Figs. 3.9 and 6.6, where we have to account for the difference in interfacial tensions between the mercury injection measurements and the field: a capillary pressure of 10^5 Pa corresponds to a mercury injection P_c of around 1.6×10^6 Pa. With L equal to a pore length, then we have in Eq. (6.68), $R_{vc} = 1.8 \times 10^9 Ca_w$. This indicates that even at the pore scale, the viscous pressure gradient in the wetting phase is more significant than the capillary pressure for all but the slowest flows.

In both the field and in laboratory measurements, the low conductance of small wetting layers results in a limiting time scale to reach capillary equilibrium, since the capillary pressure gradients required to maintain the flow are very high. In our example, the capillary pressure is equivalent to an equilibrium saturation of around 2%; see Fig. 6.6 for $J = 15$. Consider an experiment on a sample 1 cm across that attempts to measure the relationship between capillary pressure and saturation. An imposed pressure difference of 10^5 Pa, similar to the capillary pressure itself, is about the largest that could be considered reasonable to obtain representative results. By construction, this gives $R_{vc} = 1$ for $L = 1$ cm. We can use the multiphase Darcy law, Eq. (6.58), to calculate the flow rate, where $\nabla P = -10^7$ Pa/m and $k_{rw} = 4.8 \times 10^{-9}$: we find $q_w = 9 \times 10^{-11}$ m/s for $K = 1.88$ D, Table 6.3, with $\mu_w = 10^{-3}$ Pa.s. To place this in perspective, we can find the time necessary to decrease the wetting phase saturation, through layer drainage, by, say $\Delta S_w = 1\%$, assuming that the layer conductance remains approximately constant: $L^2 q_w t = \phi L^3 \Delta S_w$ or $t = 2 \times 10^5$ s or around 2.6 days for $\phi = 0.2$. While this is not a precise calculation, it indicates clearly that days are required to reach equilibrium in a small laboratory sample at low saturation; this indeed what is found for primary drainage experiments involving two fluids (Pentland et al., 2011; El-Maghraby and Blunt, 2013).

Now imagine that we attempted to increase the capillary pressure further, to drain to a saturation of 1% or less, close to the lowest saturation achieved during a mercury injection experiment. From Fig. 6.6 this requires around ten times the capillary pressure, hence an arc meniscus of one-tenth the radius, which means, from Eq. (6.70), that the relative permeability is four orders of magnitude lower. Now, for a change of saturation of only 1% the time scale for wetting layer flow is around 70 years, which is an infeasible wait for a laboratory experiment. These times would be even longer for more complex or low permeability rocks with micro-porosity.

This calculation explains why, in capillary pressure measurements using fluids, there is always an apparent irreducible wetting phase saturation, even if a very large capillary pressure is imposed: this is solely an artefact of the long times needed to allow the wetting layers to drain to a position of capillary equilibrium. In practice a high macroscopic capillary pressure is imposed on a sample, but the measured saturation does not represent a position of equilibrium, since insufficient time has been allowed for layer flow: hence for a given saturation the capillary pressure is overestimated, or – for a given capillary pressure – the measured saturation is too high. This also demonstrates the value of mercury injection measurements: here the wetting phase is a vacuum, with zero viscosity, and so it is possible to attain very low saturations rapidly.

In the field, capillary equilibrium needs to be established over an oil column which, in our example, is around 30 m high. In this case we use $L = 30$ m with an imposed pressure gradient governed by buoyancy forces $(-\nabla P = \Delta \rho g \approx 3,000$ Pa/m$)$. Now, for a saturation change of, say, 10%, the time scale is around 700,000 years, which is well within the millions of years that the hydrocarbons have resided in the field. Note, however, that were we to have a very high oil column, with lower equilibrium saturations, the time scale for wetting layer drainage is millions of years and potentially comparable with the age of the field. While it is reasonable to assume capillary equilibrium in an oilfield when it is first found, this may not always be correct.

The fourth-power scaling of relative permeability with layer curvature, Eq. (6.70), implies that macroscopic capillary equilibrium is difficult to achieve whenever flow is limited by wetting layers, particularly those with low radii of curvature compared to a typical pore size. As well as flow at the end of primary drainage, we also see thin layers at the start of waterflooding. Again this leads to a significant gradient in local capillary pressure and confines snap-off to a small region ahead of a frontal advance, as seen in Fig. 6.11.

Another situation where layer flow is important is in mixed-wet systems. Here, at the beginning of waterflooding, the wetting layers are pinned with a low conductance; see the calculation on page 232. If the water phase is only connected by these layers, then filling of pores containing water is constrained by flow through them.

The result here is to change the order in which elements are filled, similar to what we saw in the study of snap-off during Haines jumps in Chapter 4.1.1, page 125. Instead of pores and throats being filled in order of local capillary entry pressure, filling will be favoured in those elements where water can be provided through the centres of the elements, rather than by layer flow – particularly layer flow along several pores and throats. As we have demonstrated previously, the filling time for a pore where the water is provided by layers is many orders of magnitude longer than if the adjacent element is water-filled: even at low flow rates, this disparity in times will lead to large local capillary pressure gradients and a change in the sequence of pore and throat filling from that determined by local capillary pressure only. If we are considering the filling of oil-wet elements, the result is to suppress the filling of some larger, poorly-connected, elements and favour the filling and establishment of a connected pathway of smaller pores and throats where water occupies the centre. This means that at a finite flow rate, water will connect across the system through the centres of the pore space at a lower saturation than for an infinitesimally slow displacement, leading to a higher water relative permeability at low water saturation.

The final example where layer flow controls displacement is at the end of water-flooding in an oil-wet or mixed-wet system. Here it is thin layers of oil sandwiched between wetting layers in the corners and water in the centres that govern the flow of oil; Chapter 5.2. The impact of oil layers is similar to that of wetting layers at the end of primary drainage: a long time is needed to allow these layers to drain to positions of capillary equilibrium. Normally, in a waterflood experiment, oil and water are injected simultaneously and flow in the same direction; for the same imposed pressure gradient, the water flow will be much larger than that of the oil. Hence, we will need to inject a huge volume of water to allow the displacement of a relatively small amount of oil: this is evident, for instance, in Fig. 5.13, page 212, where several thousand pore volumes of water need to be injected to drive the oil saturation below 10%, while the true residual saturation may be even lower.

This is a somewhat qualitative discussion of an important effect in multiphase flow. Here we will attempt to quantify when rate-dependent layer flow is likely to be important through the generalization of some of the relationships presented previously. If we assume that the relative permeability is given by Eq. (6.70), then we substitute Eq. (6.53) for r_t and Eq. (6.56) with $P_c = \sigma/r_{AM}$. We find:

$$k_r^{layer} = \frac{a^2}{J^4} \qquad (6.71)$$

for the relative permeability for a phase (oil or water) whose flow is limited by layers: J is the J-function value that controls the size of these layers: either the prevailing value (for primary drainage, or oil layer flow in waterflooding) or the maximum reached during primary drainage (for water flow in mixed-wet or oil-wet systems).

Then from Eq. (6.68) we obtain using Eq. (6.71)

$$R_{vc}^{layer} = \frac{LCa_p J^3}{a^2 \sqrt{\phi K}},$$

(6.72)

where we use the capillary number for the phase p (oil or water) confined to a layer. Inevitably, this ratio cannot be estimated purely from macroscopically measured properties – such as J, K and ϕ alone – but also requires some pore-level information through the constant a.

We can repeat the calculation for our Bentheimer example with $J = 15$, $\sqrt{\phi K} = 0.61$ μm and $a = 1.6 \times 10^{-2}$, Table 6.3. Then in Eq. (6.72) we have $R_{vc}^{layer} = 1.3 \times 10^{13} LCa$ with L measured in metres. First, we consider the impact of viscous forces at the pore-scale and use $L = l = 79$ μm, to find $R_{vc}^{layer} \approx 10^9 Ca$. Hence, for viscous effects to be small at the pore scale for flow through layers when they are at their smallest, we require a capillary number of much less than 10^{-9} as derived previously.

There is a twist in this analysis though: the capillary number is for the phase that is in layers, which may be much less than that defined using the total, or injection, flow rate, Eq. (6.62). Moreover, in water-wet systems, the layers can swell to increase their conductance. Hence, we have a delicate balance between frontal advance and layer flow, which is controlled by both the pore-scale arrangement of fluid and the evolution of the flow pattern at the mega-scale.

One way to determine the impact of flow rate on the displacement sequence is to revert to our discussion of time scales for filling (Chapters 6.1.2 and 6.1.3) and compare these with the rate of advance of the injected water. The average speed of movement at the mega-scale, Fig. 6.10, is of order q_t/ϕ: the time taken to progress a pore-scale is $\phi l/q_t$. If we assume, for convenience, $q_t/\phi = 10^{-6}$ m/s, or one μm each second, then the water advances by a pore length in of order a minute (79 s to be specific to our Bentheimer exemplar). Of course, this is not frontal advance at this rate, but locally a gradual rise in saturation, manifest at larger scales as a progressive movement of water from injection to production wells. Filling through pore centres, Eq. (6.26), or swelling wetting layers, Eq. (6.32), takes of order ms or less. It is much faster: we attain positions of local capillary equilibrium much more rapidly than the average flow. This is yet another way of demonstrating that capillary forces dominate at the small scale: most of the time the fluid interfaces are stationary with rapid jumps from one stable arrangement to another. However, for pinned layers, Eq. (6.33), we found hour-long times: this is longer than the average speed of advance and hence insufficiently rapid to contribute significantly to displacement.

This implies that for thin layer flow the conductance is insufficient to allow much filling away from a connected front. Hence, to study rate-dependent displacement, we need to consider the behaviour of percolation-like invasion patterns filling pore and throat centres. This is treated next to address the third item in the list on page 263.

6.4.4 Correlation Lengths for Percolation-Like Displacement

The presence of a viscous pressure differential introduces a gradient in the local capillary pressure, which in turn, for a percolation-like displacement, gives a finite correlation length, conceptually similar to that derived for buoyancy forces in Chapter 3.4.2, page 108.

Percolation theory can be used to describe both the initial advance of a non-wetting or wetting phase, or the trapping of non-wetting phase. In both cases the advancing or trapped phase is at or close to its percolation threshold and poorly connected while the other phase occupies most of the pore space.

The conductance, or relative permeability, of the advancing non-wetting phase close to the percolation threshold is

$$k_{rnw} \sim (p - p_c)^t, \tag{6.73}$$

for $p - p_c \ll 1$ with $t \approx 2$ in three dimensions (Stauffer and Aharony, 1994), although there is evidence that this exponent is not universal, in that it is different for different systems (Lee et al., 1986). Then, using Eq. (3.32),

$$k_{rnw} \sim S_{nw}^{t/\beta} \tag{6.74}$$

for low saturation where the exponent $t/\beta \approx 4.8$. We observe the same scaling for the initial percolation-like advance of a wetting phase in primary imbibition where we ignore wetting layer flow, or for the filled regions of wetting phase, neglecting the contribution of layers to the conductance.

As stated previously, the viscous pressure drop in the advancing non-wetting phase leads to a gradient in local capillary pressure, and consequently the local percolation probability, giving a finite correlation length. The calculation of this correlation length is, however, complicated by the fact that the flow rate in the advancing phase depends on the macroscopic evolution of the displacement (Wilkinson, 1986). If we consider an unsteady-state flow, where one phase is injected to displace another and both phases move in the same direction, then from conservation of volume, where the injected phase fills the pore space, the fluids should move at a speed of magnitude $v = q_t/\phi$; see Chapter 9 for a longer discussion. At the advancing front – where we have the incipient percolating cluster – again conservation of volume demands that $v = q_{nw}/\phi S_{nw}$ and hence, if these speeds are of the same magnitude, $q_{nw} \sim S_{nw}q_t$ at the front, with S_{nw} given by Eq. (6.74). In terms of the percolation probability, $q_{nw} \sim \Delta p^\beta q_t$.

We use Eq. (6.64) with oil as the non-wetting phase to determine the gradient in capillary pressure:

$$\frac{\partial P_c}{\partial x} = -\frac{\sigma Ca_{nw}}{K k_{rnw}}, \tag{6.75}$$

where x is the flow direction and we have used Eq. (6.66). We have assumed that the pressure gradient is controlled by the advancing (non-wetting) phase in this case. The negative sign indicates that the capillary pressure decreases away from the inlet.

We use arguments similar to those presented on page 106 to derive an approximate expression for the change in percolation probability with distance:

$$\frac{\partial p}{\partial x} = -\frac{Ca_{nw}}{\sqrt{\phi K} J^* k_{rnw}}, \tag{6.76}$$

using J-function scaling, Eq. (6.56), with the entry capillary pressure. We now rewrite the expression in terms of a more useful capillary number, defined on the rate q_t at which the invading fluid is injected at the inlet, Eq. (6.62). Then, from Eq. (6.76) and with $Ca_{nw} \approx \Delta p^\beta Ca$ we obtain

$$\frac{\partial p}{\partial x} = -\frac{Ca \Delta p^{\beta-t}}{J^* \sqrt{\phi K}}, \tag{6.77}$$

using Eq. (6.73).

The correlation length, ξ, is such that over a distance ξ the change in percolation probability from Eq. (6.77) is also consistent with Eq. (3.30); see page 108. After some algebra we obtain (Wilkinson, 1986; Blunt and Scher, 1995):

$$\xi \sim l \left(\frac{lCa}{J^* \sqrt{\phi K}} \right)^{-\nu/(1+\nu+t-\beta)} \tag{6.78}$$

where l is a pore length. The exponent $\nu/(1 + \nu + t - \beta)$ is around 0.25, leading to relatively short correlation lengths. If we again revert to our Bentheimer example with $l = 79$ μm and $\sqrt{\phi K} = 0.61$ μm with $J^* = 0.25$ then $\xi/l \approx 0.2Ca^{-0.25}$. For a representative value of $Ca = 10^{-8}$ we find a correlation length of around 20 pore lengths or about 1.5 mm.

This result applies for both invasion and ordinary percolation. It implies that even for slow injection we see a correlation length that is only tens of pores across. If we have a substantial fraction of water-wet pores, or a high initial water saturation, invasion will be a percolation process with poor connectivity limited by layers until a high water saturation – beyond the percolation threshold – is reached. As we move to more mixed-wet conditions, the connectivity can decrease if flow is impeded by pinned corner layers; however, as discussed in the previous section, this only allows very slow filling and so cannot support any significant displacement. We therefore will see a transition to a more invasion percolation-like advance for oil-wet media. Since the correlation lengths are quite low, there is not a huge distinction between these regimes – compare, for instance, the patterns at the percolation threshold in Fig. 3.15 where the grid is ten pores across – with a smooth transition from percolation-like to invasion percolation advance as a function of wettability and initial water saturation. However, we will see a major effect on relative permeability, dependent on the flow pattern, Chapter 7.

Below this correlation length the displacement pattern is self-similar and resembles the incipient spanning cluster in percolation; if we move further towards the inlet, the displacement becomes more filled in as it departs from the percolation threshold.

If we instead consider the connected advance with a finger width W, the same expression may be used but with W replacing l in Eq. (6.78), using Eq. (4.32); see page 158.

We can also derive a correlation length when non-wetting phase is trapped: this will determine the largest size of a stranded ganglion. The expressions are different, but the derivation is simpler, since here the principal pressure gradient is in the wetting phase, which is well connected with a relative permeability of order 1. The percolation probability changes with distance as

$$\frac{\partial p}{\partial x} = \frac{Ca}{J\sqrt{\phi K}}, \tag{6.79}$$

analogous to Eq. (6.77) without the terms in Δp since $q_t \approx q_w$ and Ca is defined with the water viscosity. Here the value of J is that at the end of the displacement, just as the non-wetting phase is trapped.

We pursue similar arguments to those used in the derivation of Eqs. (3.46) and (4.33) to find (Wilkinson, 1984, 1986)

$$\xi \sim W \left(\frac{CaW}{J\sqrt{\phi K}} \right)^{-\nu/(1+\nu)}. \tag{6.80}$$

If we use the same values as before for Bentheimer, with $W = l$ we find a correlation length in this case of around 300 pore lengths or 24 mm, which is now close to the core scale. This scaling of correlation length has been observed for trapped clusters in two-dimensional micro-model experiments (Geistlinger et al., 2016).

When both viscous and gravitational effects are important we combine them since they both contribute to the gradient in local capillary pressure, as demonstrated experimentally (Zhou, 1995) to find (Blunt and Scher, 1995)

$$\xi \sim W \left(\frac{CaW}{J\sqrt{\phi K}} + \frac{BW}{l} \right)^{-\nu/(\nu+1)} \equiv WCa_{eff}^{-\nu/(\nu+1)}, \tag{6.81}$$

where B is given by Eq. (3.42) and the effective capillary number is defined by

$$Ca_{eff} = W \left(\frac{Ca}{J\sqrt{\phi K}} + \frac{B}{l} \right). \tag{6.82}$$

If we ignore gravity, say $W = l$ and use our Bentheimer values, then $Ca_{eff} \approx 500Ca$ for $J = 0.25$.

6.4.5 Correlation Length and Residual Saturation

A finite correlation length reduces the residual non-wetting phase saturation, since it truncates the distribution of trapped clusters: no ganglia larger than a length ξ are present. We could start from the scaling relationship Eq. (4.37) on page 173, used to estimate the residual saturation as a function of finger width, W. In that derivation we assumed that the correlation length was infinite. We could instead substitute ξ from Eq. (6.81) to find the trapped saturation where there are no ganglia larger than ξ: however, this is not correct, as the larger ganglia are not completely displaced. Instead, they are broken up into smaller ones: the change in the amount trapped is the saturation in these larger clusters multiplied by the shift in percolation probability from an infinite to a finite system – the rest of the non-wetting phase is trapped, albeit in smaller clusters.

The calculation of the resultant shift in saturation from finite size effects is rather subtle, but follows the argument above: we present here a simplified derivation based on Wilkinson (1984) and Blunt et al. (1992). Imagine that we are at the percolation threshold in an infinite system: at this point all the non-wetting phase is trapped. Now, consider the saturation of non-wetting phase that would span a system of size ξ: this is given by Eq. (3.32) and represents the contribution of clusters larger than ξ. This is illustrated in Fig. 6.13: in an infinite system all the non-wetting phase shown is trapped, but some larger clusters span the correlation length ξ. The shift in percolation threshold Δp represents the additional displacement that is needed to break up these larger clusters, making them trapped over a finite length. Hence the change in trapped saturation compared to an infinite system is the saturation in clusters spanning a distance ξ (which is Δp^β) times the fraction of the non-wetting phase in these clusters, Δp, that are filled when the percolation threshold shifts to render them disconnected. We obtain

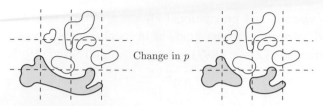

Change in p

Figure 6.13 A diagram to illustrate how a finite correlation length affects the residual non-wetting phase saturation using percolation theory. On the left, the trapped non-wetting phase in an infinite system with no gradient in capillary pressure is shown. The dotted lines represent boxes the size of the correlation length ξ. Some clusters – the one shown shaded – span one or more boxes. In a displacement with a gradient in local capillary pressure, and hence a finite correlation length, further displacement is needed, as no clusters larger than ξ remain in the system, shown in the right figure. The shift in residual saturation is the change in percolation threshold times the fraction of the non-wetting phase in the larger clusters, Eq. (6.83).

$$\frac{\Delta S_{nwr}}{S_{nwr}^\infty} \sim \Delta p^{\beta+1} \sim \xi^{-\nu(1+\beta)} \tag{6.83}$$

using Eq. (3.30). Then from Eq. (6.81) we find

$$\frac{\Delta S_{nwr}}{S_{nwr}^\infty} \sim Ca_{eff}^{(1+\beta)/(1+\nu)}. \tag{6.84}$$

The exponent $(1 + \beta)/(1 + \nu)$ has a value of approximately 0.76: for $Ca = 10^{-8}$ and hence $Ca_{eff} = 5 \times 10^{-6}$ we find a fractional change in residual saturation of only 10^{-4} for our Bentheimer example. Capillary numbers of around 10^{-5}, giving around a 2% change, or higher, are required to see a significant impact on trapping.

If we revert to the micro-model experiments of imbibition and trapping, Fig. 6.11, we see that the decrease in the amount of trapping is engendered by two effects. The first, treated here, is due to a gradient in the filling probability across the system, preventing the formation of large trapped clusters. The second, which is distinct, as it involves layer flow, is the retardation of snap-off at higher flow rates: it is this latter effect that dominates in most cases, so that the real change in residual saturation is somewhat more marked than would be predicted by Eq. (6.84).

While these percolation-type arguments are appealing, they assume that viscous forces introduce a perturbation to the local capillary pressure, with a gradient in the local filling probability, and ignore – as we have said – the impact of layer flow. They also cannot account for more radical effects, such as the movement of trapped phases induced by a pressure gradient, the final item in the list on page 263, which we treat next.

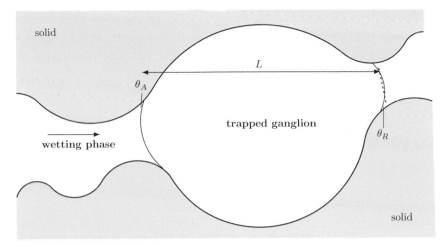

Figure 6.14 An illustration of ganglion mobilization. We find the viscous pressure drop across a trapped ganglion of length L necessary to force it through the pore space. Flow is from left to right: the wetting phase pressure is higher on the left, leading to a change in local capillary pressure across the system. The ganglion is about to pass through the throat to the right indicated by the dotted line.

6.4.6 Mobilization of Trapped Ganglia

Consider the situation shown in Fig. 6.14, where we calculate the flow rate necessary to push a trapped ganglion of non-wetting phase through the throat indicated. Hitherto, we have considered that these ganglia cannot move, since the viscous force is negligible in comparison with capillary effects. Here we test this hypothesis. Flow is from left to right: the pressure in the wetting phase is higher on the left of the figure and pushes the ganglion through the pore. The local capillary pressure varies: while we may assume that the non-wetting phase pressure is fixed (it is disconnected), the capillary pressure is lower on the left, in the pore, than on the right, where the ganglion is about the pass through the throat.

We can use Eq. (6.64) to find the difference in pressure across the ganglion; in this case $q_o = 0$, $q_w = q_t$ and $k_{rw} \approx 1$ giving:

$$\Delta P_v = \frac{L\sigma}{K} Ca \tag{6.85}$$

using Eq. (6.62) for the capillary number. Here we have used the subscript v to indicate that this is a change in local capillary pressure induced by viscous forces.

To push the ganglion through the pore space, this pressure must overcome the difference in capillary pressure DP_c necessary to enter the throat compared to movement through the pore to the left. If we have elements of circular cross-section, then $DP_c = 2\sigma (\cos \theta_R / r_t - \cos \theta_A / r_p)$. Note the different contact angles used: where the ganglion enters the throat we use the receding angle, and where it retracts from the pore we use the (water) advancing angle. Where $r_p \gg r_t$ and $\cos \theta \approx 1$ (for both advancing and receding angles), then to mobilize the trapped cluster we require:

$$\Delta P_v \geq D P_c, \tag{6.86}$$

$$Ca \geq \frac{2K}{Lr_t}. \tag{6.87}$$

We can use Eq. (6.53) to find

$$Ca \geq \frac{2a\phi r_t}{L}. \tag{6.88}$$

This represents a threshold capillary number for ganglion movement: if we have a higher flow rate, then the ganglion can be forced through the throat; if the flow is slower, the ganglion remains trapped in the pore. This threshold capillary number is not of order 1: as in all these types of calculation we need to incorporate pore-scale geometric information. For our Bentheimer example, Table 6.3, for a pore length $L = 79$ μm, we find a critical capillary number of around 2×10^{-3}: this is much higher than normally encountered in experiments or in the field, and higher than needed to see a significant shift in the amount of trapping caused by a small correlation length and the suppression of snap-off; see Fig. 6.11. This calculation is, essentially, the same as presented previously to find the ratio of viscous to capillary forces for flow through the centre of the pore space, demonstrating that in most displacements the fluid configuration remains capillary-controlled.

We can have ganglia that span several pores. Indeed, we can use our percolation theory arguments to estimate the maximum trapped cluster size, Eq. (6.81), and hence the capillary number necessary to push a blob this size through the pore space. In our example, for $Ca = 10^{-8}$, ξ was 300 pore lengths. For a ganglion this size, direct mobilization, Eq. (6.88), requires a capillary number of 6×10^{-6}. This is over two order of magnitude higher than the prevailing value. Higher flow rates lead to less trapping, not because already trapped ganglia are pushed through the pore space and flushed out of the system, but because large clusters are never completely surrounded in the first place.

This is an important result: ganglion movement is less important than the quasistatic perturbation to fluid configuration engendered by a local capillary pressure gradient. It is only significant if there is a large increase in capillary number after initial waterflooding, caused, for instance, by the injection of surfactant to lower the interfacial tension, or in model systems with rather similar pore and throat sizes such that DP_c is very small.

Once the capillary number reaches a value of around 10^{-3}, viscous and capillary forces are of a similar magnitude at the pore scale, the correlation length is therefore also the pore size, and there can be a significant decrease in residual saturation caused by the movement and recovery of non-wetting phase ganglia. Exactly how a ganglion will be mobilized by viscous forces depends on its length, the radius of the bounding throat and the wettability of the system; Eq. (6.88) simply gives an indication of the capillary number at which we expect to see geometrically trapped non-wetting phase swept through the pore space.

The decrease in residual saturation with capillary number has received extensive study in the literature (see, for instance, Lake (1989) and Dullien (1997)) as it is relevant to enhanced oil recovery processes, where the interfacial tension may be lowered by surfactants, or through the injection of a gas that is miscible or nearly miscible with the oil.

Rather than provide an exhaustive treatment here, we will instead present two instructive datasets. The first, Fig. 6.15, shows the results of the imbibition experiments in a sintered glass bead pack shown in Fig. 6.12. This is a capillary desaturation curve, where the residual saturation is plotted as a function of capillary number. We see little impact on flow rate for capillary numbers less than around 10^{-5} consistent with our estimates based on percolation theory, Eq. (6.84), and visual inspection of micro-model patterns, Fig. 6.11. We see a sudden drop in residual saturation at a critical capillary number of around 2×10^{-4}, likely caused by a significant reduction in the amount of layer flow and snap-off. The residual saturation for capillary numbers of 10^{-3} and larger is very low, declining to zero, as at these rates, ganglia can simply be flushed out of the system. These results are broadly consistent with earlier core flood experiments on bead packs, which again observed a sharp decline in residual saturation, albeit at slightly lower capillary numbers of around 10^{-3}; in these experiments there may have been less snap-off, since the porous medium was rather uniform, meaning that the residual saturation is only reduced once ganglia can be displaced directly from the pore space (Morrow et al., 1988). Trapping, ganglion formation and mobilization have also been imaged in a water-wet sintered bead pack using X-ray tomography by Armstrong et al. (2014a). At a pore level, the non-wetting phase breaks up into smaller clusters: it is the displacement of oil from some of these ganglia that causes the shift in saturation.

The second dataset is for experiments on Berea sandstone; see Fig. 6.16 (Chatzis and Morrow, 1984). Here two curves are shown for each experiment. The lower saturations are obtained for continuous oil: this means waterflood experiments performed at different flow rates. Here the principal cause of the decline in residual oil saturation is the alteration in the pore-level filling sequence caused by a gradient in local capillary pressure. The higher curve is for discontinuous oil: here a residual

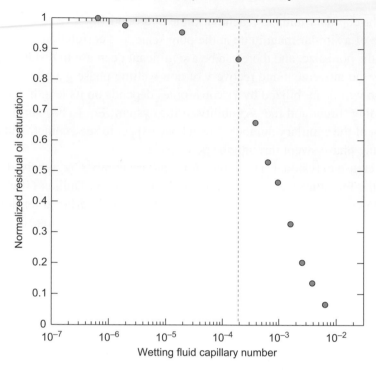

Figure 6.15 Residual oil saturation S_{or}, normalized by its maximum value, shown as a function of capillary number for the imbibition experiments illustrated in Fig. 6.12. We see that the amount of trapping does not vary significantly for small wetting fluid capillary number Ca, but decreases precipitously as Ca increases above 2×10^{-4} (dashed grey line), consistent with previous core flooding experiments on bead packs, Morrow et al. (1988). From Datta et al. (2014b).

saturation is established at a low flow rate and then the flow rate is increased. Now the only mechanism to recover more oil is direct mobilization, which – as we demonstrated above – occurs at a higher capillary number. In either case, there is a transition from a high to an almost zero value of residual oil saturation over more than an order of magnitude in flow rate, which is a reflection of the small-scale heterogeneity of the porous medium, with different throat sizes and consequently threshold flow rates for movement. Once the capillary number is 10^{-3} the residual saturation is very close to zero, since even the smallest, single-pore, blobs can be flushed from the rock.

One unsatisfactory feature of this analysis is that the critical capillary number, when viscous effects begin to dominate, is not one, but a much smaller value controlled by pore-scale geometry. This is an inevitable result of any analysis of the ratio of forces at the pore scale: it introduces characteristics of the void geometry, as well as macroscopically measurable parameters. Several authors have attempted to use a different expression for capillary number, based on the ganglion-scale ratio

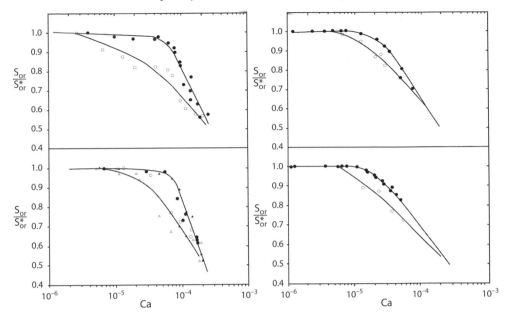

Figure 6.16 Experiments on different water-wet Berea sandstone cores showing the fractional change in residual oil saturation as a function of capillary number, Ca. The results for discontinuous oil, with more trapping (solid symbols), are experiments where waterfooding is performed at a low flow rate and then the flow rate is increased. With continuous oil (open symbols), separate experiments are performed at different flow rates to reach the residual oil saturation. Data from Chatzis and Morrow (1984).

of viscous to capillary forces, with a critical value of 1, using either theoretical arguments based on the real size of trapped ganglia or direct computation of the flow field in pore-space images (Hilfer and Øren, 1996; Anton and Hilfer, 1999; Andrew et al., 2014a; Kimbrel et al., 2015).

6.4.7 Ganglion Dynamics, Connectivity and Flow Regimes

When viscous forces become significant, displacement is governed by the flow field, which in turn is a function of the capillary number (flow rate), the viscosities of the fluids, and their density difference. The contribution of these effects on both residual saturation and the pore-scale morphology of the fluids has been observed directly by X-ray imaging (Kimbrel et al., 2015).

In some circumstances, the principal mechanism for fluid displacement is through the movement of ganglia, as opposed to flow along connected pathways of each phase, as assumed implicitly in the use of the traditional multiphase Darcy law, Eq. (6.58). This is observed at high flow rates, for capillary numbers larger than around 10^{-3} where, as we have shown above, viscous forces become important even at the pore scale. Ganglion movement is also significant in systems where there is little variation in pore and throat size, since then the difference in capillary pressure across a stranded blob necessary to allow passage through the pore space may be small. Lastly, ganglion transport may be forced through topological constraints. If, for instance, we inject both oil and water simultaneously in a two-dimensional micro-model that allows little or no layer flow, then the phases can only move through the centres of the pore space. For filling controlled by local capillary pressure, and where the pore sizes are not spatially correlated, a phase can only be connected across the system if it is above its percolation threshold, and this has to allow both site and bond filling. If this threshold is above 0.5, then it is not possible for both phases to connect across the system at the same time. For instance, for a square lattice, with a site percolation threshold of around 0.59, we cannot observe a pathway of filled junctions (pores) for both oil and water. As a consequence, continuous steady-state injection of both phases will, in almost all cases, result in ganglia of one or other phase being pushed through the system. This behaviour may not, however, be representative of better-connected three-dimensional porous media which also allow layer flow.

Payatakes and colleagues have studied and characterized the different types of ganglion transport as a function of capillary number, viscosity ratio and fractional flow of each phase through micro-model observations, and simulations using a network representation of the flow, and have quantified the macroscopic behaviour in terms of viscous coupling coefficients in Eq. (6.59) (Payatakes, 1982; Dias and Payatakes, 1986; Avraam and Payatakes, 1995b; Theodoropoulou et al., 2005). Ganglion transport has also been described and modelled using a mechanistic model by Valavanides et al. (1998) and Valavanides and Payatakes (2001). As an example, Fig. 6.17 shows different types of flow observed by Avraam and Payatakes (1995a). Four displacement patterns were seen. At low flow rates and relatively low oil (the non-wetting phase) saturation, flow occurs through the movement, break-up and coalescence of ganglia that typically span several pores. This was called large ganglion dynamics. As the flow rate increases, these ganglia become smaller, entering the small ganglion dynamics regime when the typical blob size is around that of a single pore. As the flow rate increases further, the ganglia became even smaller with several able to occupy each pore and throat – with a size that just spans a diameter of a pore – and we enter a regime termed drop-traffic flow. At the highest flow rates, the two phases are forced into approximately parallel flow channels and are connected: this is connected pathway flow. Ganglion motion was also observed experimentally in micro-model experiments by Tallakstad et al. (2009a, b) at capillary numbers greater than 10^{-3}. The behaviour in these experiments is very different from the conventional description of multiphase flow where it is assumed, instead, that connected pathways exist in the capillary-controlled limit. In the micro-model experiments of Avraam

Figure 6.17 Flow regimes observed in micro-model experiments. The non-wetting phase is shown in black. The upper right picture shows small ganglion dynamics, where the non-wetting phase moves in disconnected blobs which merge and break-up and whose size is around one, or a few, pores across. The upper left figure, at a higher magnification, shows a higher flow rate, where the ganglia break up into blobs that are around the width of a pore; this is called drop-traffic flow. The lower figure, at an even higher flow rate, showing a larger portion of the micro-model, illustrates connected pathway flow where both phases span the system, flowing in approximately parallel pathways from left to right. In all cases the distance between pores (the junctions of the network) is 1.22 mm. From Avraam and Payatakes (1995a).

and Payatakes (1995a), connected flow was the high rate regime, while ganglion transport was observed for capillary numbers around 10^{-8} to 10^{-5}: this is likely a consequence of forcing two phases through a two-dimensional model, as discussed above. Also note that this behaviour was not observed in the micro-model experiments of Lenormand and Zarcone (1984) (Fig. 6.11) since here water was injected in an unsteady-state experiment to displace oil: water and oil were never forced to flow simultaneously across the system. Furthermore, the design of the micro-models allowed more flow in wetting layers and roughness.

The appearance of small stranded droplets has also been observed in a system with a more complex pore space. Pak et al. (2015) performed waterflooding experiments in a carbonate, with a wide range of pore size, and imaged the residual non-wetting phase using X-ray tomography, as shown in Fig. 6.18. For low capillary numbers, around 2×10^{-7}, displacement was capillary-controlled, and large ganglia from a single pore-size upwards were observed, broadly consistent with the trapping experiments presented in Chapter 4, page 174. Here, the better connectivity of the phases in three dimensions allows connected pathway flow – until the non-wetting phase is trapped – and a percolation-like displacement pattern, unlike the topologically constricted micro-model experiments of Avraam and Payatakes (1995a). However, at a higher flow rate, a capillary number of approximately 10^{-5}, which, from Fig. 6.16, represents the onset of viscous effects in trapping (albeit in a sandstone with a narrower distribution of pore size), the apparent break-up of large clusters was seen, with the appearance of sub-pore-size blobs adhered to the solid surface; see Fig. 6.18. This demonstrates how, once viscous effects are observed at the pore scale in a heterogeneous rock, the resultant dynamics is not easily explained by quasi-static or percolation-like arguments.

The break-up of ganglia as the flow rate increases has been confirmed through three-dimensional imaging (Datta et al., 2014a). Fig. 6.19 shows fluid patterns as a function of time illustrating the movement and break-up of ganglia in a sintered bead pack: the system is that described in Figs. 3.3 and 6.12 except that here, as in the micro-model experiments of Avraam and Payatakes (1995a), both fluid phases are injected together, creating a steady-state flow (even if the fluid configurations continue to evolve over time). Movement and fragmentation of oil ganglia are observed at sufficiently large flow rates in either the non-wetting or wetting phases. Qualitatively we see the ganglion transport flow patterns observed by Avraam and Payatakes (1995a), but the conditions are different. As in the experiments of Pak

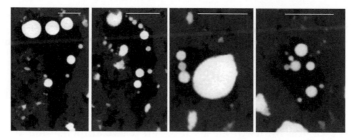

Figure 6.18 Two-dimensional cross-sections of three-dimensional images of fragmented oil droplets (white) in pores that are a few millimetres in size. Black represents the brine phase. The trapped oil phase is shown after waterflooding at a capillary number of approximately 10^{-5}. Droplets (white) that appear to be in free suspension are in contact with the pore surface when viewed in three dimensions. The scale bar is 1 mm. From Pak et al. (2015).

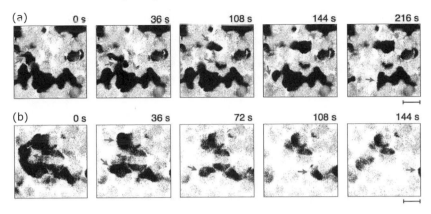

Figure 6.19 Oil ganglia break-up in a bead pack: this occurs when the capillary number, Ca_p, Eq. (6.66), exceeds around 8×10^{-4} for the wetting phase or 5×10^{-3} for the non-wetting phase. The images are shown at different times; the dark regions show oil. The imposed flow direction is from left to right. (a) At a high water flow rate, but slow oil flow, the wetting fluid breaks the oil up into smaller, discrete ganglia. (b) At lower water, but higher oil flow rates, the oil breaks up as it flows around the beads forming the porous medium. Labels show time elapsed after the first frame. Scale bars are 50 μm long. Ganglia are indicated by the right-pointing arrows, while the position of a reference bead is indicated by the left-pointing arrow. From Datta et al. (2014a).

et al. (2015), at low flow rates, connected pathway flow is seen and ganglia surrounded by wetting phase remain immobile. It is only when the capillary number exceeds 10^{-4} that these dynamic effects are apparent.

These observations can be quantified in a regime or phase diagram, delineating the conditions under which either connected flow or ganglion transport occur; see Fig. 6.20. There is a transition with increasing capillary number from the non-wetting phase flowing in a connected pathway, which is consistent with our analysis of quasi-static displacement, to first the intermittent break-up of these flow paths, followed – at the highest rates – by the advection of discrete ganglia as the principal mode of non-wetting phase movement. In these experiments, this transition is seen if either the wetting or non-wetting phase is flowing sufficiently quickly, with critical capillary numbers, Ca_p, for the onset of ganglion transport, Eq. (6.66), of 8×10^{-4} for the wetting phase and 5×10^{-3} for the non-wetting phase. These thresholds correspond approximately to the onset of a sharp decline in residual non-wetting phase saturation, Fig. 6.15. It is likely that in media with a more heterogeneous pore space, these transitions may start at lower capillary numbers: the suggestion from the experiments of Pak et al. (2015) is that at $Ca \approx 10^{-5}$ dynamic effects and ganglion fragmentation can be seen; this is also consistent with the beginning of a decrease in residual saturation at this value in Berea sandstone; see Fig. 6.16.

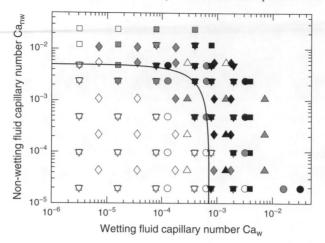

Figure 6.20 State or phase diagram of the transition from fully connected to broken-up flow for simultaneous two-phase displacement through a three-dimensional porous medium as a function of Ca_w and Ca_{nw}, Eq. (6.66), based on the results shown in Fig. 6.19. Open symbols represent flow of the non-wetting fluid through a connected three-dimensional pathway, grey symbols represent intermittent break-up of this pathway, and black symbols represent continual break-up of the oil into discrete ganglia, which are advected through the pore space. Each symbol shape represents a different viscosity, bead size or porous medium cross-sectional area. From Datta et al. (2014a).

The conclusion of this section is that while ganglion transport can be seen even at low capillary numbers in two-dimensional micro-models with restricted layer flow, in more realistic three-dimensional media, the advection of otherwise trapped clusters of non-wetting phase is only seen at capillary numbers of around 10^{-3} or higher, which is generally much larger than typically encountered in field-scale waterflooding or CO_2 storage applications, unless we consider surfactant flooding or near-miscible gas injection. These observations are consistent of our calculations of the balance between capillary and viscous forces at the pore scale.

More intriguing though is the appearance of small droplets in the pore space (Pak et al., 2015), Fig. 6.18, or the intermittent disconnection of flow pathways (Datta et al., 2014a; Rücker et al., 2015), Figs. 4.12 and 6.19, at lower capillary numbers. The break-up of ganglia at low rates has also been seen using X-ray tomography by Armstrong et al. (2014a). Here non-wetting phase flow does not occur by the transport of discrete ganglia, but then neither is the non-wetting phase always connected. A small number of snap-off and re-connection events causes a fluctuating pattern of connectivity, particularly when the non-wetting phase is close to its percolation threshold. This behaviour has an impact on the fluid distribution, and consequently on macroscopic parameters such as capillary pressure and relative permeability. There is not one constant connected path, but a more transient behaviour with flow occurring along temporarily trapped clusters that can then reconnect at critical junctions, like the movement of cars controlled by traffic lights: the road network is connected, but does not always admit flow. Since the application of pore-scale imaging in three dimensions to elucidate displacement processes is relatively recent, we do not have a full understanding of these rich phenomena. However, we can investigate the likely frequency of pore-scale re-arrangement during steady-state flow. That is, even if there is no ganglion transport, and even if there is no gradient in capillary pressure across the system, is it possible for the fluid to re-arrange in the pore space, potentially disconnecting (and then reconnecting) pathways? This cannot be analysed from a study of pressure differences, but requires a discussion of energy balance, presented below.

6.4.8 Viscous and Capillary Forces as an Energy Balance

An alternative approach to the comparison of viscous and capillary forces is to consider an energy balance. When we inject one phase to displace another, we do work in two ways. The first, which we considered explicitly for the derivation of the Young-Laplace equation, is the injection of one phase at a different pressure to another: the work done is then $P_c \Delta V$ where P_c is the pressure difference between the phases and ΔV is the change in volume of one phase as a result of displacement. This work creates interfaces between the fluid phases and changes in the solid surface energies. This process occurs regardless of the flow rate and is accounted for by considering a sequence of fluid configurations in capillary equilibrium.

There is, however, a second source of energy, which is simply the work done from injecting a fluid through the porous medium, even without displacement. This energy is usually equal to the viscous dissipation. Imagine a situation where, locally, there is no net displacement: $\Delta V = 0$. However, there is still energy conversion due to fluid flow: while most of this energy is lost as heat, it is possible that the fluid movement also allows the creation of additional surface energy, or results in fluctuations of the local fluid configurations, as seen in pore-scale imaging experiments and presented schematically on page 142, and in the previous section; see Fig. 6.19. If W is the work done then, if only viscous dissipation occurs

$$\frac{dW}{dt} = \int \mu \mathbf{v} \cdot \nabla^2 \mathbf{v} dV \equiv Q \Delta P \qquad (6.89)$$

for an incompressible fluid of constant viscosity. The integral is performed over some volume V, Q is the total flow rate through this volume, while ΔP is the average pressure change. This expression is strictly valid for single-phase flow, see, for instance, Talon et al. (2012), but can be extended to multiphase flow by considering Eq. (6.89) for each phase separately (Raeini et al., 2014a), in which case we write:

$$\frac{dW}{dt} = Q_w \Delta P_w + Q_o \Delta P_o \tag{6.90}$$

for the flow of oil and water.

While the rate of energy dissipation in Eq. (6.89) is useful to define an average pressure drop from a direct simulation of flow in the pore space, it is not possible to measure directly. However, we can estimate a likely order of magnitude and compare it with the energy needed to create a fluid interface in the pore space. Let us consider a volume that spans a representative pore length (or distance between throats) l. Rather than estimate the size of the integral in Eq. (6.89), we invoke a macroscopic description through the multiphase Darcy law, as employed previously when we introduced capillary number, to estimate the size of the left-hand-side of Eq. (6.90). $Q_p = q_p l^2$ for either phase p, while we substitute ΔP_p in Eq. (6.58) where we ignore the effects of gravity. $\nabla P_p \equiv -\Delta P_p / l$ and hence from Eq. (6.89),

$$\frac{dW}{dt} = \left(\frac{l^3}{K}\right)\left(\frac{\mu_w q_w^2}{k_{rw}} + \frac{\mu_o q_o^2}{k_{ro}}\right). \tag{6.91}$$

Imagine that the fluid flow provides energy to create an interface across the pore space, in a position of local capillary equilibrium, which is then displaced and re-formed periodically, generating an oscillating or fluctuating fluid configuration, as observed by Datta et al. (2014a) and Rücker et al. (2015). We can now compute the time needed for sufficient energy to be provided for such a process; of course, not all the energy can be used this way, as most is lost as heat. As an order of magnitude, we assume that the interface has an area of l^2 and a corresponding additional interfacial energy $W \sim \sigma l^2$. Then the minimum time required, using Eq. (6.91), if all the energy creates new interface, is

$$t = \frac{\sigma K}{l\left(\mu_w q_w^2 / k_{rw} + \mu_o q_o^2 / k_{ro}\right)} = \frac{K}{l\left(Ca_w q_w / k_{rw} + Ca_o q_o / k_{ro}\right)} \tag{6.92}$$

using Eq. (6.66). These expressions easily become sufficiently unwieldy to obscure the physical insight. To simplify, let us assume that the term in brackets, involving the flow rates of both phases, can be approximated as Caq_t using the capillary number based on the injection flow rate, Eq. (6.62). Then we have,

$$t \approx \frac{K}{Caq_t l}. \tag{6.93}$$

Using a representative Darcy velocity of 10^{-6} m/s, $Ca = 4 \times 10^{-8}$ and Bentheimer properties with $l = 79$ μm and $K = 1.88 \times 10^{-12}$ m^2, we find a time scale of approximately 600,000 s or 7 days. This is the time necessary for the viscous dissipation to be equivalent to the typical change in interfacial energy on creating a fluid interface at the pore scale.

To put this in perspective, Eq. (6.92) can be written in terms of a dimensionless time, t_D, which we will encounter again in Chapter 9; this is the number of pore volumes of fluid injected. In this case, if we have a system of volume l^3, the pore volume is ϕl^3 and the flow rate is $q_t l^2$. It therefore takes a time $\phi l/q$ to fill the pore volume with the injected phase. Then we can write, from Eq. (6.92)

$$t_D = \frac{qt}{\phi l} = \frac{K}{\phi l^2} \frac{1}{Ca}, \qquad (6.94)$$

which is the product of two dimensionless quantities: one related to macroscopic properties and one dependent on pore geometry, just as we have encountered for all other such calculations. For our example with $\phi = 0.2$, the number of pore volumes required is around 40,000. This means that an interface is created at the pore scale for every 40,000 pore volumes of fluid flowing by: to a good approximation we can consider flow at the scale of a few pores as proceeding with a fixed arrangement of fluid from which we may compute, or measure, the conductivity and hence relative permeability accurately, with little impact from fluid re-arrangement. Yes, there is displacement with interfacial fluctuations, but in a well-developed steady-state flow at representative values of the capillary number they are rare, and for most of the time we see flow of each phase in separate sub-networks consistent with the hypothesis behind the multiphase Darcy law. For slow flows, insufficient energy is introduced to allow significant fluctuations fluid configuration at the pore scale during the time it takes to flow across the system. The exception though will be cases where one phase is close to the percolation threshold: then the displacement of a single red bond, see Eq. (3.35) in Chapter 3.4.1, is sufficient to disconnect the phase. Here we may expect even infrequent fluctuations in local fluid patterns to have a large impact on connectivity.

If we have a higher capillary number, however, we rapidly see the onset of viscous effects, since the energy dissipation scales as the square of the flow rate, Eq. (6.91). For instance, in the experiments of Pak et al. (2015), with $Ca = 10^{-5}$, $K \approx 5 \times 10^{-14}$ m^2, and $\phi = 0.17$ with a characteristic length of around 10^{-4} m for the larger pores (see Fig. 6.18), then Eq. (6.94) gives $t_D \approx 3$. In the experiment 10 pore volumes were injected on a sample of length, $L = 12.5$ mm: over a length l this is equivalent to $10 \times L/l \approx 10^3$ pore volumes: more than enough to provide sufficient energy for the fragmentation of ganglia. When t_D is of order one, at capillary numbers of around 10^{-3}, then there is sufficient energy in the flowing fluid to allow the complex ganglion dynamics described previously.

6.4.9 Direct Computation of Multiphase Flow

The interaction of viscous and capillary forces is sufficiently complex that, apart from the somewhat simple theoretical arguments already presented, the only way to quantify or predict the flow behaviour is through numerical simulation. If only capillary forces act, a displacement sequence can be defined using a quasi-static network representation of the porous medium, as described in the previous chapters. To capture more accurately real pore-space geometries, other methods can be used to compute positions of capillary equilibrium, such as semi-analytical calculations on two-dimensional images (Zhou et al., 2014), or the use of level sets, see page 33 (Sussman et al., 1994; Spelt, 2005; Prodanović and Bryant, 2006; Jettestuen et al., 2013). However, once viscous effects are included, we also need to account for the time evolution of the flow field on both local capillary pressure and the rates at which the fluid may advance. The work we describe below will incorporate many of the phenomena, such as wetting layer flow, which we have already discussed, while computing the movement of a displacement pattern and finding averaged flow properties. There are a number of methods which have been employed to achieve this that we will discuss briefly here. This is not intended to be an exhaustive presentation of a very extensive literature; instead the main concepts and a few illustrative results will be presented. An excellent review of the main methods is provided by Meakin and Tartakovsky (2009).

Perturbative network models. It is assumed that flow rate produces a change in local capillary pressure, and either analytical arguments or a solution for the flow field with an instantaneously static configuration of fluid are used to estimate this change. The effect of flow rate is simply to reorder the filling sequence of pores and throats: the local capillary pressure for displacement remains the same, but this corresponds to a different injection pressure, since there is a viscous pressure drop across the system. Pores and throats are now filled in order of injection pressure, rather than the local capillary pressure. This method retains the speed and simplicity of quasi-static network modelling and can capture the perturbative influence of flow rate on displacement patterns and trapping. However, it lacks an explicit time dependence, does not accommodate different rates of filling and cannot simulate more complex dynamic phenomena, such as ganglion motion. This idea has been used by Blunt and Scher (1995) and Hughes and Blunt (2000) to identify different flow patterns as a function of capillary number, contact angle and initial wetting phase saturation, controlling the degree of layer flow. This approach included a fixed resistance to flow through wetting layers and demonstrated that, due to their low conductance, snap-off can be sufficiently suppressed to alter the flow pattern at capillary numbers as low as 10^{-8}. By varying the resistance to flow in wetting layers, a transition from frontal advance to a more percolation-like displacement

with snap-off is seen, which is qualitatively similar to that observed experimen
tally. The method was extended to allow a variable wetting layer resistance by
Idowu and Blunt (2010): the efficiency of the algorithm allowed the simulation of
a macroscopic displacement pattern at the core (cm) scale.

Dynamic network models. A numerical solution for the flow field is coupled
with interface movement and filling, governed by the local pressure gradients and
phase conductance, while obeying conservation of volume. Many researchers have
developed such models to explore displacement patterns, determine residual saturation as a function of flow rate and to compute averaged properties, such as relative
permeability. We can only highlight some of the main contributions here: reviews
have been provided in Blunt (2001b), Joekar-Niasar and Hassanizadeh (2012) and
Aghaei and Piri (2015). Chen and Koplik (1985) developed a network model of
imbibition and drainage with which they compared computed pore-scale displacement processes with those from micro-model experiments; Blunt and King (1991)
extended this work to study unsteady-state radial flow to show the impact of viscosity ratio and flow rate. Other studies have focussed on ganglion dynamics (Hashemi
et al., 1999), interpreted the micro-model studies described earlier (Dias and Payatakes, 1986) and studied the effect of viscosity ratio and local viscous stresses
on displacement patterns (Vizika et al., 1994). Other, more sophisticated models
have been developed later based on these ideas, to capture the details of meniscus
movement (Dahle and Celia, 1999; Al-Gharbi and Blunt, 2005) with successful
comparison with micro-model experiments (van der Marck et al., 1997).

Dynamic models also allow the exploration and formulation of extensions to the
conventional Darcy-like description of multiphase flow, which will be discussed in
more detail, Chapter 6.5 (Joekar-Niasar and Hassanizadeh, 2011), and to compute
non-traditional averaged properties such as interfacial area and velocity (Nordhaug
et al., 2003).

The inclusion of wetting layers is important, as we have shown, particularly
for imbibition processes. This may be studied semi-analytically to compare, for
instance, co-current and counter-current flow processes (Unsal et al., 2007b, a).
Mogensen and Stenby (1998) computed the flow field through the centres of the
pore space in a three-dimensional network, but assumed a fixed conductance in
wetting layers, which tends to overestimate the impact of flow rate on the suppression of snap-off and trapping. The swelling and flow through such layers needs to
be computed to capture accurately the filling sequence (Constantinides and Payatakes, 2000; Singh and Mohanty, 2003). Nguyen et al. (2006) developed a model
that accounted for wetting layers: they showed that if the pores are much larger
than throats (high aspect ratios), flow rate has a significant impact, since snap-
off controls the displacement in the quasi-static limit, which is then impeded as
the capillary number increases. They also noted the equivalence between capillary

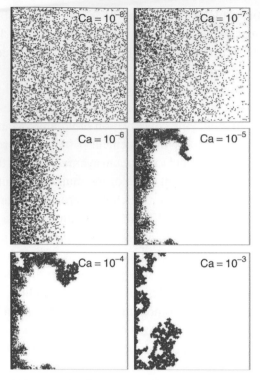

Figure 6.21 Simulation results from a two-dimensional dynamic network model at the capillary numbers, Ca, Eq. (6.62) indicated. Note the qualitative agreement with the micro-model experiments of Lenormand and Zarcone (1984) shown in Fig. 6.11. From Nguyen et al. (2006).

number and contact angle in inhibiting snap-off and showed that increasing Ca was similar to increasing θ in terms of the resultant relative permeabilities and the degree of trapping for water-wet systems. Example illustrative results are shown in Fig. 6.21: note the qualitative agreement with the micro-model experiments shown in Fig. 6.11.

Among the more sophisticated models of displacement, Aker et al. (1998) captured the change in local capillary pressure as a meniscus moved from throat to pore, while tracking multiple interfaces within each pore or throat. This work was used to study the scaling properties of the advancing interface, which were in accord with percolation theory, including burst dynamics (see Chapter 3.4.1, page 104 (Aker et al., 2000)), and to compare displacement patterns for different flow rates with experiment (Løvoll et al., 2005). Later, a similar model was used with a reconstructed sandstone network to study saturation patterns and electrical resistance (Tørå et al., 2012). A two-dimensional model with periodic boundary conditions was employed to study flow regimes during steady-state displacement, similar, conceptually, to the work of Avraam and Payatakes (1995a), and to delineate the conditions on capillary number, viscosity ratio and saturation under which there was flow of one phase, or the other, or two-phase flow (Knudsen et al., 2002; Knudsen and Hansen, 2006).

A fully dynamic network model is much slower than one that uses a perturbative approach, since the movement of fluid menisci need to be computed. In a quasi-static model with n elements (pores or throats), sorting one threshold capillary pressure in rank order takes of order $\ln n$ operations, see page 97. If the calculation of the inlet pressure needed for filling is relatively fast when the perturbative effects of viscous forces are included, or performed analytically, then it takes of order $n \ln n$ operations to complete the simulation. A dynamic simulation also requires a computation of the flow field. With the most efficient techniques, this too takes of order $n \ln n$ operations (see, for instance, Stüben (2001)). However, this is updated many times as displacement proceeds. If the number of re-computations is m, then the total simulation time scales as $mn \ln n$. m is related to the capillary number: the flow field will change significantly once a meniscus has traversed a pore. A dynamic model has to accommodate the wide range of time scales associated with filling, from the very rapid advance of a meniscus during, for instance, a Haines jump, compared to the slow overall flow, or the filling through wetting layers, which as we have shown, Eq. (6.34), may be many orders of magnitude slower. In general, the flow field needs to be recomputed frequently, and often after the filling of any pore, since this has a large effect on the movement of fluid in its neighbourhood. For fast flows, many elements are filled simultaneously, so that in the time taken for the fastest-moving meniscus to pass through a pore, many other pores may be filled, or partially filled as well, whereas in the quasi-static limit, only a single meniscus may be moving at any one time, with the rest of the fluid interfaces in static equilibrium. This implies that $m \sim 1/Ca$. Note that now the quasi-static limit or low Ca is most computationally demanding: this is because it is difficult to maintain most interfaces immobile, combined with the locally rapid movement of relatively few menisci which have reached their threshold capillary pressures. In contrast, rapid filling, with most, or all, the menisci in motion, is much easier to capture, since the flow field is more uniform and many pores are filled at once.

This constraint on computer time had, hitherto, limited dynamic simulations to rather small networks, with at most 100,000 pores and throats. However, more efficient algorithms and the advent of parallel computing do allow much larger systems to be studied now. Aghaei and Piri (2015) used a network model with 5.8 million pores and throats to represent flow at the same core scale as a suite of steady-state displacement experiments, allowing a detailed comparison between simulation and measurements. The Berea network employed is shown in Fig. 6.22:

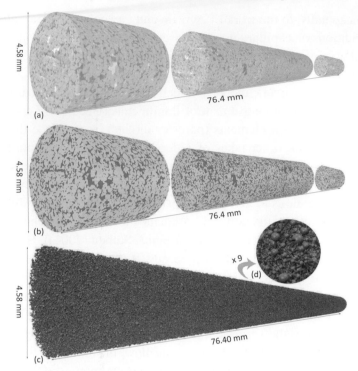

Figure 6.22 Pore network representation of a Berea sandstone core sample. (a) Greyscale images with a voxel size of 2.49 μm. (b) A segmented image where red and grey represent the pore and grain voxels, respectively. (c) The pore network generated from image (b). Red and blue represent the pores and throats in the network, respectively. For illustrative purposes only pores and throats are shown with circular cross-section. (d) A magnified image of a small section of the network. From Aghaei and Piri (2015); the network was generated by iRock Technologies, Beijing. (A black and white version of this figure will appear in some formats. For the colour version, please refer to the plate section.)

the flow model incorporated the effects of wetting layers and used contact angles consistent with direct measurements on pore-scale images of the same sample. Some of the results of this model will be shown in Chapter 7 when we describe relative permeability in detail.

Direct simulation methods on pore-space images. For single-phase flow, to compute flow fields, predict permeability and simulate transport processes, direct simulation on pore-space images has many advantages over a network approach, since it automatically captures the pore-space geometry in an image and is sufficiently fast to allow computations on domains the size of a representative elementary volume or larger for simple systems; see page 242. However, for multiphase flow, the problem becomes much more computationally demanding because of the number of times the flow field needs to be computed and the resolution required. We can use the same arguments presented above to predict an $n \ln n / Ca$ scaling of run time; however, here n is the number of grid blocks rather than the number of pores. To capture, accurately, the movement of menisci at a sub-pore resolution may require re-computation of the flow field as these menisci traverse every voxel. Furthermore, to resolve layers it is necessary to have a reasonably fine grid: as a minimum, 10 blocks need to span every pore and throat, which implies at least 10^3 voxels per pore in three dimensions. Hence, direct methods applied to study flow on the same sized sample will be many orders of magnitude slower than an equivalent network model. More significantly, in the quasi-static limit where viscous effects are unimportant, a capillary-controlled model is extremely efficient and – while the pore geometry is idealized – allows a semi-analytic computation of fluid configurations and conductance within each pore with effectively infinite resolution. A direct simulation is extremely slow and the limited resolution makes the prediction of important phenomena, such as layer flow, challenging. This is easy to appreciate with a definite example. If we return to our Bentheimer image with 10^9 voxels, then, assuming that this is sufficient resolution to capture multiphase flow (and in fact it does not properly account for wetting layers, and oil layers in mixed-wet systems), then for a slow displacement at a representative capillary number of 10^{-7}, the simulation requires of order 10^{18} operations. In contrast, for quasi-static network modelling with around 80,000 pores and throats, see Table 2.1, we have of order 10^6 operations, which is 10^{12} times faster! Now, we have not explored pre-factors in this calculation, but needless to say, a network calculation assuming capillary equilibrium takes typically less than 1 s on a standard workstation, whereas the equivalent direct simulation requires high-speed computing and many days of simulation time, yet still fails to capture properly layer flow. If we return to our discussion of length scales in more heterogeneous rocks (page 36), we estimated that typically 10^{18} voxels, at least, are required to describe the pore space from the size of the smallest pores to a representative elementary volume, making direct simulation infeasible.

The conclusion of this discussion is that at present, while direct methods can capture flow accurately for simple systems at the scale of several pores, they are mainly confined to the study of flow phenomena at the sub-pore or few-pore level. Even with the fanciest computers, it is not realistic to presume that such methods will ever be able to compute macro-scale multiphase flows on complex systems. What will be required in future is a hybrid approach, where a network model is conceived as not so much an explicit (and therefore simplified) reproduction of the pore space, but an upscaled representation, where the displacement between two regions of the pore space is parametrized in terms of a saturation-dependent conductance and capillary pressure: the parameters themselves may be derived from direct simulation at the smaller scale; see Chapter 2.2.3, page 47. These considerations are often overlooked in the hopeless quest to develop one simulation method that can capture everything from pore-scale displacement to predictions of relative permeability.

Direct simulation methods need to solve the Navier-Stokes equations while accounting for interfacial forces: essentially solutions for flow with interfaces. Scardovelli and Zaleski (1999) provides a review of grid-based methods to address such problems, but is not focussed specifically on porous media applications, where capillary effects dominate at the pore scale. In the volume of fluid method the Navier-Stokes equations are solved, usually using a finite volume approach, and, from mass conservation, the volume of each phase in each grid block is computed. In blocks containing both phases an interface is constructed and interfacial phenomena incorporated into the dynamics (Hirt and Nichols, 1981; Rudman, 1997; Gueyffier et al., 1999). Huang et al. (2005) used this approach to study multiphase flow in fractures; later the work was extended to consider slow flow in porous media (Raeini et al., 2012, 2014a, b). Ferrari and Lunati (2013) presented a series of simulations of multiphase flow using this idea to demonstrate that capillary pressure is not the difference in the average pressure between the phases in a dynamic displacement, but is better captured from a computation of interfacial energy, applying the Young-Laplace equation.

There are other methods that can be used to combine a grid-based solution of the Navier-Stokes equation with fluid interfaces. In the volume of fluid method mentioned above, a molecularly sharp interface is represented by some indicator function giving a phase boundary of finite width. It is possible instead to employ a front-tracking technique, where a separate unstructured surface mesh is used to capture the jump between phases (Unverdi and Tryggvason, 1992; Tryggvason et al., 2001). Other front-tracking, sharp interface or level set methods have been proposed which again capture these boundaries precisely (Sussman et al., 1994; Popinet and Zaleski, 1999; Liu et al., 2005), with Ding and Spelt (2007) providing a comparison of approaches. However, these methods have not yet been applied directly to slow immiscible flows in rocks.

The final grid-based method I will mention is direct hydrodynamic simulation, which employs a diffuse interface approximation at fluid/fluid boundaries using a density functional approach and incorporates a flexible thermodynamic formulation that allows, in principle, complex flow processes, such as non-Newtonian rheology and thermal effects, to be modelled. It has been applied to study pore-scale displacement phenomena and to simulate multiphase flow in pore-space images, predicting relative permeability for capillary numbers of around 10^{-5} and larger (Demianov et al., 2011; Koroteev et al., 2014).

As an example of the capability and utility of direct grid-based simulation, Fig. 6.23 shows the fluid configurations and local capillary pressures computed

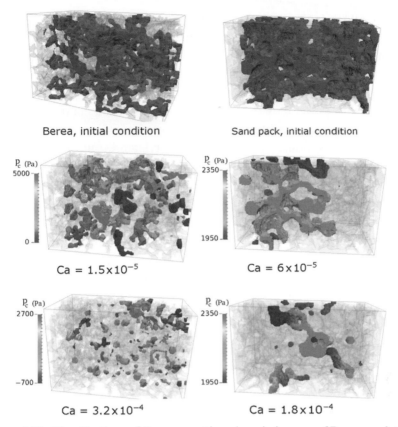

Figure 6.23 Visualizations of the non-wetting phase in images of Berea sandstone (left) and a sand pack (right) at the start (top) and at the end of water injection (middle and lower figures). Simulations are performed at the different capillary numbers shown. The colours indicate the local capillary pressure. Through capturing flow and the geometry of the pore space accurately, the residual saturation, particularly for the sand pack, can be predicted more accurately than with network modelling. From Raeini et al. (2015). (A black and white version of this figure will appear in some formats. For the colour version, please refer to the plate section.)

directly on pore-space images with around 10^6 unstructured grid blocks for different capillary numbers (Raeini et al., 2015). The fluid patterns, and local capillary pressures in the trapped ganglia, are qualitatively similar to those shown in Figs. 4.28 and 6.19. The results were also compared with those from pore-scale modelling with a network extracted from the same image and containing around 50,000 pores and throats. Using our analysis of computation times, for a capillary number of 1.5×10^{-5}, we would expect the network model to be approximately $10^6 \ln 10^6 / 1.5 \times 10^{-5} \times 50{,}000 \ln 50{,}000 \approx 1.7 \times 10^6$ times faster: the real speed-up was a factor of 5.6×10^5 on the same computer (2 s as opposed to 13 days).

Compared to macroscopic experimental measurements, these simulations gave reasonable predictions of relative permeability (Raeini et al., 2014a), the trend of residual saturation with capillary number and the dependence of residual saturation on initial saturation; see Fig. 4.33 (Raeini et al., 2015). While much slower than quasi-static network modelling, one advantage of this approach is a better representation of cooperative pore filling in homogeneous media, allowing a much more accurate prediction of trapping in a sand pack: pore-scale models use empirical approaches for pore filling (Chapter 4.2.2, page 131), rather than track properly the evolution of the meniscus through the pore space, leading to an over-prediction of the residual saturation in some cases. Other features are the ability to study the impact of viscous forces on residual saturation, as well as analyse pore-scale displacement processes impacted by viscous effects, including the suppression of snap-off discussed earlier (Raeini et al., 2014b), and, similar conceptually to the work of Ferrari and Lunati (2013), to show how considerations of energy balance can be used to determine relative permeability (Raeini et al., 2014a).

Particle-based approaches. To date, most direct simulations of multiphase flow have used particle-based methods, usually lattice Boltzmann simulation; see Chapter 6.2.4. These approaches are naturally adapted for parallel simulation and through accommodating interactions between different types of particle, elegantly and simply include interfacial forces. However, they cannot overcome the computational limitations discussed above: the scaling of run time remains as before, albeit – with the right computer and an efficient algorithm – lower pre-factors.

The first lattice Boltzmann simulations of multiphase flow on pore space images were performed by Ferréol and Rothman (1995) and Martys and Chen (1996), who studied images of Fontainebleau sandstone (see Fig. 2.10) and computed absolute and relative permeabilities. Subsequent studies have widened the range of application to predict capillary pressure (Pan et al., 2004), capillary pressure hysteresis (Ahrenholz et al., 2008), interfacial area (Porter et al., 2009) and relative permeability for different rocks (Boek and Venturoli, 2010; Hao and Cheng, 2010; Ramstad et al., 2010). The advantage of a flexible computational method is that the exact flow and fluid conditions in an experiment, even with large density contrasts (Inamuro et al., 2004), can be replicated, allowing the effects of viscous coupling or different boundary conditions (such as steady-state as opposed to unsteady-state flows) to be investigated (Li et al., 2005; Ramstad et al., 2010).

Fig. 6.24 shows computed fluid configurations for imbibition in a Bentheimer sandstone image (Ramstad et al., 2012). These simulations allow the effects of pore geometry and flow rate to be investigated, are sufficiently refined to capture layer flow and make good predictions of relative permeability, which we will present in the next chapter.

Another simulation approach, introduced in Chapter 6.2.4, is smoothed particle hydrodynamics. Beyond computations of flow and predictions of permeability, it has been employed to study miscible flow (Tartakovsky and Meakin, 2005) and

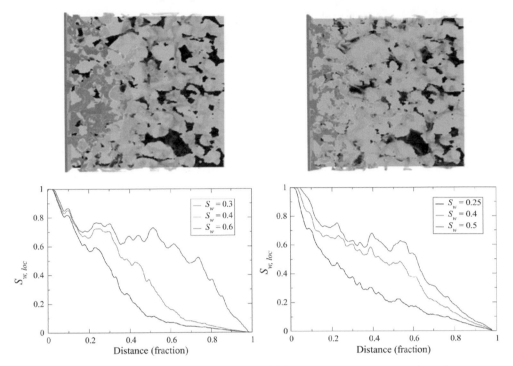

Figure 6.24 Fluid configurations for imbibition at an average saturation $S_w =$ 0.4 (top) with associated local saturation profiles (bottom) for two flow rates in a Bentheimer image with 256^3 voxels using lattice Boltzmann simulation. The wetting fluid saturation is displayed with red the largest value towards blue as zero. The pore space is black. The capillary number, $Ca \approx 5 \times 10^{-5}$ in the left picture and 5×10^{-6} in the right. From Ramstad et al. (2012). (A black and white version of this figure will appear in some formats. For the colour version, please refer to the plate section.)

solute transport (Ovaysi and Piri, 2010). This method has been applied to multiphase flow (Tartakovsky and Meakin, 2006; Tartakovsky et al., 2007, 2009a) for the study of oil-phase dissolution (Tartakovsky et al., 2009b), bubble movement in a channel (Tartakovsky et al., 2009a) and interfacial area in drainage and imbibition (Sivanesapillai et al., 2016). Liu and Liu (2010) provide a review of the method and its applications.

This whirlwind review of a fascinating zoo of numerical techniques does not, on its own, provide new insight into multiphase flow. What we will do now is synthesize the numerical results, and those from experiments, to discuss three main topics. The first is how to extend the multiphase Darcy law to better capture viscous effects, then to explore the generic flow patterns that result for displacements at different flow rates and viscosity ratios and, finally, to present relative permeability curves.

6.5 Extensions to the Multiphase Darcy Law

There have been many attempts to extend the multiphase Darcy law to provide a richer and more accurate characterization of flow. These extended formulations are developed for different reasons: they may be designed to describe displacement which is not percolation-like and for which the Darcy formulation is intrinsically inappropriate; to include rate-dependent effects; to reduce or eliminate hysteresis in the characterization of capillary pressure and relative permeability; and to incorporate rigorous considerations of energy, momentum and entropy balance.

6.5.1 Infiltration and Phase Field Models

As discussed in Chapter 4.3.6, the downwards migration of a wetting phase – rainfall moving through a dry soil – is an unstable process, characterized by persistent fingers which are nearly saturated at the tip, but with a lower saturation behind. This was explained as a fingered frontal advance of the wetting phase followed by lateral drainage. However, the conventional model of relative permeability cannot explain this behaviour: a non-monotonic saturation profile with depth cannot result from a monotonic imbibition capillary pressure curve (Eliassi and Glass, 2001). The displacement is unconditionally stable if there is no saturation overshoot: there is a uniform wetting phase saturation at the inlet whose Darcy velocity is equal to the rate of infiltration with a decrease in saturation with depth. Indeed this is what is seen if the soil is initially sufficiently damp to allow wetting layer flow and a percolation-like advance of the water.

The fingering is a consequence of the frontal advance of the wetting phase: this is not a percolation-like process and cannot be described by an equilibrium continuum model of capillary pressure and relative permeability. This is not a problem with the incorporation of viscous effects, but intrinsic to the flow regime. The water has to penetrate the porous medium at a high saturation: if the Darcy velocity for this saturation exceeds the inlet flux (the rate of rainfall), then the only possibility is to have a series of unstable fingers of wetting phase. However, this is a somewhat unsatisfactory conclusion, since it appears to preclude a macroscopic treatment of an important multiphase flow phenomenon.

This problem has been addressed using a wide variety of extensions to the conventional theory that allow a fingered profile to develop. This work has been analysed and reviewed in detail by DiCarlo et al. (2008) and DiCarlo (2013) who showed that to reproduce the fingering seen experimentally in an averaged formulation of flow, a non-monotonic capillary pressure is required, and/or a high-order dispersive contribution to the wetting phase flux; however, the capillary pressure is not physical and is qualitatively different from that encountered in other situations, and the introduction of regularization or dispersive terms in the governing flow equations is somewhat *ad hoc*.

Of the variety of approaches, the most physically appealing impose a capillary pressure for the advancing finger, consistent with a frontal pore-scale displacement. This can either be done directly (Steenhuis et al., 2013) or through the incorporation of a so-called hold-back-pile-up effect, which acts to force frontal advance in the fingers (Eliassi and Glass, 2002). It was also found that the wetting phase pressure in the finger tip increases with flow rate (and here we have local capillary numbers that can be in the viscous-controlled regime of 10^{-3} or higher). This can be predicted using a rate-dependent contact angle (Steenhuis et al., 2013). This can also be reproduced using a velocity-dependent capillary pressure instead (Hilpert, 2012), which fundamentally is trying to capture the same behaviour. In any event, there is a transition from frontal advance for the advancing finger tips to percolation-like drainage later with an inevitable jump in relative permeability (or flow conductance) and capillary pressure; such discontinuities are sufficient to allow the propagation of non-monotonic saturation profiles (Hilfer and Steinle, 2014).

Another elegant solution to this conundrum is through a modification of the Darcy velocity of the wetting phase, pursuing an analogy with film flow, such as drops of water flowing down a window pane, to provide a thermodynamically consistent treatment of unstable frontal advance (Cueto-Felgueroso and Juanes, 2008, 2009a, b). An additional flux, representing interfacial tension at the wetting front, is included in the multiphase Darcy law, Eq. (6.58):

$$\mathbf{q}_p = -\frac{k_{rp}K}{\mu_p}\left(\nabla P_p - \rho_p\mathbf{g} + \frac{\Gamma}{\rho g}\nabla(\nabla^2 S_p)\right) \qquad (6.95)$$

applied to the wetting phase (water). The additional, final, term related to the divergence of the saturation profile becomes vanishingly small for steady-state displacement, or where there is a smooth variation in saturation, and hence the model reverts to the traditional form in these cases. However, at a wetting front, the model allows the development of an instability and a non-monotonic saturation profile, as observed experimentally. The model parameter Γ can be estimated on physical grounds, leading to a predictive theory with no additional parameters (Cueto-Felgueroso and Juanes, 2009a).

While the model is compelling, providing a physical justification for an additional dispersive term in the water flux, it does not address the wider issue of the rate-dependence of relative permeability. The introduction of a third-order derivative in Eq. (6.95) also appears daunting from an analytical and numerical perspective, although it can be, and has been, successfully incorporated into simulations of fluid flow (Cueto-Felgueroso and Juanes, 2009a).

6.5.2 Accounting for Non-Equilibrium Effects

In the Russian literature, Barenblatt (1971) developed an extension to the multiphase Darcy law to accommodate non-equilibrium behaviour. This work is best described in the book Barenblatt et al. (1990) and a later pedagogical paper (Barenblatt et al., 2003). In our discussion of primary drainage and layer flow, see pages 113, 232 and 266, we suggested that there is a systematic tendency for the macroscopic capillary pressure to be overestimated, or, more specifically, for the measured wetting phase saturation to be too high, since there had been insufficient time for the local and macroscopic capillary pressures to equilibrate through slow layer flow. We mentioned that similar behaviour could be seen during waterflooding, particularly in mixed-wet or oil-wet media where flow through oil layers resulted in long time scales for capillary equilibrium to be achieved.

These considerations can be accommodated by instead of writing the relative permeabilities and capillary pressures (or Leverett J function) as functions of saturation (normally the water saturation in two-phase flow), we consider them as a functions of some effective value η. In steady state we recover $\eta = S_w$ and k_{rp} and P_c resume their equilibrium values; hence they can be measured for slow flow and then the same functions used for non-equilibrium displacement if the relationship between η and S_w is known. In general

$$\eta - S_w = \Psi\left(S_w, \ \tau\frac{\partial S_w}{\partial t}\right),\tag{6.96}$$

where Ψ is some dimensionless function and τ is a time scale for achieving porescale equilibrium. The simplest version of this model is to assume that Eq. (6.96) can be written

$$\eta - S_w = \tau\frac{\partial S_w}{\partial t}.\tag{6.97}$$

Note the sign of the terms. τ is positive. For a drainage experiment, S_w decreases with time, meaning that η is less than S_w: the capillary pressure and flow behaviour is properly captured by a lower saturation than observed macroscopically in an experiment performed at a finite rate. This is physically correct: the imposed macroscopic capillary pressure is higher ($\eta < S_w$) than its local or equilibrium value at the current wetting phase saturation. In an imbibition experiment with S_w increasing, $\eta > S_w$ and the macroscopic behaviour is representative of an equilibrium displacement at a higher wetting phase saturation.

This formulation has been used to study spontaneous imbibition, which is intrinsically a non-equilibrium process (Barenblatt et al., 2003). Here a sample initially either completely or partially saturated with the non-wetting phase (at a high capillary pressure) is placed in contact with wetting phase with a capillary pressure of zero: the wetting phase then enters the porous medium driven by capillary forces; see Chapter 9 for a fuller presentation. Experimental data could be matched using Eq. (6.97) and a relaxation time, τ of a few s to around 1,000 s, dependent on the system, which is comparable with the times necessary to fill a single pore by wetting layer flow, Eqs. (6.32)–(6.34).

While this model has a physical appeal and is relatively simple, there is no obvious way to determine τ experimentally; moreover, it is a function of saturation, since the time scales for layer flow are dependent on layer thickness, which is governed by capillary pressure. Furthermore, while spontaneous imbibition could be explained in this non-equilibrium framework, as we show in Chapter 9.3, a satisfactory quantitative interpretation and prediction of the behaviour can be provided using the conventional formulation of multiphase flow. Lastly, it is not clear how this model emerges from a rigorous averaging of pore-scale dynamics; this is discussed next.

6.5.3 Averaged Equations from Energy, Momentum and Entropy Balance

In a series of papers Hassanizadeh and Gray provided a rigorous framework for the averaging of multiphase flow from the pore to the macro-scale, including considerations of energy, momentum and entropy balance (Hassanizadeh and Gray, 1979a, b, 1980; Gray, 1983; Gray and Hassanizadeh, 1991; Hassanizadeh and Gray, 1993b; Gray and Hassanizadeh, 1998) followed by a discussion of a physically grounded representation of the average flow equations (Hassanizadeh and Gray, 1993a). A detailed review of this approach and the various models employed is provided by Miller et al. (1998) and Gray et al. (2013). The treatment is somewhat mathematically involved and will not be repeated here: instead, the key physical insights will be described briefly.

If we treat multiphase flow as an energy balance, then we inevitably require a description of interfacial area: as a consequence, a series of equations for flow are derived which contain terms related to changes in the areas of the fluid/fluid and fluid/solid boundaries. At a macroscopic level these areas, together with the conventional relative permeability and capillary pressure functions, are, in general, functions of each other and the displacement history. To arrive at a tractable set of equations it is necessary to formulate measurable constitutive relationships between these variables.

In particular, an averaged theory requires that the relative permeabilities are a function of both saturation and saturation gradient (mimicking, albeit in a simpler form, the phase field model described above), and that the capillary pressure is a function of both saturation and fluid/fluid surface area (Niessner et al., 2011). If we have a steady-state experiment, the saturation gradients are negligible and the conventional form of the equations is recovered; if interfacial area is ignored, then we have to deal with flow functions which vary with displacement path.

The approach suggested by Hassanizadeh and Gray (1993a) is to write the capillary pressure as:

$$P_c \equiv P_c(S_w, \ a_{wnw}), \tag{6.98}$$

where a_{wnw} is the specific interfacial area between the fluid phases (the area of the interface per unit volume).

The hypothesis is that while, when considered only as a function of saturation, the capillary pressure is hysteretic – that is, dependent on the direction and history of displacement; see, for instance, Fig. 5.1 – if we instead consider the pressure as a two-dimensional surface in saturation and area space, then hysteresis will be eliminated, or at least significantly reduced. This idea certainly adds to the richness of the description of multiphase flow, but suffers from three problems.

The first problem is that interfacial area is difficult to measure directly and certainly is not routinely obtained in core analysis. We mentioned the use of imaging to obtain this information in Chapter 4.5, but this is presently more of a research topic than a standard measurement with a resolution-dependent distinction between terminal and arc menisci.

The second problem is that even if interfacial area is included, hysteresis may still persist: indeed, since wettability will alter on first contact with oil, there will always be a distinction between primary drainage and subsequent flooding cycles. The evidence here is contradictory, with some direct measurements on water-wet systems seeing no hysteresis (Porter et al., 2010), while pore-scale modelling studies suggest that it is substantially reduced, but not eliminated (Reeves and Celia, 1996). However, for mixed-wet media, a significant degree of hysteresis persists, even for a model of a bundle of parallel angular tubes (Helland and Skjæveland, 2007), with even more complexity when pore-space connectivity and trapping are included (Raeesi and Piri, 2009).

Despite these caveats, including interfacial area results in a smooth functional form for the capillary pressure, without the sharp changes observed when considered as a function of saturation only (Reeves and Celia, 1996). Finally, however, the third problem is that this formulation does not explicitly address viscous effects.

Hassanizadeh and Gray (1993a) did consider rate dependence and suggested that the simplest extension of the conventional treatment is to consider a capillary pressure as a sum of its equilibrium value, P_c^{eq}, and a term proportional to the rate-of-change of saturation:

$$P_c = P_c^{eq} - \Lambda \frac{\partial S_w}{\partial t}, \tag{6.99}$$

for some coefficient Λ. This is similar to Eq. (6.97) with, to first order in the viscous perturbation, $\Lambda = \tau \partial P_c / \partial S_w$. Again, while simple and amenable to physical explanation, the model suffers from the same limitations mentioned previously, in that Λ is not easily measured, is a function of saturation, and is generally not needed to interpret and predict experimental measurements. However, this model is able to reproduce unstable infiltration fingering with overshoot, discussed above (DiCarlo, 2013). Hassanizadeh et al. (2002) showed experimental and pore-scale modelling evidence displaying a wide range of relaxation time scales, up to around 10,000 s. This approach was also employed by Helmig et al. (2007) to match experiment and incorporated into numerical models of multiphase flow with characteristic time scales τ of a few 100 s, broadly consistent with our pore-scale analysis of filling times mediated by layer flow; see Chapter 6.1.3. In further analysis, the magnitude of Λ was shown to be related to the width of the advancing front, with dynamic effects most significant where the transition from high to low saturation was sharpest (Manthey et al., 2008).

6.5.4 Consideration of Trapped Phases and Other Approaches

An alternative theoretical approach focusses on one of the most important features of multiphase flow, namely trapping, and to view macroscopic properties, such as capillary pressure and relative permeability, as the outcome of the theory, rather than empirical inputs. Here conservation equations are written for continuous and trapped phases, with only the percolating (continuous) phase flowing (Hilfer, 2006a, b). It is assumed that the rate of transfer between continuous and trapped phases is proportional to the rate of saturation change. Hysteresis naturally emerges from the theory, rather than having to be imposed. One problem that hampers routine application is that the fundamental relationships – the transfers between trapped and flowing phases – are not readily measured, and so have to be hypothesized to define the model, while the resultant formulation involves 10 coupled non-linear partial differential equations.

Numerical solutions to the flow equations have been presented (Doster et al., 2010), together with approximate analytical solutions for a variety of problems, including spontaneous imbibition and cases where both drainage and imbibition occur simultaneously (Doster et al., 2012). Finally, this theory also successfully predicts the development of non-monotonic saturation profiles during infiltration

This is not a complete review of this subject: other approaches, such as meniscus dynamics (Panfilov and Panfilova, 2005) have also been used to study multiphase flow, with the conventional model only emerging under quite restrictive conditions.

However, we will not pursue these ideas further, even though they offer a field ripe for future research. New measurement methods, particularly pore-scale imaging, may inspire fresh insights, to help guide the simplification of general averaging theories.

At present none of these models have found routine application; their value in interpreting phenomena that cannot be satisfactorily explained using conventional theory is limited (with the exception of models of unstable infiltration); no approach addresses all the problems identified – the incorporation of viscous effects, hysteresis and non-percolation or frontal displacement; and, lastly, most of the theories require additional parameters which either cannot be or are not easily measured.

Instead we will synthesize the body of numerical and experimental work to delineate flow regimes, and the conditions under which a steady-state equilibrium description of flow is appropriate.

6.6 Flow Regimes

We discussed flow regimes for displacement in Chapter 4.3 as a function of contact angle and heterogeneity, considering the absence or presence of layer flow. In Chapter 5.2.2 we showed how for waterflooding invasion percolation may be replaced by ordinary percolation. Here we extend the analysis to include the impact of viscous and buoyancy forces. As usual, for concreteness, we will by default consider a system where water displaces oil.

We will first clarify what we mean by a flow regime or pattern, with reference to Fig. 6.10. At the mega-scale we observe changes in saturation: generally the pressure drop across the system due to flow is much larger than a typical capillary pressure. Hence the large-scale displacement is viscous-controlled. However, this is not what we are interested in: if we zoom in to the macro-scale, encompassing a representative elementary volume spanning several pores, we are concerned with the arrangement of phases, since this controls how well they flow. Indeed, if we know the relative permeability and capillary pressure as a function of saturation then we have a usable averaged description, and this in turn is governed by the pore-scale configuration of fluid. If we have a percolation-like pattern, then there is a range of saturation over which both phases are interconnected and we have finite relative permeabilities. The saturation can evolve slowly in space and time, allowing a definition of multiphase flow properties consistent with Eq. (6.58), even if these properties are also functions of both saturation and saturation history. If instead we have cluster growth or frontal advance, then the average saturation changes over a pore scale and we cannot define average properties as a smooth, differentiable function of saturation. We will now delineate the conditions when we can and cannot invoke the conventional average description of multiphase flow.

6.6.1 Dimensionless Numbers

We will study the types of pattern that result as a function of the capillary number, Ca, Eq. (6.62), and the Bond number (the ratio of buoyancy to capillary forces) B, Eq. (3.42), generalized to accommodate both drainage and imbibition:

$$B = \frac{\Delta \rho g l r_t}{\sigma} \tag{6.100}$$

where l and r_t are a characteristic pore length and throat radius, respectively. Both capillary and viscous forces introduce a gradient in the local percolation probability and so, rather than treat them separately, as done, for instance, by Ewing and Berkowitz (2001), we will combine them to study flow as a function of an effective capillary number, previously introduced to quantify the effect of viscous and gravitational effects on trapping, Eq. (6.82). We simplify the expression assuming a finger width $W = 1$ and $J = 1$:

$$Ca_{eff} = \left(\frac{lCa}{\sqrt{\phi K}} + B \right). \tag{6.101}$$

This definition has the disadvantage of including pore-scale parameters, l and r_t, but any assessment of the impact of viscous forces requires this information. In addition, this expression has the appealing feature that a pore-scale transition to viscous-dominated flow occurs for $Ca_{eff} \approx 1$, rather than for some much smaller, medium-dependent value. For our Bentheimer example, with $l = 79 \ \mu$m and $\sqrt{\phi K} = 0.61 \ \mu$m, then, ignoring gravity, $Ca_{eff} \approx 130 Ca$. Hence $Ca = 10^{-3}$, which was the value when the ratio of viscous to capillary forces is around 1 corresponds to $Ca_{eff} = 0.13$. Had we retained the dependence of Ca_{eff} on J, Eq. (6.82), this transition would have occurred closer to a value of 1.

The other dimensionless variable we need to account for is M the viscosity ratio,

$$M = \frac{\mu_d}{\mu_i}, \tag{6.102}$$

where μ_d is the viscosity of the displaced fluid, while μ_i is the viscosity of the injected phase. For $M > 1$, when a less viscous fluid displaces one of higher viscosity, an instability may develop. This phenomenon, called viscous fingering, has received extensive analytical, numerical and experimental attention in the literature starting with the first quantitative analysis by Saffman and Taylor (1958). If $M < 1$ the displacement is stable. We will return to viscous fingering below.

The final variable is f_w, or the fractional flow of water at the inlet (the fraction of the injected volume that is water). In what follows we will assume that $f_w = 1$;

however, we have shown that for some steady-state experiments, where an intermediate value of f_w is imposed, we can observe a range of flow patterns featuring the movement of disconnected phases; see Chapter 6.4.7. If we have a percolation-like displacement pattern, our hypothesis is that with sufficiently slow flow it is possible to observe a smooth transition in the average saturation from a high value at the inlet ($1 - S_{or}$ for $f_w = 1$) to a lower initial value further away. In this case the flow pattern for $f_w = 1$ will display the fluid configurations for a range of saturation, as seen, for instance, in Fig. 6.11. If instead we have a more frontal advance, then an intermediate f_w is a somewhat artificial construct: in a macro-scale displacement, we will see the movement of a front with S_w close to 1 with a transition in saturation over a pore scale. For instance, in the previous discussion of fingering during infiltration, when the rainfall falls at a rate less than the Darcy velocity then $f_w < 1$: if the soil is moist, this leads to a smooth monotonic variation of saturation with depth consistent with a percolation-like displacement; if the soil is dry we have frontal advance and this leads to fingering. The water penetrates the soil at a high saturation. At no stage do we force disconnected phases through the system, as shown in Fig. 6.17: even though $f_w < 1$, wherever there is imbibition, it occurs at a high saturation. As a consequence, we will not consider separate flow patterns for steady-state flow further.

Even with $f_w = 1$ we still have four dimensionless parameters: θ (the contact angle), Ca_{eff}, M, and some measure of heterogeneity. Rather than attempt to describe, or present, a full four-dimensional phase diagram, we will discuss the behaviour in different limiting situations before outlining the conditions under which a percolation-like description is valid, allowing – under certain circumstances – a Darcy-like macroscopic description of the flow.

The limit of $Ca_{eff} \rightarrow 0$ has been described previously in Chapter 4.3; see Figs. 4.20 and 4.21. Since viscous effects are negligible, M has no impact on the displacement. We see invasion percolation or percolation with trapping regimes, both of which are percolation-like and hence amenable to a conventional multiphase Darcy law treatment for larger contact angles (drainage), and for heterogeneous systems regardless of wettability. We see frontal advance or cluster growth for small contact angles and uniform media.

6.6.2 Viscous Fingering and DLA

Now we consider the opposite extreme for $Ca_{eff} > 1$, where viscous (and buoyancy) effects dominate at the pore scale. In this limit, the contact angle has no impact on the behaviour as capillary forces are negligible. Now the pattern is controlled by the viscosity ratio. If $M < 1$, the displacement is stable, giving frontal advance. Since the injected fluid is more viscous, with a higher pressure drop than in the displaced fluid, the growth of any protuberance in the fluid advance is suppressed. Hence the flow pattern is flattened and only shows a self-affine behaviour, see page 150, over a restricted range of scale.

More interestingly, for $M > 1$ the displacement is unstable, regardless of medium heterogeneity, resulting in viscous fingering; see Homsy (1987) for a review. The pressure gradient in the injected phase is less than in the displaced phase. Hence, if a bulge develops in the pattern, the fluid within it flows more rapidly than the more viscous defending fluid around it, and the perturbation grows. The interface between the fluids is a fractal with a cascade of subsidiary digitation from the pore scale to the size of the system: the fractal dimension depends on the viscosity ratio, varying from 2.5 for $M \to \infty$ to 2 for $M = 1$ (a flat surface) (Blunt and King, 1990). However, the injected fluid itself has a mass dimension of 3: the flow pattern is filled in with little or no trapping (Frette et al., 1994), consistent with the discussion in Chapter 6.4.6. For a linear flood, the distance between the fastest and the slowest-moving fingers increases linearly with time: the average behaviour is captured accurately using empirical models based on a non-linear fractional flow, which we will present in Chapter 7.3.1.

There is one final limit, which is when $M \to \infty$. In this case not only the interface between the fluids, but the bulk of the injected phase is fractal: since there is a negligible pressure gradient in the invading phase, it can support a wispy structure. The fractal dimension in this case is approximately 1.715 ± 0.004 in two dimensions and 2.5 in three, given by an analytical expression $D = (d^2 + 1)/(d + 1)$ for $d \geq 3$ (Tolman and Meakin, 1989). The pattern is called diffusion limited aggregation, or DLA (Witten and Sander, 1981). Since there is no pressure drop in the injected phase, the fluid/fluid interface has a constant pressure (that of the injector). In a homogeneous medium with conservation of volume and flow governed by Darcy's law, in the displaced phase we obey the Laplace equation, $\nabla^2 \phi = 0$ where ϕ is the potential (equivalent to the pressure). Any process where growth is governed by the Laplace equation, with $\phi = 0$ at the boundary and a growth rate $\nabla \phi$ normal to the boundary, may be described by the DLA model. One additional, and essential, feature is some degree of noise or randomness that allows the unstable fingers to form. The easiest way to model DLA is to simulate a random walk: on average the concentration of particles (equivalent to ϕ) obeys the Laplace equation. When a particle encounters the DLA cluster it sticks and a new walker is released. Experimentally, the process is also easy to reproduce through the fast injection of a low viscosity phase (such as air) into a liquid-filled porous medium. This rather elegant and simple concept describes a number of growth processes, from snowflakes to dielectric breakdown and metal deposition at electrodes (Vicsek, 1992).

Figure 6.25 A simulated diffusion-limited aggregation (DLA) pattern. It is possible to generate clusters with many millions of particles. The infinite viscosity
contrast supports a ramified, fractal structure. Image reproduced with permission
from Paul Bourke, http://paulbourke.net/fractals/dla/.

Viscous fingering and DLA, thanks to their fractal structure, are also beautiful.
An example DLA cluster, starting from a central seed, is shown in Fig. 6.25. Notice
the branched, dendritic nature of the pattern: if the fingers extend a radius r, the
number of particles (or mass of the cluster) scales as $N \sim r^D$ where D is the fractal
dimension. The density, $\rho \sim r^{D-d}$, where d is the spatial dimension: since $d > D$,
$\rho \to 0$ as $r \to \infty$, meaning that the cluster becomes increasingly ramified as it
grows.

The visual appeal and ease with which DLA can be both simulated and demonstrated experimentally means that it is often used as the exemplar for viscous
fingering, or unstable growth. This will be evident in the figures taken from the
literature below. However, this can be misleading, as we do not, at least in field-
scale displacements, ever inject a fluid with zero (or negligible) viscosity. Typical
reservoir-condition (high temperature) water viscosities are around 5×10^{-4} Pa.s;
oil viscosities are more variable, in the range 10^{-3} to 1 Pa.s. As mentioned previously, we tend to see viscous-dominated displacement when we reduce interfacial
tension, either through surfactant injection, or the injection of a gas. Gas viscosities can be much lower than that of water (and the oil it displaces): for example,
the viscosity of CO_2 for pressures of between 10 and 20 MPa and temperatures
between 320 and 280 K vary from 2×10^{-5} to 7×10^{-5} Pa.s (Fenghour et al.,
1998). However, with gas we only reach or nearly reach miscibility with lighter,

lower viscosity oils. Therefore representative viscosity ratios, M, for field-scale viscous-dominated displacement are typically $10 - 100$: this is high, but not infinite. For a finite viscosity ratio, the fingers are, as mentioned above, more filled-in with a fractal dimension $D = d$; it is the surface which is self-similar.

An example viscous fingering pattern, simulated for a two-dimensional porous medium with no capillary forces ($Ca = \infty$) and $M = 33$, is shown in Fig. 6.26. In this case the injected and displaced fluids are miscible; there is no phase boundary between them and no capillary pressure. Rather than consider saturation, it is more natural to consider the process in terms of the concentration c of injected fluid: $c = 1$ at the inlet, while $c = 0$ corresponds to the displaced fluid. Although highly unstable, the injected phase does displace the resident fluid completely near the inlet: this is a fingered version of frontal advance. Notice the hierarchy of fingers within fingers, characteristic of a fractal surface (think of a bracken leaf as another example).

In the heyday of fractals in the 1980s, viscous fingering and DLA received obsessive attention in the literature; the author was even sufficiently misguided to complete a PhD on the topic. Rather than make a mediocre attempt to review the field further, I refer the interested reader to other books that treat the subject with the requisite enthusiasm (Vicsek, 1992; Sahini and Sahimi, 1994; Sahimi, 2011).

6.6.3 Summary of Regime Diagrams

There are four types of displacement pattern: invasion percolation, percolation with trapping, frontal advance, cluster growth and viscous fingering. For the first two percolation-like patterns, we can use the multiphase Darcy law, Eq. (6.58), as mentioned above. Viscous fingering is also amenable to a macroscopic description since

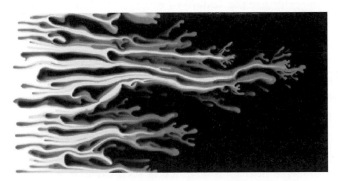

Figure 6.26 The concentration field during the unstable displacement of a more viscous fluid (dark) by a fully miscible, less viscous fluid (light). $M \approx 33$. Unlike DLA, the pattern does fill in from the inlet, but we still see a fractal interface between the fluids with a cascade of unstable fingering. From Jha et al. (2011). (A black and white version of this figure will appear in some formats. For the colour version, please refer to the plate section.)

we can average the flow in the fingers, but a formulation in terms of relative permeability, while mathematically possible, is not the most physically revealing method to describe the flow: this is discussed in more detail in Chapter 7.3.1. For frontal advance and cluster growth, the average saturation changes from an initial value to almost one across a pore length and we cannot use a conventional description of the flow. This was evident for infiltration where a plethora of putative additions to the flow equations have been proposed; the reality is that there is not a macroscopically differentiable saturation profile.

Before quantifying when each pattern is observed, we will first present some experimental and numerical evidence. Fig. 6.27 shows a series of experiments performed in a bead pack with a radial geometry. Air was injected into a mixture of water and glycerol, with $M \approx 320$ (Trojer et al., 2015). Both the flow rate and contact angle were varied: there was no wetting layer flow. The Darcy velocity is inversely proportional to radius for a constant injection rate and so the pore-level balance of viscous to capillary forces at the advancing front changes over time. Trojer et al. (2015) defined an effective capillary number

$$Ca^* = \frac{\mu_d Q R}{\sigma b d^2} = Ca_d \left(\frac{R}{d}\right)^2, \tag{6.103}$$

where Q is the injection rate (volume injected per unit time), R is the radius of the system, b is the depth and d is the grain diameter. We have defined Ca_d, for the displaced fluid, when the displacement reaches the edge of the packing.

This is a different definition of effective capillary number than we have used, Eq. (6.101): the fundamental distinction is that we rescale Ca by a ratio of length scales – a pore size to the square-root of permeability times porosity – whereas in Eq. (6.103) the scaling is the square of the system size to bead diameter. Empirically we find a transition from capillary to viscous dominance for $Ca^* \approx 1$. Similar behaviour has been observed for experiments with a smaller viscosity ratio, $M \approx 14$, but with a filled-in viscous fingering pattern at the highest capillary numbers (Frette et al., 1994).

At the lowest flow rates, there is a change from invasion percolation to frontal advance as the injected fluid becomes more wetting. This is the Cieplak-Robbins transition discussed in Chapter 4.3 with a divergent finger width apparent at a contact angle of around 120° measured through the injected phase (this corresponds to 60° in Fig. 6.27). Again, frontal advance is seen even for a drainage displacement ($\theta > 90°$).

At the highest flow rates, we see a fingered pattern, regardless of contact angle. Since the viscosity ratio, M, is very large, the structure is ramified, or DLA-like, although we do see fatter fingers when the invading phase is wetting. There is not a sharp transition between flow regimes, but we can observe, locally, a

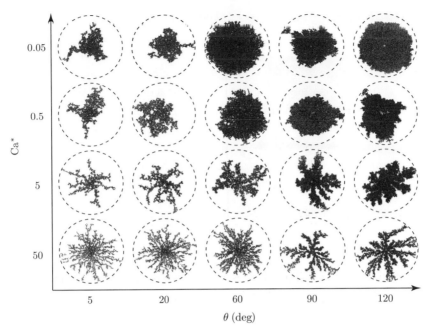

Figure 6.27 Pictures of the displacement patterns obtained during air injection (dark) into a porous medium filled with a mixture of water and glycerol (clear), as a function of the effective capillary number Ca^*, Eq. (6.103) and the static contact angle θ ranging from drainage ($\theta \approx 5°$) to imbibition ($\theta \approx 120°$). The contact angle is defined (correctly) through the denser, displaced phase: in the main text we assume that the denser phase is injected, giving an angle $180° - \theta$. The viscosity ratio, M, is around 320. From Trojer et al. (2015).

capillary-controlled pattern for non-wetting advance with $Ca^* < 1$. For stable flow, $M < 1$, in drainage, we see invasion percolation with a more frontal-type advance beyond a correlation length which decreases with increasing Ca (Lenormand et al., 1988). Similar patterns have been obtained from a numerical simulation of two-dimensional flow using a dynamic pore-scale model (Holtzman and Segre, 2015).

Fig. 6.28 shows a regime or phase diagram based on the micro-model experiments and simulations of Lenormand and Zarcone (1984), Lenormand et al. (1988), and Lenormand (1990) as a function of Ca_{eff} and M. This synthesizes the results shown previously with a cross-over from stable to unstable (viscous fingering) patterns at $M = 1$ and high flow rates, with the emergence of capillary-dominated flow for $Ca_{eff} < 1$. There are generous transitions between viscous-dominated and capillary-dominated regions, where we may see intermittent flow patterns with the disconnection and reconnection of flow paths, Chapter 6.4.7. In Fig. 6.28 there is a transition from percolation to viscous fingering, at a fixed Ca_{eff}, as M is altered: this is simply an artefact of defining the capillary number through the injected

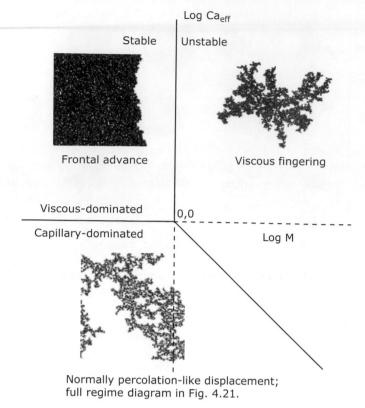

Figure 6.28 Flow regimes as a function of viscosity ratio, M, Eq. (6.102), and effective capillary number, Ca_{eff}, Eq. (6.101). In the capillary-controlled regime, for low Ca_{eff}, percolation patterns are normally observed, with the full regime diagram shown in Fig. 4.21. When flow is controlled by viscous forces, there is stable frontal advance for $M < 1$ and viscous fingering for $M > 1$: DLA is only seen in the limit $M \to \infty$.

phase. When this phase has a very low viscosity, the pressure drop within it is very low, but viscous forces are still significant, since the displaced phase has to flow: the cross-over occurs for a fixed value of Ca_d (or Ca_{eff} defined using μ_d).

We will now combine all these studies to delineate the conditions under which different generic flow regimes are observed. We will assume that layer flow is allowed, which confines the Cieplak-Robbins transition to neutrally wet and uniform systems, such as bead packs, which while they have received extensive experimental attention are not necessarily a good representation of reservoir rocks. We will also lump together all percolation-like patterns: percolation with trapping is seen with layer flow, while invasion percolation occurs during the primary invasion of a non-wetting phase, or water injection into an oil-wet medium with negligible layer flow. In the end the state or regime diagram, Fig. 6.28, is quite simple, with a more complex delineation of possible patterns in the capillary limit shown in Fig. 4.21.

(1) **Percolation-like advance**: $M < 1$ and $Ca_{eff} < 1$ or $Ca < 10^{-3}$; for $M > 1$, $Ca_{eff} < 1/M$ (or $Ca_d < 10^{-3}$); we also require some threshold pore-scale heterogeneity which is contact-angle dependent, as illustrated in Fig. 4.21. As $\theta \to 180°$ we see a percolation advance even for uniform media. These criteria encompass both invasion percolation (often termed capillary fingering) and percolation with trapping (seen when layer flow is present). It is only in a percolation-like regime that a macroscopically smooth variation in saturation can be defined and we may employ the multiphase Darcy law, Eq. (6.58), even though until $Ca_{eff} < 10^{-2}$ there will be a noticeable sensitivity in the behaviour to flow rate.

(2) **Frontal advance**: $M < 1$ and $Ca_{eff} > 1$ or $Ca_{eff} < 1$ and a contact angle and heterogeneity dependent regime, favouring neutrally wet systems without layer flow and uniform media. The finger width is infinite. The saturation changes from near 0 to 1 over a pore scale. Conventional averaged macroscopic descriptors of relative permeability and capillary pressure cannot capture this behaviour.

(3) **Cluster growth**: $Ca_{eff} < 1$, low values of the contact angle and a homogeneous system. This is the equivalent of frontal advance when layer flow is allowed and again does not permit the definition of smooth saturation-dependent capillary pressure and relative permeabilities.

(4) **Viscous fingering**: $M > 1$ and $Ca_{eff} > 1/M$. For $M \to \infty$ we see a mass-fractal DLA pattern. While viscous fingering can also be described macroscopically, the concept of relative permeability is no longer consistent with the hypothesis of different flow channels.

This is not a complete description of possible flow patterns. Hughes and Blunt (2000) identified random cluster growth when there is sufficient initial wetting phase saturation present to break up the evolution of a flat cluster; however, this is essentially a percolation-like pattern, since now the filling sequence is principally determined by pore size rather than the local meniscus configuration.

We have not considered gravitational forces separately. This has been treated by Ewing and Berkowitz (2001). In percolation theory, both buoyancy and flow provide a stabilizing gradient in the percolation probability, leading to a finite correlation length in the flow patterns and a shift in residual saturation, as addressed in Chapter 6.4.4. However, if we have a gravitationally unstable displacement, then this can impact the flow pattern, even for infinitesimal flow rates, as exemplified for infiltration in imbibition; see Chapters 4.3.6 and 6.5.1. Furthermore, for an advancing front the scaling of correlation length with capillary and Bond numbers is different: compare Eqs. (4.33) and (6.78).

Yortsos et al. (1997) used percolation theory arguments to develop a phase diagram for drainage considering gravitational and viscous forces. However, the approach was different from here: Yortsos et al. (1997) delineated the conditions under which the mega-scale displacement pattern (see Fig. 6.10) would either be compact or fingered. In this treatment we are only concerned with the macroscopic pattern over a representative elementary volume. If the pattern is percolation-like at this level, then averaged saturation-dependent quantities can be defined. We hypothesize that the mega-scale displacement can then be determined (normally numerically) using these quantities and invoking conservation of mass, see Chapter 9: the pattern will either be stable and compact at the largest scales or unstable and fingered.

In the next chapter we will assume that we have a percolation-like displacement pattern and can define average saturations, giving us macroscopic capillary pressures and relative permeabilities. We will allow the relative permeability to be a function of saturation, saturation path (displacement history), flow rate and viscosity ratio.

7

Relative Permeability

In this chapter we will discuss relative permeabilities, relating their features to the pore-scale phenomena described previously. Both experimental measurements and predictions using different modelling approaches will be shown. However, this is not intended to be a comprehensive review of relative permeability, nor will it address measurement techniques: for a discussion of the latter see Honarpour et al. (1986) and Anderson (1987b).

7.1 Water-Wet Media

We show some example relative permeability curves for, unsurprisingly, our Bentheimer sandstone exemplar in Figs. 7.1, 7.2 and 7.3 (Ramstad et al., 2012; Alizadeh and Piri, 2014a). In the experiments shown in Fig. 7.1, the contact angle between water and oil was measured directly using X-ray imaging during primary drainage and waterflooding, with average values of $22°$ and $28°$ respectively, indicating strongly water-wet conditions (Aghaei and Piri, 2015).

7.1.1 Primary Drainage

In primary drainage, the rock is initially completely saturated with water: $k_{ro} = 0$ and $k_{rw} = 1$. The oil first traverses the largest throats in an invasion percolation process, as described in Chapter 3.4: this cuts off the most conductive flow paths for the water and k_{rw} drops sharply. There is a critical saturation when the oil phase will first span the sample. In the experiments of Alizadeh and Piri (2014a) the core is 3.81 cm in diameter and our pore length is 79 μm: putting these values in Eq. (3.29) gives an estimated critical saturation of around 5%, consistent with the results shown in Fig. 7.1. Percolation theory predicts a power-law rise in k_{ro} as a function of oil saturation with an exponent of around 4.8, Eq. (6.74).

The two relative permeabilities cross, $k_{rw} = k_{ro}$, at a saturation $S_w \approx 0.55$ with a value of around 0.1 or lower. When there is a significant saturation of both

315

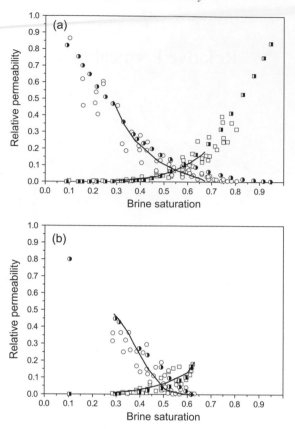

Figure 7.1 Bentheimer sandstone relative permeabilities as a function of brine saturation. The solid lines are predictions from dynamic pore-scale network modelling of steady-state flow, see Fig. 6.22, compared to experimental data from the literature, Øren et al. (1998) and Alizadeh and Piri (2014a), shown by the unfilled and half-filled symbols, respectively. (a) Primary drainage. (b) Secondary imbibition. From Aghaei and Piri (2015).

phases, the sum of the relative permeabilities is much less than 1, indicating that the total flow of oil and water is impeded. This is caused by the presence of menisci which act as flow barriers to both phases; in particular, terminal menisci block pores and throats, while arc menisci have less of an impact, since they are confined to roughness and corners of the pore space. The cross-over saturation is greater than 0.5, consistent with a water-wet system: the oil relative permeability equals that of water when the oil saturation is lower than the water saturation. If we consider the relative permeability of a phase as a function of its own saturation then k_{rw} is lower than k_{ro}: the water is wetting and confined to wetting layers and the smaller pores and throats, while the oil occupies the larger regions with a bigger conductance.

Once the oil saturation is greater than 0.5, the oil is well connected, hogging the high-flow, large-radius pathways through the rock and k_{ro} rises steeply as the

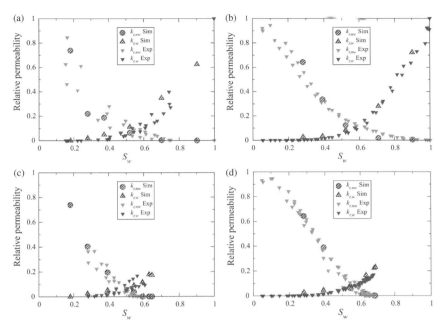

Figure 7.2 Steady-state experimental and simulation results for Bentheimer (left column) and Berea sandstones (right column). (a) and (b) show primary drainage, while (c) and (d) show secondary imbibition. Several experimental datasets are presented to demonstrate the scatter of the measurements. The simulations were performed using the lattice Boltzmann method on a 256^3 pore-space image at the same fluid and flow conditions as the experiments, see Fig. 6.24. From Ramstad et al. (2012).

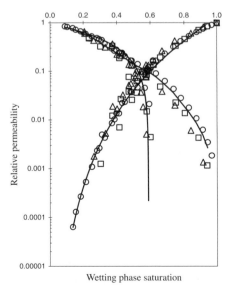

Figure 7.3 Two-phase primary drainage and secondary imbibition relative permeability data for Bentheimer sandstone. Results from oil/water (circles), gas/oil (triangles) and gas/water (squares) displacements are shown. The solid lines are curve fits. From Alizadeh and Piri (2014a).

displacement proceeds, reaching a maximum value k_{ro}^{max} close to 1. The water at low saturation does not significantly impede the flow of oil.

Below a saturation of around 0.4, k_{rw} is very low, and it requires a logarithmic plot, Fig. 7.3, to observe the sharp drop in value near the end of a primary drainage experiment. This is because the water, confined principally to layers, has a tiny conductance, as quantified in Chapter 6.1.3. Note, however, that at the end of the experiment, when the water saturation is less than 10%, k_{rw} is not zero: it simply has a very low, but finite, value. We cannot impose an infinite capillary pressure, nor wait an infinite time to allow flow at an infinitesimal rate: any experiment, and indeed any field setting, will see a fluid distribution at the end of primary drainage at a finite capillary pressure with a finite water saturation and relative permeability. It is unrealistic to expect a true irreducible saturation ever to be reached in a real rock; see Chapter 3.5.

7.1.2 Waterflooding

The relative permeabilities during waterflooding, secondary imbibition, are different. The principal reason for this is that now the non-wetting phase can be trapped, as discussed in Chapter 4.6: the residual saturations in the experiments are around 35–40%, indicative of strongly wetting conditions and a displacement process with a significant amount of snap-off. This was quantified in Table 4.1, page 136, where we recorded the displacement statistics for our Bentheimer network with different contact angles. k_{ro} for imbibition lies below k_{ro} for drainage, particularly for low oil saturations: we see an approximately linear drop from k_{ro}^{max} to zero at $S_w = 1 - S_{or}$. During imbibition, snap-off disconnects the flow pathways of the oil with relatively little change in water saturation, resulting in a rapid fall in the oil relative permeability. Of particular importance is the fact that k_{ro} decreases to zero sharply close to S_{or} as the final connected pathways of oil are disconnected: as we show later, this is a characteristic of water-wet systems, since, with layer flow in mixed-wet and oil-wet media, the approach to the residual saturation is much more gradual.

There is no evident hysteresis in k_{rw}: the water always occupies the smallest regions of the pore space, is never completely trapped and consequently has a similar conductance for the same saturation, regardless of the displacement process. This is consistent with a percolation-like imbibition pattern and a finger width that is the pore size, see Chapter 4.3, as we would expect for a strongly water-wet consolidated rock.

Fig. 7.3 shows relative permeabilities for oil/water, gas/oil and gas/water experiments. The curves for both imbibition and drainage are shown, and they overlie for the different fluid systems. In these cases, the behaviour is again suggestive of strongly wetting conditions with no impact on the displacement of the particular fluid pair used, since the contact angles in all cases are similar. Again there is no

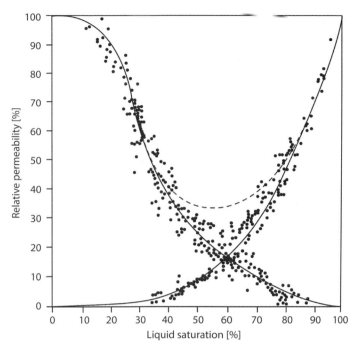

Figure 7.4 The first measurements of relative permeability, for CO_2 displacing water (liquid). Results from experiments in four sand packs are shown. The dashed line is the sum of the two relative permeabilities: note that this is higher than observed for the less well connected, consolidated sandstones shown previously. From Wyckoff and Botset (1936).

hysteresis in the wetting phase relative permeability (water in the presence of oil or gas, and oil in the presence of gas), but we do see a difference for the non-wetting phase, with a residual saturation of around 40% in imbibition.

Fig. 7.4 presents the first measurements of relative permeability on four sand packs with permeabilities of between 18 and 260 D (Wyckoff and Botset, 1936). The experiments were steady-state primary drainage displacements of water by CO_2. While we now have a much better appreciation of pore-scale displacement processes and the impact of small-scale heterogeneity – see, for instance, Reynolds and Krevor (2015) – in the subsequent 80 years we have not significantly improved the ease or accuracy of these measurements. The main difference compared to the sandstone exemplars is that, since there is a better connected pore space, the sum of the relative permeabilities is larger, with a cross-over value now greater than 0.15.

7.1.3 Predictions of Relative Permeability

As reviewed in Chapter 6.4.9, it is possible to model multiphase displacement using a variety of approaches. For relative permeability the first successful predictions based on an explicit representation of a realistic porous medium were made by Bryant and Blunt (1992) using a network model extracted from a packing of mono-disperse spheres (Finney, 1970). For a wider range of porous media, Øren et al. (1998) simulated drainage and waterflooding using networks generated with a process-based method; see Chapter 2.2.2. Specifically, good predictions were made of the primary drainage and imbibition capillary pressures and relative permeabilities for Bentheimer sandstone; similar results were also obtained by Patzek (2001).

The methodology for assigning a sequence of pore and throat filling has been described at length in the previous chapters and is only briefly reviewed here for clarity. The porous medium is represented as a network of pores connected via throats; each pore or throat has either a circular, square or triangular cross-section with a shape factor, Eq. (2.2), derived from a pore-space image or reconstruction. Displacement proceeds in order of threshold capillary pressure with local extrema defining a macroscopic capillary pressure. Fluid configurations in each pore and throat are determined from imposing capillary equilibrium: from this phase volumes and saturations may be computed. There are some subtleties here, associated with how volume is partitioned between pores and throats: ideally this is related to calculations based on the underlying pore-space images, or tuned to available data. The relative permeability is found from the conductances calculated semi-analytically for each pore and throat, accounting for layer flow, as outlined in Chapter 6.1.3. The system is treated as a random resistor network and the overall flow of each phase is computed using the conductances for each element connected together. Strictly, a Darcy or Poiseuille-like linear relationship between flow rate and pressure gradient is invoked for the flow of each phase between two pores. The constant of proportionality is found from the conductances of the individual pores and throats. Then volume conservation of each phase is imposed at every pore to derive a series of linear equations that are solved for pressure, and hence the overall flow, using standard techniques in linear analysis. This is equivalent to solving for Stokes flow through the portion of the network occupied by each phase, simplified using a series of analytical expressions for conductance in each element.

There are three important features to note. The first is that the best results are obtained assuming that there is no viscous coupling between the phases: a no-flow boundary at the fluid/fluid interface; see page 229. The second is that in many cases the immobile or connate water saturation cannot be predicted, but is inferred from the content of low-permeability minerals in the rock such as clay, or adjusted to match measured data. The third is that while, for primary drainage, good predictions are obtained for receding contact angles of zero, or close to zero, for imbibition, even in apparently strongly water-wet media, larger advancing angles are needed: for instance, Øren et al. (1998) used a range between 30° and 50°. This is because we are considering movement across a rough surface (see Fig. 2.31) and some degree of contact angle hysteresis is inevitable.

Figs. 7.1 and 7.2 show accurate predictions of the measured data. In Fig. 7.1 this is achieved using a dynamic network model which mimics the steady-state flow in the experiment. The capillary number, Eq. (6.62), was low, less than 10^{-5}, in all the experiments, so this is in the quasi-static regime. The contact angles for imbibition were between 35° and 55°: these are larger than those directly measured on pore-scale images, which were between 20° and 42°, but represent the angle when the contact line moves, rather than a – likely lower – value when the fluids are at rest (Aghaei and Piri, 2015). In Fig. 7.2, direct lattice Boltzmann simulation was used with a fixed contact angle of 35° on a 256^3 image at a capillary number of 2×10^{-6}. Notice that the modelling results lie within the scatter of the experimental data and reproduce all the main qualitative features discussed above.

Fig. 7.2 also shows measurements and successful predictions using lattice Boltzmann simulation for Berea sandstone, another exemplar quarry sample. For this rock there is a classic relative permeability dataset measured by Oak and Baker (1990) which is available (on request to the author) in spreadsheet form.

The relative permeability of Berea sandstone has also been predicted using pore-scale network modelling (Øren and Bakke, 2003; Valvatne and Blunt, 2004). Results are shown in Fig. 7.5 using, for waterflooding, average advancing contact angles around 60° (Valvatne and Blunt, 2004) and assigning an immobile water volume in clay to match the measured connate water saturation: the network model employed is shown in Fig. 2.14. The average contact angle is chosen to match the residual oil saturation: as discussed in Chapter 4.2.4 as we increase the contact angle we see less snap-off and cooperative pore filling, and trapping no longer occurs in the larger elements; see Table 4.2. This leads to a decrease in residual saturation with contact angle. Increasing the contact angle also tends to make the gradient in k_{ro} lower near S_{or} and we can see cases where $dk_{ro}/dS_w = 0$ for $S_w = 1 - S_{or}$.

The relative permeabilities for Berea display the same qualitative features as described for Bentheimer: the residual saturation is around 30% and the relative permeabilities cross at a water saturation of approximately 65% at a value of 0.1. The relative permeabilities shown here are characteristic of water-wet sandstones: the connectivity of each phase is controlled by capillarity rather than some subtle measure of the pore space itself. As we show later, in these types of rock, it is wettability, or the distribution of contact angle, that is the principal determinant of relative permeability behaviour.

These relative permeabilities can be reproduced quantitatively using the concepts of pore-space topology, quasi-static displacement, Stokes flow and capillary equilibrium developed in this book. This provides some confidence that the

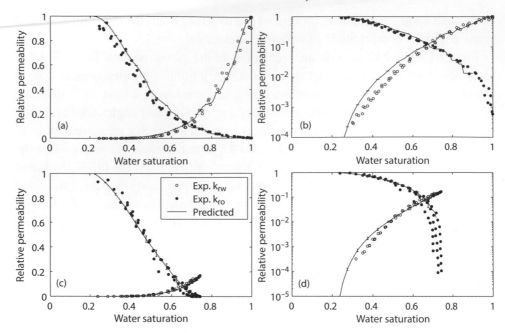

Figure 7.5 Predicted primary drainage, (a) and (b), and secondary imbibition relative permeability, (c) and (d), for water-wet Berea sandstone (lines) compared to three sets of experimental data by Oak and Baker (1990) (circles). The predicted results are the mean of 20 realizations, and the length of the error bars is twice the standard deviation. The results are plotted on a linear scale in (a) and (c), and on a semi-logarithmic scale in (b) and (d). From Valvatne and Blunt (2004).

macroscopic flow behaviour that we measure is indeed the consequence of slow flow where the microscopic configurations of the flow are governed by capillary forces with an overall displacement pattern that is percolation-like.

7.1.4 Relative Permeabilities for Different Rock Types

While it is appealing to make general statements relating rock structure to relative permeability, usually based on rather scant or poor-quality experiments, the subtlety of the multiphase connectivity within the pore-space precludes such a simplistic analysis. In the end, rather than use hand-waving correlations as a crutch, it is better to study what data there are carefully and base predictions on rigorous pore-space analysis and modelling.

As an example of the difficulty relating relative permeability to rock type, Fig. 7.6 shows the extensive relative permeability dataset on subsurface samples acquired by Bennion and Bachu (2008) for the study of CO_2 storage, with details of the rocks studied compiled in Table 7.1. The study included sandstones, carbonates, shales and an anhydrite with a variation in porosity from 0.012 to 0.195 and

Table 7.1 *Summary of rock properties for the samples whose relative permeabilities are shown in Fig. 7.6. From Bennion and Bachu (2008). Permeability is quoted in mD. 1 mD $\approx 10^{-15}$ m^2.*

Sample	Rock type	Porosity	Permeability (mD)
Viking	sandstone	0.195	21.7
Nisku	carbonate	0.114	21.0
Cardium 1	sandstone	0.153	0.356
Cardium 2	sandstone	0.161	21.2
Colorado	shale	0.044	0.0000788
Muskeg	anhydrite	0.012	0.000354
Calmar	shale	0.039	0.00000294

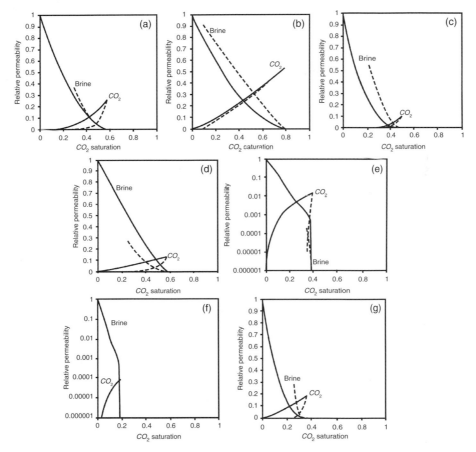

Figure 7.6 Measured relative permeabilities for primary drainage (solid lines) and secondary imbibition (dashed lines) for CO_2/brine displacement in various rock samples. (a) Viking; (b) Nisku; (c) Cardium sample 1; (d) Cardium sample 2; (e) Colorado; (f) Muskeg; and (g) Calmar. Properties of the rocks are provided in Table 7.1. The Muskeg sample was too impermeable to allow the imbibition experiment to be performed. From Bennion and Bachu (2008).

permeabilities from 20 mD to 30 nD. The first observation is that these samples are all less porous and much less permeable than the well-connected quarry systems studied hitherto, Table 6.3.

The relative permeabilities in Fig. 7.6 are plotted as functions of the non-wetting, CO_2 phase saturation, which is the opposite of the conventional formulation. First CO_2 was injected into entirely brine-saturated samples ($S_w = 1$) followed by brine flooding to residual CO_2 saturation. Experimentally it is challenging to push the wetting phase to low saturation when injecting a low viscosity fluid such as CO_2 through a low permeability sample, and so S_w at the end of primary drainage can be quite large, up to 80% in these samples. There are no error bars provided, but there will be some uncertainty in the measurements, as evident from the scatter in the data from experiments in much simpler, and easier, rocks shown previously.

The residual CO_2 saturation varies from 10% to 35%, which is in the range of values seen in more permeable rocks, Chapter 4.6: however, some of these lower values are associated with low initial non-wetting phase saturations at the beginning of imbibition. The primary indication of poor pore-space connectivity is the value at which the relative permeabilities cross: this is much lower than 0.1 (the value for the well-connected sandstones) for the lowest permeability shales and anhydrite.

Relative permeability is a dimensionless number and hence it is not directly affected by the permeability, but by how well each phase can flow through the pore space compared to when it is fully saturated. For instance, the relative permeabilities in Fig. 7.6 shown for the Cardium 1 sample, a sandstone with a permeability of 0.356 mD, is qualitatively similar to that for Berea and Bentheimer with permeabilities more than 1,000 times larger, or indeed the sand packs that are 100,000–1,000,000 times more permeable, Fig. 7.4, except for the higher apparent irreducible wetting phase saturation which is simply a consequence of the finite time and capillary pressure that can be imposed in the experiments. In contrast, the one carbonate shows much larger relative permeabilities and little trapping, suggestive of parallel pathways occupied by each phase. The three lowest porosity samples – two shales and the anhydrite – show significant trapping and only a small range of saturation where both phases can flow, particularly during imbibition: the void space is only just connected for single-phase flow, hence filling just a few pores will disconnect the other phase.

I will end this section with an instructive quotation from Muskat and Meres (1936). Commenting on relative permeabilities the authors remark that these *must be provided by empirical measurements before any particular flow problem can be discussed numerically. A satisfactory calculation of these relations theoretically would obviously be of an even higher order of complexity than that of the homogeneous fluid permeability, which already presents entirely insurmountable analytical difficulties.* Now, as demonstrated, we can predict absolute and relative permeabilities, but only based on an explicit pore-scale model of the flow and displacement: it is not reasonable to predict the complex behaviour observed in Fig. 7.6 from general principles. As a consequence, there is little point in trying to relate minutiae of some hypothesized pore structure to the average flow properties. However, it is instructive to consider how the wide variation in initial wetting phase saturation impacts the subsequent waterflood relative permeability, and this is discussed next.

7.1.5 Effect of Initial Saturation

In Chapter 4.6.2 we showed how the residual saturation depended on its initial value after primary drainage: for water-wet media there is a monotonic increase with initial saturation that can be captured accurately using empirical models. Here we study the impact on the relative permeability: specifically, in imbibition k_{ro} must drop from its value at the end of primary drainage to zero for $S_o = S_{or}(S_{oi})$. Some example results for Berea sandstone are shown in Fig. 7.7 for a CO_2/brine system: here CO_2 is the non-wetting phase. In these experiments, similar to Fig. 7.6, it is difficult to achieve high CO_2 saturations during drainage.

As mentioned already, there is relatively little hysteresis in the wetting phase relative permeability, k_{rw}; however, k_{ro} drops sharply, indeed almost vertically,

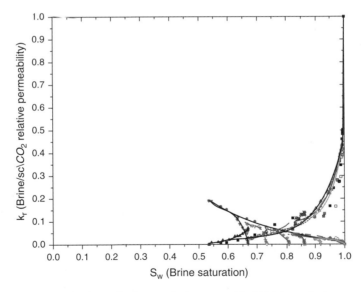

Figure 7.7 A compilation of primary drainage and imbibition relative permeabilities for Berea sandstone. The non-wetting phase is high-pressure CO_2 (labelled sc for super-critical on the graph) which is injected to reach different initial saturations, followed by brine injection. The points are the measurements, while the lines are curve fits. From Akbarabadi and Piri (2013).

with an increase in water saturation when S_{oi} is low. The reason for this is the almost linear dependence of S_{or} on S_{oi}; see Fig. 4.37: when waterflooding starts, most of the CO_2 is trapped rapidly and k_{ro} falls to zero. The non-wetting phase initially occupies a percolation-like cluster, close to threshold: a few snap-off events during imbibition are sufficient to cause disconnection; see page 180.

The rapid fall in k_{ro} close to S_{or} and a strong dependence on the initial oil saturation are characteristics of water-wet systems. However, for oilfield applications, where a wettability change occurs after primary drainage, there is a relationship between S_{oi} and wettability (see Chapter 5.3.1), since this governs how much of the pore surface is contacted by oil, and the capillary pressure, Chapter 2.3. The effect of wettability on relative permeability is discussed next.

7.2 Effect of Wettability

The simplest, and most simplistic, way to treat wettability is to consider first a completely oil-wet system as the reverse of a water-wet case, at least for waterflooding. Water is now the non-wetting phase, while oil is wetting, and so we may write $k_{rw}^{ow} = k_{ro}^{ww}$ and $k_{ro}^{ow} = k_{rw}^{ww}$ where the superscripts ow and ww refer to oil-wet and water-wet, respectively. This would mean that the relative permeabilities would tend to cross at a saturation less than 0.5, since water, being non-wetting and occupying the larger pores and throats, will have a larger conductance for the same saturation than oil in the smaller elements. The end-point value k_{rw}^{max} would be large, generally greater than 0.5, and possibly close to 1.

While appealing, and possibly correct for the trivial case of a solid surface which is treated everywhere to change the contact angle such that the wetting properties of oil and water are switched, it ignores the realities of how wettability is established in reservoirs. Firstly, k_{ro}^{max} is established after primary drainage when the system is water-wet and, as shown above, generally has a value close to 1 for low values of S_{wi} regardless of the wettability during subsequent waterflooding. If, after water injection, oil is re-injected, in a secondary process, we may see trapping of water as a non-wetting phase, a true residual S_{wr} and a lower final value of k_{ro} than after primary drainage. Secondly, and more significantly, water layers are present in the rock at the beginning of waterfooding and initially it is the conductance of these layers that will determine k_{rw} until a path of water-filled pores and throats spans the system.

Anderson (1987b) provides a review of the effect of wettability on relative permeability with an assessment of different measurement techniques. Here a selection of data will be presented and interpreted in terms of pore-scale displacement. Again, however, it will prove impossible to make quantitative predictions based on generic guidelines; the interaction of wettability, pore structure and the emergent displacement patterns reveals a rich behaviour; one that is not amenable to a purely discursive analysis.

7.2.1 Oil-Wet Media

As an example of an oil-wet case, Fig. 7.8 shows the waterflood relative permeabilities for a reservoir sample. Also shown are network modelling predictions assuming an infinitesimal flow rate (Valvatne and Blunt, 2004). Since a three-dimensional image of the rock was not available, the pore and throat sizes of a Berea network, Fig. 2.14, were tuned to match the primary drainage capillary pressure, as shown. Then, for waterflooding, all the elements containing oil became oil-wet and an accurate prediction of relative permeability was obtained. The displacement statistics for the unaltered Berea network are shown in Table 5.6.

The most evident feature of the curves in comparison to the water-wet cases shown previously is that the oil relative permeability appears close to zero (less than 0.01) for water saturations above 70%, yet displacement continues until a residual saturation less than 5% is attained. This low value of S_{or} is consistent with the data presented in Chapter 5.4. Low k_{ro} close to S_{or} is the signature of layer drainage, discussed in Chapter 5.2: layers provide continuity of the oil phase, preventing trapping, but have a very low conductance, quantified in Chapter 6.1.3 and Eq. (6.71). A model incorporating empirical expressions for layer flow in each element, Eq. (6.27), is able to reproduce the measurements. Overall the displacement is a drainage process dominated by piston-like throat filling (see Chapter 5.3.2), where continuity of the oil phase is maintained through layers.

The water relative permeability rises to almost one near S_{or}: water is the non-wetting phase and preferentially fills the larger regions of the pore space. Near the end-point, it fills the vast majority of the pore space while the oil is confined to small elements and layers.

The final, and most subtle, feature is the very low water relative permeability at low saturation. The initial water saturation is around 5%, yet k_{rw} remains lower than 0.01 until $S_w = 0.5$. The relative permeabilities cross at a saturation of close to 0.6, similar to the value seen in water-wet sands and sandstones, see Figs. 7.1 and 7.4, implying that at intermediate saturation k_{ro} is higher than k_{rw}, even if oil is the wetting phase. This seems to contradict the idea that the water fills the larger pores, with a high conductance, and hence should have a more steeply rising relative permeability, akin to the oil relative permeability in primary drainage. This apparently counter-intuitive trend has been observed in several other modelling studies (Zhao et al., 2010; Bondino et al., 2013).

The explanation for this conundrum lies in the difference between ordinary percolation and invasion percolation and the significant impact of water layers even

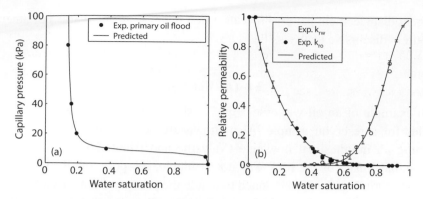

Figure 7.8 Comparison between predicted and experimental properties for an oil-wet reservoir sandstone. (a) Centrifuge primary oil flooding capillary pressures were used to modify the network until a good match – as shown – was obtained. (b) The predicted waterflood relative permeability is compared to experimental steady-state data. The very low residual oil saturation is an indication of oil-wet characteristics. From Valvatne and Blunt (2004).

if they have little conductance; see Chapter 3.4.3. Primary drainage is an invasion percolation process and displacement only occurs through throats where there is a connected path of filled elements to the inlet; see Chapter 3.4. At the percolation threshold, the saturation at which the invading phase spans the rock is very small; see Eq. (3.29), page 101. Then further displacement increases the connectivity and conductance of the non-wetting phase as it preferentially occupies the most permeable elements. Even so, for Berea and Bentheimer sandstones, the non-wetting phase saturation is between 15% and 20% before k_{ro} exceeds 0.01; see Figs. 7.3 and 7.5.

Regardless of the wettability there are water layers throughout the pore space, established at the end of primary drainage. When water is injected we can have an ordinary percolation process, as discussed in Chapter 5.2.2. In oil-wet media, these layers provide water to fill the oil-wet oil-filled pores adjacent to throats which have remained filled with water after primary drainage. The largest pores fill first. This results in a large change in saturation, since they have a significant volume, but a negligible improvement in conductance until the water-filled pores form a connected path spanning the system. This is evident in the high frequency of I_{3+} pore filling events; see Table 5.6. The water layers are pinned and cannot swell substantially, retaining, approximately, the very low conductance from the end of primary drainage; see page 232. Any appreciable connectivity requires a spanning cluster of filled pores and throats: as this is a drainage process, throat filling limits the advance of water. At the percolation threshold, the water saturation is not negligible but given approximately by Eq. (3.27). Since a fraction p_c of throats are filled, and they are the larger elements, then one would expect this saturation to exceed

p_c, assuming that initially all the water is confined to layers. For a lattice with a coordination number of 4.23 (the value for the Berea network used, Table 2.1), Eq. (3.25) gives $p_c = 0.35$. We do not expect to see a significant water relative permeability until the water saturation is significantly larger than 35%, consistent with the measurements; see Fig. 7.8.

There is a nuance to this argument, since pore filling can only proceed if a neighbouring throat is already filled with water. If a pore is surrounded by oil, then the only filling process, if oil-wet, is forced snap-off, whose threshold capillary pressure is very unfavourable, since it relies on the movement of a pinned water layer, Eq. (5.3). Hence, this low water connectivity relies on some water-filled throats being present, either at the end of primary drainage or during an initial period of water imbibition in a mixed-wet rock. If no, or negligible, initial water is present, then waterflooding reverts to being an invasion percolation process in an oil-wet medium with a more rapid rise in water relative permeability. Finally, as discussed in Chapter 6.4.3, page 269, the conductance of pinned layers is so small that even at the low flow rates encountered in the field, filling through them is effectively suppressed. This again forces a more invasion percolation advance with higher k_{rw}. This discussion illustrates that the behaviour will be very sensitive to flow rate, wettability and initial saturation, which we explore further below.

7.2.2 Cross-Over Saturation and Waterflood Recovery

While it is seductive to focus on the residual oil saturation to assess recovery, since this represents the theoretical minimum that will remain in the rock, it is an unhelpful guide for oil-wet and mixed-wet media where oil layers are present. As shown in Fig. 5.13, Chapter 5.4, hundreds or thousands of pore volumes of water may need to be injected to see residual saturations below 10%; in contrast, most waterfloods inject at most 1, or maybe 2, pore volumes. A more accurate estimate of how much oil will be displaced, after of order 1 pore volume of injection, is the saturation at which the relative permeabilities cross. When $k_{rw} < k_{ro}$ then, if we have equal oil and water viscosities, more oil than water will be produced at a well. When the relative permeabilities cross, equal amounts of oil and water flow, while when $k_{rw} > k_{ro}$ water flows at a higher rate than oil. There is some economic limit at which the cost of producing, separating and disposing of the produced water exceeds the benefit of oil production and the well is closed in; hence we cannot continue to inject hundreds of pore volumes to drive the oil saturation down to its true residual value. This limit depends on the oil price, the rate of oil production and location of the field, but the cross-over saturation gives a quick comparative guide to recovery: the higher the water saturation, S_w^{cross}; when $k_{rw} = k_{ro}$, the better the waterflood oil recovery.

It has to be emphasized that this is simply a guide to recovery and useful when comparing different relative permeability curves. It is not to be used in any sort of quantitative assessment: we will show how to compute recovery analytically from the relative permeability in Chapter 9.2. On the other hand, it does serve as a quick reference to relate recovery to relative permeability, and certainly superior to a fixation with residual saturation.

If we apply this concept to our oil-wet example, Fig. 7.8, we see $S_w^{cross} = 0.58$. In contrast $S_w^{cross} = 0.66$ for a water-wet Berea sandstone (Fig. 7.5), 0.6 for water-wet sand packs (Fig. 7.4) and 0.57 for Bentheimer (Fig. 7.3). Hence, waterflooding may be less favourable, but only slightly less, in the oil-wet rock than for the water-wet cases we have presented. This makes physical sense (again, if we stop thinking about S_{or}): water is the non-wetting phase in an oil-wet rock, fills the larger pores and flows readily, leaving the oil behind in smaller parts of the pore space. This oil clings to the surface and can only be displaced after an uneconomic amount of water injection. In contrast, for a water-wet rock, the water is held back in the smaller pores, allowing oil to flow more readily to achieve good recoveries. There is a trade-off though between the low k_{rw} and S_{or}, with the best recoveries attained for Berea, with water retained in small pores and clay, and for a sand pack, where there is little trapping.

The cross-over saturation is very sensitive to the connectivity of the water, particularly at low saturation when it is governed by layers. If we have a high initial water saturation, then we may already have a good pathway of water-filled elements which may allow k_{rw} to rise sharply. Similarly, if S_{wi} is very low, then there are no filled throats to nucleate pore filling, the displacement is an invasion percolation process, as mentioned above, and again k_{rw} may increase rapidly. In both cases S_w^{cross} is low with unfavourable recoveries. It is for intermediate values of S_{wi} that the lowest k_{rw} and largest S_w^{cross} are observed: there is some pore filling throughout the rock, but these pores only connect at high saturations. There is also an effect of flow rate, mentioned above, as pore filling through pinned layers is extremely slow; see Eq. (6.33).

Many reservoir samples are mixed-wet and allow some spontaneous imbibition of both oil and water: in waterflooding this naturally allows filling throughout the rock, providing flow paths for water but confined to the smallest elements. This will affect the behaviour and the cross-over saturation, as we demonstrate below.

7.2.3 Mixed-Wet Media

Fig. 7.9 shows the experimental waterflood capillary pressure and oil relative permeability for a reservoir sandstone compared to predictions using pore-scale network modelling, where it was assumed that 85% of the pores and throats contacted by oil at the end of primary drainage became oil-wet, and ignoring viscous effects (Øren et al., 1998). There is very little spontaneous imbibition and most displacement occurs during forced injection at a negative capillary pressure to reach a residual saturation of approximately 5%. This low value of residual oil saturation is indicative of oil-wet regions that span the system, as shown in Chapter 5.3.2. The results are qualitatively similar to those shown in Fig. 7.8, with an end-point water relative permeability close to 1, a range of low k_{ro} indicative of oil layer drainage and low k_{rw} when the flow of water is restricted by water layers.

There is, however, one surprising and significant feature of the predictions: the cross-over saturation is 0.78, when $k_{ro} = k_{rw} \approx 0.01$. This indicates that there is significant blocking of the flow of both phases, even at high water saturation, with many terminal menisci straddling throats: only when water invades almost all the pore space, confining oil to layers, does k_{rw} increase substantially. The remarkable high value of S_w^{cross} implies a favourable waterflood recovery; one that is substantially better than either the water-wet or oil-wet cases presented previously. This again is due to the low water relative permeability, limited by pinned layers: as shown in Fig. 7.9, k_{rw} only rises above 0.01 for a water saturation close to 0.8.

To illustrate the sensitivity of water connectivity to the relative permeability, Fig. 7.10 shows the network-predicted arrangement of water at a saturation of 0.5. Here we consider another reservoir sandstone where half the elements contacted by oil become oil-wet. By default, a pore or throat becomes oil-wet at random: however, we could assign the larger elements to be oil-wet, or the smaller ones

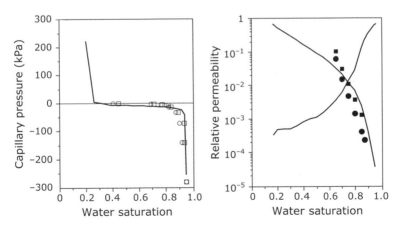

Figure 7.9 Measured waterflood capillary pressures and oil relative permeability (points) from two experiments on a clay-rich mixed-wet reservoir sandstone. The measurements are compared to predictions using pore-scale modelling (lines). To match the capillary pressure it was assumed that 85% of the pores occupied by oil after primary drainage became oil-wet. From Øren et al. (1998).

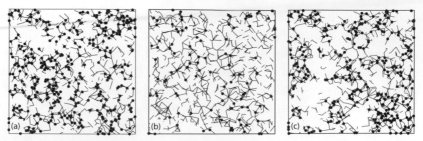

Figure 7.10 The distribution of water-filled elements during waterflooding at $S_w = 0.5$. Half the pore space is oil-wet and the water-filled elements are shown in black (small circles are the pores and lines are throats): the picture shows a projection of the three-dimensional network. The fluid pattern is shown when (a) the larger radius pores are oil-wet; (b) the smaller pores are oil-wet; (c) the wettability is spatially correlated. The saturation shown is towards the end of spontaneous displacement. Since water-wet elements are generally filled first, the phase connectivity during waterflooding is improved if there is a spatial correlation in the water-wet regions, allowing them to span the sample. From Valvatne and Blunt (2004).

(Dixit et al., 2000). For instance, Kovscek et al. (1993) suggested that the smaller pores, with a greater curvature, are more likely to become oil-wet; however direct measurements of contact angle using scanning electron microscopy suggest that, in carbonates, the larger pores become oil-wet, while in sandstones wettability is governed by the presence of clays (Robin et al., 1995; Durand and Rosenberg, 1998): the comparison of literature data and network modelling results shown in Fig. 5.10 indicated that indeed both situations are possible. In Fig. 7.10, cases where either the largest or the smallest pores become oil-wet are considered. All throats connected to an oil-wet pore were also made oil-wet. When the larger pores are oil-wet, smaller pores are filled during spontaneous water injection in preference to the larger oil-wet throats. We see a large shift in water saturation with a limited water connectivity. When the smaller pores are oil-wet, these remain unfilled until near the end of the displacement and we now see more throat filling, but the water is still poorly connected until high saturations are reached. The final case is when the wettability is spatially correlated, with patches of water-wet and oil-wet pores: this leads to better water connectivity with a pathway of water-filled water-wet elements established during spontaneous imbibition. It was this wettability pattern that was used to illustrate displacement statistics in Chapter 5.3.2.

Predictions using this model are shown in Fig. 7.11: the experiments imply a larger k_{rw} than either of the uncorrelated wettability models. Allowing spatial correlation in the wettability allows the water to span the pore space more easily and provides a better match to the experiments. However, $S_w^{cross} = 0.62$, which still indicates favourable waterflood recovery.

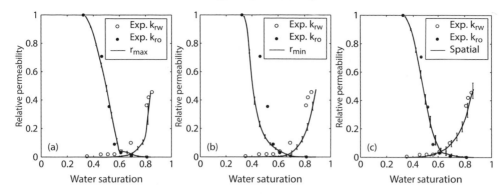

Figure 7.11 Comparison between experimental and predicted relative permeability for a reservoir sandstone. The difference between the cases is how wettability is characterized at the pore scale, see Fig. 7.10. (a) The larger pores become preferentially oil-wet. (b) The smaller pores are preferentially oil-wet. (c) The oil-wet pores are spatially correlated. From Valvatne and Blunt (2004).

The results of an extensive study of relative permeability in reservoir carbonates are shown in Fig. 7.12. The rock samples were a mix of calcite and dolomite cores with porosities between 0.17 and 0.28 and permeabilities between 1 and 46 mD. The first comment is that these low permeability samples, with extensive micro-porosity, do not show generically or qualitatively different relative permeability behaviour than sandstones with a narrower distribution of pore size, consistent with the curves shown in Fig. 7.6. This is different from capillary pressure, where there is a distinct signature of micro-porosity, see Figs. 3.8 and 3.9. These rocks are likely to be mixed- or oil-wet with little or no spontaneous imbibition of water (Dernaika et al., 2013): all the curves show an oil layer drainage regime. A high-rate, so-called bump flood, is performed at the end of waterflooding to displace any remaining mobile oil. This bump flood does indeed recover more oil, even though the oil relative permeability is too low to quantify in the measurements. The lowest oil saturation reached (and this may not be the final residual) is between around 5 and 25%, suggesting that the oil-wet patches connect across the rock.

Indeed, the somewhat low maximum or end-point values of the water relative permeability, k_{rw}^{max}, which vary from around 0.15 to 0.5, are possibly experimental artefacts. These values would indicate more water-wet conditions, with the flow of water significantly restricted by the presence of residual oil. This is likely, however, to be simply a consequence of not reaching the true residual oil saturation in these experiments.

7.2.4 Hysteresis in the Water Relative Permeability

Three types of hysteresis in the water relative permeabilities are shown in Fig. 7.12. The first two sets of curves, from samples with the highest permeability, have

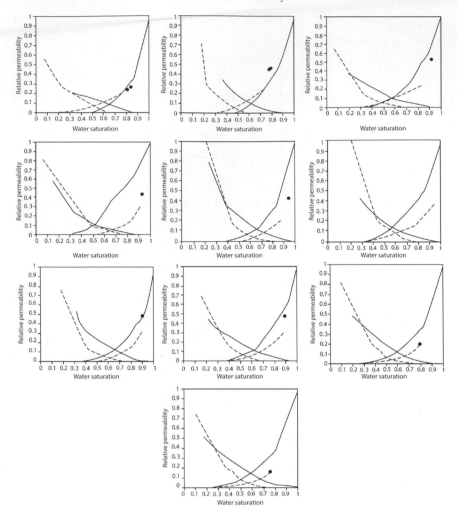

Figure 7.12 Relative permeabilities from steady-state measurements on reservoir carbonates from the Middle East. The solid lines are for primary drainage and the dashed lines for waterflooding. The isolated points indicate k_{rw} after a high-rate injection of water at the end of waterflooding that attempts to push out oil to reach its residual saturation. Three types of hysteresis behaviour in the water relative permeability are seen. (1) The first two graphs are characteristic of an oil-wet system where the water rapidly connects across the sample giving a higher water relative permeability in waterflooding than primary drainage, and cross-over saturations of 0.5 or lower indicative of poor waterflood performance. (2) The third example has a similar k_{rw} in waterflooding, but with a low k_{ro} which results in a cross-over saturation of less than 0.5, again giving poor recovery. (3) All the other examples show lower water relative permeabilities and cross-over saturations between 0.55 and 0.6, suggesting more favourable waterflood recoveries. Adapted from Dernaika et al. (2013).

relative permeabilities that might be considered characteristic of an oil-wet rock. In waterflooding, since the water is now the non-wetting phase, it occupies the larger pore spaces and hence, for the same saturation, has a higher relative permeability than in primary drainage, where it is the wetting phase and confined to smaller elements. The cross-over saturation is 0.45 and 0.5 for the two examples shown, indicative of a poor waterflood recovery. While this contrasts with the results shown hitherto, it is consistent with a rapid connection of the water through large pores in an invasion percolation-like filling sequence.

The second type of behaviour is shown by just the third graph in Fig. 7.12. Here k_{rw} for primary drainage and waterflooding are similar for $S_w \leq 0.6$. However, the oil relative permeability is lower, falling to less than 0.01 for $S_w > 0.6$. The result is that S_w^{cross} is shifted to lower values; around 0.47, giving again an unfavourable waterflood recovery. This is an intermediate behaviour, where the competition between filling large pores and poor layer connectivity conspire, coincidently, to give a similar k_{rw} to that in primary drainage.

The other cases in Fig. 7.12 show hysteresis patterns more in accord with the results presented already, with cross-over saturations between 0.55 and 0.6. The reason is that k_{rw} remains very low, lower than seen in primary drainage, or in waterflooding if the rock were water-wet, as the water connectivity is impeded by flow through thin, pinned layers in the oil-wet regions of the pore space established at the end of primary drainage; see Chapter 5.2.1. These layers have a tiny conductance, Chapter 6.4.3, hold back the water, allow oil to be displaced and lead to good waterflood recovery.

The same types of hysteresis behaviour discussed here can also be observed using pore-scale network modelling. For our example cases whose displacement statistics were presented in Tables 5.2 to 5.6, see Chapter 5.3.2, Ketton and Mount Gambier display the first type, or conventional, hysteresis, with k_{rw} for waterflooding lying above that for primary drainage with S_w^{cross} below 0.5 for most mixed-wet cases. Introducing a spatial correlation in wettability allows water to span the system early in the displacement, Chapter 5.3.2. With poorer connectivity, water flow can be more easily hampered by thin layers, resulting in very low k_{rw} and larger values of S_w^{cross} for Estaillades. Benntheimer shows the second type of hysteresis, with similar k_{rw} in drainage and waterflooding, but $S_w^{cross} < 0.5$ because of the rapid fall in k_{ro}. Berea shows both the second and third type of hysteresis behaviour, dependent on wettability, with values of $S_w^{cross} > 0.5$ for oil-wet fractions around 0.5.

Two further experimental examples are shown in Fig. 7.13 from the giant carbonate Ghawar oilfield in Saudi Arabia, which display the first type of hysteresis with unfavourable waterflood recovery (Okasha et al., 2007). The experiments were performed on core samples from the Arab-D reservoir and give results representative

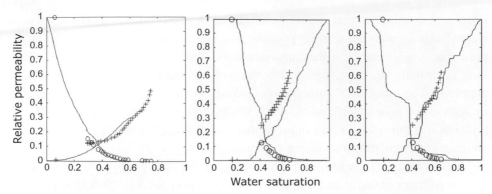

Figure 7.13 Measured waterflood relative permeabilities from the Arab-D reservoir of the giant Ghawar field, points (Okasha et al., 2007). The lines are pore-scale modelling matches. On the left the data are matched using a Mount Gambier network where 25% of the pores become oil-wet after primary drainage; see Table 5.2 for the displacement statistics. The middle and right figures show predictions of another set of experiments using two completely oil-wet networks of low connectivity (in the middle a network derived from a subsurface sample from the Middle East, and on the right from Portland limestone, see Fig. 2.3). From Gharbi and Blunt (2012).

of the field as a whole: the variation in wettability with height above the oil/water contact was shown in Fig. 5.12. The left figure is for the more water-wet sample: the results can be matched using the Mount Gambier network model, where 25% of the pores become oil-wet after primary drainage; see Table 5.2 for the displacement statistics. We see a very sharp increase in k_{rw}, suggestive of a well-connected pathway of large water-filled pores through the system. Near the end-point, k_{rw} rises almost vertically: this is seen in mixed-wet systems where the filling of a few elements leads to a very rapid improvement in connectivity. We also see evidence of a layer drainage regime, with displacement at a low k_{ro}, showing that some parts of the pore space are oil-wet. S_w^{cross} is less than 0.4, which is lower than the cases shown previously.

The second example in Fig. 7.13, appears more oil-wet with an even steeper initial rise in k_{rw}. This can be matched using completely oil-wet networks, with low coordination numbers around 2.5, representing either Portland limestone or a Middle eastern subsurface sample (Gharbi and Blunt, 2012). Again S_w^{cross} is rather low – around 0.4. Note that the modelling results are not predictions, as we know neither the pore structure nor the wettability.

7.2.5 Hysteresis in the Oil Relative Permeability

In all cases we see low values of the oil relative permeability at high water saturation, indicative of layer drainage, as discussed above. We do, however, see two types of hysteresis in k_{ro} in Fig. 7.12. This has less of an impact on the cross-over saturation than the behaviour of k_{rw}, but is again the result of the subtle interplay of displacement sequence and connectivity.

In most cases the waterflood k_{ro} lies below its value in primary drainage, since oil will now occupy some of the smaller elements. This is consistent with the results shown for sandstones. We also expect that k_{ro} in waterflooding is lower than for an equivalent water-wet sample, except when we reach low oil saturations.

The fourth and sixth cases in Fig. 7.12 show oil relative permeabilities for water-flooding that are larger than those for primary drainage, at least for low water saturation. This is another surprising and counter-intuitive trend which has also been remarked upon in modelling studies, where k_{ro} for waterflooding in an oil-wet or mixed-wet network may be larger than for the water-wet case (Zhao et al., 2010). This has also been discussed by Bondino et al. (2013) who suggested that this is the result of a competition between changes in saturation and changes in conductance.

The pore-scale interpretation is as follows. At the end of primary drainage many small throats are filled with oil; at the start of imbibition in a water-wet rock, these throats are then filled with water by snap-off. These contribute relatively little to saturation, but may disconnect major flow paths, akin to the effect of blocking some key side roads on overall traffic flow. In a mixed-wet rock the first filling is also of water-wet elements, but these leave the oil-wet patches untouched, which may continue to provide the main flow paths. Then the larger oil-wet pores are filled. While this filling will impact the conductance, they also have a large volume, so the drop in relative permeability is accompanied by a large change in saturation. This pore filling is evident in the displacement statistics in Chapter 5.3.2 where we see many more I_3 and I_{3+} events for oil-wet media, indicating the filling of isolated pores by water at the beginning of waterflooding.

The conductance of a throat scales as the fourth power of radius, see Eq. (6.9); it is this radius that limits flow and controls the relative permeability. The change in saturation is proportional to the volume displaced, which is the volume of the throat itself and, dependent on the displacement process, the adjoining pore as well. This volume scales at most with the third power of radius for a spherical element, and is often unrelated to the throat size. Hence if we fill a throat, the size of the element filled has a bigger impact on relative permeability than saturation: filling large throats gives a bigger change, as a function of saturation, than filling small ones.

However, when we fill an oil-wet rock during waterflooding, we see something different. We fill the large pores, which are those that give the largest change in saturation, while their contribution to conductance, given by the radius of the adjoining throats, may be modest: the impact depends crucially on how throat radius is correlated to pore size. If this is a weak relationship, pore filling leads to a big change in saturation with less drop in conductance than seen for a water-wet system, giving a higher oil relative permeability.

   This type of hysteresis is not observed frequently, since normally large throats are associated with large pores, but is not impossible, as Fig. 7.12 demonstrates. In modelling studies by Zhao et al. (2010) this behaviour was predicted for an oil-wet sand pack, with little variation in pore and throat size, and a poorly connected carbonate with a high aspect ratio (large pores and small throats). Our exemplars, whose displacement statistics were shown in Chapter 5.3.2, in contrast, all show the expected hysteresis of k_{ro} with lower values for mixed- and oil-wet systems.

7.2.6 Features of Relative Permeability in Mixed-Wet Rocks

It is possible to use pore-scale modelling to generate relative permeabilities more systematically, exploring the impact of structure and contact angle distribution on the behaviour (McDougall and Sorbie, 1995; Øren et al., 1998; Dixit et al., 2000; Zhao et al., 2010; Gharbi and Blunt, 2012), developing the discussion of displacement statistics in Chapter 5.3.2. We will not pursue this here, as without grounding in experimental data, the uncertainties of network extraction and the assignment of wettability, cooperative pore-filling entry pressures and microporosity make the predictions somewhat uncertain (Bondino et al., 2013). Instead, we will synthesize the results here and those of other studies to make the following general observations concerning relative permeability trends in mixed-wet and oil-wet media.

(1) There is no signature of relative permeability in carbonates with microporosity, distinct from sandstones. In general, the same trends are seen, dependent on pore-space connectivity and wettability.
(2) The defining characteristic of mixed- and oil-wet rocks is the presence of a layer drainage regime with low k_{ro} near the residual oil saturation.
(3) The oil relative permeability normally falls quickly from S_{wi} and is lower than seen in primary drainage, since oil is no longer the non-wetting phase and does not occupy the larger pores. There are exceptions, with k_{ro} larger in waterflooding for oil-wet rocks where the filling of large pores does not have a significant impact on the oil conductance.
(4) There is considerable variability in the behaviour of the water relative permeability with no obvious trend related to wettability or overall network connectivity: instead, k_{rw} depends on how well water can connect across the sample during waterflooding. In many cases, where water is confined to pinned layers in oil-filled elements and waterflooding proceeds as a percolation-like process, filling the larger oil-wet pores, the water relative

permeability remains below 0.01 until S_w exceeds 0.5. Alternatively, if there is a clear path of larger pores that rapidly fill with water, where there is negligible filling through layers, we have an invasion percolation process and k_{rw} rises more steeply and is larger than seen in primary drainage.

(5) The relative permeability is influenced principally by the fraction of pores and throats that become oil-wet after primary drainage, rather than by the contact angle in the oil-wet regions, as long as the formation of oil layers is allowed.

(6) In terms of cross-over saturation, there is a non-monotonic trend as the fraction of oil-wet pores is varied: there is normally an optimum with lower values of S_w^{cross} for strongly water-wet or strongly oil-wet systems. For carbonate networks, the most favourable behaviour (highest value of S_w^{cross}) is seen when a small fraction, around 25%, of the pores remain water-wet. This is when the displacement is most percolation-like with some filling through layers.

(7) We also observe a non-monotonic trend in cross-over saturation as we vary the initial water saturation. If S_{wi} is very low, there are few water-filled throats to nucleate pore filling, and the displacement in oil-wet regions is invasion percolation-like, facilitating rapid connection of the water. Increasing S_{wi} improves recovery, as it allows the water to access the pore space while restricting its connectivity, with a percolation-like displacement pattern, giving higher values of S_w^{cross}; as S_{wi} increases further, however, this leads to higher conductivity of the water, since it spans the rock throughout waterflooding, giving larger k_{rw}, and a decrease in S_w^{cross}.

(8) The largest values of S_w^{cross} are seen for cases with an approximately equal amount of spontaneous oil and water imbibition: for sandstones this may be when the rock is slightly water-wet overall (McDougall and Sorbie, 1995; Øren et al., 1998), whereas for carbonates the maximum is observed for more oil-wet conditions (Gharbi and Blunt, 2012).

(9) The residual oil saturation increases with oil-wet fraction, f, when f is small, since these regions become trapped after spontaneous displacement. Once the oil-wet patches span the system, S_{or} declines and reaches a minimum for strongly oil-wet conditions.

(10) The residual oil saturation can display a non-monotonic trend with initial saturation; see Chapter 5.4. As a result, scanning curves for relative permeability in mixed- and oil-wet media may cross, as do the capillary pressures (Fig. 5.17), something that is not observed in water-wet rocks.

In practical terms the most important aspect is how relative permeability affects waterflood recovery. While we have addressed recovery approximately through the use of the cross-over saturation, this will be revisited more rigorously after we present analytical solutions to the flow equations in Chapter 9.2.5. In any event, the recovery is influenced primarily by k_{rw} and its connectivity through a combination of layers, water-filled water-wet regions already present after primary drainage, or filled during the initial, spontaneous part of the displacement, and the larger oil-wet elements filled through forced injection. Unfortunately, there is no simple rule or guide to quantify this behaviour: the experimental evidence and network model predictions see a range of behaviour, subtly dependent on the distribution of wettability and the pore structure.

To make predictions, the pore structure, including micro-porosity, needs to be accommodated into a numerical model, combined with a pore-by-pore assignment of contact angle. At present this is a challenging task, and most published predictions, as we have shown, are confined to simple rocks where the wettability is known, or assumed.

At present our understanding and quantification of relative permeability is unsatisfactory. Experiments are time-consuming and prone to error, especially if saturations are not determined from *in situ* imaging, with particular problems determining accurate, or even reasonable, end points when there is flow through wetting layers of water and oil. There is no straightforward manner in which relative permeability may be predicted. Indeed, the emphasis on some sort of unilluminating black-box simulation employing a faddish new algorithm with a plethora of poorly constrained inputs is unhelpful: pore-scale models should be used principally to interpret and understand measurements while providing quantitative estimates of how properties may vary in response to changes in pore structure, displacement path and wettability.

Refinements in our measurement methods incorporating an imaginative symbiosis of imaging and modelling results may, in the near future, lead to a much better understanding than we have at present.

7.2.7 Guidelines for Assessing Wettability

It is often convenient to categorize relative permeability curves by their likely wettability, simply through inspection. This is valuable since often an independent assessment, such as a wettability index (see Chapter 5.3), is lacking, or other measurements, such as contact angles on flat surfaces, do not necessarily relate to *in situ* conditions. Furthermore, such a judgement allows a quick prediction of how the relative permeability may change with wettability, with height above the oil column; see page 113. Lastly, some features, principally the cross-over saturation, provide a rapid comparative estimate of recovery.

Unfortunately, most such guidelines are misguided, harking back to Craig (1971) and his so-called rules of thumb, from an era when mixed-wettability was not well understood, before the advent of modern pore-scale imaging and modelling. I will enumerate them here with my critique, based on the experimental evidence, before presenting my own suggestions.

(1) **Initial water saturation**. A value, at the beginning of waterflooding, of greater than 0.2–0.35 is supposed to indicate water-wetting, while less than 0.15 is oil-wet. This is not unreasonable when considering a series of cores of the same rock type from the same field. A lower initial saturation represents a sample higher in the oil column, where the imposed capillary pressure after primary drainage is larger and more of the solid surface has come in contact with oil. Hence, as S_{wi} is reduced, more oil is squeezed against the surface and the rock becomes more oil-wet, as discussed in Chapter 2.3. However, the degree of wettability alteration is dependent on the oil, brine and rock rendering a general rule impractical. As an example, Fig. 7.14 shows the variation of Amott-Harvey index, Eq. (5.7), with initial water saturation, S_{wi}, for Prudhoe Bay: the trend with depth was shown in Fig. 5.11. We do indeed see a tendency to more water-wet conditions with an increase in S_{wi}, but there is considerable scatter in the data and the division between water-wet and oil-wet conditions is not clear-cut: instead we see a variation in behaviour, with most of the cores being mixed-wet, meaning that they imbibe both oil and water (Jerauld and Rathmell, 1997). More importantly, simply knowing S_{wi} for one sample in isolation gives no indication of wettability.

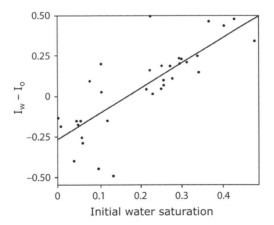

Figure 7.14 The variation of the Amott-Harvey index, $I_{AH} = I_w - I_o$, Eq. (5.7), with initial water saturation, S_{wi}, for cores from Prudhoe Bay. The points are data while the line shows the approximate trend. The corresponding variation of wettability with depth was shown in Fig. 5.11. $I_{AH} > 0$ indicates water-wet conditions, while $I_{AH} < 0$ is more oil-wet. A lower S_{wi} allows more of the oil to contact the rock surface leading to more oil-wet conditions. From Jerauld and Rathmell (1997).

(2) **Cross-over saturation**. If S_w^{cross} is greater than 0.5, the rock is water-wet; if it is less than 0.5 it is likely to be oil-wet. For the water-wet cases shown previously, we do see a cross-over value greater than 0.5, since for the same saturation of its own phase $k_{ro} > k_{rw}$. For a strongly oil-wet system, we may expect the reverse, since now the oil, as the wetting phase, is confined to the smaller pores. However, the data we have presented demonstrates that the cross-over is frequently seen for $S_w > 0.5$ even in oil-wet and mixed-wet systems. The reason for this, as discussed above, is the low conductivity of the water at the beginning of waterflooding when its spanning pathway has to include pinned water layers. This rule is unhelpful and misleading.

(3) **End-point water relative permeability**. Craig (1971) considered the ratio of the water relative permeability at the end of waterflooding to the oil relative permeability at the beginning: $k_{rw}^{max}/k_{ro}^{max}$. A value less than 0.5 indicates a water-wet rock, while greater than 0.5, and approaching 1, is oil-wet. This is a reasonable characterization, if tending to over-state the water-wet range: for strongly water-wet samples, see, for instance, Fig. 7.5, this ratio is less than 0.2, while it is approximately 0.5 for the mixed-wet sample shown in Fig. 7.11. However, this does rely on the experiments reaching a true residual oil saturation and, consequently, the maximum possible value of k_{rw} rather than an intermediate value when further displacement is possible; this can be particularly problematic for some mixed-wet rocks where k_{rw} rises very steeply near the end point, such as in Fig. 7.13.

Unfortunately, the single most important characteristic, the cross-over saturation, which gives a guide to recovery, is often misinterpreted using outdated concepts.

I will now synthesize the results of this chapter to propose my own guidelines for characterizing wettability. Here we have three conditions: water-wet, mixed-wet and oil-wet.

(1) **End-point water relative permeability**. Here I modify the rule above to suggest that $k_{rw}^{max} < 0.2$ indicates a clearly water-wet rock, a range between 0.3 and 0.6 is mixed-wet, while larger values are suggestive of oil wettability. However, as seen in Fig. 7.12 and mentioned above, we do need to reach the real residual oil saturation. As a consequence, this guideline, while reasonable, needs to be applied with caution.

(2) **Is oil layer drainage seen?** Study k_{ro} near the end-point. If it declines sharply, with a finite gradient, this is definitively water-wet; if instead there is a clear region, spanning a saturation change of 0.1 or more, where $k_{ro} < 0.05$ yet displacement continues, this is the signature of oil layers and the system is

mixed- or oil-wet. This does require an experiment where sufficient time is allowed for slow recovery. If a remaining oil saturation of 0.15 or lower is achieved in a consolidated rock, then again this strongly implies mixed-wet or oil-wet conditions.

(3) **Shape of the water relative permeability.** A very low k_{rw} at low water saturation is indicative of a mixed-wet system. This may be difficult to distinguish from a strongly water-wet case, but if $k_{rw} < 0.2$ for $S_w = S_{wi} + 0.2$ this suggests a mixed-wet rock. A sharp rise in k_{rw}, with a cross-over saturation of less than 0.5, is less usual, and is often, but not always, associated with oil-wet conditions. An almost vertical increase in k_{rw} close to the end-point shows mixed-wettability, as seen in Fig. 7.13.

7.3 Effect of Capillary Number and Viscosity Ratio

The impact of viscous forces, encapsulated in the capillary number, Ca, Eq. (6.62), on displacement and the residual oil saturation was discussed in Chapter 6.4.6. For $Ca < 10^{-6} - 10^{-5}$ capillary forces dominate the pore-scale arrangement of fluid with little effect of flow rate, interfacial tension or fluid viscosities on the behaviour. However, if we consider surfactant flooding, or gas/oil systems near miscibility, we encounter interfacial tensions that are orders of magnitude lower than between oil and water, with larger capillary numbers that will affect the relative permeabilities.

As an example, Fig. 7.15 shows primary drainage relative permeabilities for gas/vapour systems as the interfacial tension decreases when the fluid mixture approaches miscibility (Bardon and Longeron, 1980). Ca was in the range 10^{-8} to 10^{-3} with interfacial tensions as low as 10^{-6} N/m, almost five orders of magnitude smaller than typical oil/water values at laboratory conditions. For the highest capillary numbers or lowest interfacial tensions, the relative permeabilities of both phases increase, the apparent irreducible wetting phase (liquid) saturation decreases and, for nearly miscible fluids, the relative permeabilities approach a straight line.

Qualitatively similar behaviour has also been observed for oil/water systems during waterflooding: there is a reduction in both the residual oil saturation and irreducible water saturation as Ca increases, and an increase in both oil and water relative permeability consistent with asymptotic progress to straight lines for large Ca (Amaefule and Handy, 1982).

As a general guide, if $Ca < 10^{-6}$ we consider that the flow is capillary-controlled and there is no influence of flow rate, while for $Ca > 10^{-3}$ we reach a viscous-dominated regime. The exception, explored in Chapter 6.4.3, is when displacement is impeded by pinned water layers. As we have already shown, this can

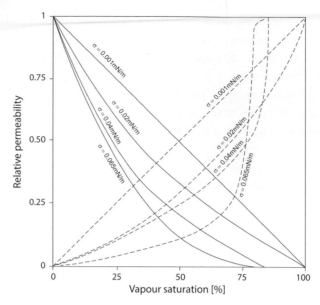

Figure 7.15 Primary drainage vapour/liquid relative permeabilities with the inter-facial tensions indicated. The lowest value corresponds, approximately, to a capillary number, Ca, Eq. (6.62), of 10^{-3} where viscous forces begin to dominate at the pore scale. The relative permeabilities are shown as functions of the vapour (non-wetting) phase saturation, which is the opposite of the conventional presen-tation in terms of the liquid (wetting phase) saturation. The viscous-dominated limit for a near-miscible gas/oil system is straight-line relative permeabilities with no trapped or irreducible saturation. From Bardon and Longeron (1980).

tip the displacement from percolation to invasion percolation with a significant impact on waterflood recovery; however, our understanding of this behaviour, based on core-scale experiments and pore-scale modelling, is insufficiently sophis-ticated to make definitive predictions of how the results presented already are affected by flow rate.

7.3.1 Relative Permeability in the Viscous Limit

It is sometimes erroneously considered that in the viscous regime, where pore-scale ganglia of oil may be displaced, the relative permeabilities approach a linear form: $k_{rp} = S_p$, whose the sum is 1 and hence capillary forces cause no impediment to the flow. This is, however, only correct when the viscosities of the two phases are the same, as we see when we approach gas/oil miscibility, see Fig. 7.15, but is not appropriate for oil/water systems during, for instance, surfactant flooding. The correct approach is to note that when two phases flow together (as in dispersed flow in a pipe) the fractional flow, f_p, of a phase is proportional to its saturation. $f_p = q_p/q_t$ where q_t is the total flow of both phases. Invoking Darcy's law, Eq. (6.42) and assuming that each phase experiences the same pressure gradient, we find for two phases labelled p and q

$$f_p = \frac{k_{rp}/\mu_p}{k_{rp}/\mu_p + k_{rq}/\mu_q} = \frac{\lambda_p}{\lambda_t}, \tag{7.1}$$

where we define a mobility $\lambda = k_r/\mu$ and a total mobility

$$\lambda_t = \lambda_p + \lambda_q \tag{7.2}$$

If the two phases flow together and we apply averaging of the mobility in parallel,

$$\lambda_t = S_p/\mu_p + S_q/\mu_q. \tag{7.3}$$

Therefore, if $f_p = S_p$ the corresponding relative permeabilities are

$$k_{rp} = S_p \left[S_p + \frac{\mu_p}{\mu_q}(1 - S_p) \right] \tag{7.4}$$

using $S_q = 1 - S_p$.

To represent the flow in a mega-scale simulation, see Fig. 6.10, it is possible to account for the viscous instability; see Chapter 6.6.2, Fig. 6.26. Although the flow pattern for viscous fingering is complex, with a fractal surface, the averaged behaviour is captured elegantly and easily using empirical models. In these cases we usually do not consider the saturation of each phase, as the fluids are miscible or nearly miscible, but the concentration, c. For a miscible gas displacing oil, there will be one hydrocarbon phase with a concentration c of gas within it: $c = 1$ is the injection condition while $c = 0$ is oil. We consider $c \equiv S_g$.

The most commonly applied, and accurate, model of fractional flow was developed by Todd and Longstaff (1972):

$$f(c) = \frac{c}{c + \frac{1-c}{M_{eff}}}, \tag{7.5}$$

where the effective viscosity ratio is given by

$$M_{eff} = M^{1-\omega} \tag{7.6}$$

with M the viscosity ratio given by Eq. (6.102): for an unstable flow $M > 1$.

In the original paper, Todd and Longstaff (1972) axiomatically, and erroneously, assumed linear relative permeabilities and then proposed a somewhat convoluted mixing model for the total mobility to recover the functional form above. Fundamentally the models for total mobility and fractional flow are independent: simulation studies suggest that the relative permeabilities are those consistent with Eqs. (7.3) and (7.5) (Blunt et al., 1994).

ω is an exponent: a value of 2/3 reproduces the average behaviour of experiments and simulations of viscous fingering: the leading edge of the fingers

$(c \rightarrow 0)$ travels at a dimensionless speed M_{eff} while the trailing edge $(c \rightarrow 1)$ has a speed $1/M_{eff}$; the experiments can also be matched using Koval's model, $M_{eff} = (0.78 + 0.22 M^{1/4})^4$ (Koval, 1963). The same concept can be used to derive the average properties for fingering in more complex compositional displacements and for waterflooding, where M represents the mobility ratio across the unstable front (Blunt et al., 1994).

7.3.2 Viscosity Ratio and Viscous Coupling

There is one other circumstance where viscous effects play a role, even for slow flow, which is when wetting layers provide a lubricating effect to the oil. This effect is particularly evident near the end of primary drainage. So far we have asserted that assuming a no-flow boundary at fluid/fluid interfaces was most consistent with experimental results; see Chapter 7.1.3, page 229. However, this is not always the case when considering arc menisci in wetting layers.

Odeh (1959) showed that in low permeability rock, the end-point oil relative permeability could exceed one, with the effect more marked for high viscosity oils. The pore-scale explanation is based on flow of the wetting phase in layers which, through viscous coupling, then enhances the movement of the non-wetting phase through the centres of the pore space. This is a more significant effect when the pores are small or the oil viscosity is large. Berg et al. (2008) compiled evidence on end-point oil relative permeability for 43 experiments that also had relative permeability values greater than 1, shown in Fig. 7.16.

Instead of the no-flow condition on solid and fluid/fluid interfaces, as we have imposed before, we may employ the Navier slip condition:

$$v = -b\frac{\partial v}{\partial r} \qquad (7.7)$$

at the boundary. v is the flow speed of the oil, r is some coordinate perpendicular to the boundary and b is the slip length. This is physically equivalent to allowing some film or layer of thickness b to be imposed between the oil and the solid, giving a finite flow speed everywhere in the oil phase. This is, in reality, provided through co-current, albeit slow, flow of wetting layers. Then if we assume that the flow of oil is governed by the Poiseuille law in a cylindrical tube, Eq. (6.9), but now with a finite velocity at the walls due to the presence of an initial water saturation:

$$k_{ro}(S_{wi}) = 1 + \frac{4b}{r_e}, \qquad (7.8)$$

where r_e is some effective capillary (throat) radius, related to the permeability by Eq. (6.49) to find:

$$k_{ro}(S_{wi}) = 1 + c\sqrt{\frac{\phi}{K}} \qquad (7.9)$$

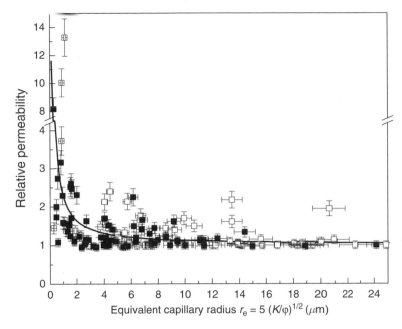

Figure 7.16 Oil end-point relative permeabilities $k_{ro}(S_{wc})$: data with $k_{ro}(S_{wc}) >$ 0.95 for aged (empty symbols) and un-aged (filled symbols) samples compiled from 43 experiments in the literature are plotted as a function of effective capillary radius, $r_e = 5\sqrt{K/\phi}$, see Eq. (6.49) (Dullien, 1997). The solid line is a best fit of the slip model with an average slip length of 318 ± 44 nm. From Berg et al. (2008).

for some constant c. Assuming that $r_e \approx 5\sqrt{K/\phi}$ (Dullien, 1997), equivalent to setting $a = 4 \times 10^{-2}$ in Eq. (6.53), see Table 6.3, and fitting to the data, Berg et al. (2008) found the slip length $b = 5c/4$. A value of around 0.3 μm matched experimental data with reasonable accuracy, as shown in Fig. 7.16. Note, however, that this does not account for the mobility ratio: one would expect more of a lubrication effect if the water in wetting layers were much less viscous than the oil.

This is an important effect as it increases the apparent permeability of the reservoir at initial saturation. At higher water saturations though, the effect of this slip condition is less noticeable and, for slow flow, there is little impact of mobility ratio or slip on the relative permeabilities (Odeh, 1959).

7.4 Empirical Models

The assessment of relative permeability is both frustrating and tantalizing. Relative permeability is difficult to measure and, for most reservoir rocks, challenging to predict using pore-scale modelling, yet the relative permeability functions themselves are not that complex and can be described by only a few key parameters that capture the end-points and shapes of the curves. In this section we derive an analytical expression for relative permeability, based on a bundle of capillary tubes before presenting the empirical models commonly applied in petroleum engineering and hydrology.

7.4.1 Relative Permeability of a Bundle of Tubes

Extending the analysis performed in Chapter 3.3, we start from the situation illustrated in Fig. 3.10 on page 89, where tubes are filled in order of size as displacement proceeds. The saturation is given by Eq. (3.19), assuming a water-wet system where wetting phase fills the smallest tubes first.

For flow, we assume the Poiseuille law in each tube, Eq. (6.9). For a series of tubes filled with water from the minimum radius r_{min} to some value r, and oil from r to the maximum r_{max} we may write, following Eq. (6.44), Chapter 6.2.1, page 238:

$$Q_w = -\frac{\pi}{8\mu_w}\nabla P \int_{r_{min}}^{r} r^4 f(r)dr, \tag{7.10}$$

and

$$Q_o = -\frac{\pi}{8\mu_o}\nabla P \int_{r}^{r_{max}} r^4 f(r)dr, \tag{7.11}$$

where we have $f(r)dr$ tubes of size between r and $r + dr$.

The relative permeability is the ratio of the flow of a given phase to the flow if all the tubes were filled with that phase. Hence

$$k_{rw} = \frac{\int_{r_{min}}^{r} r^4 f(r)dr}{\int_{r_{min}}^{r_{max}} r^4 f(r)dr}, \tag{7.12}$$

$$k_{ro} = \frac{\int_{r}^{r_{max}} r^4 f(r)dr}{\int_{r_{min}}^{r_{max}} r^4 f(r)dr}. \tag{7.13}$$

We then define, as in Chapter 3.3, a throat size distribution:

$$G(r) = \frac{r^3 f(r)}{\int_{r_{min}}^{r_{max}} r^2 f(r)dr}, \tag{7.14}$$

which allows us to rewrite Eqs. (7.12) and (7.12) as

$$k_{rw} = \frac{\int_{r_{min}}^{r} rG(r)dr}{\int_{r_{min}}^{r_{max}} rG(r)dr}, \tag{7.15}$$

$$k_{ro} = \frac{\int_{r}^{r_{max}} rG(r)dr}{\int_{r_{min}}^{r_{max}} rG(r)dr}, \tag{7.16}$$

noting that the constant integral in Eq. (7.14) cancels

The reason for this definition is that we can then relate $G(r)$ to the derivative of the primary drainage capillary pressure, Eq. (3.23): $G(r) = -P_c(dS_w/dP_c)$. The tube radius $r = 2\sigma \cos\theta/P_c$ from Eq. (3.1) and therefore $dr/dP_c = -2\sigma \cos\theta/P_c^2$. Hence with a little algebra and replacing integration over r with first P_c and then S_w we may rewrite Eqs. (7.15) and (7.16) as:

$$k_{rw} = \frac{\int_{S_{wi}}^{S_w} P_c^{-2} dS_w}{\int_{S_{wi}}^{S_w^{max}} P_c^{-2} dS_w},$$ (7.17)

$$k_{ro} = \frac{\int_{S_w}^{S_w^{max}} P_c^{-2} dS_w}{\int_{S_{wi}}^{S_w^{max}} P_c^{-2} dS_w},$$ (7.18)

where $P_c(S_w)$ is the capillary pressure and S_w^{max} is the maximum saturation: 1 for primary drainage or $1 - S_{or}$ for imbibition.

In principle, this approach can be used to predict the relative permeability from the capillary pressure in a rather elegant and simple fashion: this is potentially valuable, as relative permeability is difficult to measure and subject to significant uncertainty, while the primary drainage capillary pressure is routinely acquired using mercury injection.

However, it is apparent that the formulae in Eqs. (7.17) and (7.18) will give qualitatively the wrong shapes and magnitude to the relative permeability curves. Firstly, by definition, $k_{rw} + k_{ro} = 1$, whereas this is never the case for capillary-controlled displacement, as terminal menisci block the pore space. Secondly, in almost all the examples shown, the curves are concave, meaning that $d^2k_{rp}/dS_{wp}^2 > 0$: when plotted as a function of the saturation of its own phase, the relative permeability gets steeper – has a larger gradient – as the saturation increases. Differentiating Eqs. (7.17) and (7.18) twice gives:

$$\frac{d^2k_{rw}}{dS_w^2} = -\frac{d^2k_{rw}}{dS_w^2} \propto -\frac{1}{P_c^3}\frac{dP_c}{dS_w} > 0,$$ (7.19)

which is the correct sign for water, but wrong for oil.

The problem is, of course, that this method does not account for the tortuous connectivity of the phases, and the blocking effect of menisci, nor can it accommodate different wettabilities. Burdine (1953) did employ a modified version of this approach with an empirical saturation-dependent tortuosity factor: a prefactor was added to Eqs. (7.17) and (7.18) to produce curves that matched several experimental datasets. However, this idea was developed before the advent of advanced imaging and modelling tools, and so while a valuable pedagogic exercise, is no longer seriously applied to predict the behaviour of reservoir rocks: the relative permeability has to be measured, and if a prediction is needed, then this is best based on a physically based pore-scale model.

7.4.2 Functional Forms for Relative Permeability and Capillary Pressure

While, as just mentioned, it is not sensible to use simplistic expressions to predict relative permeability, it is convenient to have closed-form expressions that can be matched to (often rather scattered) data and then employed in analytical and numerical models.

For example, the equations used in Chapter 9 to construct analytical solutions are:

$$k_{rw} = k_{rw}^{max} S_e^a, \tag{7.20}$$

$$k_{ro} = k_{ro}^{max} (1 - S_e)^b, \tag{7.21}$$

where the effective water saturation $1 \geq S_e \geq 0$ is defined as:

$$S_e = \frac{S_w - S_{wc}}{1 - S_{or} - S_{wc}}. \tag{7.22}$$

This model has six fitting parameters: a, b, k_{rw}^{max}, k_{ro}^{max}, S_{wc} and S_{or}. We assume that S_{wc} and S_{or} are the true residual (connate) water and oil saturations, respectively, where the relative permeabilities are zero.

For capillary pressure we employ

$$P_c = P_c^{max} \frac{\left(\frac{S_w^*}{S_{wi}}\right)^{-c} - \left(\frac{S_w}{S_{wi}}\right)^{-c}}{\left(\frac{S_w^*}{S_{wi}}\right)^{-c} - 1}, \tag{7.23}$$

for $P_c \geq 0$ with fitting parameters $c > 0$, P_c^{max} and S_w^*, the saturation when $P_c = 0$. $S_w^* \leq 1 - S_{or}$ with equality only for strongly water-wet systems. S_{wi} is the lowest water saturation reached after primary drainage at the largest imposed capillary pressure P_c^{max}: $S_{wi} \geq S_{wc}$ in Eqs. (7.20) and (7.21). For the reasons discussed below, we require $a > 1 + c$.

These are simple functions and it is reasonable to consider other expressions, with possibly more parameters, if the problem or the data demand. In particular, the relative permeabilities do not allow for inflexion points, when the curvature changes, while the capillary pressure does not accommodate negative values and, for a water-wet system, assumes a zero entry pressure. There is no claim that this is correct, but for the practical purposes of Chapter 9 this is sufficient.

When using a closed-form expression for multiphase flow properties the following rules need to be obeyed, both to capture the main physical features, but also to allow reasonable solutions to the governing flow equations.

(1) The empirical forms match the data to within experimental error. This is self-evident.
(2) The relative permeability functions only reach zero at the true residual saturation. Often experiments are truncated at some minimum relative permeability: this should not be used to define the residual for mixed-wet and oil-wet media; instead the empirical model needs to match the low, but finite, relative permeability at that point.
(3) The relative permeability can have either a zero or finite gradient at the end points. A zero gradient for the oil relative permeability near S_{or} is characteristic of mixed- and oil-wet rocks, while a finite gradient indicates water-wet conditions: the model must allow either situation.
(4) The relative permeabilities and capillary pressures must be finite or zero with non-infinite first and second derivatives. This is to allow converged, and physical, solutions to the flow equations.

These criteria appear quite liberal, but item (4) is almost universally ignored for capillary pressure, with infinite values and gradients imposed at the initial water saturation and/or at the residual oil saturation. This, as already described, is unattainable.

We may relax item (4) to extrapolate capillary pressure and relative permeability to end points that were not reached experimentally. It may be argued that at the true value of S_{wc} the capillary pressure is indeed infinite. This is still a dubious assertion for waterflooding and could lead to a poor match to data at low water saturation, since P_c^{max} should represent the capillary pressure just before the first displacement event: this will always be a finite value.

If we insist on an infinite capillary pressure, we need to ensure that sensible solutions to the resultant flow equations are obtained. From the multiphase Darcy law, Eq. (6.58), the flow of a phase is proportional to the relative permeability times a pressure gradient. This pressure gradient can be caused by changes in capillary pressure: there will be a term governing flow which scales as $k_{rp} \, \partial P_p / \partial S_w \sim k_{rp} \, dP_c/dS_w \, \partial S_p/\partial x \sim k_{rp} \, dP_c/dS_w$ assuming a finite saturation gradient. This concept will be explored more rigorously in Chapter 9. On physical grounds, near end points there is very slow flow tending to zero as the end point is reached. However, if we have an infinite capillary pressure and pressure gradient, we seem to be applying an infinite force! However, the relative permeability is zero, so we apply an infinite force to an immovable flow. For physical solutions we require

$$ k_{rw} \left. \frac{dP_c}{dS_w} \right|_{\lim S_w \to S_{wc}} = 0, \tag{7.24} $$

while applying the same argument near S_{or}:

$$k_{ro} \left. \frac{dP_c}{dS_w} \right|_{\lim S_w \to 1 - S_{or}} = 0. \qquad (7.25)$$

Creating a formulation that relies on a proper limit of an infinite quantity multiplied by a zero requires careful analysis; my recommendation is still to use a finite (albeit very high) capillary pressure at end points where possible. That poorly converged coarsely gridded simulations manage to brush this problem under the numerical carpet does not make it go away.

Even with diverging capillary pressures, Eqs. (7.24) and (7.25) are normally obeyed with judicious choice of the exponents a and b: for instance, if we use Eq. (7.20) for the water relative permeability and Eq. (7.23) for capillary pressure, we obey Eq. (7.24) if $a > 1 + c$.

In the petroleum literature, power-law forms for multiphase flow properties are often termed Corey-type or Brooks-Corey after Corey (1954) and Brooks and Corey (1964), with a and b called Corey exponents. However, this is not quite correct. Corey (1954) used a Burdine-type bundle-of-tubes model to derive specific expressions for relative permeability, which for the non-wetting phase was not a power law. Brooks and Corey (1964) did allow a power-law expression with an arbitrary exponent for the wetting phase relative permeability, but the non-wetting phase relative permeability was again not a power law.

In the end it is better to abandon the somewhat shaky theoretical foundation of the models – I am not going to repeat the details here – and accept empirical equations as a curve fit, sufficient, as Burdine himself noted, for *engineering purposes* (Burdine, 1953), with a careful check that they will result in meaningful solutions to the flow equations.

7.4.3 Empirical Models in Hydrology

In the hydrology literature, a different set of functions for capillary pressure and relative permeability are almost universally applied following the work of van Genuchten (1980) and van Genuchten and Nielsen (1985). The theoretical justification is again derived from Burdine (1953) and from a similar analysis by Mualem (1976) which results in a flexible formulation with fitting parameters that can be matched honestly to data.

The formulation is written in terms of the water content, ϕS_w. Here, to be consistent, we use the terminology of the rest of the book and write relative permeability and capillary pressure as functions of saturation or normalized saturation, Eq. (7.22). The simplest model has a water relative permeability

$$k_{rw} = S_e^{1/2} \left[1 - \left(1 - S_e^{1/m} \right)^m \right]^2, \qquad (7.26)$$

and the capillary pressure is

$$P_c = \alpha \left(S_e^{-1/m} - 1\right)^{1/n}, \qquad (7.27)$$

where $m = 1 - 1/n$. $n > 1$. Here there are four matching parameters: α, m (or n), S_{wc} and S_{or}: the non-wetting phase (air) relative permeability is not specified.

Traditionally, for air/water flows, a zero air viscosity is assumed. While this is reasonable where the air is well connected in the pore space, it is not accurate near the residual saturation. The viscosity of air is around 2×10^{-5} Pa.s, giving a water/air viscosity ratio of approximately 50, which is large but certainly not infinite. Eq. (7.25) is not obeyed in the van Genuchten (1980) model, since the non-wetting phase relative permeability (air in this case) is ignored. The model has an infinite capillary pressure gradient at both end points, and it is assumed that the system is strongly water-wet. This is a particular problem for imbibition, where, invoking percolation theory, we would expect a finite capillary pressure at the end point and a zero gradient (as in primary drainage): the apparent steep downwards lurch to zero at S_{or} is simply a combination of experimental artefact (zero is the lowest pressure imposed) and finite size effects, see Fig. 4.23. In Chapter 9.3 we will show that this affects the nature of the analytical solutions for spontaneous imbibition: for reasonable results the model is limited to strongly water-wet media and air/water displacement.

This concludes the discussion of relative permeability for two-phase flow: the next chapter will introduce the fascinating topic of three-phase flow with the simultaneous movement of oil, water and gas.

8

Three-Phase Flow

This chapter will introduce the pore-scale aspects of the flow of three fluid phases: oil, water and gas. We encounter three-phase flow during gas injection into oil reservoirs, and when we have non-aqueous phase pollutants present in unsaturated soil and rock. We do not have a complete predictive understanding of two-phase flow, so it is not realistic to expect a comprehensive treatment when another phase is introduced. Instead, the pore-scale phenomena that are unique to three-phase flow will be highlighted, namely spreading layers, relationships between contact angles and multiple displacement, before presenting some example relative permeabilities and predictions using pore-scale modelling.

8.1 Spreading

Imagine spilling a drop of oil on water. Even though there is no porous medium, this is a three-phase flow process, with oil, water and gas (air) all present. What happens? There are three possibilities, Fig. 8.1, dependent on the spreading coefficient of oil, C_s (Hirasaki, 1993; Adamson and Gast, 1997):

$$C_s = \sigma_{gw} - \sigma_{go} - \sigma_{ow}, \qquad (8.1)$$

where σ is the interfacial tension while gw, go and ow label gas/water, gas/oil and oil/water, respectively. We assume that $\sigma_{gw} \geq \sigma_{ow} \geq \sigma_{go}$.

We now return to the Young equation, (1.7), and the balance of forces at a three-phase contact. This can be generalized to allow both a horizontal and vertical force balance, since the interfaces of the fluids can move.

The first possibility is when $C_s < 0$ and there is a balanced Neumann's triangle of forces (interfacial tensions) as shown, with a droplet of oil floating on water; see van Kats et al. (2001). This is non-spreading, such as many vegetable oils: place a small drop in a saucepan of water and see for yourself.

354

What happens when oil is placed on water?

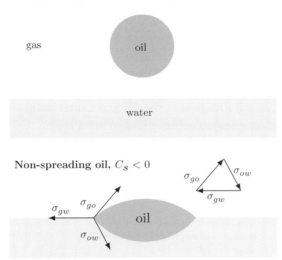

Non-spreading oil, $C_s < 0$

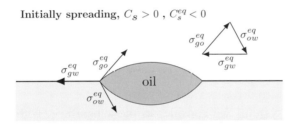

Initially spreading, $C_s > 0$, $C_s^{eq} < 0$

Spreading oil, $C_s^{eq} = 0$

$$\sigma_{gw}^{eq} = \sigma_{ow}^{eq} + \sigma_{go}^{eq}$$

Figure 8.1 An illustration of what happens when a drop of oil is placed on water (top). If we have a non-spreading oil, where C_s, Eq. (8.1), is negative, there is a Neumann's triangle of forces given by the three interfacial tensions and a stable drop floats on the surface, as shown. We can have an initially spreading oil, $C_s > 0$, which covers the gas/water interface with an oil film, shown by the bold line. This film lowers the effective, or equilibrium interfacial tension such that C_s^{eq}, Eq. (8.2), is negative. Then excess oil forms a stable drop with a triangle of forces using the equilibrium interfacial tensions. The third possibility is that oil continues to spread until a thick film of oil, shown shaded, is present, separating gas from water, where $C_s^{eq} = 0$.

The second possibility is when $C_s > 0$ and it is not possible for a droplet of oil to reside in equilibrium in the presence of gas and water, since we cannot balance the tensions where the phases meet. In this case the oil spreads: some cooking oils do this, as does petrol and crude oil (at least at ambient conditions). The oil will continue to spread until the oil film is of molecular thickness. More oil is added so that there is more than sufficient for a film to cover all the water. What happens to the excess oil? The oil film lowers the effective interfacial tension between gas and water: the film forms spontaneously and so must lessen the energy of the interface. We now define equilibrium values of the interfacial tensions, where the three phases have been in contact and films may be present. The equilibrium spreading coefficient is,

$$C_s^{eq} = \sigma_{gw}^{eq} - \sigma_{go}^{eq} - \sigma_{ow}^{eq}, \tag{8.2}$$

where $\sigma^{eq} \leq \sigma$.

If $C_s^{eq} < 0$, then excess oil forms a droplet in contact with an air/water interface covered by a film. The three equilibrium interfacial tensions now form a triangle of forces.

There is a third possibility that, as more oil is added, the film thickens. When this film is thicker than the range of intermolecular forces, we no longer really have an air/water interface, but an air/oil one followed by oil/water. Then $\sigma_{gw}^{eq} = \sigma_{ow}^{eq} + \sigma_{go}^{eq}$ and $C_s^{eq} = 0$. It is not possible to have $C_s^{eq} > 0$ as eventually the oil will enlarge the film sufficiently to drive its value to zero.

For most fluid systems we consider in the subsurface, with non-polar oil, the equilibrium spreading coefficient is either zero, or close to zero. To see this, we return to the first equation in the book, Eq. (1.1), which related an interfacial tension to its surface tensions for fluid pairs where one phase (gas or oil) was non-polar. If we apply this to our three-phase case, to find σ_{ow}, σ_{go} and σ_{gw} in terms of their surface tensions, σ_o, σ_w and σ_g, then

$$\sigma_{gw} \approx \sigma_w - \sigma_g = \sigma_w - \sigma_o + \sigma_o - \sigma_g \approx \sigma_{ow} + \sigma_{go}, \tag{8.3}$$

which gives a spreading coefficient close to zero.

We test Eq. (8.3) by studying data for fluids at ambient conditions, where the interfacial tension between air and water at 20°C is 72.8 mN/m. Table 8.1 shows the interfacial tensions and spreading coefficients for some common fluids: *n*-decane, *n*-octane, *n*-hexane, carbon tetrachloride and benzene. For alkanes, as the chain length increases, the oil becomes less spreading: decane does not spread, octane does just (but the value is 0 to within experimental error), while lighter alkanes such as hexane definitely spread, as does benzene and carbon tetrachloride. In all cases C_s^{eq} is negative, so excess oil will retract to form a droplet. While none of these values are strictly zero, C_s, and particularly C_s^{eq}, are normally much smaller

Table 8.1 *Measured interfacial tensions at $20°C$ and atmospheric pressure between liquid and air, σ_{go}, and liquid and water, σ_{ow}. Also shown are the initial, C_s, and equilibrium, C_s^{eq}, spreading coefficients. Data from Matubayasi et al. (1977), Hirasaki (1993), and Georgiadis et al. (2011).*

Liquid	σ_{go} mN/m	σ_{ow} mN/m	C_s mN/m	C_s^{eq} mN/m
n-Decane	23.8	52.0	3.0	−3.9
n-Octane	21.6	50.9	0.3	−0.9
n-Hexane	18.4	49.9	4.5	−0.2
Benzene	28.8	33.4	10.6	−1.6
Carbon tetrachloride	27.0	42.7	3.1	−1.9

Table 8.2 *Measured gas and crude oil properties, including gas/oil, gas/water and oil/water interfacial tensions (σ_{go}, σ_{gw} and σ_{ow}), and spreading coefficient, C_s, as a function of pressure and at a temperature of $82°C$. Data from Amin and Smith (1998).*

Pressure MPa	Oil density kg/m^3	Gas density kg/m^3	σ_{go} mN/m	σ_{gw} mN/m	σ_{ow} mN/m	C_s mN/m
22.6	687	159	2.71	33.0	33.9	−3.6
21.5	692	151	3.00	34.0	33.4	−2.4
20.7	695	145	3.14	35.1	33.1	−1.1
19.0	703	137	3.92	41.0	32.4	4.69
17.2	710	123	4.26	42.5	31.8	6.48
15.9	717	115	6.16	43.8	31.0	6.63
13.8	726	97	6.36	46.0	30.1	9.51
12.1	734	85	6.91	48.0	30.0	11.1
10.3	742	72	7.27	50.2	29.6	13.3
8.96	748	62	7.95	51.7	29.1	14.7
6.89	758	49	8.31	53.7	29.0	16.4
5.65	764	39	8.97	56.0	28.1	18.9
4.14	772	29	9.55	59.0	27.3	22.2
1.72	786	12	10.6	61.6	25.7	25.3

than the magnitude of any of the individual interfacial tensions in Eq. (8.1). The spreading behaviour of non-polar liquids can be predicted with reasonable accuracy from a knowledge of intermolecular forces (Hirasaki, 1993); however, for more complex crude oil/brine mixtures, direct measurement is necessary.

Table 8.2 shows data for a crude oil/natural gas mixture at reservoir conditions of temperature and pressure (Amin and Smith, 1998). Here we see that the system is spreading for most of the pressure range, only becoming non-spreading at the highest pressures when the oil/water and gas/water interfacial tensions are approximately equal. We do not know C_s^{eq} but the values are likely to be close to 0.

There is one useful limit: when gas is injected into oilfields, the injection pressure and/or the gas composition is often designed such that the oil and gas may become miscible, or at least are close to miscibility (Orr, 2007). Since the two hydrocarbon phases become indistinguishable, $\sigma_{go} \rightarrow 0$. Also, since gas and oil must therefore have similar properties, we expect $\sigma_{gw} \approx \sigma_{ow}$ and hence as we approach miscibility $C_s \rightarrow 0$ and $C_s^{eq} \rightarrow 0$.

Oil can, and will, spread between gas and water in a porous medium. This will affect the contact angles and pore-scale fluid configurations with a major influence on flow properties.

8.2 Contact Angles and the Bartell-Osterhof Equation

We now assume that the three fluids are in thermodynamic equilibrium, including the presence of any spreading films, and place them in a porous medium. We drop the superscript *eq* but in what follows assume that we are referring to equilibrium values of the interfacial tensions. Returning to the Young equation, (1.7), there are now three combinations of fluid pairs that can come into contact at the solid. We extend Fig. 1.5 to show these three possibilities in Fig. 8.2: oil/water, gas/water and gas/oil. A horizontal force balance is invoked as before to find three Young equations:

$$\sigma_{os} = \sigma_{ws} + \sigma_{ow} \cos\theta_{ow}, \tag{8.4}$$

$$\sigma_{gs} = \sigma_{ws} + \sigma_{gw} \cos\theta_{gw}, \tag{8.5}$$

$$\sigma_{gs} = \sigma_{os} + \sigma_{go} \cos\theta_{go}. \tag{8.6}$$

There are three contact angles and six interfacial tensions. However, all the solid tensions can be eliminated by re-arranging Eqs. (8.4)–(8.6). First we take Eq. (8.6) from Eq. (8.5) to obtain:

$$\sigma_{ws} + \sigma_{gw} \cos\theta_{gw} = \sigma_{os} + \sigma_{go} \cos\theta_{go}, \tag{8.7}$$

and then take away Eq. (8.4) and re-order the terms to find,

$$\sigma_{gw} \cos\theta_{gw} = \sigma_{ow} \cos\theta_{ow} + \sigma_{go} \cos\theta_{go}. \tag{8.8}$$

This is the Bartell-Osterhof equation and shows that not all the contact angles and interfacial tensions are independent (Bartell and Osterhof, 1927). The interfacial tensions are governed by intermolecular forces and can be measured separately. The contact angles are not independent though: it is not possible to have an arbitrary combination of oil, water and gas wettability in a porous medium. Knowing

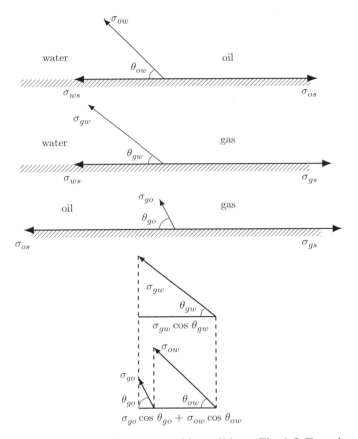

Figure 8.2 Two fluid phases in contact with a solid, see Fig. 1.5. For a three-phase system there are three possible combinations, as shown from the top: oil/water, gas/water and gas/oil. The contact angles are measured through the denser phase and may assume any value. A horizontal force balance yields the Young equations (8.4), (8.5) and (8.6) from which a constraint between the contact angles and fluid-fluid interfacial tensions may be derived, Eq. (8.8): this is illustrated geometrically in the bottom figure.

the interfacial tensions and two contact angles (generally θ_{ow} and θ_{go}) the third, θ_{gw}, may be found. Eq. (8.8) can also be derived directly from consideration of the three-fluid contact line connecting the fluid/fluid interfaces between two bulk phases and the third phase contained in a cusp near the pore wall (van Dijke and Sorbie, 2006a).

This analysis may be extended to any number of fluid phases: for an n phase system there are $n(n-1)/2$ possible contact angles but only $n-1$ are independent (Blunt, 2001a). Four fluid phases, for instance, can be encountered in low pressure oil fields undergoing CO_2 injection: there is a CO_2-rich gaseous phase, a hydrocarbon-rich gas phase, oil, and water: there are six contact angles of which three are independent.

The Bartell-Osterhof equation (8.8) is valid for equilibrium contact angles, with the fluids at rest. However, it is generally assumed that we may use the relationship for displacement processes, but this does introduce an ambiguity if there is contact angle hysteresis. For instance, imagine water is being injected into oil and gas where the appropriate gas/water and oil/water contact angles to be used are θ_{gw}^A and θ_{ow}^A, respectively. Now if one uses Eq. (8.8) to find the third angle, θ_{go}, it is not clear that the calculated value is the advancing or receding angle.

The approach used in modelling studies is to calculate θ_{gw} using θ_{ow}^A and θ_{go}^A, and then find another value employing receding values θ_{ow}^R and θ_{go}^R: the larger value of θ_{gw} is taken to be the advancing angle, and the smaller the receding angle (Piri and Blunt, 2005a).

8.2.1 Wetting and Spreading in Three-Phase Flow

We may define the wettability of a three-phase system through the oil/water contact angle, or what is generally considered to be wettability, and the spreading properties of the oil, which defines the gas/oil contact angle: then using Eq. (8.8), the gas/water contact angle is specified. To see how we relate θ_{go} to C_s consider the Young force balance, Fig. 8.2, between gas and oil on a completely water-wet surface with $\theta_{ow} = 0$. If there is a thick film of water then $\sigma_{os} \equiv \sigma_{ws}$ while $\sigma_{gs} \equiv \sigma_{gw}$ and Eq. (8.6) becomes:

$$\sigma_{gw} = \sigma_{ow} + \sigma_{go} \cos \theta_{go}, \tag{8.9}$$

or, using Eq. (8.1),

$$\cos \theta_{go} = 1 + \frac{C_s}{\sigma_{go}}. \tag{8.10}$$

For the opposite extreme of a completely oil-wet rock with $\theta_{ow} = 180°$, $\theta_{go} = 0$, as oil must also be wetting to gas.

Two approaches may now be used to assign contact angles for any wettability and spreading coefficient. The first specifies θ_{go} and θ_{ow} and then uses Eq. (8.8) to define θ_{gw}: the relationship between C_s and θ_{go} is given by Eq. (8.10) (Piri and Blunt, 2005a). The second approach instead fixes the spreading coefficient and then extrapolates θ_{go} linearly between the water-wet and oil-wet extremes outlined above (van Dijke and Sorbie, 2002b, c):

$$\cos \theta_{go} = \frac{1}{2\sigma_{go}} (C_s + C_s \cos \theta_{ow} + 2\sigma_{go}), \tag{8.11}$$

and then, using Eq. (8.8)

$$\cos \theta_{gw} = \frac{1}{2\sigma_{gw}} \left[(C_s + 2\sigma_{ow}) \cos \theta_{ow} + C_s + 2\sigma_{go} \right]. \tag{8.12}$$

This relationship has been confirmed experimentally through a series of contact angle measurements on silica surfaces whose wettability was altered systematically using different silanes (Grate et al., 2012).

In most cases the gas/oil contact angle is small: using the values for C_s^{eq} in Table 8.1, we see a range from $8°$ for hexane to $33°$ for decane for θ_{go} in Eq. (8.10).

8.2.2 Why Ducks Don't Get Wet

If we have a completely water-wet surface, where $\theta_{ow} = 0$, Eq. (8.8) gives

$$\cos\theta_{gw} = 1 - \frac{C_s + \sigma_{go}(1 - \cos\theta_{go})}{\sigma_{gw}}. \tag{8.13}$$

For a spreading system, $C_s = \theta_{go} = 0$, and we also have $\theta_{gw} = 0$. Obviously, a surface that is wetting to water in the presence of oil is also wetting to water with gas.

Now consider the opposite case of an oil-wet surface where $\cos\theta_{ow} < 0$. If $\sigma_{ow}\theta_{ow} + \sigma_{go}\cos\theta_{go} < 0$, then Eq. (8.8) gives $\cos\theta_{gw} < 0$ or a gas/water contact angle greater than $90°$. If we have a completely oil-wet case, $\cos\theta_{ow} = -1$:

$$\cos\theta_{gw} = -1 + \frac{C_s + \sigma_{go}(1 + \cos\theta_{go})}{\sigma_{gw}}. \tag{8.14}$$

Since normally $\sigma_{gw} > 2\sigma_{go}$, we must have $\cos\theta_{gw} < 0$: gas is the wetting phase in the presence of water.

Oily surfaces repel water. That water can be non-wetting when surrounded by gas is evident from every-day experience: water spilt on a varnished (or plastic) table-top forms a bead, while waterproof clothing is either plastic or waxed cloth. And the varnish and plastic are made out of oil and so, by construction, are oil-wet. Water has a contact angle of greater than $90°$ in such circumstances, has limited direct contact with the surface, and easily runs off it.

For an oil-wet porous medium there will be a finite entry capillary pressure for the water to displace gas, since invasion will be a drainage process. Water has to be forced into the material. Take, for example, the feathers of a duck. The feathers form a porous medium. The gaps between the feathers (the pore space) remain full of air; this provides an effective layer of insulation. But ducks spend a lot of time on water. The feathers are kept oil-wet by constant preening. The water therefore stays out, keeping the duck warm and dry; the water cannot force its way through the feathers. We see the same with sheep: they are covered in wool, but the wool is oily and repels the rain. A sheep with a completely sodden fleece would die of cold.

Droplets of water can also be seen on the waxy upper surfaces of leaves; see Fig. 8.3. Plants have a transpiration flow of water from uptake by water-wet roots

Figure 8.3 A photograph of water and oil on the surface of leaves. The leaves are waxy, or oil-wet. On the left water forms a non-wetting drop. Some olive oil has been placed on the leaf to the right; this has spread over the surface and soaked into the dry leaf underneath (the darker shading). In the presence of air, water is the non-wetting phase. This allows gas exchange through stomata (small holes in the leaf). Imbibition of water is prevented; were it to occur, water would saturate the pores inside the leaf, restricting the ingress of carbon dioxide for photosynthesis and the escape of water vapour for transpiration. The same phenomenon is seen with ducks: the feathers are oily and repel water, keeping the space within the feathers full of air and insulating. In a mixed-wet or oil-wet rock, this implies that gas is not necessarily the non-wetting phase. (A black and white version of this figure will appear in some formats. For the colour version, please refer to the plate section.)

(an imbibition process), up the stems, with evaporation from stomata (small holes) in the leaves. The stomata allow water vapour to escape, while also permitting the passage of oxygen and carbon dioxide for respiration and photosynthesis. To maintain this flow of gases, it cannot be thwarted by falling rain water imbibing into the leaves and saturating the pore space: hence the leaves themselves are oil-wet and repel the rain.

If we do have a water-wet pore space, then it will also imbibe water in the presence of a gas (air). For instance, when birds are affected by an oil spill at sea, a natural reaction is to rescue the birds and clean them thoroughly with soap. This renders their feathers water-wet, they retain (cold) water and the birds die of hypothermia: a more sensible (and now adopted) approach is simply to remove as much oil as possible without washing. Similarly, while sheep stay dry, a clean woollen jumper will get wet in the rain, and leave you cold, since the oils have been removed.

Unfortunately, the obvious observation that a porous medium non-wetting to water in the presence of oil is also non-wetting in the presence of gas is not appreciated in the oil industry. Traditionally, the relative permeability for water displacing oil is measured (see the waterflood relative permeabilities discussed at length in Chapter 7). If gas injection is planned, the relative permeability of gas displacing oil in the presence of an initial (generally immobile) water saturation is also found. In this experiment gas is non-wetting to oil and injection is a primary drainage process, displaying behaviour similar to that seen for initial oil invasion into water; see Chapter 7.1. However, gas is frequently injected with water, in a WAG (water alternate gas) process, to provide a more stable displacement at the field scale. The gas is no longer the non-wetting phase with water present if the rock is oil- or mixed-wet. To quantify the flow behaviour it is necessary to measure the relative permeability of gas displacing water (or water displacing gas). Sadly, this is rarely done – after all, the oil company does not want to displace water, so refuses to consider the possibility – and instead the gas relative permeability is assumed to be a function of its own saturation, measured when it displaced oil. It is then applied erroneously to situations where water is also flowing. This will tend to overestimate the gas flow and lead to pessimistic predictions of the additional recovery from gas injection. That and a wobbly oil price are sufficient to convince a conservative management to do nothing and leave the oil trapped by water underground. However, we will press on.

8.2.3 *Wettability States in Three-Phase Flow*

We may now define wettability states for three-phase flow. In the water-wet case, water is most wetting and will, consequently, occupy the smallest pores and throats, as well as wetting layers. Gas is the most non-wetting phase and will preferentially fill the larger pores. This makes oil the intermediate phase: it fills larger pores than water, but smaller ones than gas. This is a unique arrangement with a liking for the medium-sized elements that cannot be mimicked by occupancy in two-phase flow.

In a strongly oil-wet medium, oil is the most wetting phase. Since water is non-wetting to gas, this makes it the most non-wetting phase, liking the largest pores, while gas is the intermediate phase in the middle.

There is a third possibility, when $\cos\theta_{ow} < 0$ yet $\cos\theta_{gw}$ in Eq. (8.8) is still positive. We call this weakly oil-wet. This is when oil is most wetting, then water and then gas.

The order of wettability has an impact on pore occupancy and thence conductance and relative permeability. Before discussing this further we also have to consider another feature unique to three-phase flow, namely spreading layers.

8.3 Oil Layers

In a porous medium, oil does not simply spread as a molecular film between gas and water; it accumulates in layers occupying the corners, crevices and roughness of the pore space. A cartoon of such layers is shown in Fig. 8.4: there is a wedge of oil that separates water in the corner and gas in the centre of the pore space. This is different from a film of oil formed if the initial spreading coefficient is positive: this film will affect the interfacial tensions, but is of a molecular (Ångström) thickness and cannot accommodate any significant flow; a layer, in contrast, is pore-sized (microns) and will allow appreciable flow.

The most significant impact of oil layers is that they provide a mechanism for the displacement of waterflood residual oil, even for capillary-dominated flow and in the absence of phase exchange. Gas is normally injected as a tertiary process,

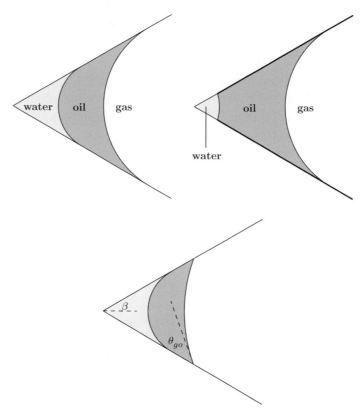

Figure 8.4 Oil layers in a single corner of the pore space. The upper left picture shows a strongly water-wet spreading system with $\theta_{go} = \theta_{ow} = 0$. The upper right figure is for a spreading oil-wet medium: there is a pinned water layer. The thick bold line indicates a surface of altered wettability. The lower figure shows an oil layer in a non-spreading system where $\theta_{go} > 0$. In this example, β, the half angle of the corner, is 30°, while $\theta_{go} = 40°$.

meaning that it is implemented after waterflooding. Therefore, gas is injected into a pore space containing water and oil at, or close to, its residual saturation. Whenever the gas contacts oil, the oil spreads as a layer between the gas and water. If the gas is continuous – and it has to be, as it is injected – then so is the oil. This enables the recovery of oil down to very low saturations; indeed to zero if $C_s = 0$, as we show later.

A common mode of gas injection is as a gas gravity drainage process: gas is injected at the top of a field forming a gas cap that overlies the oil. As more gas is injected, the gas cap increases in volume, displacing oil and water downwards. This is a gravity stable process. Imagine that the gas has contacted two hitherto trapped ganglia of oil: they are now connected by an oil layer. Under gravity, the oil from the upper ganglion can drain to the lower one. As the gas progresses further downwards, more oil may drain below it, forming a bank of mobile oil that can be produced. This is an important recovery process, allowing very high recoveries to be achieved, albeit limited by the slow flow of oil layers. In core samples, remaining saturations of only 3% have been achieved (Dumoré and Schols, 1974; Kantzas et al., 1988), with values as low as 0.1% in sand packs (Zhou and Blunt, 1997).

Oil layers have been seen in micro-model experiments (Kantzas et al., 1988; Øren et al., 1992; Øren and Pinczewski, 1994; Keller et al., 1997): an example is given in Fig. 8.5 (Feali et al., 2012). For fluids with an initially positive spreading coefficient, oil surrounds the gas, preventing the contact of gas by water. If the spreading coefficient is negative, layers are not always seen and direct gas/water contacts are formed.

In a rock, the direct observation of oil layers *in situ* is more challenging, since we need to distinguish three fluid phases at high resolution (Brown et al., 2014). However, qualitatively similar results have been obtained, showing layers of oil surrounding gas in the pore space of Bentheimer sandstone, see Fig. 8.6 (Kumar et al., 2009; Feali et al., 2012). Layers have also been observed during carbonated water injection, where the mobilization of CO_2 as the gas phase is facilitated by the formation of oil layers (Alizadeh et al., 2014).

8.3.1 Mixed-Wettability and Layer Stability

Oil layers can also form in an oil-wet or mixed-wet medium, as illustrated in Fig. 8.4. That the rock is oil-wet does not hinder the formation of layers: the key feature is the contact angle between gas and oil, which represents the degree of spreading. A necessary condition for the existence of a layer is that $\beta + \theta_{go} < \pi/2$ where β is the corner half angle, which disfavours layers for non-spreading systems with large θ_{go}. However, this is not sufficient: as in two-phase flow an energy balance needs to be performed to find the most stable configuration of the fluids (van Dijke et al., 2007).

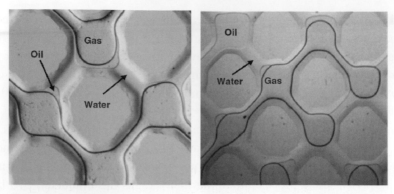

Figure 8.5 Three-phase flow in a micro-model experiment. On the left, the fluid system is spreading and we see oil accumulating in layers between gas and water: these layers prevent direct contact of gas and water. In contrast, the figure on the right shows a non-spreading system where there is contact between gas and water, although there are also narrower regions of the pore space where oil resides as a lens between the other two phases. From Feali et al. (2012).

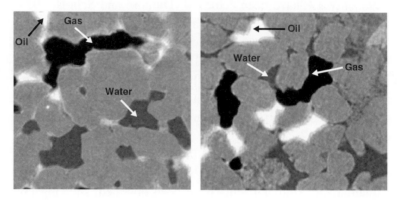

Figure 8.6 Pore-scale X-ray images of oil, water and gas in Bentheimer sandstone for the same fluid systems as in Fig. 8.5. On the left, the spreading oil surrounds the gas, while for the non-spreading system (right) there is direct contact between gas and water, similar to the observations in a micro-model. The image is approximately 1 mm across with a voxel size of around 3 μm. From Feali et al. (2012).

8.3.2 Three Phases in Capillary/Gravity Equilibrium

If there is no direct contact of gas and water, there is a simple way to estimate the capillary pressure in three-phase flow (Leverett, 1941). We measure the capillary pressure between gas and oil (normally with immobile water present), $P_{cgo}(S_g)$, and the capillary pressure between oil and water $P_{cow}(S_w)$. With three phases present, the gas/oil capillary pressure, $P_{cgo}(S_g)$, is given by the same function where now we use the gas saturation in the three-phase situation. Similarly, since only oil contacts water, $P_{cow}(S_w)$ is the same function with the three-phase water saturation. The capillary pressure between gas and water is not found from considering the curvature of their interfaces, as there are none, but is $P_{cgw} = P_{cgo} + P_{cow}$. This last relationship must always be true in capillary equilibrium since:

$$P_{cgw} = P_g - P_w = P_g - P_o + P_o - P_w = P_{cgo} + P_{cow}. \qquad (8.15)$$

Now imagine that we have a large spill of oil, say at a refinery site; this oil will infiltrate underground until it reaches the water table on which it will float. Here we will calculate the distribution of the three phases – oil, water and air – in capillary/gravity equilibrium. The other application is to determine the distribution of gas, oil and water in a reservoir upon discovery, extending the treatment in Chapter 3.5, Fig. 3.18, to where a gas cap is present. This problem was first considered by Leverett (1941) when the concept of capillary pressure was introduced into the petroleum literature.

We start with Eq. (3.38), page 106, and integrate to find how the capillary pressures between the phases vary with depth, z:

$$P_{cow} = \Delta\rho_{ow} g \left(z_{fwl} - z\right), \qquad (8.16)$$

which is the same as Eq. (3.39) where $\Delta\rho_{ow} = \rho_w - \rho_o$ is the density difference between the water and oil. z_{fwl} is the depth of the free water level, defined as where the oil/water capillary pressure is zero.

$$P_{cgo} = \Delta\rho_{go} g \left(z_{fol} - z\right), \qquad (8.17)$$

where $\Delta\rho_{go} = \rho_o - \rho_g$ is the density difference between gas and oil while z_{fol} is the depth of the free oil level where the gas/oil capillary pressure is zero. We use Eq. (8.15) to find the gas/water capillary pressure.

If we assume that the saturation distribution has been established by drainage, we may invoke Leverett J-function scaling, Chapter 6.2.3 and Eq. (6.55), to write:

$$P_{cow} = \sqrt{\frac{\phi}{K}} \sigma_{ow} \cos\theta_{ow} J(S_w), \qquad (8.18)$$

$$P_{cgo} = \sqrt{\frac{\phi}{K}} \sigma_{go} \cos\theta_{go} J(S_w + S_o), \qquad (8.19)$$

with P_{cgw} given by Eq. (8.15), and we assume in all cases that the contact angles are for the denser phase receding. In Eq. (8.19) we write the capillary pressure as a function of the apparent wetting phase to gas: $S_w + S_o$. This scaling of three-phase capillary pressures with the interfacial tensions has been confirmed experimentally for drainage in water-wet media (Lenhard and Parker, 1987; Ferrand et al., 1990).

We can now substitute Eqs. (8.16) and (8.17) into Eqs. (8.18) and (8.19) respectively and re-arrange to find expressions for the saturations as a function of height:

$$S_w = J^{-1}\left(\sqrt{\frac{K}{\phi}}\frac{\Delta\rho_{ow}g(z-z_{fwl})}{\sigma_{ow}\cos\theta_{ow}}\right),\tag{8.20}$$

$$S_o = J^{-1}\left(\sqrt{\frac{K}{\phi}}\frac{\Delta\rho_{go}g(z-z_{fol})}{\sigma_{go}\cos\theta_{go}}\right) - S_w.\tag{8.21}$$

Eq. (8.21) predicts that a zero oil saturation may be reached, when the first term is equal to the expression (8.20) at a depth, z_c where

$$z_c = z_{fwl} - \frac{\alpha H}{\alpha - 1}.\tag{8.22}$$

$H = z_{fwl} - z_{fol}$ is the effective height of oil floating on the water table. Note that a lower depth corresponds to being higher in the rock. We define

$$\alpha = \frac{\sigma_{ow}\cos\theta_{ow}\Delta\rho_{go}}{\sigma_{go}\cos\theta_{go}\Delta\rho_{ow}}\tag{8.23}$$

where, for most fluid systems, $\alpha > 1$.

If there is no oil above z_c (that is for $z \leq z_c$) then we must have a fluid distribution governed by direct gas/water contact with

$$P_{cgw} = \sqrt{\frac{\phi}{K}}\sigma_{gw}\cos\theta_{gw}J(S_w).\tag{8.24}$$

To have a continuous saturation distribution as $S_o \rightarrow 0$, we need to obey both Eqs. (8.24) and (8.15) with Eqs. (8.18) and (8.19): an inspection of the expressions indicates that this is only possible if the Bartell-Osterhof relation, (8.8), is satisfied.

This distribution of oil resting above water, established by drainage of oil and water, with transition zones for both water and oil, has been confirmed experimentally; see Fig. 8.7 (Blunt et al., 1995; Zhou and Blunt, 1997). The measured three-phase saturation distribution corresponds accurately to that predicted using only the two-phase capillary pressure (or J function). The oil saturation declines to less than 1% after a prolonged period of drainage.

8.3.3 *Layer Conductance and Relative Permeability*

If we have layer-dominated flow, then we may derive a simple relationship for the oil relative permeability (Fenwick and Blunt, 1998b). If all the oil is confined to

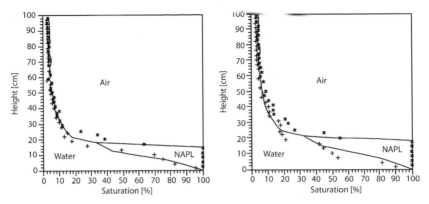

Figure 8.7 Predicted and measured three-phase saturations as a function of height. The solid lines are the predictions using Eqs. (8.20), (8.21) and (8.24), while the points are measurements. NAPL is non-aqueous phase liquid (oil). The crosses represent the NAPL/water distribution and the stars are for air/NAPL. The NAPL saturation at any given height is given by the difference in saturation between the crosses and the stars. Left, with hexane as the oil (NAPL) phase. Right, with octane as the oil. From Zhou and Blunt (1997).

layers, the oil saturation is proportional to the cross-sectional area, A_{ol}, of these layers. The oil relative permeability is proportional to their flow conductance.

We may adapt the discussion for layers in two-phase flow; see Chapter 6.1.3 and Eq. (6.27)

$$Q_{ol} = -\frac{A_{ol}r^2}{\mu \beta_R} \nabla P,$$ (8.25)

where we have replaced the area of the arc meniscus for a wetting layer with that of the oil layer in, for instance, Fig. 8.4. r is the radius of curvature of the layer and Q_{ol} is the flow rate. β_R is an empirical coefficient that can be found either experimentally or through solutions of the Stokes equation. For example, Zhou et al. (1997) presented closed-form expressions for β_R based on experiment and an approximate analytical analysis as a function of contact angles, the corner half angle, and the fluid/fluid boundary conditions.

For a corner of half angle β we adapt Eq. (3.8), Chapter 3.1.2, to find the layer area and hence the conductance:

$$A_{ol} = r^2 \left(\frac{\cos \theta_{go} \cos(\theta_{go} + \beta)}{\sin \beta} - \frac{\pi}{2} + \theta_{go} + \beta \right) - A_{AM},$$ (8.26)

where A_{AM} is the area of the water in the corner given by Eq. (3.8) with the radius of curvature of the oil/water interface. If we assume that we have a low initial water saturation established by primary drainage, then $A_{ol} \gg A_{AM}$ and we see from Eq. (8.26) that $A_{ol} \sim r^2$. Hence in Eq. (8.25) we have

$$Q_{ol} \sim A_{ol}^2$$ (8.27)

from which using $k_{ro} \sim Q_{ol}$ and $S_o \sim A_{ol}$

$$k_{ro} \sim S_o^2. \tag{8.28}$$

The relative permeability of the oil is proportional to the square of the oil saturation. This simple expression, implying no residual saturation in a spreading system, is indicative of layer drainage. Note that this is *not* consistent with some mistaken notion of thin film flow, which would give a cubic scaling.

Evidence of a quadratic dependence of relative permeability on saturation is shown in Fig. 8.8 from steady-state measurements on water-wet Bentheimer sandstone (Alizadeh and Piri, 2014a); see Figs. 7.1 and 7.3 for the two-phase relative permeabilities. At high oil saturation the relative permeability is governed by both the size of the pores occupied by oil and their connectivity with a scaling with saturation to the power 4. However, at lower saturation, approximately for $S_o < S_{or(w)}$, below the waterflood residual saturation, we see a slower decrease in relative permeability with a parabolic behaviour consistent with our hypothesis of oil layer drainage, Eq. (8.28).

A compilation of other data is shown in Fig. 8.9. Again at low saturation, behaviour consistent with layer drainage is seen in three of the experiments; for the fourth the oil used is decane, which is more non-spreading, Table 8.1, and we suggest that oil layers do not form in this case.

A final set of oil drainage measurements is shown in Fig. 8.10 (DiCarlo et al., 2000a, b). Here, as in Fig. 8.9, layer drainage is seen for an octane system in a water-wet sand pack. The same behaviour is also observed if the medium is mixed-wet: as shown in Fig. 8.4, we can have oil layers in such cases for a spreading system. Again, with non-spreading decane, there is no layer drainage and oil can

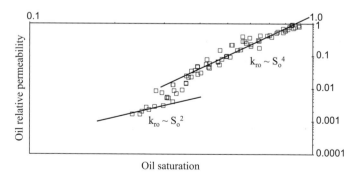

Figure 8.8 Three-phase oil relative permeability data in the direction of decreasing oil saturation in a water-wet Bentheimer core sample; see also Fig. 7.3. At high oil saturations, the oil relative permeability is governed by the network of oil-filled elements with $k_{ro} \sim S_o^4$. At low oil saturations where flow is believed to be controlled by layer drainage, $k_{ro} \sim S_o^2$. From Alizadeh and Piri (2014a).

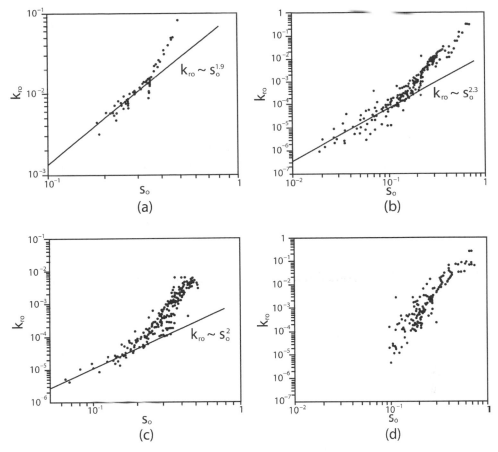

Figure 8.9 Oil relative permeabilities measured in water-wet media. (a) Data from experiments on a bead pack by Grader and O'Meara (1988) with a best-fit straight line on the doubly logarithmic plot at low oil saturation. (b) Gravity drainage measurements in a water-wet sand pack using an air/brine/octane fluid system (Sahni et al., 1998). (c) Measurements from gas injection into a consolidated sandstone (Goodyear and Jones, 1993). (d) Gravity drainage measurements in a sand pack, as in part (b), but with non-spreading decane, see Table 8.1, as the oil. Here we do not see a layer drainage regime. From Fenwick and Blunt (1998a).

be trapped by gas. The last example is the water relative permeability for an oil-wet medium. One might naively expect in this case that we can swap oil and water so that water spreads as a layer between oil nearer the corner, and gas in the centre: indeed this assumption has been proposed to predict three-phase relative permeabilities for oil-wet media (Stone, 1970, 1973).

The hypothetical arrangement of phases at the pore scale is illustrated in Fig. 8.11; however, this does not occur, since for this we require that water spreads on oil – it does not. The spreading coefficient for water, is, by analogy to Eq. (8.1)

$$C_s^w = \sigma_{go} - \sigma_{ow} - \sigma_{gw}. \qquad (8.29)$$

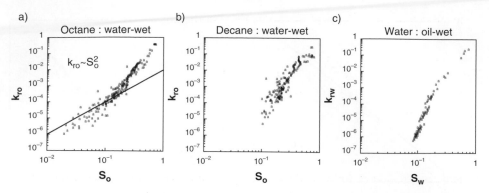

Figure 8.10 Measurements during gravity drainage in sand packs. (a) The oil relative permeability with octane as the oil phase in a water-wet system; as shown in Fig. 8.9 we see a layer drainage. (b) With decane, which is non-spreading, as the oil phase, we do not observe the effect of oil layers. (c) Water is also non-spreading on oil, see Eq. (8.29) and Fig. 8.11: as a result we cannot swap oil and water when we consider water-wet versus oil-wet media: there is no water drainage regime. From DiCarlo et al. (2000a).

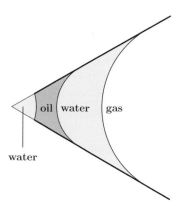

Figure 8.11 A water layer in an oil-wet corner. The thick line represents a surface of altered wettability: water remains in the corner from primary drainage; there is an oil layer followed by a layer of water and gas in the centre of the pore space. This arrangement is not possible, however, as the Bartell-Osterhof equation, (8.8) does not allow gas to be non-wetting to water in a strongly oil-wet medium for typical interfacial tensions.

In oilfield applications, see Table 8.2, $\sigma_{gw} > \sigma_{ow} > \sigma_{go}$ and hence $C_s^w \ll 0$. Water does not spread on oil. This is evident from our discussion of oil-wet media above, in Chapter 8.2: water becomes the most non-wetting phase and therefore would rather occupy the centre of the pore space. Hence, in an oil-wet medium, water can be trapped in the presence of gas: there is no water-drainage regime corresponding to what we see for oil.

Similarly, we may define a gas spreading coefficient

$$C_s^g = \sigma_{ow} - \sigma_{go} - \sigma_{gw}. \tag{8.30}$$

This is also negative for fluids at ambient conditions; see Table 8.1. However, in oilfields where σ_{go} tends to zero and $\sigma_{ow} \approx \sigma_{gw}$ then $C_s^g \to 0$. In a strongly oil-wet system, then from Eq. (8.14)

$$\cos \theta_{gw} = -1 + \frac{C_s}{\sigma_{gw}}, \tag{8.31}$$

which gives $\theta_{gw} = \pi$ for an (oil) spreading system: we certainly can see gas layers in near-miscible gas injection. Both oil and gas become spreading in the presence of water. This is considered further below in Chapter 8.4.1.

8.4 Displacement Processes

8.4.1 Fluid Configurations

The putative arrangements of three phases shown in Fig. 8.4 are not the only possible ways in which oil, water and gas may reside in a single pore. The are many configurations (Zhou and Blunt, 1998; Hui and Blunt, 2000; Piri and Blunt, 2005a; van Dijke et al., 2007). Rather than attempt a comprehensive discussion of this — it is simply a matter of finding what sequence of menisci are consistent with capillary equilibrium, the displacement sequence, and assigned contact angles — Fig. 8.12 illustrates some of the two- and three-phase arrangements that have been considered (Helland and Skjæveland, 2006a, b). Table 8.3 lists the ranges of contact angles for which the configurations are seen. It is assumed that oil is always wetting to gas. Hence only three orders of wetting state are allowed, from most non-wetting to most wetting discussed previously, namely: (i) gas, oil, water; (ii) gas, water, oil; and (iii) water, gas, oil.

The two-phase configurations shown in Fig. 8.12 have already been treated in Chapter 5.2.2; see Fig. 5.6 for instance. These were considered with reference for oil/water systems, but this is trivially extended to gas/water arrangements as well. There is one novel pattern though: F for oil and K for gas. In configuration F we have three layers: water in the corner, then oil, then water, with oil in the centre. This seems bizarre, but we can accommodate multiple layers in corners if we have repeated cycles of water and oil injection with significant contact angle hysteresis, such that during waterflooding water is non-wetting, but the system appears water-wet during oil invasion. We can see the same for gas if we have cycles of gas and water flooding and again significant hysteresis in the gas/water contact angle.

Of the three-phase arrangements, M and N are shown in Fig. 8.4. Pattern O is shown in Fig. 8.11: while it is unusual for hydrocarbon fluids, as already discussed,

Table 8.3 *Constraints on contact angles for the different fluid configurations shown in Fig. 8.12. β is the corner half angle. Empty spaces indicate that any contact angle is allowed, as long as $\theta_A \geq \theta_R$. From Helland and Skjæveland (2006b).*

	θ_{owA}	θ_{owR}	θ_{gwA}	θ_{gwR}	θ_{goA}	θ_{goR}
A	n/a	n/a	n/a	n/a	n/a	n/a
B		$< \pi/2 - \beta$		$\leq \pi/2$	$\leq \pi/2$	$\leq \pi/2$
C					$\leq \pi/2$	$\leq \pi/2$
D					$\leq \pi/2$	$\leq \pi/2$
E	$> \pi/2 + \beta$				$\leq \pi/2$	$\leq \pi/2$
F	$> \pi/2 + \beta$	$< \pi/2 - \beta$		$\leq \pi/2$	$\leq \pi/2$	$\leq \pi/2$
G	$> \pi/2 + \beta$	$< \pi/2 - \beta$		$\leq \pi/2$	$\leq \pi/2$	$\leq \pi/2$
H			$< \pi/2 - \beta$	$\leq \pi/2$	$\leq \pi/2$	$\leq \pi/2$
I					$\leq \pi/2$	$\leq \pi/2$
J	$> \pi/2$		$> \pi/2 + \beta$		$\leq \pi/2$	$\leq \pi/2$
K	$> \pi/2$		$> \pi/2 + \beta$	$< \pi/2 - \beta$	$\leq \pi/2$	$\leq \pi/2$
L	$> \pi/2$		$> \pi/2 + \beta$	$< \pi/2 - \beta$	$\leq \pi/2$	$\leq \pi/2$
M	$> \pi/2$		$> \pi/2 + \beta$	$< \pi/2 - \beta$	$\leq \pi/2$	$\leq \pi/2 - \beta$
N					$\leq \pi/2$	$< \pi/2 - \beta$
O	$> \pi/2 + \beta$			$< \pi/2 - \beta$	$\leq \pi/2$	$< \pi/2$
P	$> \pi/2$		$> \pi/2 + \beta$		$\leq \pi/2$	$< \pi/2 - \beta$
Q	$> \pi/2$	$< \pi/2 - \beta$	$> \pi/2 + \beta$	$\leq \pi/2$	$\leq \pi/2$	$\leq \pi/2$

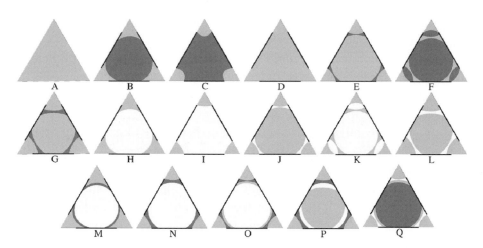

Figure 8.12 A collection of possible fluid configurations for water (blue), oil (red) and gas (yellow) in a triangular pore or throat. The bold line indicates a surface of altered wettability, Fig. 2.29, but this is not necessarily oil-wet. From Helland and Skjæveland (2006b). (A black and white version of this figure will appear in some formats. For the colour version, please refer to the plate section.)

it is not impossible to construct a fluid system that allows water layers. Configuration P has gas layers; this is less surprising, since gas has the intermediate wetting state in a strongly oil-wet medium. To allow a gas layer we require, see Table 8.3, $\theta_{gwA} > \pi/2+\beta$. Returning to Eq. (8.31), this is possible when oil and gas approach miscibility.

The final configuration, Q, has both gas and water layers. This is again unusual because of the presence of water layers and in any event requires repeated cycles of gas, water and oil injection to be observed.

8.4.2 Configuration Changes and Layer Stability

The geometric constraints that allow a fluid configuration are necessary but not sufficient conditions for it to be seen. As presented in the context of two-phase flow, Chapter 5.2.2, the range of capillary pressures for which a given arrangement of fluid is permitted, and the values at which there is a transition to a new pattern, are controlled by an energy balance. Dong et al. (1995) first used thermodynamic considerations to find when oil layers form in a water-wet capillary tube. This work has been extended to treat layer stability and displacement for systems of arbitrary wettability (Helland and Skjæveland, 2006a, b; van Dijke and Sorbie, 2006b; van Dijke et al., 2007).

Conceptually the approach is straightforward, applying the principles outlined in Chapter 2.3.1 extended to three fluid phases. Then, the transition from one stable configuration to another can be computed as a function of imposed capillary pressure and contact angles. This can be considered a generalization of the diagrams shown in Fig. 5.7 for two-phase flow, indicating when there is displacement combined with the formation and collapse of layers.

In practice, however, the calculations are involved, since we need to treat any arbitrary sequence of water, gas and/or oil invasion, any permissible combination of contact angles and interfacial tensions and a large number of possible fluid configurations; see Fig. 8.12. The details lie outside the scope of the discussion here, but are provided in the references above. This method can be further extended to consider pores of arbitrary cross-section, derived from images (Zhou et al., 2014, 2016).

8.4.3 Multiple Displacement

There is one type of displacement unique to three-phase flow that allows the movement of trapped ganglia even in the limit of capillary-controlled flow. Imagine that gas contacts a pore containing oil that had been trapped during waterflooding. So far we have allowed that the oil spreads between gas and water, allowing it to drain away slowly. But there is another, more local, process that facilitates its removal, even if there are no layers. The gas pushes the oil into a neighbouring water-filled pore. This means that gas displaces oil that displaces water. This is allowable in a capillary-controlled displacement governed by the gas/water capillary pressure: this event will occur if the pressure difference between gas and water for the displacement (the sum of the threshold invasion pressures for gas displacing oil, and oil displacing water) is more favoured than any other, specifically others where gas can displace water directly. Since in spreading systems there is no direct contact of gas and water, a double displacement process is the only way that gas may remove water.

This type of double displacement was first observed in micro-model experiments, as shown in Fig. 8.13 (Øren et al., 1992; Øren and Pinczewski, 1994). We have a water-wet system, so the process is double drainage (the gas and oil invasion processes are both drainage). In this example, two double drainage events push oil that was originally stranded in a large pore into a smaller throat and layers. This is a mechanism for the mobilization of a residual phase.

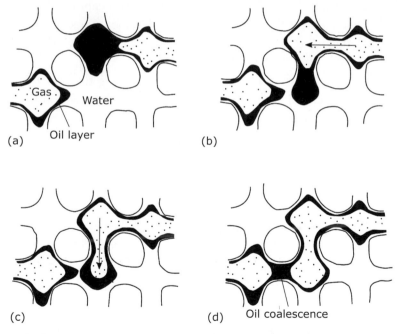

Figure 8.13 Double drainage in a water-wet micro-model. (a) Oil, water and gas in a spreading system with layers of oil surrounding the gas phase. (b) Gas invades a pore through a throat, as shown. To conserve volume, oil is pushed ahead of the gas and displaces water. This is a double drainage event. (c) Gas now invades the pore just filled with oil. (d) Oil displaces water from a neighbouring throat and coalesces with oil surrounding another gas-filled pore. The result of these events is to have removed oil from a pore, placing it in a smaller throat and allowing drainage through layers. From Øren et al. (1992).

Combinatorics show that there are six double displacement processes: gas-oil-water, gas-water-oil, oil-gas-water, oil-water-gas, water-gas-oil and water-oil-gas (double imbibition in a water-wet medium) (Fenwick and Blunt, 1998b). These other processes have also been observed in micro-model experiments (Keller et al., 1997).

However, these six new modes of displacement are not all. We can have multiple chains of displacement, with one phase displacing another and then another, with any number of intermediate steps. These are multiple displacements, first described by van Dijke and Sorbie (2003) and incorporated into a network model to reproduce micro-model experiments (van Dijke et al., 2006). Such chains are important where water and gas are injected alternatively into a reservoir containing waterflood residual oil. In these cases both the oil and gas can be trapped by the water and their mobilization is initiated through the movement of several ganglia simultaneously. This is most prevalent in rocks with a low connectivity, or in two-dimensional micro-models. We can see the same phenomenon in mixed-wet or oil-wet rocks, where now water can be trapped by gas.

As an example, Fig. 8.14 shows the relative frequency of different filling events for the simulation of gas and water injection in an oil-wet micro-model (van Dijke and Sorbie, 2002b; van Dijke et al., 2006). Gas 1 refers to the injection of gas after waterflooding. Gas directly displaces both oil and water, but there are also double or multiple displacement chains. When water is re-injected (water 1) we see mainly direct displacement of gas, since this is the connected phase. However, subsequent gas and water injection (gas 2 and water 2) leads to more multiple displacement than before, as we have poor connectivity of all three phases. This leads to further recovery of oil. Later cycles experience more single displacement, as there is less oil remaining.

An illustration of the pattern of fluids is shown in Fig. 8.15, where network model predictions (Al-Dhahli et al., 2013) are compared with experimental measurements (Sohrabi et al., 2004). The model incorporates thermodynamic criteria for layer formation and collapse, discussed above, and accommodates multiple displacement chains (Al-Dhahli et al., 2012). In these examples, the flow paths are ramified and easily disconnected, resulting in many trapped ganglia of all three phases.

Even in a capillary-controlled displacement there is movement, disconnection and re-connection of trapped phases. This is addition to the local dynamic effects that impact connectivity discussed in Chapter 6.4.7. The resultant relative permeabilities therefore have contributions from both connected-phase flow and multiple displacement: the quantitative implications for this, particularly where these events are frequent, have not been explored and offer an opportunity for further study. In what follows we will ignore this complication and present below predictions of

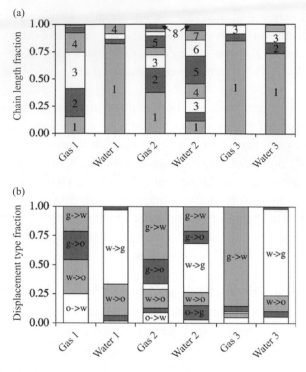

Figure 8.14 Displacement statistics for simulations of WAG floods in an oil-wet micro-model starting with an initial waterflood. (a) The fraction of the total number of displacement chains with the indicated length: 1 is single (normal) displacement, 2 is double displacement, 3 indicates a chain of three displacements, and so on. (b) The fraction of the total number of chains with the indicated displacement type (e.g. g → o indicates gas displacing oil). From van Dijke et al. (2006).

relative permeability which have ignored the contribution of trapped phases to the overall flow.

8.4.4 Displacement Paths

One other subtlety associated with three-phase flow is that the displacement paths can be complex and not necessarily imposed externally. In two-phase flow, a displacement proceeds as a monotonic change in the macroscopic capillary pressure. Of course, the direction of the displacement may change, as discussed previously; see Fig. 5.2. However, in most oilfield applications we only need consider primary drainage to some initial saturation followed by waterflooding.

Now we consider gas injection. Gas is often injected together with water, typically as alternating slugs. Hence, we have a repeated sequence of gas and water injection, as shown above in Fig. 8.15: the injection can begin and end over any

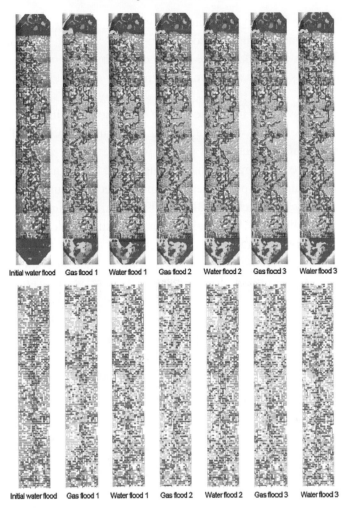

Figure 8.15 The distribution of three phases (water in blue, oil in red and gas in yellow) in an oil-wet micro-model for waterflooding followed by repeated cycles of gas and water injection. Injection occurs from the bottom of the system with recovery from the top. Experimental results from Sohrabi et al. (2004) (top) are compared with the network model of Al-Dhahli et al. (2013) (bottom). Similar displacement patterns are observed: these are the result of multiple displacement processes at the pore scale with a continual decline in remaining oil saturation, Fig. 8.14. Water is the most non-wetting phase, then gas and then oil. (A black and white version of this figure will appear in some formats. For the colour version, please refer to the plate section.)

range of mobile water, oil and gas saturations. At the mega-scale, see Fig. 6.10, where does the displaced oil go? It must be pushed ahead of the advancing gas front. Hence, at some location in the middle of the field, the first impact of gas injection is to see, locally, an increase in oil saturation or, apparently, oil injection. The same is seen at a production well: first an increase in oil rate is observed, with gas production coming later. As the oil bank moves further through the field, in the rock we see displacement by gas and water, with possibly additional cycles of oil, water and gas invasion associated with the alternate injection.

As described above, the filling sequence does not just involve an additional phase, but also has to accommodate arbitrary changes in flow direction with repeated invasions of oil, water and gas. If we have a pore-scale model, we can reproduce a sequence of saturation changes, imposed, for instance, during a steady-state experiment, using the principles of capillary equilibrium and displacement presented previously (Fenwick and Blunt, 1998b; Piri and Blunt, 2005a).

At the field scale, however, the local sequence of saturation changes is not known *a priori*. Imagine, for instance, that we inject water and gas together at an injection well. Some way from the well, we see first the formation of an oil bank. What is the increase in oil saturation as a result of this bank moving through the pore space? This will depend on how effectively the oil is displaced, which in turn is controlled by the relative permeabilities. However, the relative permeabilities are themselves a function of the displacement sequence. For instance, we know that oil layers have a major impact on oil phase connectivity. Oil layers are most stable and conductive if the oil/water capillary pressure is high (pushing water into the corners) while the gas/oil capillary pressure is low (keeping gas back from the corners and accommodating a thick oil layer). So, we may anticipate that a larger oil bank forms if we have thicker oil layers. But this is controlled by the capillary pressures of the injected water and gas, which we do not know until we can specify the saturations associated with a given injection rate of the phases, which in turn is controlled by relative permeability. This argument is confusing, because the underlying problem is somewhat tortuous: to predict relative permeability we need to know the macro-scale sequence of saturation (or capillary pressure) changes, but we cannot determine this until we know the relative permeabilities.

To reproduce a dynamic or non-steady-state displacement, we need to find self-consistent relative permeabilities (Fenwick and Blunt, 1998a): the process is shown in Fig. 8.16. We first guess a displacement path (a local sequence of saturation changes) and find the relative permeabilities associated with it from pore-scale modelling. We use these functions in a field-scale solution to the flow equations, with the imposed boundary conditions: this will predict a different sequence of saturation changes. Then we recompute the relative permeability for this new path and iterate until we find a set of relative permeabilities where the changes in saturation are consistent with a solution of the flow equations with the specified boundary conditions.

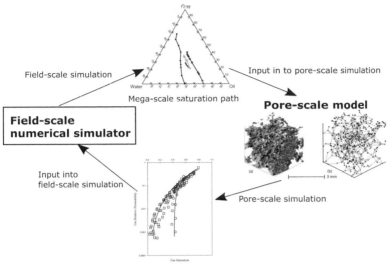

Field-scale simulation

Input in to pore-scale simulation

Mega-scale saturation path

Pore-scale model

Field-scale numerical simulator

Input into field-scale simulation

Pore-scale simulation

Predicted relative permeabilities

Figure 8.16 The iterative procedure to find self-consistent relative permeabilities. A saturation path is guessed (top) and followed in a pore-scale model (right). Note how we represent saturation changes on a ternary diagram: the top of the triangle represents $S_g = 1$, the lower right vertex $S_o = 1$ and the lower left $S_w = 1$. The relative permeabilities are predicted (bottom) and input into a field-scale numerical simulator (left). The simulation model predicts a new saturation path which is again input into the pore-scale model. The process continues until we converge on a path and predicted relative permeabilities.

8.5 Three-Phase Relative Permeability

With all the complexities associated with layers, wettability, multiple displacement and self-consistency, it is not possible to provide a comprehensive analysis of how pore-scale processes impact relative permeability. There is a large body of literature on three-phase relative permeability measurements, starting with the pioneering work of Leverett and Lewis (1941). Alizadeh and Piri (2014b) provide an extensive review of all the experiments, which highlights where information is still lacking, namely data for carbonates and systematic studies of the effects of wettability.

As examples, Fig. 8.17 shows the measured three-phase relative permeabilities for water-wet Bentheimer sandstone (Alizadeh and Piri, 2014a). In this case we can ignore many of the details described previously. For water, we see little hysteresis and the two-phase and three-phase results are similar, as we would expect, with water residing in the smaller pores and in wetting layers. For gas, the most non-wetting phase, occupying the larger pores, we again see no significant difference between two- and three-phase flow. We do, however, see a difference between gas injection, where the gas is always continuous with a behaviour matched by a

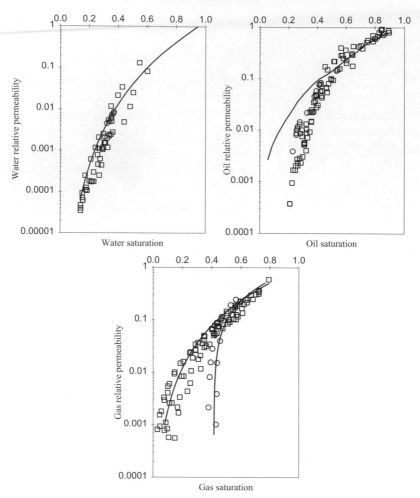

Figure 8.17 Three-phase relative permeabilities for water-wet Bentheimer sand-stone. The squares show experiments with a decreasing oil saturation and the circles are where the oil saturation increases. The solid curves represent the two-phase relative permeabilities, see Fig. 7.3. All three relative permeabilities appear to be functions only of their own saturation, although we do see the effect of trapping of gas. For oil, the relative permeability is similar to that for gas injection into oil and immobile water. At low saturation there is a signature of oil layer drainage, evident when plotted on logarithmic axes, Fig. 8.8. From Alizadeh and Piri (2014a).

primary drainage displacement, and water (or oil) flooding, where gas is trapped. For oil, again the relative permeability is, to a good approximation, a function only of the oil saturation and lies close to that observed for gas injection into oil and immobile water, as long as continuous gas is present. This somewhat simple behaviour was ascribed to the relatively narrow pore (or throat) size distribution, see Fig. 3.14, but is more likely to reflect that only a restricted range of saturation paths was explored.

The saturation paths for a series of three-phase displacements in water-wet Berea sandstone and a sand pack are shown in Fig. 8.18. Here gas and water are injected to displace oil and gas (Kianinejad et al., 2015; Kianinejad and DiCarlo, 2016). As with the measurements on Bentheimer, the water relative permeability is a function only of its own phase and similar to that measured in two-phase flow. However, now we observe an oil relative permeability that is clearly a function of saturation path, Fig. 8.19. The obvious reason, based on the arguments in Chapter 8.5.1, is that the relative permeability is controlled by the sizes of the pores occupied by oil. However, an alternative explanation is to assume that the differences are due

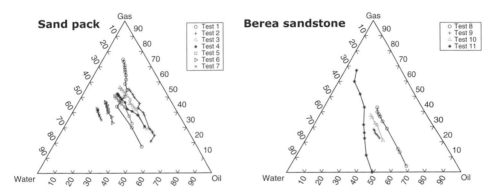

Figure 8.18 Saturation paths taken in gravity drainage experiments on a sand pack (left) and Berea sandstone (right). Through changing the amount of water present in the experiments, the trapped oil at the end of the experiment was varied. From Kianinejad and DiCarlo (2016).

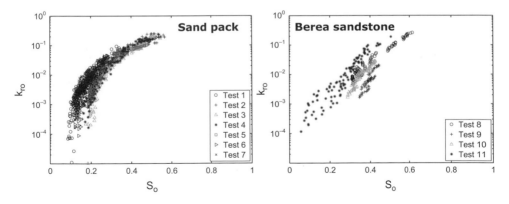

Figure 8.19 The oil relative permeabilities corresponding to the saturation paths shown in Fig. 8.18. Note that k_{ro} is not a unique function of oil saturation. From Kianinejad and DiCarlo (2016).

to changes in the amount of trapped oil. This is sensitive to the presence of water: with a low water saturation and gas present, a layer drainage regime is observed with a very low residual saturation. In contrast, if there is more water, this leads to the collapse of oil layers as the oil/water capillary pressure decreases and to direct trapping of oil through water snap-off. When the oil relative permeability is plotted as a function of the mobile saturation, $S_o - S_{or}$, the data collapse onto a single curve; see Fig. 8.20. This result implies that the mobile oil is confined to flow in a similar subset of the pore space in all cases: in pores and throats smaller than those occupied by gas, and larger than those filled with water. The differences are simply due to differing amounts of trapping by water in the larger pores not filled with gas.

Rather than attempt a more exhaustive review, we will now briefly mention the main features of three-phase relative permeability observed in the literature.

(1) **Oil layer drainage.** This was discussed in Chapter 8.3 above. For saturations below the waterflood residual, during gas injection at a low water saturation, we see approximately $k_{ro} \sim S_o^2$ with no apparent residual.

(2) **Saturation dependence.** While this is discussed further below, in water-wet media, the gas and water relative permeabilities appear to be functions only of their own saturation, while oil, being intermediate-wet, has a relative permeability that depends on the amount of water and gas (Oak and Baker, 1990). However, some experiments have shown relative permeabilities that appear to be functions only of their own phases, for a restricted range of saturation paths (Grader and O'Meara, 1988; Dria et al., 1993); also see the results for Bentheimer shown above (Alizadeh and Piri, 2014a). In the mixed-wet Prudhoe Bay reservoir, where the gas is the most non-wetting phase, the oil relative permeability is approximately only a function of its own saturation, implying that

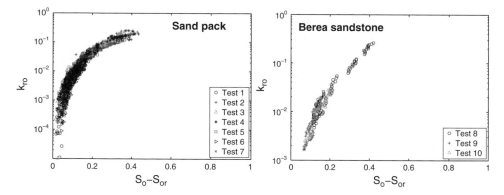

Figure 8.20 When the relative permeabilities in Fig. 8.19 are plotted as a function of $S_o - S_{or}$ they lie on a single curve, implying that k_{ro} is a function only of its mobile saturation. From Kianinejad and DiCarlo (2016).

here it tends to reside in the smaller pores; in contrast, the water relative permeability is affected by the presence of gas, as expected for the intermediate-wet phase (Jerauld, 1997).

(3) **Effect of spreading coefficient.** For a fluid system with a large and negative spreading coefficient, layer drainage is not seen, see Fig. 8.10, with lower oil recovery from gas injection than for a spreading oil. Vizika and Lombard (1996) studied the effect of spreading coefficient and wettability for water-wet, oil-wet and fractionally wet media: here sand grains of different wettability were mixed together. In water-wet and fractionally wet systems evidence of layer drainage was seen for spreading systems. For a non-spreading oil, both the oil and gas relative permeabilities were lower. They showed that the spreading coefficient had less impact in oil-wet systems although here the recovery was lower due to capillary retention of the oil as the most wetting phase. In general the oil relative permeability decreased as the porous medium became more oil-wet. Kalaydjian et al. (1997) also studied three-phase relative permeabilities for spreading and non-spreading systems. They too showed less oil recovery for non-spreading systems and a smaller gas relative permeability: a gas blocking effect caused by trapping of oil.

(4) **Effect of wettability.** Since water does not spread on oil, we do not see layer drainage of water in an oil-wet medium; see Fig. 8.10. The gas relative permeability in an oil-wet medium is lower than for a water-wet system, since the gas no longer occupies the larger pores (DiCarlo et al., 2000a, b). Oak (1991) measured three-phase relative permeabilities for a sandstone treated with silane to render all the surfaces oil-wet. He found that gas injection did not reduce the remaining oil saturation found from waterflooding, consistent with oil being the wetting phase. However, the saturation dependence of the relative permeabilities was more similar to that expected for a water-wet system, with the oil relative permeability remaining a function of two saturations.

(5) **Viscous coupling at the fluid/gas interface.** In two-phase flow we have asserted that, to a reasonable approximation, the fluid/fluid interfaces are static, representing a no-flow boundary: the flow of one phase does not alter the flow of the others and we can ignore cross-terms in the relative permeabilities; see Eq. (6.59). Although we see exceptions, where viscous coupling between phases impacts flow near end-points (see Chapter 7.3.2), this is ignored for most oil/water flows. We cannot make the same approximation in three-phase flow, where the boundary condition at a gas/fluid interface is approximately that of a free surface, Eq. (6.30), and we also see significant coupling between the flow of oil and water. This is in accord with experiments in angular capillary tubes (Zhou et al., 1997; Firincioglu et al., 1999). In gravity drainage, the oil relative permeability is strongly influenced by the flow rates of water

and gas (Dehghanpour et al., 2011a). The viscous coupling between the phases can be modelled at the pore scale and incorporated into an empirical model (Dehghanpour et al., 2011b). However, the full implications of this complexity on three-phase flow have not been explored.

We will now amplify these remarks by touching upon four topics: saturation dependence, predictions using pore-scale modelling, trapping and empirical models.

8.5.1 *Pore Occupancy and Saturation Dependence*

Before three-phase relative permeability is presented in more detail, it is instructive to consider what saturations it depends on. This is illustrated in Figs. 8.21 and 8.22 (van Dijke et al., 2001a, b). For a rock of uniform wettability we have three possibilities associated with the order of wettability, or contact angle, discussed in Chapter 8.2.3 above. In a water-wet medium, water is most wetting and preferentially occupies the smallest pores, gas is non-wetting and resides in the larger pores, while oil is intermediate-wet and is found in pores of medium size.

In a water-wet medium, the water relative permeability is a function only of the water saturation and is unaffected by the relative amount of gas and oil present, since in all cases it has a conductance controlled by the connectivity of the smallest pores and throats that it occupies. Similarly, the gas relative permeability is a function only of the gas saturation: it preferentially fills the larger pores and throats and its flow is indifferent to how oil and water are arranged in the smaller ones. Now, this does not preclude a dependence on saturation path and the effects of trapping (see Chapter 7.1.5), but for a given displacement the relative permeabilities of water and gas are functions only of their own saturations. Similarly we may write, as in Chapter 8.3.2, that $P_{cow}(S_w)$ and $P_{cgo}(S_g)$.

Since oil has the intermediate wettability state, its relative permeability will depend on both the amount of water and gas present. For a given oil saturation, with more oil present, the oil occupies larger pores with a larger relative permeability than for a large gas saturation when it is confined to small pores and a lower relative permeability. Now it is the capillary pressure excluding the intermediate phase: $P_{cgw} = P_{cgo} + P_{cow}$ that depends on two saturations.

If we have a strongly oil-wet medium and gas is the intermediate-wet phase, then its relative permeability is a function of two saturations; now the oil and water relative permeabilities are functions only of their own saturations: also we may write $P_{cow}(S_o)$ and $P_{cgw}(S_w)$, where the gas/oil capillary pressure is a function of two saturations. In a weakly oil-wet medium, where water is intermediate-wet, now this relative permeability is a function of two saturations.

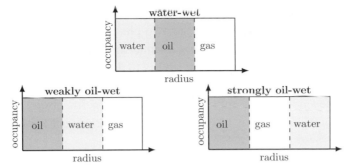

Figure 8.21 Possible pore occupancies in a uniformly wet system. Here we ignore complexities associated with pinned layers and filling that is not strictly in order of size. We also ignore trapped phases. In a water-wet rock (top) gas fills the largest pores, water the smallest and oil the medium-sized ones. In a weakly oil-wet medium (lower left), oil fills the smallest pores, while gas is most non-wetting and fills the largest ones. Water has the intermediate wetting state and occupies the medium-sized pores. If the system is strongly oil-wet (lower right) water is non-wetting and preferentially resides in the larger pores, oil in the smallest and now gas occupies the pores in the middle. The relative permeabilities of the most wetting and most non-wetting phases are functions of only their own saturations, while the occupancy – and hence relative permeability – of the phase of intermediate wettability depends on two saturations.

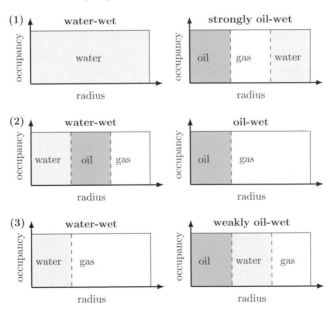

Figure 8.22 Possible connected phase occupancies for mixed-wet rocks. Each row indicates an allowed pattern of occupancy. We must preserve the same order of wettability in the water-wet and oil-wet regions, since we have the same capillary pressures throughout the connected phases. There are also special cases where one or more of the phases are completely displaced from either region. Configuration (2) applies to both weakly and strongly oil-wet cases. Based on the work of van Dijke et al. (2001b).

This argument has important consequences, since we typically extrapolate three-phase relative permeabilities from two-phase measurements. Most empirical models (see Chapter 8.5.5 below) assume that the system is water-wet, so correctly allow for the oil relative permeability to depend on two saturations. However, it is also assumed that gas is still the most-non-wetting phase in an oil-wet rock, and so they admit that only the water relative permeability can be a function of two saturations (Stone, 1970, 1973): this is not correct for strongly oil-wet media.

For mixed-wet rocks, Fig. 8.22, there are more complex arrangements of filling: each row in the figure represents a possible configuration of oil, water and gas in pores of different size (van Dijke et al., 2001a, b; van Dijke and Sorbie, 2002a, c). Here we do not make any assumptions on the relative sizes of the oil-wet and water-wet pores. However, while the pictures may appear to offer several possibilities, there is a major constraint: the capillary pressures must be the same in the water-wet and oil-wet regions. Regardless of wettability, the phase in the largest pores has the highest pressure, and the lowest pressure is found for the phase in the smallest pores. Now imagine that we have all three phases present in the water-wet regions. This implies that $P_g > P_o > P_w$ or that $P_{cgw} > P_{cgo}$, $P_{cow} > 0$. Therefore, in the oil-wet regions, gas must occupy the largest pores, then oil, and then water. But this is inconsistent with the wetting order, which has to have oil in smaller pores than water. Hence, we cannot have three connected phases present: we may just have gas, or gas in the larger pores and then oil (occupancy 2), but we cannot have connected water, as the water must have a higher pressure than the oil, $P_{cow} < 0$, which cannot be achieved. We can continue this line of argument to determine the three allowed arrangements shown in Fig. 8.22.

One consequence of the constraints placed on capillary pressure is that for a given wettability state and saturation, there is only ever one intermediate-wet phase. The relative permeability of this phase will depend on the saturations of the other two; the remaining relative permeabilities will be a function of only their own saturations. This is not true for the entire saturation range: which relative permeability is a function of both saturations will change dependent on the relative amounts of the three phases present. In Fig. 8.22, gas is intermediate-wet in pattern (1), oil in (2) and water in (3). If we consider, for instance, gas injection into oil and water, we start with the first pattern where gas is intermediate-wet, assuming that we have some strongly oil-wet pores. When the gas has removed all the water from the oil-wet pores it will begin to invade the water-wet regions and we transition to the second configuration. Note that it is possible for oil to be displaced into the water-wet pores by double displacement processes. Now oil is the intermediate-wet phase.

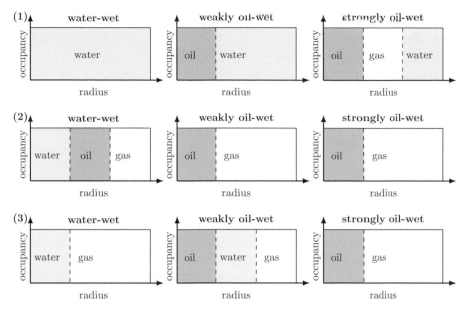

Figure 8.23 Possible connected phase occupancies for mixed-wet rocks where we allow all three possible orders of wettability to be present simultaneously. Each row indicates an allowed pattern of occupancy. We must preserve the same order of wettability in the water-wet, weakly oil-wet and strongly oil-wet regions, since we have the same capillary pressures throughout the connected phases. This is an extension of the occupancies shown in Fig. 8.22 based on the work of van Dijke and Sorbie (2002a).

This analysis can be extended to cases where all three wettability orders – water-wet, weakly oil-wet and strongly oil-wet – co-exist, as shown in Fig. 8.23. In pattern (2), oil is the intermediate-wet phase. However, in patterns (1) and (3) it is evident that changing the saturation of any one phase affects the phase occupancy of the other two. In these cases all the capillary pressures and relative permeabilities are functions of the saturation of two phases (van Dijke and Sorbie, 2002a).

This is an elegant conceptualization that helps frame a rational discussion of three-phase capillary pressure and capillary pressure. However, it does not account for trapping, pinned layers and cases where the local geometry combined with contact angles close to 90° means that filling is not strictly in order of size.

8.5.2 Predictions of Three-Phase Relative Permeability

First principles volume averaging (see Chapter 6.5), automatically incorporating the momentum transfer at fluid interfaces, can be used to study three-phase displacement. However, at present, the work is limited to rather simple pore-space geometries, although it offers the possibility of a more rigorous foundation for the description of three-phase flow (Bianchi Janetti et al., 2015, 2016).

Pore-scale network models of increasing complexity have been developed to predict three-phase relative permeability and to explore behaviour outside the somewhat limited range of three-phase experiments or to provide a physically-based interpolation from two-phase measurements. Algorithmically, the principles are the same as in two-phase flow, with sorted lists of capillary pressures for every possible displacement, with the complication that water, oil and gas invasion all need to be considered in any arbitrary sequence. In addition, sophisticated clustering algorithms are needed to account for trapping and multiple displacement (van Dijke and Sorbie, 2003; Piri and Blunt, 2005a).

The first three-phase network model assumed water-wet conditions and studied a branched, or Bethe, lattice to compute pore occupancies and relative permeabilities (Heiba et al., 1984). Soll and Celia (1993) studied capillary pressure hysteresis and trapping in a three-dimensional percolation-type model. Layer flow and double displacement based on micro-model observations (Øren et al., 1992; Øren and Pinczewski, 1994, 1995) were incorporated into a three-phase network model (Øren et al., 1994) that was later extended to account for viscous pressure gradients in wetting and spreading layers (Pereira et al., 1996; Pereira, 1999). Double displacement processes and different saturation paths were included by Fenwick and Blunt (1998a, b), multiple displacement by van Dijke and Sorbie (2003), the effect of spreading coefficient was studied by Mani and Mohanty (1997), while van Dijke and Sorbie (2002b, c) explored the effects of wettability.

The first successful predictions of three-phase relative permeability were made for Berea sandstone by Lerdahl et al. (2000). This was followed by the work of Piri and Blunt (2005a, b) who also used a Berea network to predict experimental data and assess the effect of wettability. More recent models can accommodate multiple displacement, thermodynamically consistent criteria for layer formation and collapse, while solving for displacement in networks based on pore-space images (Al-Dhahli et al., 2012, 2013).

We begin, in Fig. 8.24, not with a three-phase relative permeability, but a two-phase prediction using the Bartell-Osterhof equation, (8.8). For water-wet Berea sandstone we have already shown good predictions of the waterflood relative permeabilities measured by Oak and Baker (1990), Fig. 7.5. However, the advancing contact angles, θ_{ow}^A, had to be adjusted to match the observed residual oil saturation, $S_{or(w)}$: as shown in Table 4.2 and Fig. 4.10, increasing the contact angles leads to less snap-off and a filling sequence that is less strictly in order of size, giving a lower residual saturation. The measured relative permeabilities for oil displacing gas in the presence of immobile water were also accurately predicted (Piri and Blunt, 2005b): here a spreading system was assumed, but, for a rough surface, θ_{go}^A is not zero, although lower than θ_{ow}^A to match the measured residual gas saturation, $S_{gr(o)}$. The final two-phase measurement is for water displacing gas: here we apply

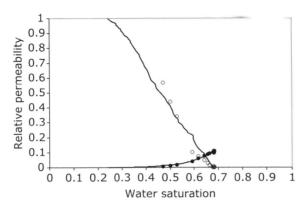

Figure 8.24 Predicted relative permeabilities for water injection into gas and immobile water. Network model results (lines) are compared to steady-state measurements by Oak and Baker (1990) (points). The gas/water contact angles, θ_{gw}^A are determined from the oil/water and gas/oil values using the Bartell-Osterhof relation, Eq. (8.8). From Piri and Blunt (2005b).

Eq. (8.8) to find θ_{gw}^A and so there are no tuning parameters. As shown in Fig. 8.24, a good agreement with experiment is obtained with an accurate match to the residual gas saturation, $S_{gr(w)}$.

As examples of three-phase predictions, Fig. 8.25 shows computed oil, water and gas relative permeabilities compared to the experiment measurements of Oak and Baker (1990) on Berea sandstone (Al-Dhahli et al., 2013). A good agreement is obtained in all cases: in particular, the correct energy-balance criteria for oil layer formation allows an accurate prediction of k_{ro} in the layer drainage regime.

While these results are promising, there is still a lack of data and model predictions for mixed-wet rocks and for repeated cycles of gas and water injection. We have also not addressed one very important aspect of three-phase flow, namely trapping.

8.5.3 Trapping in Three-Phase Flow

In three-phase flow we may see trapping of oil, water and gas, with the residual saturations controlled by wettability and saturation path. Again, like relative permeability, a comprehensive synthetic understanding of trapping is not yet available. Instead, the discussion will be framed around the study of some representative experimental datasets. We will not consider the residual oil saturation encountered in gravity drainage or through gas injection: in this case, as discussed in Chapter 8.3, the residual saturation for a spreading system is close to zero. Instead we will focus on water displacing gas and oil, trapping both hydrocarbon phases.

The first series of measurements we present were performed on water-wet sand packs where a period of gravity drainage was followed by waterflooding (Amaechi

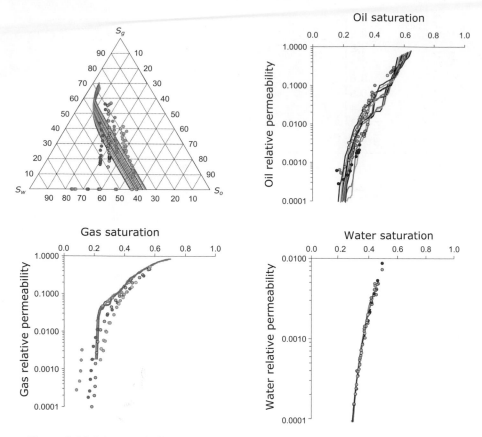

Figure 8.25 Measured oil relative permeability for tertiary gas injection (Oak and Baker, 1990) (points) compared to network model predictions (lines). The saturation paths are shown on the upper left figure; the others show the oil, gas and water relative permeabilities. From Al-Dhahli et al. (2012).

et al., 2014); the results are shown in Fig. 8.26. We might expect, since gas is the most non-wetting phase, that the amount of trapping is the same in two- and three-phase flow, as water and oil combined act as a single wetting phase. However, we see that more gas can be trapped under three-phase conditions when S_{gi} is high. In such situations, we start with a long period of gravity drainage, leaving water in the smallest pores, with oil in some of the smaller pores and in layers. Then water is injected. In two-phase flow, in a sand pack, the displacement is dominated by piston-like advance with relatively little snap-off and trapping compared to consolidated rocks; see Chapter 4.6. In a three-phase displacement in a spreading system, water cannot contact gas directly. Instead, water displaces oil out of the smaller pores, and oil, in turn, displaces gas. Since the oil is at low saturation, confined mainly to layers, it is not able to push out gas by piston-like advance: when the layers swell, the only available displacement is snap-off.

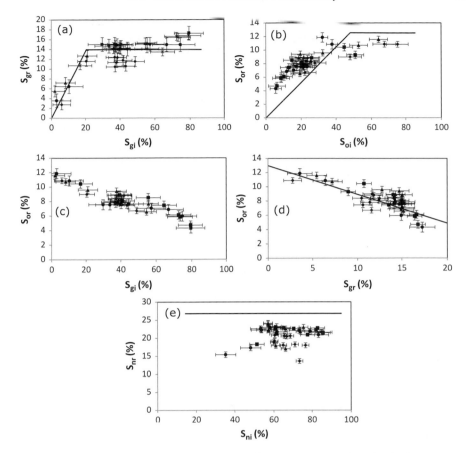

Figure 8.26 Three-phase trapping measured in water-wet sand packs during gravity drainage followed by waterflooding. The symbols represent different drainage times before water injection. (a) Plot of residual gas saturation, S_{gr}, against initial gas saturation, S_{gi}. The solid line represents the trend observed in gas/water systems (Al Mansoori et al., 2009, 2010). (b) Plot of the residual oil saturation, S_{or}, against initial oil saturation, S_{oi}. The solid line shows the trapping curve for a two-phase oil/water system. (c) The residual oil saturation, S_{or}, as a function of initial *gas* saturation, S_{gi}. (d) The residual oil saturation, S_{or}, plotted against the residual gas saturation, S_{gr}. The line here is the trend, Eq. (8.32) with $a \approx 0.4$. (e) The total residual saturation of oil plus gas, S_{nr}, against the total initial saturation of oil and gas, S_{ni}. The solid line is the sum of the maximum residual oil and gas saturations in two-phase flow. Replotted from Amaechi et al. (2014).

This therefore leads to a higher residual gas saturation than in a similar gas/water displacement.

In consolidated media, where two-phase waterflooding is dominated by snap-off with higher residual saturations than in sand packs, we see the same amount, or less, residual gas in three-phase flow (Jerauld, 1997). While direct contact of gas by water is still prevented, forcing oil to snap-off gas in the smaller throats, this has to be compared with two-phase flow where snap-off would occur anyway in these throats. Now oil and gas are trapped in the larger pores: if there is trapped oil, this limits the pores available for gas, and there is less trapping. Network modelling of trapping in Berea sandstone has seen similar behaviour (Suicmez et al., 2007, 2008).

For oil, there is more trapping under three-phase conditions at low initial saturation with slightly less trapping for higher S_{oi}. We see less trapping of oil when the initial gas saturation is higher, simply because there is less oil present initially to be trapped. However, this does illustrate an advantage of gas injection combined with waterflooding. If the oil saturation after primary drainage is high, it is advantageous to allow some displacement by gas such that for subsequent waterflooding the apparent initial oil saturation is lower, leading to a smaller residual. In experiments where gas remains a continuous phase, shown in Fig. 8.18 above, we can again see less trapping than when oil is displaced by water alone (Kianinejad et al., 2015).

The presence of residual gas does limit the amount of trapping of oil, as first noted by Holmgren and Morse (1951). We observe the following empirical relation (Kyte et al., 1956):

$$S_{or}^{3p} = S_{or(w)} - a S_{gr}^{3p}, \qquad (8.32)$$

where the superscript $3p$ indicates three-phase conditions and a is an empirical constant. Land (1968) hypothesized that the total amount of oil and gas trapped would be the same as encountered in two-phase flow, implying $a = 1$. Experiments in water-wet sandstone and limestone cores found lower values, approximately in the range 0.4–1 (Kyte et al., 1956). This is broadly consistent with the results in Fig. 8.26 which show a value of around 0.4, suggesting that under three-phase conditions there is more trapping of both non-wetting phases, but less trapping of oil. Oil and gas are trapped in the larger pores: having a large amount of trapped gas reduces the number of large pores in which oil may be trapped, and hence lowers its residual saturation. However, since we allow more snap-off in three-phase flow, as discussed above, the total trapping of oil and gas combined may exceed that seen in two-phase flow. This explains why the value of a found in a sand pack, where there is a large increase in the amount of snap-off under three-phase conditions, is at the lower end of that observed for consolidated rocks.

The total amount of hydrocarbon (oil and gas) that may be trapped can exceed the two-phase values, but the sum $S_{gr}^{3p} + S_{or}^{3p} \equiv S_{nr} < S_{or(w)} + S_{gr(w)}$, or the sum of the two-phase values, as shown in Fig. 8.26.

The number of studies that consider the effect of wettability is limited. Jerauld (1997) presented results from the mixed-wet Prudhoe Bay reservoir, Fig. 5.11. Here we appear to have a weakly oil-wet system, since gas behaved as the most non-wetting phase; see Chapter 8.5.1, page 386. The amount of trapped gas was largely independent of the presence of the other phases and could be predicted by a Land-type model, Eq. (4.39). Trapped gas lowered the water relative permeability, implying that it displaced water from the larger oil-wet pores, which water would preferentially occupy if only oil were present. Trapped gas also had a weak effect on the residual oil saturation, as would be expected if oil is the most wetting phase: the value of a in Eq. (8.32) was found to be around 0.13. This is also in accord with the results of Kyte et al. (1956) on cores which had been chemically treated to render them oil-wet. This implies that the sum of the three-phase residual saturations can exceed either two-phase value, as we saw for a water-wet sand pack. Here the reason is somewhat different though, in that the trapping of oil, as the most wetting phase, and gas, as the most non-wetting phase, are independent.

Caubit et al. (2004) performed a series of gravity drainage experiments followed by waterflooding on cores that had been aged in crude oil to replicate water-wet, neutrally wet and oil-wet conditions. The work also reviewed literature data. Here again, gas remained the most non-wetting phase with similar trapping characteristics in two- and three-phase flow in all cases. The behaviour of the oil phase was, however, rather different than discussed already. For the water-wet cores, the residual saturation was similar in two- and three-phase flow, with little apparent effect of trapped gas. For neutrally wet cores, the residual oil saturation was higher than in two-phase flow; for the oil-wet cores the residual oil saturation was lower.

The explanation for this curious wettability dependence depends on the presence of oil layers in two- and three-phase flow; see Figs. 5.16 and 8.4, respectively. In a neutrally wet system with oil/water contact angles around 90°, it is not possible to have oil layers in two-phase flow as oil is not sufficiently wetting. However, there is still relatively little trapping, where displacement proceeds as a frontal advance with no snap-off; see Chapter 4.2.3. In three-phase waterflooding, oil is strongly wetting to gas in a spreading system, and so we can form layers as shown in Fig. 8.4: the oil traps gas by snap-off, which prevents the piston-like invasion of gas by water. Since water is not clearly wetting to oil, the oil/water capillary pressure decreases rapidly, causing oil layers to collapse: water readily removes oil from a layer by a type of forced snap-off process; see Chapter 5.2.2. This traps the oil surrounding already trapped gas clusters.

For an oil-wet system, the lower residual oil saturation under three-phase conditions is at first sight unexpected; as oil is clearly the most wetting phase, its trapping should be unaffected by the presence of gas, as observed by Jerauld (1997). However, again the explanation lies in the fact that oil is more strongly wetting to gas than water. With gas present, oil layers, as illustrated in Fig. 8.4, are more stable than when oil is sandwiched between water; see Fig. 5.16. If oil is more wetting to gas than water, $\theta_{go} < \pi - \theta_{ow}$. Low values of θ_{go} mean thicker, more conductive

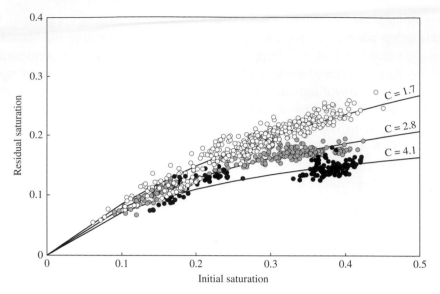

Figure 8.27 Measurements of trapping in Estaillades limestone. The unfilled symbols indicate the residual saturations measured under water-wet conditions. The other symbols show the results after the rock had been aged in crude oil: the grey points are for subsequent displacement of nitrogen by brine, while the black points are for brine displacing CO_2. There is less trapping of gas when the rock becomes more oil-wet, consistent with Eq. (8.8), since the gas/water contact angle increases. The curves are fitted using a Land model with a trapping constant C indicated, Eq. (4.39). Adapted with permission from Al-Menhali and Krevor (2016). Copyright (2016) American Chemical Society.

and stabler layers for a given imposed capillary pressure; these layers allow flow and prevent the trapping of oil, leading to lower residual saturations.

Finally, Fig. 8.27 shows the trapped saturation of nitrogen and CO_2 (representing the gaseous phase) in Estaillades limestone aged in crude oil to render it mixed-wet. Primary drainage of gas into a water-saturated core is followed by waterflooding. These experiments appear to have resulted in a stronger wettability alteration than encountered in the sandstones mentioned above, in that now the gas is intermediate-wet with significantly less trapping than for an equivalent water-wet sample (Al-Menhali and Krevor, 2016). This is consistent with a larger gas/water contact angle in an oil-wet rock and the resultant suppression of any snap-off by water. The lower residuals for CO_2 compared to nitrogen are due to the lower gas/water interfacial tension leading, in an oil-wet system, to a larger value of θ_{gw} from Eq. (8.8): CO_2 is more wetting than nitrogen, leading to less trapping. In pore-scale imaging experiments in oil-wet Bentheimer, a halving of residual saturation was also seen compared to water-wet conditions (Rahman et al., 2016).

While many of the results in the literature may be explained through consideration of pore-scale occupancy and displacement, a complete understanding is lacking. It is puzzling why gas appears to remain the most non-wetting phase in many experiments, even for cores that can spontaneously imbibe oil, and so are clearly oil-wet. Furthermore, it is not straightforward to predict the degree of trapping, as opposed to developing *post hoc* interpretations of the results: why, for instance, was the trapping of oil, as the wetting phase, affected by gas in the experiments of Caubit et al. (2004) but not in Jerauld (1997)? This is an area ripe for further careful experimental, numerical and theoretical study.

8.5.4 Direct Imaging of Trapped Phases in Three-Phase Flow

The trapping of three phases has been observed using X-ray imaging. Iglauer et al. (2013) studied the trapping of oil and gas after waterflooding in a water-wet sandstone. As seen in core-scale experiments, the presence of gas reduced the residual oil saturation. More gas was trapped when it was initially injected into a high oil saturation than when gas displaced water and waterflood residual oil. The distribution of ganglion size for both oil and gas clusters was broadly consistent with percolation theory, Eq. (4.34), Chapter 4.6.1; however, the interfacial area did not scale with volume, as expected, which implies a more connected displacement of oil and gas combined.

Alizadeh et al. (2014) observed the trapping of oil and gas during carbonated water injection through Berea sandstone, where oil was mobilized through the movement of ganglia of CO_2. Oil formed stable layers around the gas; the oil was then fragmented during water injection. Gas was also trapped in significant quantities. Some ganglia are shown in Fig. 8.28, indicating that large clusters of trapped oil and gas are seen in the pore space. There are some sheet-like structures in the oil phase, consistent with the formation of layers.

There is opportunity for future work in this area, to observe three-phase trapping at the pore scale for more displacement paths, wettabilities and pore structures to rationalize the core-scale results presented previously.

8.5.5 Empirical Models in Three-Phase Flow

As mentioned previously, the waterflood relative permeabilities of water and oil are routinely measured, see Chapter 7; we call these $k_{rw(o)}$ and $k_{ro(w)}$ respectively, indicating in brackets the other phase present. For gas injection relative permeabilities are usually found for a primary drainage displacement of gas into oil with irreducible water present. From this experiment $k_{rg(o)}$ and $k_{ro(g)}$ are obtained. Since it is unusual to have direct measurements of relative permeability when all three phases are flowing and it is, in any event, infeasible to account for all possible displacement paths, empirical models are used to interpolate the two-phase measurements to three-phase conditions. These are quite different from the empirical

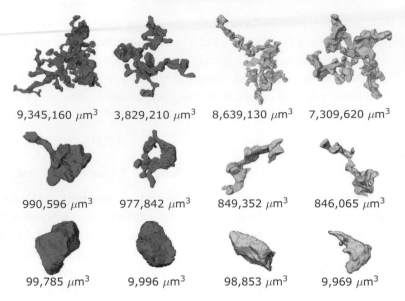

9,345,160 μm^3 3,829,210 μm^3 8,639,130 μm^3 7,309,620 μm^3

990,596 μm^3 977,842 μm^3 849,352 μm^3 846,065 μm^3

99,785 μm^3 9,996 μm^3 98,853 μm^3 9,969 μm^3

Figure 8.28 Ganglia of oil (two left-hand columns) and gas (right-hand columns) after carbonated water injection in a water-wet Berea sandstone. The volume of the ganglia are shown: the pictures are not all to the same scale. From Alizadeh et al. (2014). (A black and white version of this figure will appear in some formats. For the colour version, please refer to the plate section.)

expressions for two-phase flow presented in Chapter 7.4, where the models are used simply to fit data. For three-phase flow, however, we are making predictions in the absence of data.

The most commonly employed models in the petroleum literature were introduced by Stone (1970) (Stone 1) and Stone (1973) (Stone 2). The models view the flow of oil as being blocked by both water and gas: k_{ro} is related to the product of $k_{ro(w)}(S_w)$ (in the absence of gas) and $k_{ro(g)}$ (in the absence of mobile water). The common implementation, Aziz and Settari (1979), ensures that the three-phase values tend to the two-phase limits when one phase is absent. For the Stone 1 model:

$$k_{ro} = \frac{S_{oe} k_{ro(w)}(S_w) k_{ro(g)}(S_g)}{k_{ro}^{max}(1 - S_{we})(1 - S_{ge})}, \tag{8.33}$$

where

$$S_{oe} = \frac{S_o - S_{om}}{(1 - S_{wc} - S_{om})}, \tag{8.34}$$

$$S_{we} = \frac{S_w - S_{wc}}{(1 - S_{wc} - S_{om})}, \tag{8.35}$$

$$S_{ge} = \frac{S_g}{(1 - S_{wi} - S_{om})}. \tag{8.36}$$

S_{om} is a user-defined residual oil saturation under three-phase flow conditions.

The Stone 2 model is similar conceptually, based on a picture of segregated flow:

$$k_{ro} = \left(k_{ro(w)}(S_w) + k_{ro}^{max} k_{rw(o)}(S_w)\right) \times \tag{8.37}$$
$$\left(k_{ro(g)}(S_w) + k_{ro}^{max} k_{rg(o)}(S_q)\right) - k_{ro}^{max}\left(k_{rw(o)}(S_w) + k_{rg(o)}(S_g)\right).$$

Here the residual oil saturation is predicted when $k_{ro} = 0$. Note that in both models the relative permeabilities are evaluated as functions of the water and gas saturations. k_{rg} and k_{rw} are assumed to be functions of only their own saturations and equal to their two-phase values.

While these models are reasonable extrapolations of two-phase values they suffer from three main problems. Firstly, when tested against data the predictions are inaccurate, tending to overestimate k_{ro} for Stone 1 and exaggerating the amount of trapping for Stone 2 (Fayers and Matthews, 1984; Baker, 1988; Fayers, 1989; Blunt, 2000; Kianinejad et al., 2015). Secondly, the models cannot account for the saturation-dependence of relative permeability when the rock is no longer water-wet. Stone (1973) suggested that for an oil-wet rock, the water and oil phases are simply swapped. As discussed in Chapter 8.5.1 above, this is not necessarily correct and certainly cannot accommodate the subtleties associated with mixed-wet rocks. Thirdly, the model cannot distinguish different saturation paths and the resultant changes in the amount of trapping. The Stone 1 model does allow the separate assignment of residual saturation, which can, for instance, be assumed to vary between the two-phase water and gasflood limits with a linear, quadratic or cubic dependence to match experimental data (Fayers and Matthews, 1984; Fayers, 1989). In contrast, the uncontrolled prediction of trapping in the Stone 2 model is a major limitation.

A simpler yet more general model is to apply a saturation-weighted interpolation: this is often the default model in reservoir simulators, since it is easily coded from a table of relative permeability values. Here any wettability can be accounted for if there is a separate measurement of gas injection into water to find $k_{rg(w)}$ and $k_{rw(g)}$. This model was first proposed by Baker (1988), who showed that it provided more accurate predictions than the Stone models:

$$k_{ro} = \frac{(S_w - S_{wi})k_{ro(w)}(S_o) + (S_g - S_{gr})k_{ro(g)}(S_o)}{(S_w - S_{wi}) + (S_g - S_{gr})}. \tag{8.38}$$

S_{gr} is the residual gas saturation in the gas/oil displacement, which is normally zero if gas is injected in a primary drainage process. Note that here the two-phase relative permeabilities are calculated at the three-phase oil saturation. For the other phases:

$$k_{rw} = \frac{(S_o - S_{oi})k_{rw(o)}(S_w) + (S_g - S_{gr})k_{rw(g)}(S_w)}{(S_o - S_{oi}) + (S_g - S_{gr})}, \tag{8.39}$$

$$k_{rg} = \frac{(S_w - S_{wi})k_{rg(w)}(S_g) + (S_o - S_{oi})k_{rg(o)}(S_g)}{(S_w - S_{wi}) + (S_o - S_{oi})}, \qquad (8.40)$$

where S_{wi} is the initial water saturation in the gas/oil experiment and S_{oi} (normally zero) is the initial oil saturation in the gas/water flood. Again the relative permeabilities are computed as a function of their own saturation in three-phase flow.

While this model may be considered an improvement, it cannot capture the complex transition of saturation dependence in mixed-wet media explained in Chapter 8.5.1. It also assigns a residual saturation which is the minimum of the two-phase values, which, as we showed in Chapter 8.5.3, may not be accurate.

In the hydrology literature, empirical models have been developed based on van Genuchten or Mulahem-type expressions, see Chapter 7.4.3 (Parker et al., 1987). Water-wet conditions are assumed and like the Stone and Baker models, they do not automatically accommodate repeated flooding cycles.

These models may, however, be used as the basis for a more sophisticated characterization of three-phase flow. The important features to capture are as follows.

(1) **Saturation dependence of all three relative permeabilities.** As discussed in Chapter 8.5.1, for mixed-wet rocks it is necessary to allow all three relative permeabilities to depend on two saturations.

(2) **Trapping and hysteresis.** A model also needs to accommodate multiple sequences of saturation changes and the trapping of all three phases that may result. One approach, consistent with the trapping models advanced by Hilfer (2006a), see Chapter 6.5.4, but originating in the work of Carlson (1981), is to consider that there is no hysteresis if we consider the relative permeabilities as functions of the connected saturation. Then a separate model, based, for instance, on an extension of that proposed by Land (1968), Eq. (4.39), is used to relate total to continuous saturation (Blunt, 2000). This approach has been used to predict a series of three-phase datasets in the literature, including the results shown in Figs. 8.19 and 8.20. The dependence of the residual oil saturation on the water and gas saturations was found from the data: k_{ro} was then found to be a unique function of mobile oil saturation only, with no further dependence on saturation path.

(3) **Dependence on capillary number.** Many gas injection projects are designed such that the oil and gas are miscible, or nearly miscible, as mentioned previously. When complete miscibility is achieved, there is only one hydrocarbon phase and we have two-phase flow. Near miscibility, the interfacial tension between gas and oil may be orders of magnitude lower than seen at ambient conditions, such that the capillary number for flow, Ca, Eq. (6.62), may

approach values where viscous and capillary forces are comparable at the pore scale, and oil may be swept from the pore space, resulting in very low residual saturations; see Chapter 7.3. In a three-phase relative permeability model, these effects need to be included: specifically allowing the residual oil saturation to decline to zero for large Ca, accommodating the correct viscous limit on the gas/oil relative permeabilities, Eq. (7.4), and making the gas/water and oil/water relative permeabilities identical.

As examples, I will briefly mention two sophisticated models, based on a keen appreciation of the pore-scale physics and which match experiments (Jerauld, 1997; Larsen and Skauge, 1998). The expressions for relative permeability have been implemented into field-scale simulators for the evaluation of gas injection projects. Jerauld (1997) used an extensive set of measurements of two-phase properties from the Prudhoe Bay oilfield (Jerauld and Rathmell, 1997) to constrain a three-phase relative permeability model that was used to design the world's largest gas injection project. It was an extension of the Stone model to incorporate the trend in wettability with initial water saturation, see Fig. 5.11, as well as hysteresis and trapping of all phases, and the capillary-number dependence of relative permeability and residual oil saturation as the oil and gas neared miscibility. Larsen and Skauge (1998) developed a model for repeated cycles of water and gas injection, accounting for trapping and the observed decrease in end-point relative permeabilities.

At present, however, our understanding of three-phase relative permeability, both in terms of experimental measurements and empirical models, is unsatisfactory. Although we now have sophisticated pore-scale models incorporating the unique features of three-phase flow discussed in this chapter, namely multiple displacement, different saturation paths and spreading layers, we are still some way from a comprehensive understanding of the flow behaviour.

9

Solutions to Equations for Multiphase Flow

In this chapter we will introduce the macroscopic averaged properties discussed previously, namely relative permeability and capillary pressure, into equations for conservation of volume using the multiphase Darcy law. We will then present analytical solutions for waterflooding in one dimension, governed either by viscous or capillary forces. This exercise is presented to relate pore-scale phenomena to large-scale flow patterns and recovery.

9.1 Conservation Equations for Multiphase Flow

We start with the Darcy equations for multiphase flow, invoking conservation of mass (or volume). In Chapter 6.1 we considered mass conservation for flow in a free fluid. Here we modify this treatment for a porous medium, where we perform an average in some representative elementary volume. We start from Eq. (6.2), page 220, written for a phase p. We assume completely immiscible flow with no exchange of components between phases.

Consider some arbitrary volume V of the porous medium bounded by a surface $\mathbf{S}$. The mass per unit volume of a phase is $\rho_p \phi S_p$, while the normal component of the Darcy velocity is the volume of that phase flowing per unit area per unit time. Imposing conservation of mass by equating the flux of each phase out of the volume to the change in mass within it:

$$\int \frac{\partial \rho_p \phi S_p}{\partial t} dV = -\int \rho_p \mathbf{q}_p \cdot d\mathbf{S}, \qquad (9.1)$$

where the minus sign indicates that a net flow out of the volume leads to a decrease in mass. From Gauss' theorem, the surface integral is converted into one over the same volume V,

$$\int \frac{\partial \rho_p \phi S_p}{\partial t} dV = -\int \nabla \cdot (\rho_p \mathbf{q}_p) dV. \qquad (9.2)$$

Then, since this relation holds for any arbitrary volume of space, the integrands must be equal and

$$\frac{\partial \rho_p \phi S_p}{\partial t} = -\nabla \cdot (\rho_p \mathbf{q}_p). \tag{9.3}$$

For the Navier-Stokes equations, it was assumed that at the pore scale the fluid densities were constant; see page 220. This is a reasonable approximation for slow flows, even for gases, as the change in fluid pressure is small. Here we will consider flow at the mega-scale, see Fig. 6.10, where the fluid moves 100s or 1,000s m. In these cases the assumption of constant density is still reasonable for the flow of oil and water in cases where, at the field scale, we do not allow a continuous drop in pressure over time. In oilfields, water is injected to replace the volume of oil produced, which maintains the average pressure. The water will also displace the oil, of course, which is the subject of this chapter. There is still a pressure change between injection and production wells, usually a few MPa, but since water, oil and rock are relatively incompressible with total compressibilities of all three phases order 10^{-8} Pa^{-1}, the change in density across the field is typically only around 1%.

If we assume that the fluid densities and porosity are constant, representing incompressible flow, we simplify Eq. (9.3) to:

$$\phi \frac{\partial S_p}{\partial t} + \nabla \cdot \mathbf{q}_p = 0. \tag{9.4}$$

We can sum Eq. (9.4) over all phases (generally oil and water, but we could include oil, water and gas, if we ignore compressibility). The sum of the saturations is 1 and so the time derivative is zero. We find:

$$\nabla \cdot \mathbf{q}_t = 0, \tag{9.5}$$

where $\mathbf{q}_t = \sum_p \mathbf{q}_p$ is the total velocity.

Eq. (9.5) is the porous medium counterpart to Eq. (6.5): rather than the velocity, it is the total Darcy velocity that is divergence free for incompressible multiphase flow.

The multiphase Darcy law, Eq. (6.58), is substituted into Eq. (9.4) to derive equations for saturation and pressure. We will restrict ourselves to two-phase flow. For completeness, Eq. (6.58) is written for both for oil and water:

$$\mathbf{q}_w = -\frac{k_{rw} K}{\mu_w} (\nabla P_w - \rho_w \mathbf{g}), \tag{9.6}$$

$$\mathbf{q}_o = -\frac{k_{ro} K}{\mu_o} (\nabla P_o - \rho_o \mathbf{g}) = -\frac{k_{ro} K}{\mu_o} (\nabla P_w + \nabla P_c - \rho_o \mathbf{g}), \tag{9.7}$$

where we have used $P_c = P_o - P_w$ to have equations that only involve the water pressure. Then adding Eqs. (9.6) and (9.7) to find the total velocity:

$$\mathbf{q}_t = -K\lambda_t \nabla P_w - \lambda_o \nabla P_c + \lambda_w \rho_w \mathbf{g} + \lambda_o \rho_o \mathbf{g}, \qquad (9.8)$$

defining mobilities $\lambda = k_r/\mu$ and the total mobility $\lambda_t = \lambda_w + \lambda_o$, Eq. (7.2). Eq. (9.8) can be re-arranged to find the gradient of the water pressure, which is then substituted back into Eq. (9.6). With a few steps of algebra we obtain:

$$\mathbf{q}_w = \frac{\lambda_w}{\lambda_t}\mathbf{q}_t + K\frac{\lambda_w\lambda_o}{\lambda_t}(\rho_w - \rho_o)\mathbf{g} + K\frac{\lambda_w\lambda_o}{\lambda_t}\nabla P_c. \qquad (9.9)$$

The water Darcy velocity has three components. The first is advection, or the contribution to flow driven by the total movement of both phases: the relative contribution of water is simply the ratio of the water mobility to the total mobility. The second term is gravitational segregation: water will tend to move downwards under gravity in the presence of oil, while oil, being buoyant, will move upwards. This term is proportional to the density difference between the phases, and is dependent on the mobility of both oil and water, since for segregation to occur, both phases need to move. The third term is the contribution of capillary forces. For water injection, this also adds to the flow. $dP_c/dS_w < 0$, but ∇S_w is also negative (in waterflooding the water saturation is highest at an injection well and decreases with distance away from it), and hence $\nabla P_c = dP_c/dS_w \, \nabla S_w$ is positive: this is the effect of water imbibing into the pore space.

Eq. (9.3) (for compressible flow) or Eq. (9.4) (for incompressible flow) combined with the multiphase Darcy law, Eq. (6.58), saturation-dependent descriptions of relative permeability and capillary pressure and appropriate boundary conditions (initial conditions in the reservoir and imposed flow rates or pressures at wells), provide a complete mathematical description of multiphase flow. From these equations solutions for the evolution of saturation and pressure as functions of space and time can be obtained. In field-scale application, with orders of magnitude variation in permeability over several spatial scales, this problem can only be tackled numerically. Indeed, a whole industry is devoted to simulating fluid flow and failing to match predicted behaviour to what is measured. The subject is riddled with dodgy data, even dodgier models and cumbersome black-box commercial codes stuck in a 70s time warp. But that's the topic for another book.

The complexity of numerical solutions with a bewildering stack of poorly constrained inputs often serves to obscure the key physical insights, namely how the pore-scale configuration of phases affects field-scale recovery. This is important to understand multiphase flow processes and to design injection that ensures optimal recovery, CO_2 storage security or minimal pollution risk, depending on application.

To help make the link between the pore and field scales, it is instructive to consider analytical solutions, and for this we need to simplify the treatment to one-dimensional flows.

9.1.1 Equations in One Dimension and the Fractional Flow

In one dimension, x, Eq. (9.5) becomes $\partial q_t / \partial x = 0$, which is integrated to find $q_t(t)$ only. The total velocity can vary in time, but not in space, for incompressible one-dimensional flow. This should be intuitively obvious: the same volume that enters the system has to leave. q_t is usually imposed externally, by well rates in the field, or a pump in a laboratory experiment. We then write Eq. (9.9) as:

$$q_w = \frac{\lambda_w}{\lambda_t} q_t + K \frac{\lambda_w \lambda_o}{\lambda_t} (\rho_w - \rho_o) g_x + K \frac{\lambda_w \lambda_o}{\lambda_t} \frac{dP_c}{dS_w} \frac{\partial S_w}{\partial x} \equiv f_w q_t, \tag{9.10}$$

where g_x is the component of gravity in the flow direction. For flow tilted at an angle θ to the horizontal, $g_x = g \sin \theta$. f_w is the fractional flow, defined by

$$f_w = \frac{\lambda_w}{\lambda_t} + \frac{K}{q_t} \frac{\lambda_w \lambda_o}{\lambda_t} (\rho_w - \rho_o) g_x + \frac{K}{q_t} \frac{\lambda_w \lambda_o}{\lambda_t} \frac{dP_c}{dS_w} \frac{\partial S_w}{\partial x}. \tag{9.11}$$

As for Eq. (9.9) for the water Darcy velocity, the fractional flow has three components: advection, gravity and capillarity. By definition $f_w + f_o = 1$.

The conservation equation for water, Eq. (9.4), becomes, in one dimension:

$$\phi \frac{\partial S_w}{\partial t} + \frac{\partial q_w}{\partial x} = 0, \tag{9.12}$$

or, for a finite total velocity with $q_w = f_w q_t$:

$$\phi \frac{\partial S_w}{\partial t} + q_t \frac{\partial f_w}{\partial x} = 0, \tag{9.13}$$

since q_t is not a function of x.

We cannot solve Eq. (9.13) for general saturation-dependent relative permeabilities and capillary pressures, since we have a second-order non-linear partial differential equation thanks to the inclusion of the capillary term in Eq. (9.11). However, we can construct solutions in two physically meaningful limits.

9.1.2 Waterflooding and Spontaneous Imbibition

There are two generic recovery processes in oil reservoirs when water is injected: direct viscous displacement (simply termed waterflooding) and spontaneous imbibition. Fig. 9.1 provides a cartoon to describe these two phenomena. As mentioned above, typically the pressure difference between injection and production wells is of order a few 10s MPa: this contrasts with representative capillary pressures of

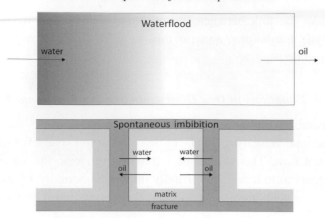

Figure 9.1 A cartoon illustrating waterflooding (top) and spontaneous imbibition (bottom). In waterflooding, we have a displacement dominated by viscous forces at the mega-scale; see Fig. 6.10. While capillary forces control the local distribution of fluids, and hence the relative permeability and capillary pressure, the capillary pressure term in the governing flow equations is relatively small and only serves to smear out the fluid front. In contrast, spontaneous imbibition is entirely controlled by capillary forces. It occurs when we cannot exert a significant viscous pressure drop across a region of rock. Generally this is seen in fractured reservoirs, where the high-conductance fractures effectively short-circuit the flow and surround the lower permeability matrix with water.

around 10 kPa (see, for instance, Figs. 4.24 and 4.25 and do not fixate on dubious end-point values). Hence at the mega- or field scale, the viscous pressure difference is much larger than capillary pressure. Buoyancy forces are also significant: the pressure change caused by differences in density is of order $(\rho_w - \rho_o)gh$, where h is the height of the oil column. Taking representative values of $\rho_w - \rho_o = 300$ kg/m^3 and $h = 30$ m, the buoyancy pressure is around 0.9 MPa: less than the viscous pressure, but more than capillary pressures for the mid-saturation range. Hence in Eq. (9.13) and (9.11) the term involving capillary pressure is relatively small. Therefore, to describe waterflooding at the large scale, we need principally to consider viscous and buoyancy effects, and can ignore the explicit effect of capillary pressure on the fractional flow. This is the first case we will treat analytically: viscous displacement or waterflooding. Water pushes out oil in a process controlled by the imposed injection rate and density difference between oil and water.

In fractured media, or reservoirs with an extreme contrast in permeability, we observe the other recovery process: spontaneous imbibition. Injected water rapidly fills the high permeability pathways through the field: the fractures or other fast-flow streaks. This strands the majority of the oil in the matrix – the lower permeability rock. The fractures provide a short-circuit for flow: it is not possible to impose a significant viscous pressure drop across a single region of rock surrounded by the fractures without applying infeasible and uneconomic flow rates.

Oil can only be recovered if water imbibes into the matrix. There are two types of spontaneous imbibition: co-current and counter-current. In the former process, oil and water flow in the same direction. In counter-current flow, in contrast, the matrix is initially completely surrounded by water and oil can only be displaced by moving in the opposite direction: the total velocity is zero.

Spontaneous imbibition, when we do not inject or force water into a porous medium but allow the invasion to be controlled entirely by capillary forces, is also seen in many other situations: the examples given at the beginning of Chapter 4, namely rainfall soaking into soil, plant roots taking up water, a paper tissue mopping up a spill and filling a baby's nappy, are all spontaneous imbibition processes.

We showed a picture of water soaking into a tissue in Fig. 4.15, a frontal advance that cannot be properly described by the conventional model of multiphase flow; another example is shown in Fig. 9.2 showing the ingress of water into an initially dry sample of Ketton limestone imaged using CT scanning. Here the pore scale displacement is likely to be a percolation-like, and we observe significant trapping. The speed of advance slows over time, consistent with the mathematical description we provide later in Chapter 9.3.

In what follows we will only consider counter-current imbibition. We will ignore viscous and buoyancy effects in Eqs. (9.9) and (9.12) with $q_t = 0$. First though, to make the analysis more concrete, we will present the relative permeabilities and capillary pressures that we will consider.

9.1.3 Exemplar Relative Permeabilities and Capillary Pressures

To illustrate the analysis in this chapter, solutions using some exemplar relative permeabilities and capillary pressures will be presented. The parameters used are

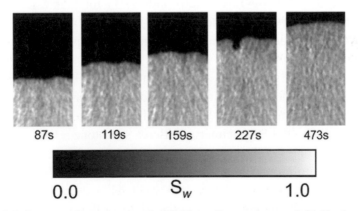

Figure 9.2 Images of spontaneous imbibition of water into an initially dry Ketton limestone core 7.7 cm high. The images were acquired using CT X-ray scanning at the different times indicated. Water enters the core from the bottom. From Alyafei (2015).

Table 9.1 *The parameters used for the examples presented in this chapter. We use Eqs. (7.20) and (7.21) for the water and oil relative permeabilities, respectively, and Eq. (7.23) for capillary pressure with the parameters listed below. Also shown are the properties used for the van Genuchten (1980) model, Eqs. (7.26) and (7.27), and the viscosity ratios,* $M = \mu_o/\mu_w$.

	Strongly water-wet	Weakly water-wet	Mixed-wet	Oil-wet
S_{wc}	0.1	0.1	0.1	0.1
S_{or}	0.4	0.3	0.15	0.05
k_{rw}^{max}	0.1	0.2	0.5	0.95
a	2	2	8	1.5
k_{ro}^{max}	1	1	1	1
b	1	1.5	2.5	4
S_w^{cross}	0.55	0.55	0.64	0.39
c	0.3	0.3	0.3	0.3
S_w^*	0.6	0.6	0.5	0.1
P_c^{max} (kPa)	200	100	100	100

Van Genuchten model	n	m	α (kPa)	S_{wc}	S_{or}
	2	0.5	30	0.1	0.4

Viscosity ratio, M	0.005	0.05	1	20	200

Permeability, K	10^{-12}	m^2
Injected (water) viscosity, μ_w	10^{-3}	Pa.s
Porosity, ϕ	0.25	

listed in Table 9.1. We use Eqs. (7.20) and (7.21) for the water and oil relative permeabilities, respectively, and Eq. (7.23) for capillary pressure (we assume $S_{wi} = S_{wc}$). We also consider a van Genuchten model, Eqs. (7.27) and (7.26), for a strongly water-wet case.

The relative permeability curves are shown in Fig. 9.3: this covers the range of typical multiphase flow properties discussed in Chapter 7, although we do not attempt to match any particular model or experimental dataset. The strongly water-wet case is similar to the Bentheimer and Berea sandstone relative permeabilities shown in Figs. 7.1 and 7.5. The weakly water-wet case represents an example with contact angles up to around 90° and some displacement by forced water injection. The mixed-wet case has a very low water relative permeability until high water saturations, representing the low connectivity of flow restricted to layers in a percolation-like displacement; see Chapter 7.2.1 and Fig. 7.8, for instance. The oil-wet example is, however, more representative of cases with a larger water relative

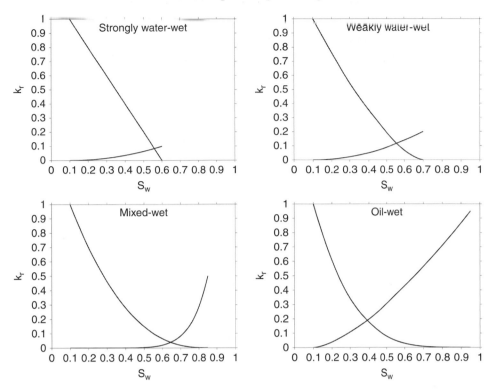

Figure 9.3 The exemplar relative permeability curves for which analytical solutions for waterflooding and spontaneous imbibition will be constructed. Table 9.1 lists the parameters used. The relative permeabilities display the range of behaviour discussed in Chapter 7.

permeability, indicating rapid connectivity of the larger oil-wet pores through invasion percolation, Fig. 7.13.

As described in Chapter 7.2.2, the saturation when the oil and water relative permeabilities are equal, S_w^{cross}, gives a quick indication of likely waterflood recovery. From Table 9.1 the mixed-wet case gives the most favourable behaviour, since water is held back in the pore space, followed by the water-wet examples, with the oil-wet exemplar giving the worst recovery, since water readily spans the pore space through the larger pores and throats, leaving unrecovered oil in the smaller elements. However, this analysis is only approximately true and limited to viscosity ratios around 1: as we show later, this is a rather simplistic analysis. The solution of the flow equations provides a much richer and more accurate guide to recovery. Which wettability type is best depends on the mobility ratio: we will see, for instance, that the oil-wet case is most favourable in some circumstances.

We consider a range of the ratio of the displaced to injected phase viscosities, M, Eq. (6.102). $M = 0.005 = 1/200$ is representative of polymer injection into a light oil, where the water is made highly viscous. $M \approx 1/50$ represents water displacing air. $M = 0.05 = 1/20$ is typical of brine displacing CO_2 in storage applications (Fenghour et al., 1998). $M = 1$ indicates the waterflood of a light (relatively low viscosity) oil, whereas $M = 20$ and 200 are cases with an increasingly viscous (heavy) oil. The corresponding fractional flows are shown in Fig. 9.4 using Eq. (9.11), ignoring capillary and buoyancy effects (only the first term is considered).

If instead of waterflooding, where water is injected to push out oil and viscous forces dominate, we have a spontaneous imbibition process, then water imbibes to reach the saturation at which the capillary pressure is zero, S_w^*. In this case, the water-wet cases give the highest ultimate recovery since S_w^* is largest, while there is no displacement at all in the oil-wet case (there is no imbibition). However, the rate of recovery is controlled by the relative permeabilities as discussed later, Chapter 9.3.

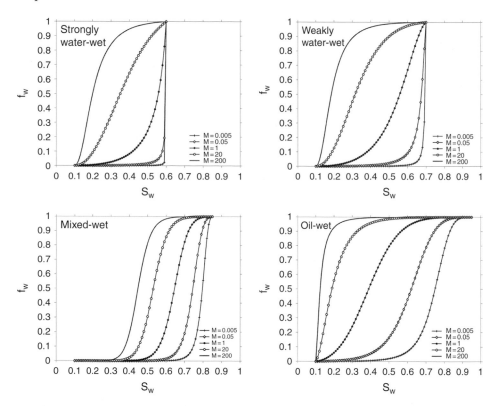

Figure 9.4 The fractional flow curves, Eq. (9.21), for the relative permeabilities corresponding to Fig. 9.3. The effects of buoyancy and capillary pressure have been ignored. We consider a range of mobility ratios, M, Eq. (6.102), from $M = 200$ representing, for instance, water displacing a heavy oil, to $M = 0.005$, which corresponds to a polymer solution displacing light oil. Note that as M increases the curves shift to the left, indicating that the injected aqueous phase becomes more mobile relative to the displaced phase (oil, CO_2 or air).

The water flow rate is proportional to the absolute permeability and inversely proportional to the water viscosity, but for given values of K and μ_w the differences in behaviour are represented by the multiphase flow properties and M. By default we take $K = 10^{-12}$ m^2 and $\mu_w = 10^{-3}$ Pa.s.

9.1.4 Boundary Conditions for One-Dimensional Flow Problems

We assume that initially the rock is full of oil and irreducible water ($S_w = S_{wc}$, where $k_{rw} = 0$). Now, we could consider cases with a mobile initial water saturation ($k_{rw} > 0$), and indeed we have placed some emphasis on the fact that we never reach a true irreducible saturation in field or laboratory settings. However, in any event the initial water relative permeability is very low and the assumption made here has a negligible impact on the solutions. We also assume that we have a medium of constant permeability and porosity, with relative permeabilities and capillary pressures that are functions only of water saturation.

In waterflooding we inject at some known flow rate (total velocity) at an injection well that we place at $x = 0$; the production well is at $x = L$. At the injection well, but within the porous medium, we will drive the oil saturation down to its minimum or residual value.

In spontaneous imbibition, the total velocity is zero and the saturation at the inlet is S_w^* corresponding to $P_c = 0$. This analysis ignores capillary back-pressure, which is seen if the system is strongly water-wet and $S_w^* = S_{or}$: for oil to flow out, it needs to be mobile with a finite capillary pressure (Unsal et al., 2007b, a; Haugen et al., 2014). Instead we will allow flow through having an infinite saturation gradient at the inlet, discussed later. There is no production well or outlet; we consider an infinite system where the oil escapes out of the inlet boundary ($x = 0$).

In mathematical form the boundary conditions are as follows:

$$S_w = S_{wc}, \qquad t = 0, \quad x > 0, \tag{9.14}$$

and, for water injection:

$$S_w = 1 - S_{or}, \qquad t > 0, \quad x = 0, \tag{9.15}$$

with some known injection rate $q_t(t)$, while for spontaneous imbibition $q_t = 0$ and

$$S_w = S_w^*, \qquad t > 0, \quad x = 0. \tag{9.16}$$

9.2 Buckley-Leverett Analysis for Two-Phase Flow

Here we ignore capillary effects explicitly in the flow equations. The analysis presented below was first derived by Buckley and Leverett (1942) and is named after these authors. The presentation will be brief, as it is already well covered in other books, see Craig (1971), Dake (1983) and Lake (1989); our principal interest is to relate pore-scale fluid displacement to large-scale recovery. We will skip a lengthy presentation of the traditional graph-paper-and-pencil methodology for constructing solutions by hand.

We rewrite Eq. (9.13) as:

$$\phi \frac{\partial S_w}{\partial t} + q_t \frac{d f_w}{d S_w} \frac{\partial S_w}{\partial x} = 0, \tag{9.17}$$

and solve this subject to Eqs. (9.14) and (9.15). The fractional flow is given by

$$f_w = \frac{\lambda_w}{\lambda_t} + \frac{K}{q_t} \frac{\lambda_w \lambda_o}{\lambda_t} (\rho_w - \rho_o) g_x. \tag{9.18}$$

Eq. (9.18) is traditionally written in terms of a gravity-to-viscous ratio:

$$N_{gv} = \frac{K (\rho_w - \rho_o) g_x}{\mu_o q_t}, \tag{9.19}$$

and then we have from Eq. (9.18):

$$f_w = \frac{1 + k_{ro} N_{vg}}{1 + \mu_w k_{ro} / \mu_o k_{rw}}. \tag{9.20}$$

From the discussion above, in most field-scale displacements $|N_{gv}| < 1$. For instance, if we take representative values: $K = 10^{-13} - 10^{-12}$ m^2, $\rho_w - \rho_o = 300$ kg/m^3, $g_x = 9.81$ m/s^2, $\mu_o = 5 \times 10^{-3}$ Pa.s and $q_t = 10^{-6}$ m/s, we have from Eq. (9.19), $N_{gv} = 0.06 - 0.6$; gravity has a relatively small, but not negligible, impact on the flow. N_{gv} may be positive or negative, dependent on whether the flow is downhill (N_{gv} is positive as this adds to the flow of water) or uphill. Furthermore, in cases where the injection rate, q_t, is low, f_w may be either negative or greater than 1. This is a result of counter-current flow, where the oil and water move in opposite directions. The sign of q_w differs from q_o; hence f_w is either > 1 or < 0.

For the remainder of the chapter we will ignore gravity: Eq. (9.20) becomes

$$f_w = \frac{1}{1 + \mu_w k_{ro} / \mu_o k_{rw}} \tag{9.21}$$

for the curves shown in Fig. 9.4. However, it is straightforward to include buoyancy effects in the analysis (Lake, 1989).

9.2.1 Dimensionless Variables and Wavespeeds

It is instructive to rewrite Eq. (9.17) in terms of dimensionless variables indicated by a subscript D. The dimensionless distance,

$$x_D = \frac{x}{L},\tag{9.22}$$

while the dimensionless time is,

$$t_D = \frac{\int_0^t q(t')dt'}{\phi L}.\tag{9.23}$$

t_D is the pore volumes of water injected. Eq. (9.17) then becomes, noting that $dt_D = (q_t/\phi L)dt$:

$$\frac{\partial S_w}{\partial t_D} + \frac{df_w}{dS_w}\frac{\partial S_w}{\partial x_D} = 0.\tag{9.24}$$

This is a non-linear first-order partial differential equation that we will solve assuming that we can write $S_w(x,t)$ as a function of a single variable

$$v_D = \frac{x_D}{t_D},\tag{9.25}$$

where v_D is the dimensionless wavespeed. $v_D = (q_t/\phi)v$ where v is the wavespeed with dimensions length/time.

Eq. (9.24) is rewritten as an ordinary differential equation as a function of v_D. We use Eq. (9.25) to find

$$\frac{\partial S_w}{\partial t_D} = \frac{\partial v_D}{\partial t_D}\frac{dS_w}{dv_D} = -\frac{v_D}{t_D}\frac{dS_w}{dv_D},\tag{9.26}$$

$$\frac{\partial S_w}{\partial x_D} = \frac{\partial v_D}{\partial x_D}\frac{dS_w}{dv_D} = \frac{1}{t_D}\frac{dS_w}{dv_D}.\tag{9.27}$$

Then substituting Eqs. (9.26) and (9.27) into Eq. (9.24)

$$\frac{dS_w}{dv_D}\left(\frac{df_w}{dS_w} - v_D\right) = 0.\tag{9.28}$$

There are two solutions to Eq. (9.28): the first, $dS_w/v_D = 0$, is a constant saturation, or a constant state; the second is

$$v_D = \frac{df_w}{dS_w}.\tag{9.29}$$

The dimensionless wavespeed is the gradient of the fractional flow.

To construct a solution $S_w(v_D)$ must be single-valued: $S_w(v_D)$ varies from $1-S_{or}$ at $v_D = 0$ to S_{wc} as $v_D \to \infty$, consistent with the boundary conditions, Eqs. (9.14) and (9.15). From the example fractional flow functions shown in Fig. 9.4 this is not possible, since the wavespeed, the gradient of the curve, is typically zero at $S_w = S_{wc}$, with a maximum at some intermediate saturation.

We cannot solve the partial differential equation: the problem is using a differential equation in the first place (this limitation, since the time of Newton and Leibnitz, is rarely admitted in applied mathematics textbooks): instead we have to take a step back and invoke conservation of volume as a difference equation, which is, after all, what we do in numerical simulations, and abandon the notion that the saturation has to have continuous derivatives.

9.2.2 Shocks

When water is injected to displace oil, a bank or shock develops with a sharp rise in water saturation from its initial, irreducible value. If we return to Fig. 6.10, we are considering a mega-scale displacement; at the macro-scale there will be locally an approximately constant saturation varying slowly in space and time, where fluid displacement is controlled by capillary pressure. At low water saturation, the water flow is low, and hence injected water behind it, at a higher saturation, will catch up. However, water cannot overtake itself, since the wavespeed is only a function of saturation. What happens is that the water builds up at the leading front. We can reach a stable state where capillary forces ensure that the water across this front moves sufficiently fast to avoid further accumulation of water: in Eq. (9.10), the capillary contribution to the water Darcy velocity is proportional to the saturation gradient and hence can, and will, be significant if these gradients are sufficiently large. Interfacial forces will smear the front over a distance such that the gradients in pressure caused by capillary and viscous forces approximately balance.

If we ignore capillary pressure in the flow equations, we predict instead a shock, which is a discontinuity in saturation. This is only meaningful at the mega, km, scale: capillary pressure will still ensure a smooth change in saturation, but over smaller distances, of order 1 m.

We can derive the shock speed through explicit conservation of volume. Consider the situation shown in Fig. 9.5, focussing on the region where there is a discontinuity in saturation. In a time Δt there is a change in volume due to flow of $A\phi(S_w^L - S_w^R)v_s\Delta t$ where L and R refer to the conditions at the left and right of the shock, respectively. A is the cross-sectional area and v_s is the shock speed. This must be balanced by the net Darcy flow and hence

$$A\phi(S_w^L - S_w^R)v_s\Delta t = A(q_w^L - q_w^R)\Delta t, \qquad (9.30)$$

which can be written as

$$v_s = \frac{q_t}{\phi}\frac{\Delta f_w}{\Delta S_w}, \qquad (9.31)$$

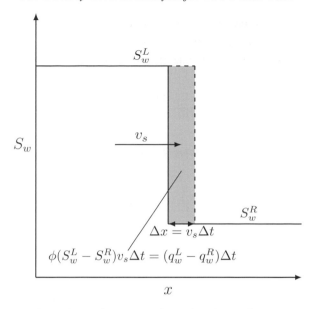

Figure 9.5 An illustration of conservation of volume for a shock: there is a jump in saturation from S_w^L to S_w^R that moves with speed v_s. In a time Δt the shock moves a distance $\Delta x = v_s \Delta t$. The volume of water represented by the shaded area must equal the net change in volume per unit area caused by flow: $(q_w^L - q_w^R)\Delta t$, where q_w is the Darcy velocity (volume per unit area per unit time), Eq. (9.30). This relation may also be expressed using dimensionless units and the fractional flows, Eq. (9.32).

where $\Delta S_w = S_w^L - S_w^R$ and $q_w^L - q_w^R = q_t(f_w^L - f_w^R) = q_t \Delta f_w$. In terms of the dimensionless speed:

$$v_{Ds} = \frac{\Delta f_w}{\Delta S_w}. \tag{9.32}$$

9.2.3 Constructing a Solution for Saturation

We will now construct solutions for $S_w(v_D)$ involving constant states (S_w is a constant), rarefactions (a smooth variation in saturation obeying Eq. (9.29)) and shocks. There will be a monotonic decrease in S_w with increasing v_D from $1 - S_{or}$ to S_{wc}.

The approach to this problem was first presented by Welge (1952). We first find the shock from $S_w^R = S_{wc}$ to $S_w^L = S_{ws}$ such that the speed of the shock, Eq. (9.32) and the rarefaction are equal:

$$v_{Ds}(S_{ws}) = \frac{f_w(S_{ws})}{S_{ws} - S_{wc}} = \left. \frac{df_w}{dS_w} \right|_{S_w = S_{ws}}. \tag{9.33}$$

This defines a unique value of the shock saturation S_{ws} and hence the shock speed. This solution makes physical sense: it is the limit of the solution to Eq. (9.13) with a the fractional flow given by Eq. (9.11) when the capillary pressure becomes negligible.

The solution has the following components.

(1) For $v_D > v_{Ds}$, $\quad S_w = S_{wc}$.
(2) For $v_D = v_{Ds}$ given by Eq. (9.33), S_w increases from S_{wc} to S_{ws}.
(3) For $v_D^{min} \geq v_D \geq v_{Ds}$, v_D is given by Eq. (9.29), which also defines the corresponding saturation where $1 - S_{or} \geq S_w \geq S_{ws}$.
(4) v_D^{min} is defined as $df_w/dS_w|_{S_w=1-S_{or}}$. If $v_D^{min} = 0$, this completes the solution. If $v_D^{min} > 0$ we have a constant state, or a fixed saturation of $1 - S_{or}$ from $v_D = 0$ to v_D^{min}.

Solutions are shown in Fig. 9.6 for our examples. By hand, the shock is found graphically from the tangent to the fractional flow for a straight line starting at $S_w = S_{wc}$ and $f_w = 0$ using the curves in Fig. 9.4.

For the water-wet cases, at the lowest mobility ratios (water is much more viscous than the oil or gas it displaces) the solution is all shock, in that the fractional flow, see Fig. 9.4, is concave ($d^2 f_w/dS_w^2 > 0$): the water is held back in the pore space, since it is confined to the smaller pores and layers, and only advances at the maximum saturation. This is favourable for oil recovery, as only oil is produced until breakthrough: the final recovery is therefore controlled by the residual oil saturation. At higher mobility ratios, a rarefaction and a faster-moving shock develops. We see this behaviour whenever there is an inflexion point in the fractional flow where $d^2 f_w/dS_w^2 = 0$. The water is now more mobile compared to the oil. In all cases, though, there is a finite wavespeed for $S_w = 1 - S_{or}$, which means that at some time all the recoverable oil will be displaced. The weakly water-wet case has a lower residual saturation, thanks to less snap-off and a less strict filling in order of pore size, see Chapter 4.2.4, which tends to shift the saturation profiles to higher saturation.

The mixed-wet case combines a high-saturation shock front, indicating good recovery at water breakthrough, with a low residual oil saturation. Here the water mobility is even lower than in the water-wet cases as it is hampered by flow through pinned water layers; see Chapter 7.2.4. Note, however, that the flow speeds for $M > 1$ near $S_w = 1 - S_{or}$ are extremely slow and tend to zero for $S_w = 1 - S_{or}$: it will take an infinite time to reach the true residual saturation; we never even get close in a field setting, as this entails the sluggish drainage of oil layers; see Chapter 5.2.

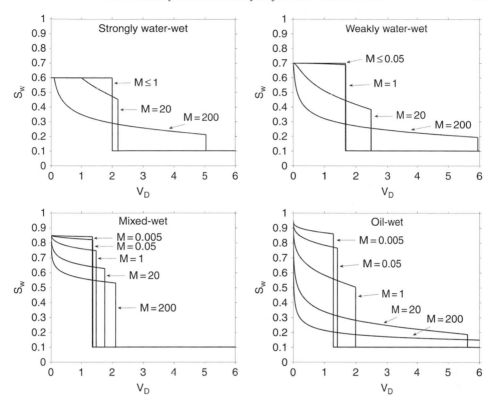

Figure 9.6 Water saturation, S_w, as a function of dimensionless wavespeed, v_D, for the exemplar cases whose fractional flow curves are shown in Fig. 9.4, see also Table 9.1. The viscosity ratio, M, is indicated. In the oil-wet case for $M = 200$ there is no shock at all, and the maximum speed of the rarefaction is approximately 20.

The oil-wet case has the highest water mobility, since the water rapidly spans the rock through larger pores; see Chapter 7.2.1. This can lead to very rapid movement of the water front for $M > 1$, leading to early breakthrough and unfavourable recoveries. For the most unfavourable mobility ratio there is no shock at all (note the convex shape of the fractional flow curve in Fig. 9.4; $d^2 f_w / dS_w^2 \leq 0$ for $1 - S_{or} \geq S_w \geq S_{wc}$), giving a fast-moving rarefaction and a continuous increase in water production after breakthrough. Again, the wavespeeds for high water saturations are very low; it takes an infinite time to reach the true residual oil saturation. However, if the water is rendered much more viscous than the oil (with a polymer, for instance) giving $M < 1$, then a rather favourable displacement is possible with a high shock saturation, since the water is held back until the water saturation is close to its maximum value and the oil is rapidly displaced to near its very low residual saturation.

9.2.4 Recovery Calculations

The final step in the analysis is to compute recovery, defined here as the pore volumes of oil produced, N_{pD}, as a function of pore volumes of water injected, t_D. Before water breaks through at the production well, conservation of volume means that $N_D = t_D$: the volume of oil recovered is equal to the volume of water injected. This continues until the shock reaches $x_D = 1$, or for $t_D \leq 1/v_{Ds}$.

Once water is produced, N_{pD} is found from integrating the saturation profile.

$$N_{pD}(t_D) = \bar{S}_w - S_{wc}, \tag{9.34}$$

where $\bar{S}_w$ is the average saturation given by

$$\bar{S}_w = \int_0^1 S_w(x_D, t_D)dx_D = t_D \int_0^{1/t_D} S_w(v_D)dv_D. \tag{9.35}$$

We perform an integration by parts (1 and S_w) in Eq. (9.35) to obtain,

$$\bar{S}_w = t_D \left[v_D S_w(v_D)\right]_0^{1/t_D} - t_D \int_0^{1/t_D} v_D \frac{dS_w}{dv_D}dv_D. \tag{9.36}$$

Now we invoke Eq. (9.29) to convert the integral to one in f_w, and evaluate the limits in the first term of Eq. (9.36):

$$\bar{S}_w = S_w(v_D) - t_D \int_1^{f_w} df_w, \tag{9.37}$$

and evaluating the integral, using Eq. (9.34) to find N_{pD}

$$N_{pD}(t_D) = S_w(v_D) - S_{wc} + \frac{1 - f_w}{v_D}, \tag{9.38}$$

since $1/t_D = v_D = df_w/dS_w$ at $x_D = 1$. Eq. (9.38) is evaluated by fixing a value of v_D and then finding the corresponding value of f_w and S_w for which $v_D = df_w/S_w$. Graphically this is obtained using the Welge (1952) construction. A tangent to the fractional flow at some saturation above S_{ws} is drawn: the saturation, S_w, where this line hits $f_w = 1$ is the average saturation behind the front, $\bar{S}_w$.

9.2.5 Example Recovery Curves

The computed recovery curves for our exemplars are shown in Fig. 9.7: for reference the relative permeabilities, fractional flows and saturation profiles are shown in Figs. 9.3, 9.4 and 9.6, respectively: the parameters used are shown in Table 9.1.

While, experimentally, it is possible to inject hundreds or thousands of pore volumes of water to drive the average saturation down close to its residual value (see Fig. 5.13 for instance), in field settings this is infeasible. Since there is a cost

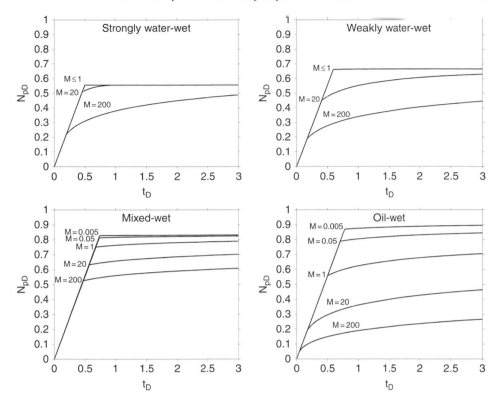

Figure 9.7 Exemplar recovery curves for different wettabilities and viscosity ratios, M. The pore volumes of oil produced, N_{pD}, is shown as a function of pore volumes of water injected, t_D. The relative permeabilities, fractional flows and saturation profiles are shown in Figs. 9.3, 9.4 and 9.6, respectively: the parameters used are listed in Table 9.1. The most favourable wettability state is governed by both M and t_D. As we consider lower M and higher t_D, the recovery is controlled by the residual oil saturation, which is lowest for the oil-wet case; for high M and economically constrained values of t_D (around 1), the water mobility is the key determinant of recovery.

associated with water injection, and with the production, separation and disposal or re-injection of produced water, the amount that it is possible to inject is limited. As a general rule, approximately the same amount of water is injected as oil originally present in the reservoir, which means that to assess waterflood recovery we should be comparing the curves for $t_D \approx 1$ and not obsess over the value for $t_D \to \infty$ which is only of theoretical (or experimental) interest.

If we have a very stable displacement with a low viscosity ratio ($M = 0.005$ and 0.05 in Fig. 9.7) – water displacing air, a low-pressure gas (in a natural gas field), CO_2 migration, or polymer flooding – then the water is held back during waterflooding and the recovery is controlled by the residual saturation. In this case, oil-wettability appears most favourable for recovery, since this state has the lowest residual oil saturation: it is least favourable for CO_2 storage, though, since we wish, in this application, to leave as much behind as possible, see Fig. 8.27 and page 396 for a fuller discussion. The worst for recovery – and best for storage – is the strongly water-wet rock with the highest residual non-wetting phase saturation. These results seem to contradict the simple cross-over saturation analysis presented in Chapter 7.2.2, which would indicate that the oil-wet case is the worst (see Table 9.1). However, this was a rather approximate assessment that did not account for viscosity contrasts.

For $M = 1$, the high water mobility in the oil-wet case causes a significant drop in recovery. Now the mixed-wet rock gives the best recoveries for $t_D \approx 1$, since we combine a low water relative permeability, suppressing rapid water flow, with a low residual saturation. The weakly water-wet is next best in terms of recovery through the same combination of relatively low residual saturation and water confined to the smaller pores. The strongly water-wet case is still worst, since the recovery is limited by the high residual saturation.

As the mobility ratio increases beyond 1, representing the typical case of an oil that is more viscous than water, the oil-wet recovery declines even further and is the least favourable wettability: this now agrees with the cross-over saturation analysis and common oilfield experience. An oil-wet reservoir generally gives rapid water breakthrough (when the recovery curve departs from a straight line of unit slope) and poor ultimate recoveries.

A representative viscosity ratio for an oil reservoir is around 20. In this case the mixed-wet case has clearly the best recovery, for $t_D = 1$. The weakly and strongly water-wet cases are similar, but in field operations, the strongly water-wet case (not that this is ever seen in reality) would be more favourable, since the maximum recovery is reached shortly after breakthrough. This is better than having to recycle large quantities of water for a long time after breakthrough. The oil-wet case is worst, foiled by the high water mobility at the pore scale.

For the largest viscosity ratio, for a heavier oil, the mixed-wet case is best, followed by the strongly water-wet example, then weakly water-wet and with the oil-wet rock worst. This seems a rather curious trend, but is an indication that recovery is now dominated by the water mobility, specifically the water relative permeability at intermediate saturation. This is lowest (giving best recoveries) for mixed-wet rocks with pinned layers; next best are strongly water-wet cases where, while wetting layers can swell, filling is confined strictly to the smaller regions of the pore space; then comes the weakly water-wet case where, while we tend to fill smaller pores, there is a less strict size-dependent order of filling (see Chapter 4.2.4). Oil-wet rocks are, again, worst, since the water preferentially fills the larger elements and quickly spans the pore space.

This exercise demonstrates how pore-scale configurations, their manifestation in the relative permeability and mobility effects all combine to control waterflood recovery.

Note that the recovery trend with wettability is non-monotonic: there is no obvious change from good to bad, or bad to good as we progress from water-wet to oil-wet. For the highest mobility ratios there is not even one local wettability maximum: strongly water-wet and mixed-wet reservoirs are favourable with lower recoveries for wettability states in between. Attempting to grasp qualitative rules based on *ad hoc* and often misleading experience is unwise. For instance, if we revert to Craig's rules (see Chapter 7.2.7), asserting a sheep-like *oil-wet bad, mixed-wet better, water-wet best* mantra is incorrect, limited and unhelpful. Instead, we have, properly, to consider the pore-scale arrangement of fluid and mobility ratio. If a simple rule is needed, then we see that for low M the residual saturation controls recovery (a low value is favourable), while, as M increases, the water relative permeability controls the behaviour (low values at intermediate saturation are best). However, do not obsess over rules: there is a clear methodology for the rapid assessment of recovery and the impact of wettability. Construct Buckley-Leverett solutions and recovery curves for relative permeabilities of interest.

This is a one-dimensional analysis that takes no account of buoyancy and heterogeneity in the absolute and relative permeabilities. If we have a gravity-stable displacement (water is injected from the bottom of the field or we consider the natural ingress of an expanding aquifer) then the shock front saturation is higher, with more favourable recovery than a horizontal or gravity unstable flood. This will also tend to make the residual saturation, rather than the water mobility, the determinant of recovery. In contrast, for horizontal flooding with $M > 1$ we may have a viscously unstable displacement, see Chapter 6.6, with channelling through high-permeability streaks and generally much less favourable recoveries than shown here; this effect will tend to exaggerate the impact of wettability and water mobility, making the oil-wet case even less advantageous. A fuller discussion of heterogeneity and field-scale recovery is beyond the scope of this book and requires a numerical solution to the flow equations; however, the general trends presented here still give a representative indication of relative recovery, even though the absolute answers will be case-dependent and lower than shown here.

This analysis is only for synthetic data: we will present some experimental confirmation of the trends shown here compared to imbibition recoveries at the end of the chapter, but to do this we need first to present an analogous analytical solution for imbibition.

9.3 Analysis of Imbibition

9.3.1 Capillary Dispersion and Fractional Flow

The second process considered is spontaneous imbibition, shown schematically in Fig. 9.1. Using Eq. (9.12) with q_w given by Eq. (9.10), and ignoring viscous and buoyancy effects, results in

$$\phi \frac{\partial S_w}{\partial t} = \frac{\partial}{\partial x} \left(D(S_w) \frac{\partial S_w}{\partial x} \right), \tag{9.39}$$

where $D(S_w)$ is the capillary dispersion defined as

$$D(S_w) = -K \frac{\lambda_w \lambda_o}{\lambda_t} \frac{dP_c}{dS_w}, \tag{9.40}$$

which has a positive sign since dP_c/dS_w is negative. For the van Genuchten model, that assumes that $\lambda_t = \lambda_o$ (the displaced mobility is assumed to be infinite compared to water),

$$D(S_w) = -K \lambda_w \frac{dP_c}{dS_w}. \tag{9.41}$$

The water Darcy velocity is, from Eq. (9.10)

$$q_w = -D(S_w) \frac{\partial S_w}{\partial x}. \tag{9.42}$$

The example capillary pressures and capillary dispersion, Eq. (9.40), are shown in Figs. 9.8 and 9.9, respectively. We use the parameters listed in Table 9.1 and the relative permeabilities shown in Fig. 9.3. For the capillary pressure only the

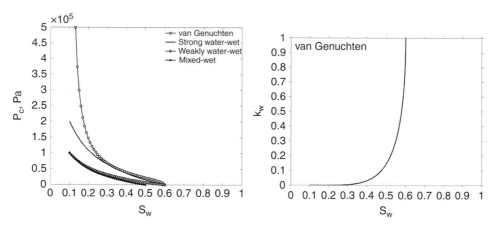

Figure 9.8 The left-hand figure shows the capillary pressures used in our exemplar solutions: Eqs. (7.23) and (7.27) with the parameters listed in Table 9.1. The right-hand figure shows the relative permeability for the van Genuchten model, Eq. (7.26).

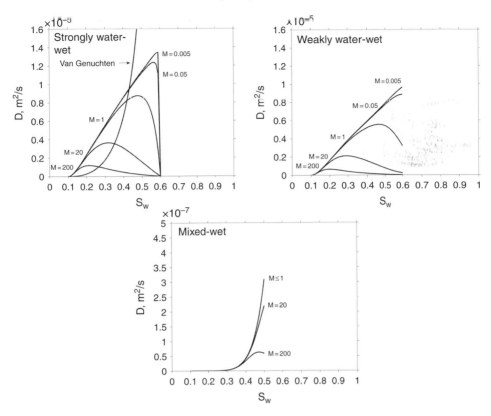

Figure 9.9 The capillary dispersion, $D(S_w)$, Eq. (9.40), with the parameters listed in Table 9.1. The relative permeabilities are shown in Fig. 9.3, and the capillary pressures in Fig. 9.8. For the van Genuchten model Eq. (9.41) is used with the relative permeability and capillary pressure shown in Fig. 9.8. Note the limits: $D(S_{wc}) = 0$ in all cases, as here the water mobility is zero. $D(S_w^*) = \infty$ for the van Genuchten model, since the capillary pressure gradient is infinite and it is assumed that the displaced phase has no flow resistance. For the other strongly water-wet cases, $D(S_w^*) = 0$ since the oil is immobile at the end of imbibition; for the weakly water-wet and mixed-wet cases, where some mobile oil remains at the end of imbibition, $D(S_w^*)$ is finite. These different behaviours will manifest themselves in generically distinct saturation profiles. Note the different scale for the mixed-wet case, indicating a slower rate of imbibition.

positive parts of the curves, $S_w \leq S_w^*$, are shown, since there is no spontaneous imbibition for $P_c < 0$. There is a finite maximum value of the capillary pressure at the initial water saturation, using Eq. (7.23). The exception is the van Genuchten curve, where $P_c \to \infty$ for $S_w = S_{wc}$, Eq. (7.27); the gradient of P_c is also infinite at $S_w^* = 1 - S_{or}$ (see Chapter 7.4), which gives an infinite value of $D(S_w^*)$.

In all cases $D(S_{wc}) = 0$, since the water relative permeability is zero and, except for the van Genuchten case, the capillary dispersion has a maximum value at a saturation that approaches S_w^* as the mobility ratio decreases (the water viscosity increases in comparison to the oil or gas displaced). As we move to less strongly wetting conditions, the magnitude of D decreases, which implies from Eq. (9.42) that the imbibition rate will be slower. In the strongly water-wet cases, all the mobile oil is displaced by imbibition, and, with the exception of the van Genuchten model noted above, $D(S_w^*) = 0$. For weakly water-wet and mixed-wet rocks, there is still mobile oil present at the end of imbibition and consequently $D(S_w^*)$ is finite. These limits – $D(S_w^*)$ may be infinite, zero or finite – will manifest themselves in different shapes of the resultant saturation profiles.

We will solve Eq. (9.39) subject to the boundary conditions Eqs. (9.14) and (9.16). The solution was first proposed by McWhorter and Sunada (1990); however, the authors did not appreciate that the boundary condition they used was generally applicable for counter-current imbibition. Independently, Chen et al. (1990) developed a similar solution and exploited the analogy with the Buckley-Leverett analysis presented above: unfortunately, though, the work was only published in an internal report and consequently was ignored in the open literature. Schmid et al. (2011) rediscovered the McWhorter and Sunada (1990) solution, showed that it applied to spontaneous imbibition and used the analysis to derive scaling groups to estimate recovery (Schmid and Geiger, 2012, 2013).

There are two key steps in the acquisition of a solution. The first is to note that since we have a diffusion-type equation to solve, albeit one that is non-linear, we should look for solutions that are a function of the variable:

$$\omega = \frac{x}{\sqrt{t}}. \tag{9.43}$$

This is the counterpart to v_D in the Buckley-Leverett analysis. We first consider this transformation on Eq. (9.12) involving the water velocity, Eq. (9.42). Substituting $\partial/\partial x = 1/\sqrt{t}\, d/d\omega$ and $\partial/\partial t = -\omega/2t\, d/d\omega$, we find

$$\frac{\omega\phi}{2t}\frac{dS_w}{d\omega} = \frac{1}{\sqrt{t}}\frac{dq_w}{d\omega}, \tag{9.44}$$

which can be rewritten as

$$\frac{dq_w}{dS_w} = \omega\frac{\phi}{2\sqrt{t}}. \tag{9.45}$$

The second key insight is to look for a solution analogous to Eq. (9.29) with a capillary fractional flow $F_w(S_w)$, which we define by (McWhorter, 1971)

$$F_w(S_w) = \frac{q_w(S_w)}{q_w(S_w^*)}, \tag{9.46}$$

where $S_w = S_w^*$ at $x = 0$. Unlike the Buckley-Leverett problem, where the inlet (injection well) flow rate, q_t, is imposed, here $q_w(x = 0)$ is a rock-and-fluid dependent quantity that cannot be specified *a priori*. Furthermore, the water flow is a function of time. Implicit in the hypothesis that the saturation profile is a function only of ω, Eq. (9.43), is that the flow rate scales as $1/\sqrt{t}$: indeed, this was the principal assumption made in the original work of McWhorter and Sunada (1990). Hence we write

$$q_w(x = 0, t) = q_w(S_w^*) = \frac{C}{\sqrt{t}},$$ (9.47)

for some constant C. The oil recovered is equal to the total amount of water invasion: per unit cross-sectional area, this is

$$V(t) = \int_0^t q_w(t')dt' = 2C\sqrt{t}.$$ (9.48)

Returning to Eq. (9.45) we substitute Eqs. (9.46) and (9.47) to find

$$\omega = \frac{2C}{\phi}\frac{dF_w}{dS_w}.$$ (9.49)

C determines how much water imbibes: it has the same role as q_t in the Buckley-Leverett analysis. The distance moved by a saturation S_w can then be written, using Eqs. (9.48) and (9.49):

$$x(S_w) = \frac{V}{\phi}\frac{dF_w}{dS_w}.$$ (9.50)

Hence, as in the Buckley-Leverett analysis, the distance moved by a given saturation value is proportional to the gradient of the fractional flow. However, we need to find F_w: how this is done is presented below.

Eq. (9.39) is rewritten as a function of ω to obtain an ordinary differential equation:

$$\phi\omega\frac{dS_w}{d\omega} + 2\frac{d}{d\omega}\left(D(S_w)\frac{dS_w}{d\omega}\right) = 0.$$ (9.51)

Eq. (9.51) is integrated to find,

$$\int \omega dS_w = -\frac{2D(S_w)}{\phi}\frac{dS_w}{d\omega}.$$ (9.52)

Eq. (9.49) is applied to transform the left-hand-side of Eq. (9.52) to an integral over F_w,

$$F_w = -\frac{D(S_w)}{C}\frac{dS_w}{d\omega}.$$ (9.53)

The integration constant is zero, since when $S_w = S_{wc}$, there is no water flow and $F_w = D = 0$.

The final step to derive a non-linear ordinary differential equation for F_w is to differentiate Eq. (9.49),

$$\frac{d\omega}{dS_w} = \frac{2C}{\phi}\frac{d^2 F_w}{dS_w^2} \equiv \frac{2C}{\phi}F_w''. \tag{9.54}$$

We substitute this into Eq. (9.53) to obtain

$$F_w F_w'' = -\frac{\phi D(S_w)}{2C^2}. \tag{9.55}$$

This is the principal equation governing spontaneous imbibition: once we solve Eq. (9.55) for F_w we can construct a solution, analogous to the Buckley-Leverett case, simply by differentiating F_w to find the saturation profile, Eq (9.49).

Formally Eq. (9.55) can be solved using an implicit integral. There are three conditions that need to be specified: the two constants of integration and the value of C. There are three constraints: $F_w(S_{wc}) = 0$, $F_w(S_w^*) = 1$, and $\omega(S_w^*) = 0$, which comes from the inlet boundary condition Eq. (9.16).

Eq. (9.55) is integrated twice in a form that ensures $F_w(S_w^*) = 1$ and $dF/dS_w|_{S_w^*} = 0$, equivalent to stating $\omega(S_w^*) = 0$, Eq. (9.49):

$$F_w(S_w) = 1 - \frac{\phi}{2C^2}\int_{S_w}^{S_w^*}\int_{\beta}^{S_w^*}\frac{D(\alpha)}{F_w(\alpha)}d\alpha d\beta, \tag{9.56}$$

where α and β are dummy integration variables. The expression is simplified by integrating by parts (1 and D/F) to obtain,

$$F_w(S_w) = 1 - \frac{\phi}{2C^2}\int_{S_w}^{S_w^*}\frac{(\beta - S_w)D(\beta)}{F_w(\beta)}d\beta. \tag{9.57}$$

The last constraint is $F_w(S_{wc}) = 0$ which in Eq. (9.57) leads to an expression for C:

$$C^2 = \frac{\phi}{2}\int_{S_{wc}}^{S_w^*}\frac{(\beta - S_{wc})D(\beta)}{F_w(\beta)}d\beta, \tag{9.58}$$

and hence Eq. (9.57) becomes

$$F_w(S_w) = 1 - \int_{S_w}^{S_w^*}\frac{(\beta - S_w)D(\beta)}{F_w(\beta)}d\beta \left/ \int_{S_{wc}}^{S_w^*}\frac{(\beta - S_{wc})D(\beta)}{F_w(\beta)}d\beta \right. . \tag{9.59}$$

Eq. (9.59) is an implicit integral, since to obtain F_w we need F_w in the integrand. This equation is solved iteratively: a good initial guess is $F_w = (S_w - S_{wc})/(S_w^* - S_{wc})$ (Schmid et al., 2011).

Table 9.2 *The calculated imbibition rates, C, Eq. (9.47), computed for the example cases whose properties are listed in Table 9.1.*

Wettability	Viscosity ratio, M	C, 10^{-5}m/s$^{1/2}$
van Genuchten	$\to 0$	87.5
Strongly water-wet	0.005	41.3
Strongly water-wet	0.05	40.4
Strongly water-wet	1	34.5
Strongly water-wet	20	19.6
Strongly water-wet	200	8.92
Weakly water-wet	0.005	34.5
Weakly water-wet	0.05	34.0
Weakly water-wet	1	28.6
Weakly water-wet	20	14.8
Weakly water-wet	200	6.36
Mixed-wet	0.005	2.73
Mixed-wet	0.05	2.73
Mixed-wet	1	2.72
Mixed-wet	20	2.51
Mixed-wet	200	1.77

9.3.2 Example Solutions

The capillary fractional flows for our example cases are shown in Fig. 9.10; the corresponding derivatives, providing the saturation as a function of ω, are given in Fig. 9.11. The imbibition rates, C, are listed in Table 9.2.

The overall imbibition rates, C, are, as expected, largest for the most strongly water-wet conditions and much smaller, by an order of magnitude or more, for the mixed-wet examples, since the water relative permeabilities and capillary pressures are lower; see Table 9.2. The use of power-law relative permeabilities overly simplifies the problem, however, and in experiments an even more dramatic difference between mixed-wet and water-wet media is observed, as shown later. The rates are also sensitive to the mobility ratio. This is most marked for the strongly water-wet cases, where oil has to escape through the inlet in the limit that the oil relative permeability tends to zero. For the mixed-wet media, the effect is less evident, as at S_w^* the oil is still reasonably well connected and can easily be removed from the inlet: in this limit, the rate of imbibition is almost entirely controlled by the much lower mobility of water.

The more interesting observations come from a study of the shape of the water saturation profiles; see Fig. 9.11. Firstly, we can see that in all cases the leading edge of the water has an infinite gradient. We do not see a shock, but there is a sharp decline in saturation to the initial, irreducible, value. This behaviour can be

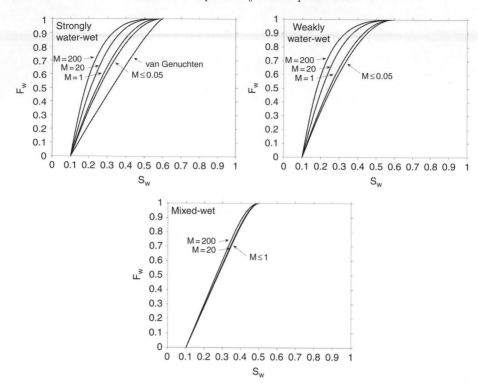

Figure 9.10 The capillary fractional flows, $F_w(S_w)$, for the example cases. The capillary dispersion, Fig. 9.9, is used to evaluate the implicit integral, Eq. (9.59) to find F_w.

elucidated analytically: define $S_w = S_{wc} + \epsilon$, where $\epsilon \ll 1$. Then, as $\epsilon \to 0$, λ_o is finite, while $\lambda_w \sim \epsilon^a$, using the power-law scaling of water relative permeability, Eq. (7.20). Then from Eq. (9.40), $D \sim \epsilon^{a_p}$. $a_p = a$ for all our examples, bar the van Genuchten model, as we assume that the capillary pressure and its gradient are finite at S_{wc}: however, many models do have a divergent P_c and hence we may have $a_p < a$; generally though, we always have $a_p > 1$. For instance, in the van Genuchten model, near S_{wc}, $k_{rw} \sim \epsilon^{1/2+2/m}$ and $dP_c/dS_w \sim -\epsilon^{-1/m}$, Eqs. (7.26) and (7.27). Hence $a_p = 1/2 + 1/m = 2.5$ in our examples.

From Eq. (9.55), we may write $F_w F_w'' \sim -\epsilon^{a_p}$. Now $F_w(S_{wc}) = 0$, so using a Taylor series expansion: $F_w = F_w(S_{wc}) + \epsilon F_w' + O(\epsilon^2) = \epsilon F_w' + O(\epsilon^2)$. Hence to leading order we obtain:

$$F_w' F_w'' \sim -\epsilon^{a_p-1}. \tag{9.60}$$

The gradient F_w' is clearly finite near S_{wc}, since this is proportional to the maximum rate of ingress of the water, Eq. (9.49); therefore $F_w'' \sim -\epsilon^{a_p-1}$: in the limit $\epsilon \to 0$ for $a_p > 1$, the second derivative of the capillary fractional flow must also be zero.

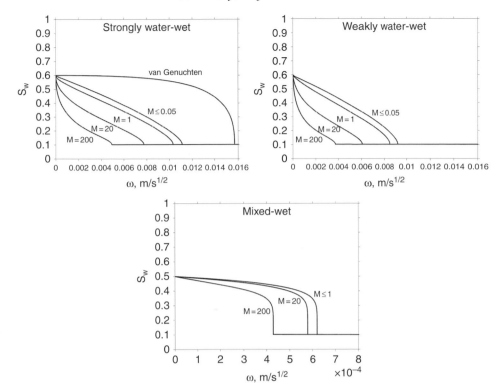

Figure 9.11 The water saturation as a function of ω, Eq. (9.43), for our example cases. $\omega(S_w)$ is proportional to the derivative of the capillary fractional flow, $F_w(S_w)$ shown in Fig. 9.10. Notice that at the leading edge of the water front, the saturation gradient is infinite. At the trailing edge ($\omega = 0$) note how the different models have disparate behaviour, with either a zero (van Genuchten), infinite (the other strongly water-wet cases) or a finite (weakly water-wet and mixed-wet) saturation gradient. This is related to the functional form of the capillary dispersion, $D(S_w)$, Fig. 9.9.

But, from Eq. (9.54) this also means that $d\omega/dS_w \rightarrow 0$, or $dS_w/d\omega \rightarrow \infty$: in the plot of $S_w(\omega)$ we have an infinite gradient at the leading nose of the imbibition front, in all cases, which may be viewed as an incipient shock. This makes physical sense: we have a dispersive flow, but at S_{wc} the capillary dispersion D becomes zero, and so the spreading is impeded, leading to an infinite slope.

We see different behaviour though at the inlet: a zero slope, for the van Genuchten model; an infinite slope, for the other strongly water-wet cases; and a finite slope for the weakly water-wet and mixed-wet examples. Here F_w is close to 1, and so from Eq. (9.55) we may write $F_w'' \sim -D$. From the discussion above, $dS_w/d\omega \sim 1/F_w''$ and hence $dS_w/d\omega|_{S_w^*} \sim 1/D(S_w^*)$.

In the van Genuchten model, $D(S_w^*) = \infty$, giving $dS_w/d\omega|_{S_w^*} = 0$ or a zero gradient at the inlet. Since the displaced phase is assumed to have infinite mobility (it can always escape even if there is no mobile saturation for it to flow), and the capillary pressure gradient is infinite, water may move at finite speed in response to an infinitesimal saturation gradient.

In the other strongly water-wet cases, we require an infinite saturation gradient to force a finite flow of displaced fluid through the inlet where it has, asymptotically, no mobility: $D(S_w^*) = 0$, and so $dS_w/d\omega|_{S_w^*} = \infty$. This implies that a long time is required to completely saturate the rock by spontaneous imbibition. This is the opposite behaviour that we observed for the water injection, or Buckley-Leverett, problem, where for low values of M the displacement was all shock with complete recovery at breakthrough; see Figs. 9.6 and 9.7.

For weakly water-wet or mixed-wet rocks, the displaced phase mobility is finite at the inlet. Thus there is flow in response to a finite saturation gradient. The magnitude of this gradient depends on the mobility ratio: it is lower for smaller M, where the displaced fluid flows more readily, while for larger M there is a larger gradient indicating a slower approach to the ultimate recovery.

While this analysis has a certain mathematical elegance, it is somewhat speculative: how do these results compare with measurements? This is addressed next.

9.3.3 *Experimental Analysis of Spontaneous Imbibition*

Several experiments have confirmed that for spontaneous imbibition, before the advancing water front encounters a boundary, the saturation profiles are indeed a function of ω, Eq. (9.43), only, while the overall recovery increases as $\sqrt{t}$, Eq. (9.48); see, for instance, Morrow and Mason (2001), Zhou et al. (2002) and Mason and Morrow (2013). This scaling, though, is limited to counter-current imbibition with no impediments to flow at the inlet, and co-current flows where the displaced fluid is of low viscosity (such as air) (Mason and Morrow, 2013). Quantitative analysis of the behaviour in terms of the semi-analytical formulation presented above is, however, more limited.

McWhorter (1971) performed imbibition experiments in a sand and then used independently measured capillary pressures and relative permeabilities to predict the recovery: the capillary pressure was shown in Fig. 5.3. The results in Fig. 9.12 demonstrate good agreement between theory and the measurements. In particular, the amount imbibed scales as $\sqrt{t}$ with a constant that can be derived from the capillary dispersion. However, there was no monitoring of the saturation profiles.

The results of a more recent series of experiments on Ketton limestone were illustrated in Fig. 9.2: the averaged saturation profiles are shown in Fig. 9.13 (Alyafei, 2015). Again, we see a scaling with ω. Qualitatively, the profile is most similar to the van Genuchten case in Fig. 9.11. In this experiment, water imbibed

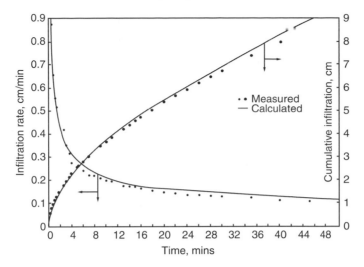

Figure 9.12 The measured recovery and recovery rate for spontaneous imbibition into a sand. The experimental data (points) are compared to theoretical predictions using independently measured relative permeabilities and capillary pressure. The capillary pressure was shown in Fig. 5.3. Data from McWhorter (1971).

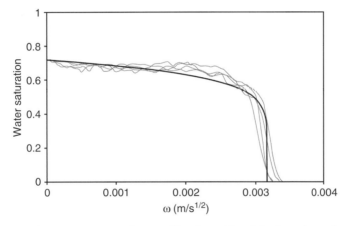

Figure 9.13 The average saturation profiles from Fig. 9.2 plotted as a function of ω, Eq. (9.43). The results at different times fall, approximately, onto one universal curve. The thick line is an analytical prediction using the method outlined above: a very low water relative permeability is needed to reproduce the behaviour. From Alyafei (2015).

into a dry core, displacing air. The results imply a strongly water-wet rock with little or no influence of a finite air mobility. The profiles could also be matched reasonably well using power-law relative permeabilities and capillary pressures, as in our examples. However, to obtain the frontal-type profile a water relative permeability exponent a of 10 was needed (Alyafei, 2015).

In both the experiments presented, a wetting phase was introduced into a completely dry porous medium. A more usual situation is to have some wetting phase present; indeed, as discussed in Chapter 4.3, the absence of wetting layers may result in a frontal advance, rather than a percolation-like displacement, which means that we cannot employ a traditional Darcy-type description of flow. Also, air was the displaced phase with co-current, rather than counter-current, flow. If we assume that the air is infinitely mobile, then co-current and counter-current flow are the same: there is no difference between air moving back through the inlet, or flowing to the outlet. However, this is not necessarily true near S_w^* if $S_w^* \approx 1 - S_{or}$. There are therefore opportunities to extend this work to study the *in situ* saturation profiles where water is initially present, for counter-current flow, and for the displacement of oil.

One valuable application of the analytical theory is to use measured saturation profiles $S_w(\omega)$ to determine or constrain relative permeability and capillary pressure. Using Eq. (9.55) we can find the capillary dispersion $D(S_w)$ from multiplying the gradient, F_w'', and the integral F_w of the saturation profile, since $\omega \propto F_w'$. If we have independently measured the relative permeabilities (from, for instance, a Buckley-Leverett style displacement), then the capillary pressure may be found.

9.4 Recovery, Imbibition and the Trillion-Barrel Question

We end this chapter, and the main material of the book, with a brief synthesis of two experiments with associated pore-scale modelling, which study recovery as a function of wettability. We will then discuss the implications for field-scale reservoir management.

The first series of experiments is presented in Fig. 9.14 (Jadhunandan and Morrow, 1995). Here, waterflood recovery was measured on Berea sandstone, after ageing in crude oil at different initial water saturations. As discussed in Chapter 3.5, this provides a transition from more water-wet conditions at high S_{wi} and more mixed- to oil-wet wettability as S_{wi} decreases. As S_{wi} decreases, more of the pore space comes in contact with oil, while the imposed capillary pressure increases, promoting the collapse of wetting films and engendering a wettability change.

We observe a non-monotonic trend in recovery: the highest recoveries are observed for an intermediate value of S_{wi}. The lowest recoveries are found for the smallest S_{wi}, representing an oil-wet system, and for the highest S_{wi}, which is more water-wet. In these experiments the viscosity ratio M is around 5. The reasons for this have been discussed previously in detail for our example cases; see Chapter 9.2.5. A mixed-wet rock, if it combines a low water relative permeability with a low residual saturation, can give the highest waterflood recoveries.

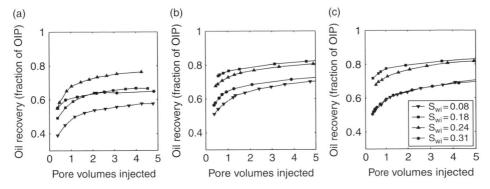

Figure 9.14 Oil recovery for different initial water saturations. (a) Experimental data from Jadhunandan and Morrow (1995). (b) Predicted recoveries using pore-scale modelling with a fixed distribution of intrinsic contact angles between $85°$ and $120°$. (c) Predictions where fewer pores become oil-wet for S_{wi} above 0.20. Here OIP stands for oil initially in place. The recovery, R_f, is related to pore volumes produced N_{pD} by $R_f = N_{pD}/(1 - S_{wi})$. From Valvatne and Blunt (2004).

Also shown in Fig. 9.14 are predictions using pore-scale network modelling (Valvatne and Blunt, 2004). The purpose here is not to worry about a quantitative match, since the *in situ* wettability of the rocks is not known, but to explore what conditions are necessary to reproduce similar trends. A non-monotonic dependence of recovery on initial saturation is seen if we apply, in the model, either reason for the wettability dependence of S_{wi}: assigning the same contact angles to those portions of the rock contacted by oil, or through allowing more pores to have an altered wettability for higher imposed capillary pressures after primary drainage. While there is no claim that recovery can be predicted *a priori* without a knowledge of both the pore structure and contact angles, the hypothesis, based on our example cases, that mixed-wet rocks may provide a better waterflood recovery than either water-wet or oil-wet media is supported by these experiments.

These results have implications for transition-zone reservoirs (see Fig. 3.18), where there is a variation in S_{wi} above the free water level. Cores taken from near the top of the oil column, or where a low initial water saturation is established in flooding experiments, will display oil-wet characteristics, indicating unfavourable waterflood recovery, whereas in most of the oil column, a better mixed-wet performance will be observed (Jackson et al., 2003). It is therefore important to capture the variation in wettability and consequently relative permeability in flow models of such reservoirs.

The second dataset is from Zhou et al. (2000), where both imbibition and waterflood recoveries are compared for sandstone cores exposed to crude oils for different ageing times. The rock became more mixed-wet as the ageing time

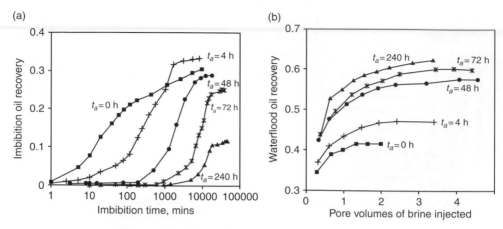

Figure 9.15 A comparison of oil recovery from spontaneous imbibition (left) and waterflooding (right) on cores exposed to different ageing times, t_a (in hours, h), as indicated. For waterflooding we see an increase in recovery, with t_a consistent with a transition from water-wet to more mixed-wet conditions. Note, however, that for imbibition, not only does the final recovery decrease with t_a, as less of the pore space is water-wet, but the time scales for recovery stretch by almost four orders of magnitude: this is a more marked impact of wettability than observed in our example cases, although it can be mimicked in pore-scale models. Data from Zhou et al. (2000).

increased. Fig. 9.15 compares the recovery by spontaneous counter-current imbibition and waterflooding. The waterflood recovery increases with ageing time, indicating a transition from water-wet (no ageing) to more mixed-wet conditions, consistent with Fig. 9.14. Since all the samples did imbibe water, a fully oil-wet state was never reached: some oil-filled pores always remained water-wet.

The imbibition recoveries show, as expected, that as the ageing time increases and the samples are less water-wet, the final recovery drops, as water can only imbibe into the decreasing fraction of the pore space that remains water-wet. The exception is for an ageing time of only four hours; this is likely to represent weakly water-wet conditions, where all the pore space can still imbibe water, but the residual oil saturation has decreased.

The most remarkable feature is the time scale of recovery. The virgin sample imbibes water almost 10,000 times faster than the mixed-wet core with the longest ageing time: the main feature for field-scale operations is not how much can be recovered, but the rate, and this is exceptionally sensitive to wettability. Indeed, our rather limited power-law relative permeability examples shown in Fig. 9.11 and Table 9.2 only showed around a 10-fold decrease in rate as we moved from strongly water-wet to mixed-wet conditions. These results imply that the water relative permeability in the mixed-wet cores is extremely low at low water saturation, hampering the influx of water and leading to recovery that proceeds at a glacial pace. While this dramatic effect is not captured in our synthetic examples, it can be reproduced using pore-scale modelling. Here the contact angles were adjusted to match the waterflood data, predicting, as expected, a change from strongly water-wet to a mixed-wet state with ageing time. Then the imbibition recovery was predicted using the same waterflood relative permeabilities and capillary pressures: a numerical approach was used, as the authors were not aware of the analytic formulation at the time (Behbahani and Blunt, 2005). The comparison is shown in Fig. 9.16: while an exact match is not achieved, the most important generic feature, which is the huge range of time scale, is captured.

In mixed-wet systems, pinned water layers greatly impede the water flow, with a percolation-like displacement pattern resulting in very low relative permeabilities during the initial stages of water displacement; see Chapter 7.2.4. This is advantageous for waterflooding, since the water is held back, allowing oil to be displaced to close to its low residual saturation. In spontaneous imbibition it is very unfavourable, since the ingress of water is controlled by its mobility: recovery is agonizingly slow and the final value is low since not all the pore space is water-wet.

Which leads us to the trillion-barrel question ... Most of the world's conventional oil resides in huge, fractured carbonate reservoirs in the Middle East. While

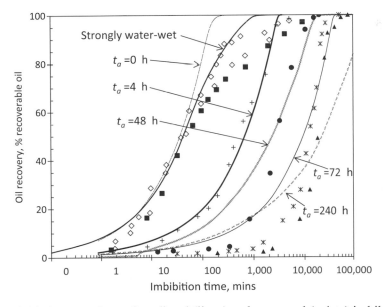

Figure 9.16 A comparison of predicted (lines) and measured (points) imbibition recoveries. The experimental results are shown in Fig. 9.15 with additional data on water-wet samples. The important feature to note is that a pore-scale model can replicate the enormous range of time-scale for recovery: this is dependent on the water relative permeability at low water saturation. Data from Behbahani and Blunt (2005).

estimates of reserves are notoriously unreliable, it is fair to say that possibly around one trillion barrels of oil *may* be recoverable from these fields. The emphasis on the word 'may' is because it all depends on the efficiency of displacement by water, either through a natural aquifer drive or by water injection. These fields often have only, to date, recovered a small fraction of the reserves, and so well-head production alone does not yet indicate the likely ultimate recovery. For waterflooding, the evidence is conflicting: the relative permeabilities from the United Arab Emirates presented in Fig. 7.12 suggested favourable mixed-wet conditions in most cases, while those from the giant Ghawar field in Saudi Arabia, Fig. 7.13, implied rather less optimistic recoveries, with rapid water breakthrough and behaviour more representative of oil-wet rocks. Moreover, these fields are fractured. If fracture flow dominates, then the recovery mechanism is now spontaneous imbibition and even mixed-wet conditions give poor, and very slow, recovery. At the pore scale, it all hinges on, well, hinging contacts, pinned or swelling water layers and whether we have a percolation-like or invasion percolation-like displacement pattern.

In the future it may be possible to manipulate the wettability of the rock *in situ* during waterflooding, through the use of controlled salinity water injection, or the addition of surface-active surfactants, as discussed in Chapter 1.3.2. The analysis of oil recovery shows that the optimal wettability state is dependent on the pore structure, displacement process and mobility ratio; a simple progression towards, for instance, more water-wet conditions is unlikely to be successful in all cases.

We do not know the characteristic wettability state and corresponding relative permeabilities for many of these fields, nor the importance of fracture flow and spontaneous imbibition to recovery. It is unlikely that there will be a simple answer. The aspiration though is that through a combination of careful experiment, intelligent pore-scale modelling and the application of some of the ideas presented here, a better understanding, and better management, of these fields can be achieved.

Appendix

Exercises

Here I present a series of questions based on past examination papers set at Imperial College London and Politecnico di Milano. I also provide solutions online. These are not model answers, but give the correct numerical answers and, in places, a guide on how to derive them.

(1) Capillary pressure and Leverett J function. (25 marks total)

You measure the following primary drainage capillary pressure in the laboratory using mercury. The core has a permeability of 600 mD, a porosity of 0.20, the interfacial tension is 487 mN/m and the contact angle is 140°.

Pressure (Pa)	Saturation
0	1
50,000	1
74,000	0.6
150,000	0.4
250,000	0.3
300,000	0.3

  (i) Write an equation that relates the capillary pressure to the Leverett J function. Define all the terms and give appropriate units. (5 marks)

  (ii) In the field the average permeability is 200 mD, the porosity is 0.15 and the interfacial tension is 25 mN/m. Plot a graph of water saturation against height above the free water level in the reservoir. The oil density is 700 kg/m^3 and the brine density is 1,050 kg/m^3. g=9.81 ms^{-2}. (15 marks)

  (iii) What approximations have you made in this analysis and what did you have to assume? (5 marks)

(2) Relative permeability. (25 marks total)

  (i) Write down the multiphase Darcy equation, define all terms and give them suitable units. (5 marks)

(ii) Draw a schematic of the waterflood (water displacing oil) and gasflood (gas displacing oil and connate water) relative permeabilities for a water-wet sandstone. Label the graph and comment on the values given and any differences between the relative permeability functions. (6 marks)

(iii) Discuss briefly how the three-phase oil relative permeability can be estimated when all three phases – oil, water and gas – are flowing. (6 marks)

(iv) Estimate the oil production rate from a reservoir of cross-sectional area 1 km by 5 km. The oil drains under gas gravity drainage: oil has a density of 700 kg/m^3 and gas a density of 300 kg/m^3. The permeability is 50 mD, the oil viscosity is 1.5 mPa.s and the relative permeability is 0.001. $g = 9.81$ m/s^2. Comment on your result. Why is the oil relative permeability so low? (8 marks)

(3) Wettability and contact angle. (25 marks total)

(i) Define intrinsic, advancing and receding contact angles. (3 marks)

(ii) Give all the reasons why the advancing contact angle is typically significantly higher than the receding contact angle. (3 marks)

(iii) Draw a graph of primary drainage capillary pressure for a fine sand pack. Also show the waterflood capillary pressure if the sand pack becomes oil-wet after drainage. Explain why the waterflood capillary pressure is lower than the drainage capillary pressure. (4 marks)

(iv) What is the capillary pressure for invasion through a tube of circular cross-section, but which has sides sloping at an angle α and a contact angle θ (see Fig. A.1)? The interfacial tension is σ. (10 marks)

(v) Comment on your answer to part (iv): how can it be used to explain capillary pressure hysteresis (consider a porous medium with diverging and converging pores)? (5 marks)

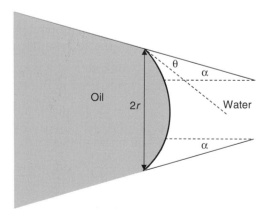

Figure A.1 Diagram for invasion in a conical geometry.

(4) Gas storage (25 marks total)

You are asked to consider using a depleted gas field to store CO_2 as part of a carbon-capture-and-storage project.

   (i) The natural gas originally in the reservoir had a density of 300 kg/m^3. The brine in the formation has a density of 1,100 kg/m^3. The cap-rock has a porosity of 0.1 and a permeability of 0.01 mD. The gas/brine interfacial tension is 50 mN/m. Make an approximate estimate of the capillary pressure necessary for the gas to enter the cap-rock. Explain your calculation carefully. (10 marks)

  (ii) Use the answer to part (i) to estimate the maximum height of a gas column that could be sustained under the cap rock. (7 marks)

 (iii) If CO_2 were stored in the same formation, what would the maximum height of the CO_2 be? The CO_2 density at reservoir conditions is 600 kg/m^3 and the interfacial tension is 20 mN/m. Comment on the result. Can CO_2 be safely stored in this field if it were to replace the gas? (8 marks)

(5) Conservation equations for a partitioning tracer. (25 marks total)

After waterflooding, there is a non-flowing residual oil saturation S_{or}. A tracer is injected in the water that partitions (dissolves) in the oil. If the concentration in the water is C, then the concentration in oil is aC.

   (i) Derive a conservation equation for the tracer concentration for one-dimensional incompressible flow. Explain all the terms carefully. (10 marks)

  (ii) With what speed does the tracer move? What is the speed of a conservative tracer that does not dissolve in oil? (5 marks)

 (iii) A conservative tracer and a partitioning tracer ($a = 2$) are injected into a waterflooded oilfield. The conservative tracer breaks through at a production well after 100 days, while the partitioning tracer breaks through at 150 days. Estimate the residual oil saturation. (10 marks)

(6) Relative permeability. (25 marks total)

   (i) Write down the multiphase Darcy equation, define all terms and give them suitable units. (5 marks)

  (ii) Draw a schematic of the waterflood (water displacing oil) relative permeabilities for water-wet, oil-wet and mixed-wet rock, noting differences between them. (7 marks)

 (iii) Comment on the implications for waterflood recovery in an oilfield. For a light oil with a viscosity similar to water, which wettability type gives the most favourable recovery? (6 marks)

(iv) Polymer flooding is being proposed for an oilfield. Polymer is injected with the water to increase the water viscosity. This method only works if it improves recovery beyond normal waterflooding. Why does polymer flooding work? For what wettability type(s) is this likely to be most favourable? Explain your answer carefully. (7 marks)

(7) Pore-scale displacement. (25 marks total)

(i) Define snap-off and piston-like advance. What is meant by pore-filling? Explain the processes that control the degree of non-wetting phase trapping. (7 marks)

(ii) Derive an equation for the entry pressure for piston-like advance through a cylindrical throat of inscribed radius r with contact angle θ. (4 marks)

(iii) Derive an equation for the threshold capillary pressure for filling by snap-off for a throat with an equilateral triangular cross-section of inscribed radius r and contact angle θ. What is the ratio of the threshold pressures for snap-off divided by piston-like advance? (8 marks)

(iv) Comment on your answer to part (iii). Use this result to explain how the residual non-wetting phase saturation varies with contact angle for contact angles less than 90°. (6 marks)

(8) Gravity drainage and three-phase flow (25 marks total)

(i) Explain the concept of layer drainage in three-phase flow and explain carefully why the oil relative permeability is proportional to the square of the oil saturation in the layer drainage regime. (7 marks)

(ii) In an oilfield that is being produced by gravity drainage, the oil relative permeability is $k_{ro} = 0.1S_o^2$. Write down the conservation equation for oil saturation, putting in the expression for the oil Darcy velocity for vertical flow under gravity. Use this to find the speed with which a saturation S_o travels. (9 marks)

(iii) Estimate the time needed to drain the oil saturation to 20%. The reservoir has an oil column of height 50 m, the oil density is 850 kg/m^3, the gas density is 350 kg/m^3, the oil viscosity is 0.3 mPa.s and the vertical permeability is 50 mD. The porosity is 0.2. Comment on your answer. (9 marks)

(9) Conservation equations for CO_2 storage in a fractured medium. (25 marks total)

Huge fractured aquifers are possible storage locations for CO_2 collected from power stations and other industrial plants. The CO_2 flows through the fractures. CO_2 also dissolves in brine; this CO_2 saturated brine can enter the matrix.

(i) The CO_2 in its own phase remains in the fractures. Explain physically what prevents the CO_2 in its own phase entering the matrix. (5 marks)

(ii) By what physical mechanism does the CO_2 dissolved in brine move through the matrix? (4 marks)

(iii) If the fractures are closely spaced, then all the water in the matrix in contact with a fracture containing CO_2 will have the same dissolved CO_2 concentration, equal to the solubility. If this solubility is C_s, write down a conservation equation for the flow of CO_2. You may assume that the only Darcy flow is in the fractures and can ignore water in the fractures themselves. To simplify the analysis, you can assume the Darcy flow of CO_2 is $S_c q_t$, where S_c is the saturation and q_t is the total (Darcy) velocity. ϕ_f is the porosity of the fractures and ϕ_m is the porosity of the matrix. Derive an expression for the speed of the CO_2 if the fracture saturation is 1. Draw a sketch to illustrate how the CO_2 moves that explains your answer. (13 marks)

(iv) Find the speed of the CO_2 in the fractures with dissolution if the total Darcy velocity is 10^{-7} m/s, $\phi_f = 0.0005$, $\phi_m = 0.3$, $C_s = 40$ kg/m^3 and the density of CO_2 in its own phase is 600 kg/m^3. (3 marks)

(10) Pore-scale displacement. (25 marks total)

(i) Define and explain the pore-scale filling processes that govern trapping in porous media when water displaces oil in a water-wet porous medium. Under what circumstances do you expect to see a significant amount of capillary trapping (a high residual non-wetting phase saturation)? Draw pictures to help illustrate your explanation. (7 marks)

(ii) Explain clearly what is meant by an oil layer in two-phase flow. Under what circumstances are oil layers observed? How do they affect the degree of trapping of oil? (4 marks)

(iii) Derive an equation for the threshold capillary pressure for filling by snap-off for a throat with a square cross-section of inscribed radius r and contact angle θ. What is the largest contact angle possible for snap-off to occur (for the threshold capillary pressure to be positive)? (10 marks)

(iv) Comment on your answer to part (iii). What happens to the amount of trapping for larger contact angles than the value found in part (iii)? (4 marks)

(11) Relative permeability. (25 marks total)

(i) In words explain what is meant by the concept of relative permeability. (4 marks)

(ii) Explain what is meant by wettability. Why do we often encounter oil-wet surfaces in oilfields? (5 marks)

Relative permeabilities

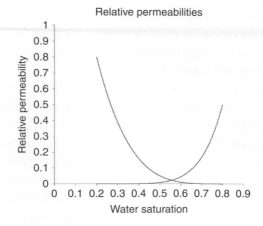

Figure A.2 Relative permeabilities.

(iii) Define the terms water-wet, oil-wet and mixed-wet. (4 marks)

(iv) The relative permeability in Fig. A.2 is measured on a core sample from a giant oilfield in the Middle East. What is the likely wettability of the sample? Explain your answer carefully. (5 marks)

 (v) In this field, water is injected to displace oil. If the oil and water viscosities are similar, estimate approximately the water saturation at which more water than oil will be produced from the field. What fraction of the oil that was originally in the reservoir will be produced? (7 marks)

(12) Capillary-controlled displacement. (25 marks total)

  (i) Write down the Young-Laplace equation. Define all the terms and provide units for each term. (4 marks)

 (ii) On a rock sample, a typical pore radius is 1 μm and the interfacial tension is 25 mN/m. What approximately is a typical capillary pressure? (4 marks)

(iii) If I have another rock sample where all the pores are twice the size as before, by what amount does the typical capillary pressure change? By what factor does the permeability change? (5 marks)

(iv) It takes 1,000 s for water to imbibe into a core of radius 1 cm, how long – all else being equal – does it take for water to imbibe into a matrix block in a reservoir that is around 1 m in radius? Explain your answer. (6 marks)

 (v) For the reservoir-scale matrix block, how long will it take for imbibition if all the pore sizes are now half the size than in the core-scale experiment? (6 marks)

(13) Pore-scale displacement. (25 marks total)
   (i) Define and explain the different pore-filling mechanisms in imbibition. Draw pictures to help illustrate your explanation. (7 marks)
   (ii) Explain how these pore-filling processes alone lead to little trapping and a flat frontal advance. (4 marks)
   (iii) Derive an equation for the threshold capillary pressure for filling by snap-off for a throat with an equilateral triangular cross-section of inscribed radius r and contact angle θ. What is the largest contact angle possible for snap-off to occur (for the threshold capillary pressure to be positive)? (10 marks)
   (iv) Explain how snap-off can lead to trapping. Use the answers to parts (ii) and (iii) to comment on what happens when the contact angle increases. (4 marks)

(14) Multiphase flow and oil layers. (25 marks total)
   (i) Write down the multiphase Darcy equation. Define all the terms and provide units. (5 marks)
   (ii) Write down the equation for the Darcy flow, under gravity only, of oil in the presence of a gas. You may assume that the gas has negligible viscosity and density. (5 marks)
   (iii) Explain why, in three-phase flow, with oil layer drainage the relative permeability is proportional to the square of the oil saturation. (8 marks)
   (iv) Calculate the oil Darcy velocity if $k_{ro} = 0.1S_o^2$, $K = 100$ mD, $\mu_o = 0.001$ Pa.s, $\rho_o = 800$ kg/m^3 and $S_o = 0.2$. Comment on the result. (7 marks)

(15) Relative permeability. (25 marks total)
   Fig. A.3 shows field data of relative permeability from a Middle Eastern carbonate core. The solid line labelled is the primary drainage relative permeability; this is before wettability alteration and the core is water-wet. The dashed line is the waterflood relative permeability after wettability alteration. The circle is the final waterflood relative permeability after flooding at a high flow rate (a so-called bump flood).
   (i) Describe, briefly, what is meant by wettability alteration at the end of primary drainage. (3 marks)
   (ii) Explain the likely wettability of the core during waterflooding. Explain your answer carefully. (7 marks)
   (iii) Frequently waterflood curves are referred to as imbibition. Is this correct? Explain using your answer to parts (i) and (ii). (5 marks)
   (iv) Explain why the water relative permeability during waterflooding is below the relative permeability for primary drainage. (10 marks)

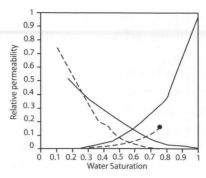

Figure A.3 Relative permeabilities for a Middle Eastern carbonate (Dernaika et al., 2013).

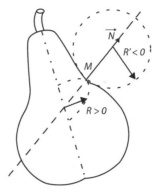

Figure A.4 The curvature of a pear (de Gennes et al., 2003).

(16) Capillary pressure and contact angles. (25 marks total)

  (i) Write down the Young-Laplace equation. Define all the terms and provide units for each term. (4 marks)

 (ii) Calculate the capillary pressure for the pear-shaped surface at the point M indicated, Fig. A.4. The interfacial tension is 60 mN/m and the radii have sizes 0.5 mm and 2 mm for R and R', respectively. (10 marks)

(iii) Write down the Bartell-Osterhof equation, the relationship between contact angles and interfacial tensions in three-phase flow. (4 marks)

(iv) Compute the contact angle between gas and water: the interfacial tension is 70 mN/m. The oil/water contact angle is 150° and the interfacial tension is 50 mN/m, the gas/oil contact angle is 20° and the interfacial tension is 20 mN/m. Comment on your result: can water be non-wetting to gas? (7 marks)

(17) You measure the following from appraisal wells in an oilfield. (50 marks total) The acceleration due to gravity = 9.81 m/s². Depths are measured from the surface.

Depth (m)	Pressure (MPa)	Fluid and density (kg/m^3)
2,250	17.51	Gas, 305
2,285	17.63	Oil, 650
2,327	18.01	Water, 1,040

(i) Explain physically why gas and oil pressures in a reservoir are typically higher than the water pressure in the surrounding aquifer. (8 marks)

(ii) Is the reservoir normally pressured, over pressured or under pressured? Explain your answer. (4 marks)

(iii) Find the depths of the free water and free oil levels. Hence find the height of the oil column. (18 marks)

(iv) Log measurements place the oil/water contact at a depth of 2,290 m. Explain the discrepancy with the answer in part (iii). Which value would you take to calculate the original oil in place? What is your estimate now of the height of the oil column? (12 marks)

(v) Find the original oil in place, measured at surface conditions. The reservoir has an area of 3.6×10^7 m^2, an average porosity of 0.22, an initial water saturation of 0.25, a net-to-gross of 0.85 (the fraction of the rock volume that contains producible hydrocarbon) and the oil formation volume factor is 1.41 (the ratio of reservoir oil volume to oil volume at the surface). You may quote the answer in m^3. (8 marks)

(18) Buckley-Leverett analysis. (50 marks total)

(i) Write down the multiphase extension of Darcy's law, explain all the terms with units. What does relative permeability mean physically? Draw typical curves for a water-wet and a mixed-wet rock. (13 marks)

(ii) Plot the relative permeability curves shown below, as well as the corresponding water fractional flow. What is the likely wettability of the rock? (12 marks)

$$k_{rw} = \frac{(S_w - 0.2)^4}{0.7^3}, \tag{A.1}$$

$$k_{ro} = \frac{(S_o - 0.2)^4}{0.7^2}, \tag{A.2}$$

where $\mu_o = 0.002$ Pa.s and $\mu_w = 0.001$ Pa.s.

(iii) Calculate the water saturation as a function of dimensionless velocity and the pore volumes produced as a function of pore volumes injected. Plot your answers on a graph. (15 marks)

(iv) Plot the oil produced as a function of time if $B_o = 1.5$ (the ratio of reservoir to surface oil volume) and $B_w = 1.1$ (the ratio of reservoir to surface water volume), and the total water injection rate is 10^5 stb/day (stb stands for stock tank, or surface, barrels). The pore volume of the reservoir is 1.5×10^8 rb (rb is reservoir barrels). (10 marks)

References

Adamson, A. W., and Gast, A. P. 1997. *Physical Chemistry of Surfaces*. John Wiley & Sons, Inc., New York.

Adler, P. M. 1990. Flow in simulated porous media. *International Journal of Multiphase Flow*, **16**(4), 691–712.

Adler, P. M. 2013. *Porous Media: Geometry and Transports*. Elsevier, Amsterdam.

Aghaei, A., and Piri, M. 2015. Direct pore-to-core up-scaling of displacement processes: Dynamic pore network modeling and experimentation. *Journal of Hydrology*, **522**, 488–509.

Ahrenholz, B., Tolke, J., Lehmann, P., Peters, A., Kaestner, A., Krafczyk, M., and Durner, W. 2008. Prediction of capillary hysteresis in a porous material using lattice-Boltzmann methods and comparison to experimental data and a morphological pore network model. *Advances in Water Resources*, **31**(9), 1151–1173.

Aissaoui, A. 1983. *Etude théorique et expérimentale de lhystérésis des pressions capillaries et des perméabilitiés relatives en vue du stockage souterrain de gaz*. PhD thesis, Ecole des Mines de Paris, Paris.

Akbarabadi, M., and Piri, M. 2013. Relative permeability hysteresis and capillary trapping characteristics of supercritical CO_2/brine systems: An experimental study at reservoir conditions. *Advances in Water Resources*, **52**, 190–206.

Akbarabadi, M., and Piri, M. 2014. Nanotomography of Spontaneous Imbibition in Shale. URTeC: 1922555, Unconventional Resources Technology Conference.

Aker, E., Måløy, K. J., Hansen, A., and Batrouni, G. G. 1998. A two-dimensional network simulator for two-phase flow in porous media. *Transport in Porous Media*, **32**(2), 163–186.

Aker, E., Måløy, K. J., Hansen, A., and Basak, S. 2000. Burst dynamics during drainage displacements in porous media: Simulations and experiments. *EPL (Europhysics Letters)*, **51**(1), 55–61.

Al-Dhahli, A., Geiger, S., and van Dijke, M. I. J. 2012. Three-phase pore-network modeling for reservoirs with arbitrary wettability. *SPE Journal*, **18**(2), 285–295.

Al-Dhahli, A., van Dijke, M. I. J., and Geiger, S. 2013. Accurate modelling of pore-scale films and layers for three-phase flow processes in clastic and carbonate rocks with arbitrary wettability. *Transport in Porous Media*, **98**(2), 259–286.

Al-Futaisi, A., and Patzek, T. W. 2003a. Extension of Hoshen-Kopelman algorithm to non-lattice environments. *Physica A: Statistical Mechanics and Its Applications*, **321**(3–4), 665–678.

Al-Futaisi, A., and Patzek, T. W. 2003b. Impact of wettability alteration on two-phase flow characteristics of sandstones: A quasi-static description. *Water Resources Research*, **39**(2), 1042.

Al-Gharbi, M. S., and Blunt, M. J. 2005. Dynamic network modeling of two-phase drainage in porous media. *Physical Review E*, **71**(1), 016308.

Al Mansoori, S. K., Iglauer, S., Pentland, C. H., and Blunt, M. J. 2009. Three-phase measurements of oil and gas trapping in sand packs. *Advances in Water Resources*, **32**(10), 1535–1542.

Al Mansoori, S. K., Itsekiri, E., Iglauer, S., Pentland, C. H., Bijeljic, B., and Blunt, M. J. 2010. Measurements of non-wetting phase trapping applied to carbon dioxide storage. *International Journal of Greenhouse Gas Control*, **4**(2), 283–288.

Al-Menhali, A., Niu, B., and Krevor, S. 2015. Capillarity and wetting of carbon dioxide and brine during drainage in Berea sandstone at reservoir conditions. *Water Resources Research*, **51**(10), 7895–7914.

Al-Menhali, A. S., and Krevor, S. 2016. Capillary trapping of CO_2 in oil reservoirs: observations in a mixed-wet carbonate rock. *Environmental Science & Technology*, **50**(5), 2727–2734.

Al-Raoush, R. I. 2009. Impact of wettability on pore-scale characteristics of residual nonaqueous phase liquids. *Environmental Science & Technology*, **43**(13), 4796–4801.

Al-Raoush, R. I., and Willson, C. S. 2005. Extraction of physically realistic pore network properties from three-dimensional synchrotron X-ray microtomography images of unconsolidated porous media systems. *Journal of Hydrology*, **300**(1–4), 44–64.

Alava, M., Dubé, M., and Rost, M. 2004. Imbibition in disordered media. *Advances in Physics*, **53**(2), 83–175.

Alizadeh, A. H., and Piri, M. 2014a. The effect of saturation history on three-phase relative permeability: An experimental study. *Water Resources Research*, **50**(2), 1636–1664.

Alizadeh, A. H., and Piri, M. 2014b. Three-phase flow in porous media: A review of experimental studies on relative permeability. *Reviews of Geophysics*, **52**(3), 468–521.

Alizadeh, A. H., Khishvand, M., Ioannidis, M. A., and Piri, M. 2014. Multi-scale experimental study of carbonated water injection: An effective process for mobilization and recovery of trapped oil. *Fuel*, **132**, 219–235.

Alyafei, N. 2015. *Capillary Trapping and Oil Recovery in Altered-Wettability Carbonate Rock*. PhD thesis, Imperial College London.

Alyafei, N., and Blunt, M. J. 2016. The effect of wettability on capillary trapping in carbonates. *Advances in Water Resources*, **90**, 36–50.

Alyafei, N., Raeini, A. Q., Paluszny, A., and Blunt, M. J. 2015. A sensitivity study of the effect of image resolution on predicted petrophysical properties. *Transport in Porous Media*, **110**(1), 157–169.

Amaechi, B., Iglauer, S., Pentland, C. H., Bijeljic, B., and Blunt, M. J. 2014. An experimental study of three-phase trapping in sand packs. *Transport in Porous Media*, **103**(3), 421–436.

Amaefule, J. O., and Handy, L. L. 1982. The effect of interfacial tensions on relative oil/water permeabilities of consolidated porous media. *SPE Journal*, **22**(3), 371–381.

Amin, R., and Smith, T. N. 1998. Interfacial tension and spreading coefficient under reservoir conditions. *Fluid Phase Equilibria*, **142**(12), 231–241.

Amott, E. 1959. Observations relating to the wettability of porous rock. *Petroleum Transaction of the AIME*, **216**, 156–162.

Anderson, W. G. 1986. Wettability literature survey – Part 2: Wettability measurement. *Journal of Petroleum Technology*, **38**(11), 1246–1462.

Anderson, W. G. 1987a. Wettability literature survey – Part 4: Effects of wettability on capillary pressure. *Journal of Petroleum Technology*, **39**(10), 1283–1300.

Anderson, W. G. 1987b. Wettability literature survey – Part 5: The effects of wettability on relative permeability. *Journal of Petroleum Technology*, **39**(11), 1453–1468.

Andrä, H., Combaret, N., J., Dvorkin, Glatt, E., Han, J., Kabel, M., Keehm, Y., Krzikalla, F., Lee, M., Madonna, C., Marsh, M., Mukerji, T., Saenger, E. H., Sain, R., Saxena, N., Ricker, S., Wiegmann, A., and Zhan, X. 2013a. Digital rock physics benchmarks – Part I: Imaging and segmentation. *Computers & Geosciences*, **50**, 25–32.

Andrä, H., Combaret, N., J., Dvorkin, Glatt, E., Han, J., Kabel, M., Keehm, Y., Krzikalla, F., Lee, M., Madonna, C., Marsh, M., Mukerji, T., Saenger, E. H., Sain, R., Saxena, N., Ricker, S., Wiegmann, A., and Zhan, X. 2013b. Digital rock physics benchmarks – Part II: Computing effective properties. *Computers & Geosciences*, **50**, 33–43.

Andrew, M. G. 2014. *Reservoir-Condition Pore-Scale Imaging of Multiphase Flow*. PhD thesis, Imperial College London.

Andrew, M. G., Bijeljic, B., and Blunt, M. J. 2013. Pore-scale imaging of geological carbon dioxide storage under in situ conditions. *Geophysical Research Letters*, **40**(15), 3915–3918.

Andrew, M. G., Bijeljic, B., and Blunt, M. J. 2014a. Pore-by-pore capillary pressure measurements using X-ray microtomography at reservoir conditions: Curvature, snap-off, and remobilization of residual CO_2. *Water Resources Research*, **50**(11), 8760–8774.

Andrew, M. G., Bijeljic, B., and Blunt, M. J. 2014b. Pore-scale contact angle measurements at reservoir conditions using X-ray microtomography. *Advances in Water Resources*, **68**, 24–31.

Andrew, M. G., Bijeljic, B., and Blunt, M. J. 2014c. Pore-scale imaging of trapped supercritical carbon dioxide in sandstones and carbonates. *International Journal of Greenhouse Gas Control*, **22**, 1–14.

Andrew, M. G., Menke, H., Blunt, M. J., and Bijeljic, B. 2015. The imaging of dynamic multiphase fluid flow using synchrotron-based x-ray microtomography at reservoir conditions. *Transport in Porous Media*, **110**(1), 1–24.

Anton, L., and Hilfer, R. 1999. Trapping and mobilization of residual fluid during capillary desaturation in porous media. *Physical Review E*, **59**(6), 6819–6823.

Armstrong, R. T., and Berg, S. 2013. Interfacial velocities and capillary pressure gradients during Haines jumps. *Physical Review E*, **88**(4), 043010.

Armstrong, R. T., Porter, M. L., and Wildenschild, D. 2012. Linking pore-scale interfacial curvature to column-scale capillary pressure. *Advances in Water Resources*, **46**, 55–62.

Armstrong, R. T., Georgiadis, A., Ott, H., Klemin, D., and Berg, S. 2014a. Critical capillary number: Desaturation studied with fast X-ray computed microtomography. *Geophysical Research Letters*, **41**(1), 55–60.

Armstrong, R. T., Ott, H., Georgiadis, A., Rücker, M., Schwing, A., and Berg, S. 2014b. Subsecond pore-scale displacement processes and relaxation dynamics in multiphase flow. *Water Resources Research*, **50**(12), 9162–9176.

Arns, C. H., Knackstedt, M. A., Pinczewski, W. V., and Lindquist, W. B. 2001. Accurate estimation of transport properties from microtomographic images. *Geophysical Research Letters*, **28**(17), 3361–3364.

Arns, C. H., Knackstedt, M. A., Pinczewski, W. V., and Martys, N. S. 2004. Virtual permeametry on microtomographic images. *Journal of Petroleum Science and Engineering*, **45**(1–2), 41–46.

Arns, C. H., Knackstedt, M. A., and Martys, N. S. 2005. Cross-property correlations and permeability estimation in sandstone. *Physical Review E*, **72**(4), 046304.

Arns, C. H., Knackstedt, M. A., and Mecke, K. R. 2009. Boolean reconstructions of complex materials: Integral geometric approach. *Physical Review E*, **80**(5), 051303.

Avraam, D. G., and Payatakes, A. C. 1995a. Flow regimes and relative permeabilities during steady-state two-phase flow in porous media. *Journal of Fluid Mechanics*, **293**, 207–236.

Avraam, D. G., and Payatakes, A. C. 1995b. Generalized relative permeability coefficients during steady-state two-phase flow in porous media, and correlation with the flow mechanisms. *Transport in Porous Media*, **20**(1–2), 135–168.

Aziz, K., and Settari, A. 1979. *Petroleum Reservoir Simulation*. Chapman & Hall, London.

Baker, L. E. 1988. Three-phase relative permeability correlations. SPE 17369, proceedings of the SPE Enhanced Oil Recovery Symposium, Tulsa, Oklahoma, 16–21 April.

Bakke, S., and Øren, P. E. 1997. 3-D pore-scale modelling of sandstones and flow simulations in the pore networks. *SPE Journal*, **2**(2), 136–149.

Baldwin, C. A., Sederman, A. J., Mantle, M. D., Alexander, P., and Gladden, L. F. 1996. Determination and characterization of the structure of a pore space from 3D volume images. *Journal of Colloid and Interface Science*, **181**(1), 79–92.

Bardon, C., and Longeron, D. G. 1980. Influence of very low interfacial tensions on relative permeability. *SPE Journal*, **20**(5), 391–401.

Barenblatt, G. I. 1971. Filtration of two non-mixing fluids in a homogeneous porous medium. *Fluid Dynamics*, **6**(5), 857–864.

Barenblatt, G. I., Entov, V. M., and Ryzhik, V. M. 1990. *Theory of Fluid Flows Through Natural Rocks*. Kluwer Academic Publishers, Dordrecht, The Netherlands.

Barenblatt, G. I., Patzek, T. W., and Silin, D. B. 2003. The mathematical model of nonequilibrium effects in water-oil displacement. *SPE Journal*, **8**(4), 409–416.

Bartell, F. E., and Osterhof, H. J. 1927. Determination of the wettability of a solid by a liquid. *Industrial and Engineering Chemistry*, **19**, 1277–1280.

Batchelor, G. K. 1967. *An Introduction to Fluid Dynamics*. Cambridge Mathematical Library. Cambridge University Press, Cambridge.

Bauters, T. W. J., Steenhuis, T. S., Parlange, J. Y., and DiCarlo, D. A. 1998. Preferential flow in water-repellent sands. *Journal of the Soil Science Society of America*, **62**(5), 1185–1190.

Bauters, T. W. J., Steenhuis, T. S., DiCarlo, D. A., Nieber, J. L., Dekker, L. W., Ritsema, C. J., Parlange, J. Y., and Haverkamp, R. 2000. Physics of water repellent soils. *Journal of Hydrology*, **231–232**, 233–243.

Bear, J. 1972. *Dynamics of Fluids in Porous Media*. Dover Publications, Mineola, New York.

Behbahani, H., and Blunt, M. J. 2005. Analysis of imbibition in mixed-wet rocks using pore-scale modeling. *SPE Journal*, **10**(4), 466–474.

Ben-Avraham, D., and Havlin, S. 2000. *Diffusion and Reactions in Fractals and Disordered Systems*. Cambridge University Press, Cambridge.

Bennion, B., and Bachu, S. 2008. Drainage and imbibition relative permeability relationships for supercritical CO_2/brine and H_2S/brine systems in intergranular sandstone, carbonate, shale, and anhydrite rocks. *SPE Reservoir Evaluation & Engineering*, **11**(3), 487–486.

Berg, S., Cense, A. W., Hofman, J. P., and Smits, R. M. M. 2008. Two-phase flow in porous media with slip boundary condition. *Transport in Porous Media*, **74**(3), 275–292.

Berg, S., Ott, H., Klapp, S. A., Schwing, A., Neiteler, R., Brussee, N., Makurat, A., Leu, L., Enzmann, F., Schwarz, J. O., Kersten, M., Irvine, S., and Stampanoni, M. 2013. Real-time 3D imaging of Haines jumps in porous media flow. *Proceedings of the National Academy of Sciences*, **110**(10), 3755–3759.

Berkowitz, B., Cortis, A., Dentz, M., and Scher, H. 2006. Modeling non-Fickian transport in geological formations as a continuous time random walk. *Reviews of Geophysics*, **44**(2), RG2003.

Bianchi Janetti, E., Riva, M., and Guadagnini, A. 2015. Three-phase permeabilities: Upscaling, analytical solutions and uncertainty analysis in elementary pore structures. *Transport in Porous Media*, **106**(2), 259–283.

Bianchi Janetti, E., Riva, M., and Guadagnini, A. 2016. Analytical expressions for three-phase generalized relative permeabilities in water- and oil-wet capillary tubes. *Computational Geosciences*, **20**, 555–565.

Bijeljic, B., Mostaghimi, P., and Blunt, M. J. 2011. Signature of non-Fickian solute transport in complex heterogeneous porous media. *Physical Review Letters*, **107**(20), 204502.

Bijeljic, B., Mostaghimi, P., and Blunt, M. J. 2013a. Insights into non-Fickian solute transport in carbonates. *Water Resources Research*, **49**(5), 2714–2728.

Bijeljic, B., Raeini, A., Mostaghimi, P., and Blunt, M. J. 2013b. Predictions of non-Fickian solute transport in different classes of porous media using direct simulation on pore-scale images. *Physical Review E*, **87**(1), 013011.

Birovljev, A., Furuberg, L., Feder, J., Jøssang, T., Måløy, K. J., and Aharony, A. 1991. Gravity invasion percolation in two dimensions: Experiment and simulation. *Physical Review Letters*, **67**(5), 584–587.

Biswal, B., Manwart, C., Hilfer, R., Bakke, S., and Øren, P. E. 1999. Quantitative analysis of experimental and synthetic microstructures for sedimentary rock. *Physica A: Statistical Mechanics and Its Applications*, **273**(3–4), 452–475.

Biswal, B., Øren, P. E., Held, R. J., Bakke, S., and Hilfer, R. 2007. Stochastic multiscale model for carbonate rocks. *Physical Review E*, **75**(6), 061303.

Biswal, B., Held, R. J., Khanna, V., Wang, J., and Hilfer, R. 2009. Towards precise prediction of transport properties from synthetic computer tomography of reconstructed porous media. *Physical Review E*, **80**(4), 041301.

Blunt, M., and King, P. 1990. Macroscopic parameters from simulations of pore scale flow. *Physical Review A*, **42**(8), 4780–4787.

Blunt, M., and King, P. 1991. Relative permeabilities from two- and three-dimensional pore-scale network modelling. *Transport in Porous Media*, **6**(4), 407–433.

Blunt, M., King, M. J., and Scher, H. 1992. Simulation and theory of two-phase flow in porous media. *Physical Review A*, **46**(12), 7680–7699.

Blunt, M., Zhou, D., and Fenwick, D. 1995. Three-phase flow and gravity drainage in porous media. *Transport in Porous Media*, **20**(1–2), 77–103.

Blunt, M. J. 1997a. Effects of heterogeneity and wetting on relative permeability using pore level modeling. *SPE Journal*, **2**(4), 70–87.

Blunt, M. J. 1997b. Pore level modeling of the effects of wettability. *SPE Journal*, **2**(1), 494–510.

Blunt, M. J. 1998. Physically-based network modeling of multiphase flow in intermediate-wet porous media. *Journal of Petroleum Science and Engineering*, **20**(3–4), 117–125.

Blunt, M. J. 2000. An empirical model for three-phase relative permeability. *SPE Journal*, **5**(4), 435–445.

Blunt, M. J. 2001a. Constraints on contact angles for multiple phases in thermodynamic equilibrium. *Journal of Colloid and Interface Science*, **239**(1), 281–282.

Blunt, M. J. 2001b. Flow in porous media – pore-network models and multiphase flow. *Current Opinion in Colloid and Interface Science*, **6**(3), 197–207.

Blunt, M. J., and Bijeljic, B. 2016. *Imperial College Consortium on Pore-scale Modelling*. http://www.imperial.ac.uk/engineering/departments/earth-science/research/research-groups/perm/research/pore-scale-modelling/.

Blunt, M. J., and Scher, H. 1995. Pore-level modeling of wetting. *Physical Review E*, **52**(6), 6387–6403.

Blunt, M. J., Barker, J. W., Rubin, B., Mansfield, M., Culverwell, I. D., and Christie, M. A. 1994. Predictive theory for viscous fingering in compositional displacement. *SPE Reservoir Engineering*, **9**(1), 73–80.

Blunt, M. J., Bijeljic, B., Dong, H., Gharbi, O., Iglauer, S., Mostaghimi, P., Paluszny, A., and Pentland, C. 2013. Pore-scale imaging and modelling. *Advances in Water Resources*, **51**, 197–216.

Boek, E. S., and Venturoli, M. 2010. Lattice-Boltzmann studies of fluid flow in porous media with realistic rock geometries. *Computers & Mathematics with Applications*, **59**(7), 2305–2314.

Bondino, I., Hamon, G., Kallel, W., and Kachuma, D. 2013. Relative permeabilities from simulation in 3D rock models and equivalent pore networks: Critical review and way forward. *Petrophysics*, **54**(6), 538–546.

Bourbiaux, B. J., and Kalaydjian, F. J. 1990. Experimental study of cocurrent and countercurrent flows in natural porous media. *SPE Reservoir Engineering*, **5**(3), 361–368.

Bourbie, T., and Zinszner, B. 1985. Hydraulic and acoustic properties as a function of porosity in Fontainebleau sandstone. *Journal of Geophysical Research*, **90**(B13), 11524–11532.

Broadbent, S. R., and Hammersley, J. M. 1957. Percolation processes. I. Crystals and mazes. *Proceedings of the Cambridge Philosophical Society*, **53**, 629–641.

Brooks, R. H., and Corey, T. 1964. *Hydraulic properties of porous media*. Hydrology Papers, Colorado State University.

Brown, K., Schülter, S., Sheppard, A., and Wildenschild, D. 2014. On the challenges of measuring interfacial characteristics of three-phase fluid flow with x-ray microtomography. *Journal of Microscopy*, **253**(3), 171–182.

Brown, R. J. S., and Fatt, I. 1956. Measurements Of Fractional Wettability Of Oil Fields' Rocks By The Nuclear Magnetic Relaxation Method. SPE 743-G, proceedings of the Fall Meeting of the Petroleum Branch of AIME, Los Angeles, California, 14–17 October.

Brun, F., Mancini, L., Kasae, P., Favretto, S., Dreossi, D., and Tromba, G. 2010. Pore3D: A software library for quantitative analysis of porous media. *Nuclear Instruments and Methods in Physics Research Section A: Accelerators, Spectrometers, Detectors and Associated Equipment*, **615**(3), 326–332.

Brusseau, M. L., Peng, S., Schnaar, G., and Murao, A. 2007. Measuring air-water interfacial areas with X-ray microtomography and interfacial partitioning tracer tests. *Environmental Science & Technology*, **41**(6), 1956–1961.

Bryant, S., and Blunt, M. 1992. Prediction of relative permeability in simple porous media. *Physical Review A*, **46**(4), 2004–2011.

Bryant, S. L., King, P. R., and Mellor, D. W. 1993a. Network model evaluation of permeability and spatial correlation in a real random sphere packing. *Transport in Porous Media*, **11**(1), 53–70.

Bryant, S. L., Mellor, D. W., and Cade, C. A. 1993b. Physically representative network models of transport in porous media. *AIChE Journal*, **39**(3), 387–396.

Buckingham, E. 1907. Studies on the movement of soil moisture. *USDA Bureau of Soils Bulletin*, **38**, 61–70.

Buckley, J. S., and Liu, Y. 1998. Some mechanisms of crude oil/brine/solid interactions. *Journal of Petroleum Science and Engineering*, **20**(3–4), 155–160.

Buckley, J. S., Takamura, K., and Morrow, N. R. 1989. Influence of electrical surface charges on the wetting properties of crude oils. *SPE Reservoir Engineering*, **4**(3), 332–341.

Buckley, J. S., Bousseau, C., and Liu, Y. 1996. Wetting alteration by brine and crude oil: From contact angles to cores. *SPE Journal*, **1**(3), 341–50.

Buckley, S. E., and Leverett, M. C. 1942. Mechanism of fluid displacement in sands. *Petroleum Transactions of the AIME*, **146**(1), 107–116.

Bultreys, T., Van Hoorebeke, L., and Cnudde, V. 2015a. Multi-scale, micro-computed tomography-based pore network models to simulate drainage in heterogeneous rocks. *Advances in Water Resources*, **78**, 36–49.

Bultreys, T., Boone, M. A., Boone, M. N., De Schryver, T., Masschaele, B., Van Loo, D., Van Hoorebeke, L., and Cnudde, V. 2015b. Real-time visualization of Haines jumps in sandstone with laboratory-based microcomputed tomography. *Water Resources Research*, **51**(10), 8668–8676.

Bultreys, T., Boone, M. A., Boone, M. N., De Schryver, T., Masschaele, B., Van Hoorebeke, L., and Cnudde, V. 2016a. Fast laboratory-based micro-computed tomography for pore-scale research: Illustrative experiments and perspectives on the future. *Advances in Water Resources*, **95**, 341–351.

Bultreys, T., De Boever, W., and Cnudde, V. 2016b. Imaging and image-based fluid transport modeling at the pore scale in geological materials: A practical introduction to the current state-of-the-art. *Earth-Science Reviews*, **155**, 93–128.

Burdine, N. T. 1953. Relative permeability calculations from pore size distribution data. *Journal of Petroleum Technology*, **5**(3), 71–78.

Caers, J. 2005. *Petroleum Geostatistics*. Society of Petroleum Engineers, Richardson, Texas.

Carlson, F. M. 1981. Simulation of Relative Permeability Hysteresis to the Nonwetting Phase. SPE 10157, proceedings of the SPE Annual Technical Meeting and Exhibition, San Antonio, Texas, 4–7 October.

Carman, P. C. 1956. *Flow of Gases Through Porous Media*. Butterworths, London.

Casado, C. M. M. 2012. *Laplace. La mecánica celeste*. RBA Publishers, Barcelona.

Caubit, C., Bertin, H., and Hamon, G. 2004. Three-Phase Flow in Porous Media: Wettability Effect on Residual Saturations During Gravity Drainage and Tertiary Waterflood. SPE 90099, proceedings of the SPE Annual Technical Conference and Exhibition, Houston, Texas, 26–29 September.

Cazabat, A. M., Gerdes, S., Valignat, M. P., and Villette, S. 1997. Dynamics of wetting: From theory to experiment. *Interface Science*, **5**(2–3), 129–139.

Chandler, R., Koplik, J., Lerman, K., and Willemsen, J. F. 1982. Capillary displacement and percolation in porous media. *Journal of Fluid Mechanics*, **119**, 249–267.

Chatzis, I., and Morrow, N. R. 1984. Correlation of capillary number relationships for sandstone. *SPE Journal*, **24**(5), 555–562.

Chatzis, I., Morrow, N. R., and Lim, H. T. 1983. Magnitude and detailed structure of residual oil saturation. *SPE Journal*, **23**(2), 311–326.

Chaudhary, K., Bayani Cardenas, M., Wolfe, W. W., Maisano, J. A., Ketcham, R. A., and Bennett, P. C. 2013. Pore-scale trapping of supercritical CO_2 and the role of grain wettability and shape. *Geophysical Research Letters*, **40**(15), 3878–3882.

Chen, J. D., and Koplik, J. 1985. Immiscible fluid displacement in small networks. *Journal of Colloid and Interface Science*, **108**(2), 304–330.

Chen, S., and Doolen, G. D. 1998. Lattice Boltzmann method for fluid flows. *Annual Review of Fluid Mechanics*, **30**(1), 329–364.

Chen, Z. X., Zimmerman, R. W., Bodvarsson, G. S., and Witherspoon, P. A. 1990. A new formulation for one-dimensional horizontal imbibition in unsaturated porous media. *Lawrence Berkeley Preprint LBL-28638*.

Cieplak, M., and Robbins, M. O. 1988. Dynamical transition in quasistatic fluid invasion in porous media. *Physical Review Letters*, **60**(20), 2042–2045.

Cieplak, M., and Robbins, M. O. 1990. Influence of contact angle on quasistatic fluid invasion of porous media. *Physical Review B*, **41**(16), 11508–11521.

Cnudde, V., and Boone, M. N. 2013. High-resolution X-ray computed tomography in geosciences: A review of the current technology and applications. *Earth-Science Reviews*, **123**, 1–17.

Coker, D. A., Torquato, S., and Dunsmuir, J. H. 1996. Morphology and physical properties of Fontainebleau sandstone via a tomographic analysis. *Journal of Geophysical Research: Solid Earth*, **101**(B8), 17497–17506.

Coles, M. E., Hazlett, R. D., Spanne, P., Soll, W. E., Muegge, E. L., and Jones, K. W. 1998. Pore level imaging of fluid transport using synchrotron X-ray microtomography. *Journal of Petroleum Science and Engineering*, **19**(1–2), 55–63.

Constantinides, G. N., and Payatakes, A. C. 2000. Effects of precursor wetting films in immiscible displacement through porous media. *Transport in Porous Media*, **38**(3), 291–317.

Corey, A. T. 1954. The interrelation between gas and oil relative permeabilities. *Producers Monthly*, **19**(1), 38–41.

Costanza-Robinson, M. S., Harrold, K. H., and Lieb-Lappen, R. M. 2008. X-ray microtomography determination of air-water interfacial area-water saturation relationships in sandy porous media. *Environmental Science & Technology*, **42**(8), 2949–2956.

Craig, F. F. 1971. *The Reservoir Engineering Aspects of Waterflooding*. Monograph Series 3, Society of Petroleum Engineers, Richardson, Texas.

Crocker, M. E. 1986. Wettability. *Enhanced Oil Recovery, Progress Review for the Quarter Ending Sept. 30, DOE/BC-86/3*, **47**, 100–110.

Cueto-Felgueroso, L., and Juanes, R. 2008. Nonlocal interface dynamics and pattern formation in gravity-driven unsaturated flow through porous media. *Physical Review Letters*, **101**(24), 244504.

Cueto-Felgueroso, L., and Juanes, R. 2009a. A phase field model of unsaturated flow. *Water Resources Research*, **45**(10), W10409.

Cueto-Felgueroso, L., and Juanes, R. 2009b. Stability analysis of a phase-field model of gravity-driven unsaturated flow through porous media. *Physical Review E*, **79**(3), 036301.

Cueto-Felgueroso, L., and Juanes, R. 2016. A discrete-domain description of multiphase flow in porous media: Rugged energy landscapes and the origin of hysteresis. *Geophysical Research Letters*, **43**(4), 1615–1622.

Culligan, K. A., Wildenschild, D., Christensen, B. S. B., Gray, W. G., Rivers, M. L., and Tompson, A. F. B. 2004. Interfacial area measurements for unsaturated flow through a porous medium. *Water Resources Research*, **40**(12), W12413.

Culligan, K. A., Wildenschild, D., Christensen, B. S. B., Gray, W. G., and Rivers, M. L. 2006. Pore-scale characteristics of multiphase flow in porous media: A comparison of air-water and oil-water experiments. *Advances in Water Resources*, **29**(2), 227–238.

Dahle, H. K., and Celia, M. A. 1999. A dynamic network model for two-phase immiscible flow. *Computational Geosciences*, **3**(1), 1–22.

Dake, L. P. 1983. *Fundamentals of Reservoir Engineering*. Elsevier, Amsterdam, 2nd Edition.

Darcy, H. 1856. *Les fontaines publiques de la ville de Dijon: English Translation, The Public Fountains of the City of Dijon, translated by P. Bobeck, Kendall/Hunt Publishing Company, Dubuque, Iowa, 2004*. Victor Dalmont, Paris.

Datta, S. S., Dupin, J. B., and Weitz, D. A. 2014a. Fluid breakup during simultaneous two-phase flow through a three-dimensional porous medium. *Physics of Fluids*, **26**(6), 062004.

Datta, S. S., Ramakrishnan, T. S., and Weitz, D. A. 2014b. Mobilization of a trapped non-wetting fluid from a three-dimensional porous medium. *Physics of Fluids*, **26**(2), 022002.

de Gennes, P. G. 1985. Wetting: statics and dynamics. *Reviews of Modern Physics*, **57**(3), 827–863.

de Gennes, P. G., Brochard-Wyatt, F., and Quérédef, D. 2003. *Capillarity and Wetting Phenomena: Drops, Bubbles, Pearls and Waves*. Springer, New York.

Dehghanpour, H., Aminzadeh, B., Mirzaei, M., and DiCarlo, D. A. 2011a. Flow coupling during three-phase gravity drainage. *Physical Review E*, **83**(6), 065302.

Dehghanpour, H., Aminzadeh, B., and DiCarlo, D. A. 2011b. Hydraulic conductance and viscous coupling of three-phase layers in angular capillaries. *Physical Review E*, **83**(6), 066320.

Demianov, A., Dinariev, O., and Evseev, N. 2011. Density functional modelling in multiphase compositional hydrodynamics. *The Canadian Journal of Chemical Engineering*, **89**(2), 206–226.

Deng, W., Cardenas, M. Bayani, and Bennett, P. C. 2014. Extended Roof snap-off for a continuous non-wetting fluid and an example case for supercritical CO_2. *Advances in Water Resources*, **64**, 34–46.

Dernaika, M. R., Basioni, M. A., Dawoud, A. M., Kalam, M. Z., and Skjæveland, S. M. 2013. Variations in bounding and scanning relative permeability curves with different carbonate rock type. *SPE Reservoir Evaluation & Engineering*, **16**(3), 265–280.

Dias, M. M., and Payatakes, A. C. 1986. Network models for two-phase flow in porous media Part 2. Motion of oil ganglia. *Journal of Fluid Mechanics*, **164**, 337–358.

Dias, M. M, and Wilkinson, D. 1986. Percolation with trapping. *Journal of Physics A: Mathematical and General*, **19**(15), 3131–3146.

DiCarlo, D. A. 2004. Experimental measurements of saturation overshoot on infiltration. *Water Resources Research*, **40**(4), W04215.

DiCarlo, D. A. 2013. Stability of gravity-driven multiphase flow in porous media: 40 years of advancements. *Water Resources Research*, **49**(8), 4531–4544.

DiCarlo, D. A., Bauters, T. W. J., Darnault, C. J. G., Steenhuis, T. S., and Parlange, J. 1999. Lateral expansion of preferential flow paths in sands. *Water Resources Research*, **35**(2), 427–434.

DiCarlo, D. A., Sahni, A., and Blunt, M. J. 2000a. The effect of wettability on three-phase relative permeability. *Transport in Porous Media*, **39**(3), 347–366.

DiCarlo, D. A., Sahni, A., and Blunt, M. J. 2000b. Three-phase relative permeability of water-wet, oil-wet, and mixed-wet sandpacks. *SPE Journal*, **5**(1), 82–91.

DiCarlo, D. A., Cidoncha, J. I. G., and Hickey, C. 2003. Acoustic measurements of pore-scale displacements. *Geophysical Research Letters*, **30**(17), 1901.

DiCarlo, D. A., Juanes, R., LaForce, T., and Witelski, T. P. 2008. Nonmonotonic traveling wave solutions of infiltration into porous media. *Water Resources Research*, **44**(2), W02406.

Ding, H., and Spelt, P. D. M. 2007. Inertial effects in droplet spreading: A comparison between diffuse-interface and level-set simulations. *Journal of Fluid Mechanics*, **576**, 287–296.

Dixit, A. B., Buckley, J. S., McDougall, S. R., and Sorbie, K. S. 1998a. Core wettability: should IAH equal IUSBM? SCA-9809, proceedings of the International Symposium of the Society of Core Analysts, The Hague.

Dixit, A. B., McDougall, S. R., and Sorbie, K. S. 1998b. A pore-level investigation of relative permeability hysteresis in water-wet systems. *SPE Journal*, **3**(2), 115–123.

Dixit, A. B., Buckley, J. S., McDougall, S. R., and Sorbie, K. S. 2000. Empirical measures of wettability in porous media and the relationship between them derived from pore-scale modelling. *Transport in Porous Media*, **40**(1), 27–54.

Donaldson, E. C., Thomas, R. D., and Lorenz, P. B. 1969. Wettability determination and its effect on recovery efficiency. *SPE Journal*, **9**(1), 13–30.

Dong, H., and Blunt, M. J. 2009. Pore-network extraction from micro-computerized-tomography images. *Physical Review E*, **80**(3), 036307.

Dong, M., Dullien, F. A. L., and Chatzis, I. 1995. Imbibition of oil in film form over water present in edges of capillaries with an angular cross section. *Journal of Colloid and Interface Science*, **172**(1), 21–36.

Doster, F., Zegeling, P. A., and Hilfer, R. 2010. Numerical solutions of a generalized theory for macroscopic capillarity. *Physical Review E*, **81**(3), 036307.

Doster, F., Hönig, O., and Hilfer, R. 2012. Horizontal flow and capillarity-driven redistribution in porous media. *Physical Review E*, **86**(1), 016317.

Dria, D. E., Pope, G. A., and Sepehrnoori, K. 1993. Three-phase gas/oil/brine relative permeabilities measured under CO_2 flooding conditions. *SPE Reservoir Engineering*, **8**(2), 143–150.

Dullien, F. A. L. 1997. *Porous Media: Fluid Transport and Pore Structure*. Academic Press, San Diego. 2nd Edition.

Dullien, F. A. L, Zarcone, C., Macdonald, I. F., Collins, A., and Bochard, R. D. E. 1989. The effects of surface roughness on the capillary pressure curves and the heights of capillary rise in glass bead packs. *Journal of Colloid and Interface Science*, **127**(2), 362–372.

Dumoré, J. M., and Schols, R. S. 1974. Drainage capillary-pressure functions and the influence of connate water. *SPE Journal*, **14**(5), 437–444.

Durand, C., and Rosenberg, E. 1998. Fluid distribution in kaolinite- or illite-bearing cores: Cryo-SEM observations versus bulk measurements. *Journal of Petroleum Science and Engineering*, **19**(1-2), 65–72.

Dussan, E. B. 1979. On the spreading of liquids on solid surfaces: static and dynamic contact lines. *Annual Review of Fluid Mechanics*, **11**(1), 371–400.

El-Maghraby, R. M. 2012. *Measurements of CO_2 Trapping in Carbonate and Sandstone Rocks*. PhD thesis, Imperial College London.

El-Maghraby, R. M., and Blunt, M. J. 2013. Residual CO_2 trapping in Indiana limestone. *Environmental Science & Technology*, **47**(1), 227–233.

Eliassi, M., and Glass, R. J. 2001. On the continuum-scale modeling of gravity-driven fingers in unsaturated porous media: The inadequacy of the Richards Equation with standard monotonic constitutive relations and hysteretic equations of state. *Water Resources Research*, **37**(8), 2019–2035.

Eliassi, M., and Glass, R. J. 2002. On the porous-continuum modeling of gravity-driven fingers in unsaturated materials: Extension of standard theory with a hold-back-pile-up effect. *Water Resources Research*, **38**(11), 1234.

Ewing, R. P., and Berkowitz, B. 2001. Stochastic pore-scale growth models of DNAPL migration in porous media. *Advances in Water Resources*, **24**(3–4), 309–323.

Fatt, I. 1956a. The network model of porous media I. Capillary pressure characteristics. *Petroleum Transactions of the AIME*, **207**, 144–159.

Fatt, I. 1956b. The network model of porous media II. Dynamic properties of a single size tube network. *Petroleum Transactions of the AIME*, **207**, 160–163.

Fatt, I. 1956c. The network model of porous media III. Dynamic properties of networks with tube radius distribution. *Petroleum Transactions of the AIME*, **207**, 164–181.

Fayers, F. J. 1989. Extension of Stone's method 1 and conditions for real characteristics in three-phase flow. *SPE Reservoir Engineering*, **4**(4), 437–445.

Fayers, F. J., and Matthews, J. D. 1984. Evaluation of normalized Stone's methods for estimating three-phase relative permeabilities. *SPE Journal*, **24**(2), 224–232.

Feali, M., Pinczewski, W., Cinar, Y., Arns, C. H., Arns, J. Y., Francois, N., Turner, M. L., Senden, T., and Knackstedt, M. A. 2012. Qualitative and quantitative analyses of the three-phase distribution of oil, water, and gas in Bentheimer sandstone by use of micro-CT imaging. *SPE Reservoir Evaluation & Engineering*, **15**(6), 706–711.

Fenghour, A., Wakeham, W. A., and Vesovic, V. 1998. The viscosity of carbon dioxide. *Journal of Physical and Chemical Reference Data*, **27**(1), 31–44.

Fenwick, D. H., and Blunt, M. J. 1998a. Network modeling of three-phase flow in porous media. *SPE Journal*, **3**(1), 86–96.

Fenwick, D. H., and Blunt, M. J. 1998b. Three-dimensional modeling of three phase imbibition and drainage. *Advances in Water Resources*, **21**(2), 121–143.

Fernø, M. A., Torsvik, M., Haugland, S., and Graue, A. 2010. Dynamic laboratory wettability alteration. *Energy & Fuels*, **24**(7), 3950–3958.

Ferrand, L. A., Milly, P. C. D., Pinder, G. F., and Turrin, R. P. 1990. A comparison of capillary pressure-saturation relations for drainage in two- and three-fluid porous media. *Advances in Water Resources*, **13**(2), 54–63.

Ferrari, A., and Lunati, I. 2013. Direct numerical simulations of interface dynamics to link capillary pressure and total surface energy. *Advances in Water Resources*, **57**, 19–31.

Ferréol, B., and Rothman, D. H. 1995. Lattice-Boltzmann simulations of flow through Fontainebleau sandstone. *Transport in Porous Media*, **20**(1–2), 3–20.

Finney, J. L. 1970. Random packings and the structure of simple liquids. I. The geometry of random close packing. *Proceedings of the Royal Society of London. Series A, Mathematical and Physical Sciences*, **319**(1539), 479–493.

Firincioglu, T., Blunt, M. J., and Zhou, D. 1999. Three-phase flow and wettability effects in triangu-lar capillaries. *Colloids and Surfaces A: Physicochemical and Engineering Aspects*, **155**(2–3), 259–276.

Flannery, B. P., Deckman, H. W., Roberge, W. G., and D'Amico, K. L. 1987. Three-dimensional X-ray microtomography. *Science*, **237**(4821), 1439–1444.

Fleury, M., Ringot, G., and Poulain, P. 2001. Positive imbibition capillary pressure curves using the centrifuge technique. *Petrophysics*, **42**(4), 344–351.

Frette, V., Feder, J., Jøssang, T., and Meakin, P. 1992. Buoyancy-driven fluid migration in porous media. *Physical Review Letters*, **68**(21), 3164–3167.

Frette, V., Feder, J., Jøssang, T., Meakin, P., and Måløy, K. J. 1994. Fast, immiscible fluid-fluid displacement in three-dimensional porous media at finite viscosity contrast. *Physical Review E*, **50**(4), 2881–2890.

Furuberg, L., Måløy, K. J., and Feder, J. 1996. Intermittent behavior in slow drainage. *Physical Review E*, **53**(1), 966–977.

Garcia, X., Akanji, L. T., Blunt, M. J., Matthai, S. K., and Latham, J. P. 2009. Numerical study of the effects of particle shape and polydispersity on permeability. *Physical Review E*, **80**(2), 021304.

Gauglitz, P. A., and Radke, C. J. 1990. The dynamics of liquid film breakup in constricted cylindrical capillaries. *Journal of Colloid and Interface Science*, **134**(1), 14–40.

Geistlinger, H., and Mohammadian, S. 2015. Capillary trapping mechanism in strongly water wet systems: Comparison between experiment and percolation theory. *Advances in Water Resources*, **79**, 35–50.

Geistlinger, H., Ataei-Dadavi, I., Mohammadian, S., and Vogel, H. J. 2015. The impact of pore struc-ture and surface roughness on capillary trapping for 2-D and 3-D-porous media: Comparison with percolation theory. *Water Resources Research*, **51**(11), 9094–9111.

Geistlinger, H., Ataei-Dadavi, I., and Vogel, H. J. 2016. Impact of surface roughness on capillary trapping using 2D micromodel visualization experiments. *Transport in Porous Media*, **112**(1), 207–227.

Georgiadis, A., Maitland, G., Trusler, J. P. M., and Bismarck, A. 2011. Interfacial tension measure-ments of the (H_2O + n-Decane + CO_2) ternary system at elevated pressures and temperatures. *Journal of Chemical & Engineering Data*, **56**(12), 4900–4908.

Georgiadis, A., Berg, S., Makurat, A., Maitland, G., and Ott, H. 2013. Pore-scale micro-computed-tomography imaging: Nonwetting-phase cluster-size distribution during drainage and imbibi-tion. *Physical Review E*, **88**(3), 033002.

Geromichalos, D., Mugele, F., and Herminghaus, S. 2002. Nonlocal dynamics of spontaneous imbibition fronts. *Physical Review Letters*, **89**(10), 104503.

Gharbi, O. 2014. *Fluid-Rock Interactions in Carbonates: Applications to CO_2 Storage*. PhD thesis, Imperial College London.

Gharbi, O., and Blunt, M. J. 2012. The impact of wettability and connectivity on relative permeability in carbonates: A pore network modeling analysis. *Water Resources Research*, **48**(12), W12513.

Glass, R. J., and Nicholl, M. J. 1996. Physics of gravity fingering of immiscible fluids within porous media: An overview of current understanding and selected complicating factors. *Geoderma*, **70**(24), 133–163.

Glass, R. J., Steenhuis, T. S., and Parlange, J. Y. 1989a. Mechanism for finger persistence in homogeneous unsaturated porous media: Theory and verification. *Soil Science*, **148**(1), 60–70.

Glass, R. J., Parlange, J. Y., and Steenhuis, T. S. 1989b. Wetting front instability: 1. Theoretical discussion and dimensional analysis. *Water Resources Research*, **25**(6), 1187–1194.

Glass, R. J., Steenhuis, T. S., and Parlange, J. Y. 1989c. Wetting front instability: 2. Experimental determination of relationships between system parameters and two-dimensional unstable flow field behavior in initially dry porous media. *Water Resources Research*, **25**(6), 1195–1207.

Goodyear, S. G., and Jones, P. I. R. 1993. Relative permeabilities for gravity stabilised gas injection. Proceedings of the 7th European Symposium on Improved Oil Recovery, Moscow.

Grader, A. S., and O'Meara, D. J. Jr. 1988. Dynamic Displacement Measurements of Three-Phase Relative Permeabilities Using Three Immiscible Liquids. SPE 18293, proceedings of the SPE Annual Technical Conference and Exhibition, Houston, Texas, 2–5 October.

Grate, J. W., Dehoff, K. J., Warner, M. G., Pittman, J. W., Wietsma, T. W., Zhang, C., and Oostrom, M. 2012. Correlation of oil-water and air-water contact angles of diverse silanized surfaces and relationship to fluid interfacial tensions. *Langmuir*, **28**(18), 7182–7188.

Graue, A., Viksund, B. G., Eilertsen, T., and Moe, R. 1999. Systematic wettability alteration by aging sandstone and carbonate rock in crude oil. *Journal of Petroleum Science and Engineering*, **24**(2–4), 85–97.

Gray, W. G. 1983. General conservation equations for multi-phase systems: 4. Constitutive theory including phase change. *Advances in Water Resources*, **6**(3), 130–140.

Gray, W. G., and Hassanizadeh, S. M. 1991. Unsaturated flow theory including interfacial phenomena. *Water Resources Research*, **27**(8), 1855–1863.

Gray, W. G., and Hassanizadeh, S. M. 1998. Macroscale continuum mechanics for multiphase porous-media flow including phases, interfaces, common lines and common points. *Advances in Water Resources*, **21**(4), 261–281.

Gray, W. G., and O'Neill, K. 1976. On the general equations for flow in porous media and their reduction to Darcy's Law. *Water Resources Research*, **12**(2), 148–154.

Gray, W. G., Miller, C. T., and Schrefler, B. A. 2013. Averaging theory for description of environmental problems: What have we learned? *Advances in Water Resources*, **51**, 123–138.

Groenzin, H., and Mullins, O. C. 2000. Molecular size and structure of asphaltenes from various sources. *Energy & Fuels*, **14**(3), 677–684.

Gueyffier, D., Li, J., Nadim, A., Scardovelli, R., and Zaleski, S. 1999. Volume-of-fluid interface tracking with smoothed surface stress methods for three-dimensional flows. *Journal of Computational Physics*, **152**(2), 423–456.

Guises, R., Xiang, J., Latham, J. P., and Munjiza, A. 2009. Granular packing: Numerical simulation and the characterisation of the effect of particle shape. *Granular Matter*, **11**(5), 281–292.

Haines, W. B. 1930. Studies in the physical properties of soil. V. The hysteresis effect in capillary properties, and the modes of moisture distribution associated therewith. *The Journal of Agricultural Science*, **20**(1), 97–116.

Hammervold, W. L, Knutsen, Ø., Iversen, J. E., and Skjæveland, S. M. 1998. Capillary pressure scanning curves by the micropore membrane technique. *Journal of Petroleum Science and Engineering*, **20**(3–4), 253–258.

Hao, L., and Cheng, P. 2010. Pore-scale simulations on relative permeabilities of porous media by lattice Boltzmann method. *International Journal of Heat and Mass Transfer*, **53**(9–10), 1908–1913.

Harvey, R. R., and Craighead, E. M. 1965. *Aqueous displacement of oil*. US Patent 3,170,514.

Hashemi, M., Sahimi, M., and Dabir, B. 1999. Monte Carlo simulation of two-phase flow in porous media: Invasion with two invaders and two defenders. *Physica A: Statistical Mechanics and Its Applications*, **267**(1–2), 1–33.

Hassanizadeh, M., and Gray, W. G. 1979a. General conservation equations for multi-phase systems: 1. Averaging procedure. *Advances in Water Resources*, **2**, 131–144.

Hassanizadeh, M., and Gray, W. G. 1979b. General conservation equations for multi-phase systems: 2. Mass, momenta, energy, and entropy equations. *Advances in Water Resources*, **2**, 191–203.

Hassanizadeh, M., and Gray, W. G. 1980. General conservation equations for multi-phase systems: 3. Constitutive theory for porous media flow. *Advances in Water Resources*, **3**(1), 25–40.

Hassanizadeh, S. M., and Gray, W. G. 1993a. Thermodynamic basis of capillary pressure in porous media. *Water Resources Research*, **29**(10), 3389–3405.

Hassanizadeh, S. M., and Gray, W. G. 1993b. Toward an improved description of the physics of two-phase flow. *Advances in Water Resources*, **16**(1), 53–67.

Hassanizadeh, S. M., Celia, M. A., and Dahle, H. K. 2002. Dynamic effect in the capillary pressure-saturation relationship and its impacts on unsaturated flow. *Vadose Zone Journal*, **1**(1), 38–57.

Hassenkam, T., Skovbjerg, L. L., and Stipp, S. L. S. 2009. Probing the intrinsically oil-wet surfaces of pores in North Sea chalk at subpore resolution. *Proceedings of the National Academy of Sciences*, **106**(15), 6071–6076.

Haugen, Å., Fernø, M. A., Mason, G., and Morrow, N. R. 2014. Capillary pressure and relative permeability estimated from a single spontaneous imbibition test. *Journal of Petroleum Science and Engineering*, **115**, 66–77.

Hazen, A. 1892. *Some physical properties of sands and gravels, with special reference to their use in filtration*. 24th Annual Report, Massachusetts State Board of Health, Pub. Doc. No. 34.

Hazlett, R. D. 1993. *On surface roughness effects in wetting phenomena*. VSP, Utrecht, the Netherlands. In: Contact Angle, Wettability and Adhesion, Ed. Mittal, K. L. Pages 173–181.

Hazlett, R. D. 1995. Simulation of capillary-dominated displacements in microtomographic images of reservoir rocks. *Transport in Porous Media*, **20**(1–2), 21–35.

Hazlett, R. D. 1997. Statistical characterization and stochastic modeling of pore networks in relation to fluid flow. *Mathematical Geology*, **29**(6), 801–822.

Heiba, A. A., Davis, H. T., and Scriven, L. E. 1984. Statistical Network Theory of Three-Phase Relative Permeabilities. SPE 12690, proceedings of the SPE Enhanced Oil Recovery Symposium, Tulsa, Oklahoma, 15–18 April.

Helland, J. O., and Skjæveland, S. M. 2006a. Physically based capillary pressure correlation for mixed-wet reservoirs from a bundle-of-tubes model. *SPE Journal*, **11**(2), 171–180.

Helland, J. O., and Skjæveland, S. M. 2006b. Three-phase mixed-wet capillary pressure curves from a bundle of triangular tubes model. *Journal of Petroleum Science and Engineering*, **52**(1–4), 100–130.

Helland, J. O., and Skjæveland, S. M. 2007. Relationship between capillary pressure, saturation, and interfacial area from a model of mixed-wet triangular tubes. *Water Resources Research*, **43**(12), W12S10.

Helmig, R., Weiss, A., and Wohlmuth, B. I. 2007. Dynamic capillary effects in heterogeneous porous media. *Computational Geosciences*, **11**(3), 261–274.

Herring, A. L., Harper, E. J., Andersson, L., Sheppard, A., Bay, B. K., and Wildenschild, D. 2013. Effect of fluid topology on residual nonwetting phase trapping: Implications for geologic CO_2 sequestration. *Advances in Water Resources*, **62**, 47–58.

Herring, A. L., Andersson, L., Schlüter, S., Sheppard, A., and Wildenschild, D. 2015. Efficiently engineering pore-scale processes: The role of force dominance and topology during nonwetting phase trapping in porous media. *Advances in Water Resources*, **79**, 91–102.

Herring, A. L., Sheppard, A., Andersson, L., and Wildenschild, D. 2016. Impact of wettability alteration on 3D nonwetting phase trapping and transport. *International Journal of Greenhouse Gas Control*, **46**, 175–186.

Hilfer, R. 1991. Geometric and dielectric characterization of porous media. *Physical Review B*, **44**(1), 60–75.

Hilfer, R. 2002. Review on scale dependent characterization of the microstructure of porous media. *Transport in Porous Media*, **46**(2–3), 373–390.

Hilfer, R. 2006a. Capillary pressure, hysteresis and residual saturation in porous media. *Physica A: Statistical Mechanics and Its Applications*, **359**, 119–128.

Hilfer, R. 2006b. Macroscopic capillarity and hysteresis for flow in porous media. *Physical Review E*, **73**(1), 016307.

Hilfer, R., and Lemmer, A. 2015. Differential porosimetry and permeametry for random porous media. *Physical Review E*, **92**(1), 013305.

Hilfer, R., and Øren, P. E. 1996. Dimensional analysis of pore scale and field scale immiscible displacement. *Transport in Porous Media*, **22**(1), 53–72.

Hilfer, R., and Steinle, R. 2014. Saturation overshoot and hysteresis for twophase flow in porous media. *The European Physical Journal Special Topics*, **223**(11), 2323–2338.

Hilfer, R., and Zauner, Th. 2011. High-precision synthetic computed tomography of reconstructed porous media. *Physical Review E*, **84**(6), 062301.

Hilfer, R., Armstrong, R. T., Berg, S., Georgiadis, A., and Ott, H. 2015a. Capillary saturation and desaturation. *Physical Review E*, **92**(6), 063023.

Hilfer, R., Zauner, T., Lemmer, A., and Biswal, B. 2015b. *Institute for Computational Physics*, *http://www.icp.uni-stuttgart.de/microct/*.

Hill, S. 1952. Channeling in packed columns. *Chemical Engineering Science*, **1**(6), 247–253.

Hilpert, M. 2012. Velocity-dependent capillary pressure in theory for variably-saturated liquid infiltration into porous media. *Geophysical Research Letters*, **39**(6), L06402.

Hilpert, M., and Miller, C. T. 2001. Pore-morphology-based simulation of drainage in totally wetting porous media. *Advances in Water Resources*, **24**(3–4), 243–255.

Hiorth, A., Cathles, L. M., and Madland, M. V. 2010. The impact of pore water chemistry on carbonate surface charge and oil wettability. *Transport in Porous Media*, **85**(1), 1–21.

Hirasaki, G. J. 1991. Wettability: Fundamentals and surface forces. *SPE Formation Evaluation*, **6**(2), 217–226.

Hirasaki, G. J. 1993. *Structural interactions in the wetting and spreading of van der Waals fluids*. VSP, Utrecht, the Netherlands. In: Contact Angle, Wettability and Adhesion, Ed. Mittal, K. L. Pages 183–220.

Hirasaki, G. J., Rohan, J. A., Dubey, S. T., and Niko, H. 1990. Wettability evaluation during restored state core analysis. SPE 20506, proceedings of the Annual Technical Conference and Exhibition of the SPE, New Orleans, September 23–26.

Hirt, C. W., and Nichols, B. D. 1981. Volume of fluid (VOF) method for the dynamics of free boundaries. *Journal of Computational Physics*, **39**(1), 201–225.

Holmgren, C. R., and Morse, R. A. 1951. Effect of free gas saturation on oil recovery by water flooding. *Journal of Petroleum Technology*, **3**(5), 135–140.

Holtzman, R., and Segre, E. 2015. Wettability stabilizes fluid invasion into porous media via nonlocal, cooperative pore filling. *Physical Review Letters*, **115**(16), 164501.

Homsy, G. M. 1987. Viscous fingering in porous media. *Annual Review of Fluid Mechanics*, **19**(1), 271–311.

Honarpour, M. M., Koederitz, F., and Herbert, A. 1986. *Relative Permeability of Petroleum Reservoirs*. CRC Press, Boca Raton, Florida.

Hoshen, J., and Kopelman, R. 1976. Percolation and cluster distribution. I. Cluster multiple labeling technique and critical concentration algorithm. *Physical Review B*, **14**(8), 3438–3445.

Huang, H., Meakin, P., and Liu, M. B. 2005. Computer simulation of two-phase immiscible fluid motion in unsaturated complex fractures using a volume of fluid method. *Water Resources Research*, **41**(12), W12413.

Hughes, R. G., and Blunt, M. J. 2000. Pore scale modeling of rate effects in imbibition. *Transport in Porous Media*, **40**(3), 295–322.

Hui, M. H., and Blunt, M. J. 2000. Effects of wettability on three-phase flow in porous media. *The Journal of Physical Chemistry B*, **104**(16), 3833–3845.

Humphry, K. J., Suijkerbuijk, B. M. J. M., van der Linde, H. A., Pieterse, S. G. J., and Masalmeh, S. K. 2013. Impact of wettability on residual oil saturation and capillary desaturation curves. SCA2013-025, proceedings of Society of Core Analysts Annual Meeting, Napa Valley, California.

Hunt, A., and Ewing, R. 2009. *Percolation Theory for Flow in Porous Media*. Vol. 771. Springer Science & Business Media, New York.

Hunt, A. G. 2001. Applications of percolation theory to porous media with distributed local conductances. *Advances in Water Resources*, **24**(3), 279–307.

Idowu, N., Long, H., Øren, P. E., Carnerup, A. M., Fogden, A., Bondino, I., and Sundal, L. 2015. Wettability analysis using micro-CT, FESEM and QEMSCAN and its applications to digital rock physics. SCA2015-010, proceedings of the International Symposium of the Society of Core Analysts, St. Johns Newfoundland and Labrador, Canada, 16–21 August.

Idowu, N. A., and Blunt, M. J. 2010. Pore-scale modelling of rate effects in waterflooding. *Transport in Porous Media*, **83**(1), 151–169.

Iglauer, S., Favretto, S., Spinelli, G., Schena, G., and Blunt, M. J. 2010. X-ray tomography measurements of power-law cluster size distributions for the nonwetting phase in sandstones. *Physical Review E*, **82**(5), 056315.

Iglauer, S., Paluszny, A., Pentland, C. H, and Blunt, M. J. 2011. Residual CO_2 imaged with X-ray micro-tomography. *Geophysical Research Letters*, **38**(21), L21403.

Iglauer, S., Fernø, M. A., Shearing, P., and Blunt, M. J. 2012. Comparison of residual oil cluster size distribution, morphology and saturation in oil-wet and water-wet sandstone. *Journal of Colloid and Interface Science*, **375**(1), 187–192.

Iglauer, S., Paluszny, A., and Blunt, M. J. 2013. Simultaneous oil recovery and residual gas storage: A pore-level analysis using in situ X-ray micro-tomography. *Fuel*, **103**, 905–914.

Iglauer, S., Pentland, C. H., and Busch, A. 2015. CO_2 wettability of seal and reservoir rocks and the implications for carbon geo-sequestration. *Water Resources Research*, **51**(1), 729–774.

Inamuro, T., Ogata, T., Tajima, S., and Konishi, N. 2004. A lattice Boltzmann method for incompressible two-phase flows with large density differences. *Journal of Computational Physics*, **198**(2), 628–644.

Issa, R. I. 1986. Solution of the implicitly discretized fluid flow equations by operator-splitting. *Journal of Computational Physics*, **62**(1), 40–65.

Jackson, M. D., and Vinogradov, J. 2012. Impact of wettability on laboratory measurements of streaming potential in carbonates. *Colloids and Surfaces A: Physicochemical and Engineering Aspects*, **393**, 86–95.

Jackson, M. D., Valvatne, P. H., and Blunt, M. J. 2003. Prediction of wettability variation and its impact on flow using pore- to reservoir-scale simulations. *Journal of Petroleum Science and Engineering*, **39**(3–4), 231–246.

Jadhunandan, P. P., and Morrow, N. R. 1995. Effect of wettability on waterflood recovery for crude-oil/brine/rock systems. *SPE Reservoir Engineering*, **10**(1), 40–46.

Jain, V., Bryant, S., and Sharma, M. 2003. Influence of wettability and saturation on liquid-liquid interfacial area in porous media. *Environmental Science & Technology*, **37**(3), 584–591.

Jerauld, G. R. 1997. General three-phase relative permeability model for Prudhoe Bay. *SPE Reservoir Engineering*, **12**(4), 255–263.

Jerauld, G. R., and Rathmell, J. J. 1997. Wettability and relative permeability of Prudhoe Bay: A case study in mixed-wet reservoirs. *SPE Reservoir Engineering*, **12**(1), 58–65.

Jerauld, G. R., and Salter, S. J. 1990. The effect of pore-structure on hysteresis in relative permeability and capillary pressure: Pore-level modeling. *Transport in Porous Media*, **5**(2), 103–151.

Jerauld, G. R., Scriven, L. E., and Davis, H. T. 1984. Percolation and conduction on the 3D Voronoi and regular networks: A second case study in topological disorder. *Journal of Physics C: Solid State Physics*, **17**(19), 3429–3439.

Jettestuen, E., Helland, J. O., and Prodanović, M. 2013. A level set method for simulating capillary-controlled displacements at the pore scale with nonzero contact angles. *Water Resources Research*, **49**(8), 4645–4661.

Jha, B., Cueto-Felgueroso, L., and Juanes, R. 2011. Fluid mixing from viscous fingering. *Physical Review Letters*, **106**(19), 194502.

Jiang, F., Oliveira, M. S. A., and Sousa, A. C. M. 2007a. Mesoscale SPH modeling of fluid flow in isotropic porous media. *Computer Physics Communications*, **176**(7), 471–480.

Jiang, Z., Wu, K., Couples, G., van Dijke, M. I. J., Sorbie, K. S., and Ma, J. 2007b. Efficient extraction of networks from three-dimensional porous media. *Water Resources Research*, **43**(12). W12S03.

Jiang, Z., van Dijke, M. I. J., Sorbie, K. S., and Couples, G. D. 2007c. Representation of multiscale heterogeneity via multiscale pore networks. *Water Resources Research*, **49**(9), 5437–5449.

Jiang, Z., van Dijke, M. I. J., Wu, K., Couples, G. D., Sorbie, K. S., and Ma, J. 2012. Stochastic pore network generation from 3D rock images. *Transport in Porous Media*, **94**(2), 571–593.

Jin, C., Langston, P. A., Pavlovskaya, G. E., Hall, M. R., and Rigby, S. P. 2016. Statistics of highly heterogeneous flow fields confined to three-dimensional random porous media. *Physical Review E*, **93**(1), 013122.

Joekar-Niasar, V., and Hassanizadeh, S. M. 2011. Effect of fluids properties on non-equilibrium capillarity effects: Dynamic pore-network modeling. *International Journal of Multiphase Flow*, **37**(2), 198–214.

Joekar-Niasar, V., and Hassanizadeh, S. M. 2012. Analysis of fundamentals of two-phase flow in porous media using dynamic pore-network models: A review. *Critical Reviews in Environmental Science and Technology*, **42**(18), 1895–1976.

Johannesen, E. B., and Graue, A. 2007. Mobilization of remaining oil – emphasis on capillary number and wettability. SPE 108724, proceedings of the International Oil Conference and Exhibition, Mexico, 27–30 June.

Johns, M. L., and Gladden, L. F. 2001. Surface-to-volume ratio of ganglia trapped in small-pore systems determined by pulsed-field gradient nuclear magnetic resonance. *Journal of Colloid and Interface Science*, **238**(1), 96–104.

Juanes, R., Spiteri, E. J., Orr, F. M., and Blunt, M. J. 2006. Impact of relative permeability hysteresis on geological CO_2 storage. *Water Resources Research*, **42**(12), W12418.

Kaimourgiakis, M. E., S., Kikkinides E., A., Galani, C., Charalambopoulou G., and K., Stubos A. 2005. Digitally reconstructed porous media: Transport and sorption properties. *Transport in Porous Media*, **58**(1–2), 43–62.

Kalaydjian, F. 1990. Origin and quantification of coupling between relative permeabilities for two-phase flows in porous media. *Transport in Porous Media*, **5**(3), 215–229.

Kalaydjian, F. J. M., Moulu, J. C., Vizika, O., and Munkerud, P. K. 1997. Three-phase flow in water-wet porous media: Gas/oil relative permeabilities for various spreading conditions. *Journal of Petroleum Science and Engineering*, **17**(3–4), 275–290.

Kang, Q., Lichtner, P. C., and Zhang, D. 2006. Lattice Boltzmann pore-scale model for multicomponent reactive transport in porous media. *Journal of Geophysical Research: Solid Earth*, **111**(B5), B05203.

Kantzas, A., Chatzis, I., and Dullien, F. A. L. 1988. Mechanisms of capillary displacement of residual oil by gravity-assisted inert gas injection. SPE 17506, proceedings of the SPE Rocky Mountain Regional Meeting, Casper, Wyoming, 11–13 May.

Karabakal, U., and Bagci, S. 2004. Determination of wettability and its effect on waterflood performance in limestone medium. *Energy & Fuels*, **18**(2), 438–449.

Karpyn, Z. T., Piri, M., and Singh, G. 2010. Experimental investigation of trapped oil clusters in a water-wet bead pack using X-ray microtomography. *Water Resources Research*, **46**(4), W04510.

Kazemifar, F., Blois, G., Kyritsis, D. C., and Christensen, K. T. 2016. Quantifying the flow dynamics of supercritical CO_2-water displacement in a 2D porous micromodel using fluorescent microscopy and microscopic PIV. *Advances in Water Resources*, **95**, 325–368.

Keehm, Y., Mukerji, T., and Nur, A. 2004. Permeability prediction from thin sections: 3D reconstruction and Lattice-Boltzmann flow simulation. *Geophysical Research Letters*, **31**(4), L04606.

Keller, A. A., Blunt, M. J., and Roberts, P. V. 1997. Micromodel observation of the role of oil layers in three-phase flow. *Transport in Porous Media*, **26**(3), 277–297.

Kianinejad, A., and DiCarlo, D. A. 2016. Three-phase relative permeability in water-wet media: A comprehensive study. *Transport in Porous Media*, **112**(3), 665–687.

Kianinejad, A., Chen, X., and DiCarlo, D. A. 2015. The effect of saturation path on three-phase relative permeability. *Water Resources Research*, **51**(11), 9141–9164.

Killins, C. R., Nielsen, R. F., and Calhoun, J. C. 1953. Capillary desaturation and imbibition in porous rocks. *Producers Monthly*, **18**(2), 30–39.

Killough, J. E. 1976. Reservoir simulation with history-dependent saturation functions. *SPE Journal*, **16**(1), 37–48.

Kim, H., Rao, P. S. C., and Annable, M. D. 1997. Determination of effective air-water interfacial area in partially saturated porous media using surfactant adsorption. *Water Resources Research*, **33**(12), 2705–2711.

Kimbrel, E. H., Herring, A. L., Armstrong, R. T., Lunati, I., Bay, B. K., and Wildenschild, D. 2015. Experimental characterization of nonwetting phase trapping and implications for geologic CO_2 sequestration. *International Journal of Greenhouse Gas Control*, **42**, 1–15.

Klise, K. A., Moriarty, D., Yoon, H., and Karpyn, Z. 2016. Automated contact angle estimation for three-dimensional X-ray microtomography data. *Advances in Water Resources*, **95**, 152–160.

Kneafsey, T. J., Silin, D., and Ajo-Franklin, J. B. 2013. Supercritical CO_2 flow through a layered silica sand/calcite sand system: Experiment and modified maximal inscribed spheres analysis. *International Journal of Greenhouse Gas Control*, **14**, 141–150.

Knudsen, H. A., and Hansen, A. 2006. Two-phase flow in porous media: Dynamical phase transition. *The European Physical Journal B–Condensed Matter and Complex Systems*, **49**(1), 109–118.

Knudsen, H. A., Aker, E., and Hansen, A. 2002. Bulk flow regimes and fractional flow in 2D porous media by numerical simulations. *Transport in Porous Media*, **47**(1), 99–121.

Koiller, B., Ji, H., and Robbins, M. O. 1992. Fluid wetting properties and the invasion of square networks. *Physical Review B*, **45**(14), 7762–7767.

Koroteev, D., Dinariev, O., Evseev, N., Klemin, D., Nadeev, A., Safonov, S., Gurpinar, O., Berg, S., van Kruijsdijk, C., and Armstrong, R. 2014. Direct hydrodynamic simulation of multiphase flow in porous rock. *Petrophysics*, **55**(4), 294–303.

Koval, E. J. 1963. A method for predicting the performance of unstable miscible displacements in heterogeneous media. *Petroleum Transactions of the AIME*, **228**(2), 143–150.

Kovscek, A. R., and Radke, C. J. 1996. Gas bubble snap-off under pressure-driven flow in constricted noncircular capillaries. *Colloids and Surfaces A: Physicochemical and Engineering Aspects*, **117**(1–2), 55–76.

Kovscek, A. R., Wong, H., and Radke, C. J. 1993. A pore-level scenario for the development of mixed wettability in oil reservoirs. *AIChE Journal*, **39**(6), 1072–1085.

Kozeny, J. 1927. Über kapillare Leitung des Wassers im Boden. *Proceedings of the Royal Academy of Sciences, Vienna*, **136**(2a), 271–280.

Krevor, S., Blunt, M. J., Benson, S. M., Pentland, C. H., Reynolds, C., Al-Menhali, A., and Niu, B. 2015. Capillary trapping for geologic carbon dioxide storage – From pore scale physics to field scale implications. *International Journal of Greenhouse Gas Control*, **40**, 221–237.

Krummel, A. T., Datta, S. S., Münster, S., and Weitz, D. A. 2013. Visualizing multiphase flow and trapped fluid configurations in a model three-dimensional porous medium. *AIChE Journal*, **59**(3), 1022–1029.

Kumar, M., Senden, T., Knackstedt, M. A, Latham, S. J., Pinczewski, V., Sok, R. M, Sheppard, A. P, and Turner, M. L. 2009. Imaging of pore scale distribution of fluids and wettability. *Petrophysics*, **50**(4), 311–323.

Kyte, J. R., Stanclift, R. J. Jr., Stephan, S. C. Jr., and Rapoport, L. A. 1956. Mechanism of water flooding in the presence of free gas. *Petroleum Transactions of the AIME*, **207**, 215–221.

Lago, M., and Araujo, M. 2001. Threshold pressure in capillaries with polygonal cross section. *Journal of Colloid and Interface Science*, **243**(1), 219–226.

Lake, L. W. 1989. *Enhanced Oil Recovery*. Prentice Hall Inc., Englewood Cliffs, New Jersey.

Land, C. S. 1968. Calculation of imbibition relative permeability for two- and three-phase flow from rock properties. *SPE Journal*, **8**(2), 149–156.

Laplace, P. S. 1805. Traite de Mecanique Celeste (Gauthier-Villars, Paris, 1839), suppl. au livre X, 1805 and 1806, resp. *Oeuvres compl*, **4**.

Larsen, J. A., and Skauge, A. 1998. Methodology for numerical simulation with cycle-dependent relative permeabilities. *SPE Journal*, **3**(2), 163–173.

Larson, R. G., Scriven, L. E., and Davis, H. T. 1981. Percolation theory of two phase flow in porous media. *Chemical Engineering Science*, **36**(1), 57–73.

Latham, S., Varslot, T., and Sheppard, A. 2008. Image Registration: Enhancing and Calibrating X-Ray Micro-CT Imaging. SCA2008-35, proceedings of the International Symposium of the Society of Core Analysts, Abu Dhabi, 29 October–2 November.

Latief, F. D. E., Biswal, B., Fauzi, U., and Hilfer, R. 2010. Continuum reconstruction of the pore scale microstructure for Fontainebleau sandstone. *Physica A: Statistical Mechanics and Its Applications*, **389**(8), 1607–1618.

Lee, S. I., Song, Y .and Noh, T. W., Chen, X. D., and Gaines, J. R. 1986. Experimental observation of nonuniversal behavior of the conductivity exponent for three-dimensional continuum percolation systems. *Physical Review B*, **34**(10), 6719–6724.

Lemaitre, R., and Adler, P. M. 1990. Fractal porous media IV: Three-dimensional stokes flow through random media and regular fractals. *Transport in Porous Media*, **5**(4), 325–340.

Lenhard, R. J., and Parker, J. C. 1987. Measurement and prediction of saturation-pressure relationships in three-phase porous media systems. *Journal of Contaminant Hydrology*, **1**(4), 407–424.

Lenormand, R. 1990. Liquids in porous media. *Journal of Physics: Condensed Matter*, **2**(S), SA79–SA88.

Lenormand, R., and Bories, S. 1980. Description d'un mecanisme de connexion de liaision destine a l'etude du drainage avec piegeage en milieu poreux. *CR Acad. Sci*, **291**, 279–282.

Lenormand, R., and Zarcone, C. 1984. Role of roughness and edges during imbibition in square capillaries. SPE 13264, proceedings of the 59th SPE Annual Technical Conference and Exhibition, Houston, Texas, 16–19 September.

Lenormand, R., and Zarcone, C. 1985. Invasion percolation in an etched network: Measurement of a fractal dimension. *Physical Review Letters*, **54**(20), 2226–2229.

Lenormand, R., Zarcone, C., and Sarr, A. 1983. Mechanisms of the displacement of one fluid by another in a network of capillary ducts. *Journal of Fluid Mechanics*, **135**, 337–353.

Lenormand, R., E., Touboul, and Zarcone, C. 1988. Numerical models and experiments on immiscible displacements in porous media. *Journal of Fluid Mechanics*, **189**, 165–187.

Lerdahl, T. R., Øren, P. E., and Bakke, S. 2000. A Predictive Network Model for Three-Phase Flow in Porous Media. SPE 59311, proceedings of the SPE/DOE Improved Oil Recovery Symposium, Tulsa, Oklahoma, 3–5 April.

Leverett, M. C. 1939. Flow of oil-water mixtures through unconsolidated sands. *Petroleum Transactions of the AIME*, **132**(1), 150–172.

Leverett, M. C. 1941. Capillary behavior in porous sands. *Petroleum Transactions of the AIME*, **142**(1), 152–169.

Leverett, M. C. 1987. Trends and needs in reactor safety improvement. Pages 219–227 of: Lave, L. B. (ed), *Risk Assessment and Management*. Advances in Risk Analysis, vol. 5. Springer US.

Leverett, M. C., and Lewis, W. B. 1941. Steady flow of gas-oil-water mixtures through unconsolidated sands. *Petroleum Transactions of the AIME*, **142**(1), 107–116.

Li, H., Pan, C., and Miller, C. T. 2005. Pore-scale investigation of viscous coupling effects for two-phase flow in porous media. *Physical Review E*, **72**(2), 026705.

Lindquist, W. B., Lee, S. M., Coker, D. A., Jones, K. W., and Spanne, P. 1996. Medial axis analysis of void structure in three-dimensional tomographic images of porous media. *Journal of Geophysical Research: Solid Earth*, **101**(B4), 8297–8310.

Liu, H., Krishnan, S., Marella, S., and Udaykumar, H. S. 2005. Sharp interface Cartesian grid method II: A technique for simulating droplet interactions with surfaces of arbitrary shape. *Journal of Computational Physics*, **210**(1), 32–54.

Liu, M. B., and Liu, G. R. 2010. Smoothed particle hydrodynamics (SPH): An overview and recent developments. *Archives of Computational Methods in Engineering*, **17**(1), 25–76.

Longeron, D., Hammervold, W. L., and Skjæveland, S. M. 1994. Water-oil capillary pressure and wettability measurements using micropore membrane technique. SCA9426, proceedings of the Society of Core Analysts International Symposium, Stavanger, Norway, September 12–14.

Lorenz, C. D., and Ziff, R. M. 1998. Precise determination of the bond percolation thresholds and finite-size scaling corrections for the sc, fcc, and bcc lattices. *Physical Review E*, **57**(1), 230–236.

Løvoll, G., Méheust, Y., Måløy, K. J., Aker, E., and Schmittbuhl, J. 2005. Competition of gravity, capillary and viscous forces during drainage in a two-dimensional porous medium, a pore scale study. *Energy*, **30**(6), 861–872.

Ma, S., Mason, G., and Morrow, N. R. 1996. Effect of contact angle on drainage and imbibition in regular polygonal tubes. *Colloids and Surfaces A: Physicochemical and Engineering Aspects*, **117**(3), 273–291.

Mani, V., and Mohanty, K. K. 1997. Effect of the spreading coefficient on three-phase flow in porous media. *Journal of Colloid and Interface Science*, **187**(1), 45–56.

Manthey, S., Hassanizadeh, S. M., Helmig, R., and Hilfer, R. 2008. Dimensional analysis of two-phase flow including a rate-dependent capillary pressure-saturation relationship. *Advances in Water Resources*, **31**(9), 1137–1150.

Manwart, C., Torquato, S., and Hilfer, R. 2000. Stochastic reconstruction of sandstones. *Physical Review E*, **62**(1), 893–899.

Manwart, C., Aaltosalmi, U., Koponen, A., Hilfer, R., and Timonen, J. 2002. Lattice-Boltzmann and finite-difference simulations for the permeability for three-dimensional porous media. *Physical Review E*, **66**(1), 016702.

Martys, N., Cieplak, M., and Robbins, M. O. 1991a. Critical phenomena in fluid invasion of porous media. *Physical Review Letters*, **66**(8), 1058–1061.

Martys, N., Robbins, M. O., and Cieplak, M. 1991b. Scaling relations for interface motion through disordered media: Application to two-dimensional fluid invasion. *Physical Review B*, **44**(22), 12294–12306.

Martys, N. S., and Chen, H. 1996. Simulation of multicomponent fluids in complex three-dimensional geometries by the lattice Boltzmann method. *Physical Review E*, **53**(1), 743–750.

Masalmeh, S. K. 2002. The effect of wettability on saturation functions and impact on carbonate reservoirs in the Middle East. SPE 78517, proceedings of the Abu Dhabi International Petroleum Exhibition and Conference, Abu Dhabi, UAE.

Masalmeh, S. K, and Oedai, S. 2000. Oil mobility in the transition zone. SCA 2000-02, proceedings of Society of Core Analysts Annual Meeting, Abu Dhabi, UAE.

Mason, G., and Morrow, N. R. 1984. Meniscus curvatures in capillaries of uniform cross-section. *Journal of the Chemical Society, Faraday Transactions 1: Physical Chemistry in Condensed Phases*, **80**(9), 2375–2393.

Mason, G., and Morrow, N. R. 1991. Capillary behavior of a perfectly wetting liquid in irregular triangular tubes. *Journal of Colloid and Interface Science*, **141**(1), 262–274.

Mason, G., and Morrow, N. R. 2013. Developments in spontaneous imbibition and possibilities for future work. *Journal of Petroleum Science and Engineering*, **110**, 268–293.

Masson, Y., and Pride, S. R. 2014. A fast algorithm for invasion percolation. *Transport in Porous Media*, **102**(2), 301–312.

Matubayasi, N., Motomura, K., Kaneshina, S., Nakamura, M., and Matuura, R. 1977. Effect of pressure on interfacial tension between oil and water. *Bulletin of the Chemical Society of Japan*, **50**(2), 523–524.

Mayer, A. S., and Miller, C. T. 1993. An experimental investigation of pore-scale distributions of nonaqueous phase liquids at residual saturation. *Transport in Porous Media*, **10**(1), 57–80.

Mayer, R. P., and Stowe, R. A. 1965. Mercury porosimetry-breakthrough pressure for penetration between packed spheres. *Journal of Colloid Science*, **20**(8), 893–911.

McDougall, S. R., and Sorbie, K. S. 1995. The impact of wettability on waterflooding: Pore-scale simulation. *SPE Reservoir Engineering*, **10**(3), 208–213.

McWhorter, D. B. 1971. Infiltration affected by flow of air. *Hydrology Papers, Colorado State University, Fort Collins*, **49**.

McWhorter, D. B., and Sunada, D. K. 1990. Exact integral solutions for two-phase flow. *Water Resources Research*, **26**(3), 399–413.

Meakin, P. 1993. The growth of rough surfaces and interfaces. *Physics Reports*, **235**(4–5), 189–289.

Meakin, P., and Tartakovsky, A. M. 2009. Modeling and simulation of pore-scale multiphase fluid flow and reactive transport in fractured and porous media. *Reviews of Geophysics*, **47**(3), RG3002.

Meakin, P., Feder, J., Frette, V., and Jøssang, T. 1992. Invasion percolation in a destabilizing gradient. *Physical Review A*, **46**(6), 3357–3368.

Meakin, P., Wagner, G., Vedvik, A., Amundsen, H., Feder, J., and Jøssang, T. 2000. Invasion percolation and secondary migration: Experiments and simulations. *Marine and Petroleum Geology*, **17**(7), 777–795.

Mecke, K., and Arns, C. H. 2005. Fluids in porous media: A morphometric approach. *Journal of Physics: Condensed Matter*, **17**(9), S503–S534.

Mehmani, A., and Prodanović, M. 2014a. The application of sorption hysteresis in nano-petrophysics using multiscale multiphysics network models. *International Journal of Coal Geology*, **128–129**, 96–108.

Mehmani, A., and Prodanović, M. 2014b. The effect of microporosity on transport properties in porous media. *Advances in Water Resources*, **63**, 104–119.

Miller, C. T., Christakos, G., Imhoff, P. T., McBride, J. F., Pedit, J. A., and Trangenstein, J. A. 1998. Multiphase flow and transport modeling in heterogeneous porous media: Challenges and approaches. *Advances in Water Resources*, **21**(2), 77–120.

Mogensen, K., and Stenby, E. H. 1998. A dynamic two-phase pore-scale model of imbibition. *Transport in Porous Media*, **32**(3), 299–327.

Mohammadian, S., Geistlinger, H., and Vogel, H. J. 2015. Quantification of gas-phase trapping within the capillary fringe using computed microtomography. *Vadose Zone Journal*, **14**(5). 10.2136/vzj2014.06.0063.

Mohanty, K. K., Davis, H. T., and Scriven, L. E. 1987. Physics of oil entrapment in water-wet rock. *SPE Reservoir Engineering*, **2**(1), 113–128.

Morrow, N., and Buckley, J. 2011. Improved oil recovery by low-salinity waterflooding. *Journal of Petroleum Technology*, **63**(5), 106–112.

Morrow, N. R. 1970. Physics and thermodynamics of capillary action in porous media. *Industrial & Engineering Chemistry*, **62**(6), 32–56.

Morrow, N. R. 1976. Capillary pressure correlations for uniformaly wetted porous media. *Journal of Canadian Petroleum Technology*, **15**(4), 49–69.

Morrow, N. R. 1990. Wettability and its effect on oil recovery. *Journal of Petroleum Technology*, **42**(12), 1476–1484.

Morrow, N. R., and Mason, G. 2001. Recovery of oil by spontaneous imbibition. *Current Opinion in Colloid and Interface Science*, **6**(4), 321–337.

Morrow, N. R., Chatzis, I., and Taber, J. J. 1988. Entrapment and mobilization of residual oil in bead packs. *SPE Reservoir Engineering*, **3**(3), 927–934.

Mostaghimi, P., Bijeljic, B., and Blunt, M. J. 2012. Simulation of flow and dispersion on pore-space images. *SPE Journal*, **17**(4), 1131–1141.

Mostaghimi, P., Blunt, M. J., and Bijeljic, B. 2013. Computations of absolute permeability on micro-CT images. *Mathematical Geosciences*, **45**(1), 103–125.

Mousavi, M., Prodanović, M., and Jacobi, D. 2012. New classification of carbonate rocks for process-based pore scale modeling. *SPE Journal*, **18**(2), 243–263.

Mualem, Y. 1976. A new model for predicting the hydraulic conductivity of unsaturated porous media. *Water Resources Research*, **12**(3), 513–522.

Muljadi, B. P., Blunt, M. J., Raeini, A. Q., and Bijeljic, B. 2016. The impact of porous media heterogeneity on non-Darcy flow behaviour from pore-scale simulation. *Advances in Water Resources*, **95**, 329–340.

Murison, J., Semin, B., Baret, J. C. Herminghaus, S., Schröter, M., and Brinkmann, M. 2014. Wetting heterogeneities in porous media control flow dissipation. *Physical Review Applied*, **2**(3), 034002.

Muskat, M. 1949. *Physical Principles of Oil Production*. McGraw-Hill, New York.

Muskat, M., and Meres, M. W. 1936. The flow of heterogeneous fluids through porous media. *Journal of Applied Physics*, **7**, 346–363.

Navier, C. L. M. H. 1823. Mémoire sur les lois du mouvement des fluides. *Mémoires de l'Académie Royale des Sciences de l'Institut de France*, **6**, 389–416.

Neto, C., Evans, D. R., Bonaccurso, E., Butt, H. J., and Craig, V. S. J. 2005. Boundary slip in Newtonian liquids: A review of experimental studies. *Reports on Progress in Physics*, **68**(12), 2859–2897.

Ngan, C. G., and Dussan, V. E. B. 1989. On the dynamics of liquid spreading on solid surfaces. *Journal of Fluid Mechanics*, **209**, 191–226.

Nguyen, V. H., Sheppard, A. P., Knackstedt, M. A., and Pinczewski, W. V. 2006. The effect of displacement rate on imbibition relative permeability and residual saturation. *Journal of Petroleum Science and Engineering*, **52**(1–4), 54–70.

Niessner, J., Berg, S., and Hassanizadeh, S. M. 2011. Comparison of two-phase Darcy's law with a thermodynamically consistent approach. *Transport in Porous Media*, **88**(1), 133–148.

Niu, B., Al-Menhali, A., and Krevor, S. C. 2015. The impact of reservoir conditions on the residual trapping of carbon dioxide in Berea sandstone. *Water Resources Research*, **51**(4), 2009–2029.

Nono, F., Bertin, H., and Hamon, G. 2014. Oil recovery in the transition zone of carbonate reservoirs with wettability change: hysteresis models of relative permeability versus experimental data. SCA2014-007, proceedings of Society of Core Analysts Annual Meeting, Avignon, France.

Nordhaug, H. F., Celia, M., and Dahle, H. K. 2003. A pore network model for calculation of interfacial velocities. *Advances in Water Resources*, **26**(10), 1061–1074.

Nutting, P. G. 1930. Physical analysis of oil sands. *AAPG Bulletin*, **14**(10), 1337–1349.

Oak, M. J. 1991. Three-Phase Relative Permeability of Intermediate-Wet Berea Sandstone. SPE 22599, proceedings of the SPE Annual Technical Conference and Exhibition, Dallas, Texas, 6–9 October.

Oak, M. J., and Baker, L. E. 1990. Three-phase relative permeability of Berea sandstone. *Journal of Petroleum Technology*, **42**(8), 1054–1061.

Odeh, A. S. 1959. Effect of viscosity ratio on relative permeability. *Petroleum Transactions of the AIME*, **216**, 346–353.

Okabe, H., and Blunt, M. J. 2004. Prediction of permeability for porous media reconstructed using multiple-point statistics. *Physical Review E*, **70**(6), 066135.

Okabe, H., and Blunt, M. J. 2005. Pore space reconstruction using multiple-point statistics. *Journal of Petroleum Science and Engineering*, **46**(1–2), 121–137.

Okasha, T. M., Funk, J. J., and Rashidi, H. N. 2007. Fifty years of wettability measurements in the Arab-D carbonate reservoir. SPE 105114, proceedings of the SPE Middle East Oil and Gas Show and Conference, Manama, Kingdom of Bahrain, 11–14 March.

OpenFOAM. 2010. *OpenFOAM Programmers guide, http://foam.sourceforge.net/doc/Guides-a4 /ProgrammersGuide.pdf*. OpenCFD Limited.

Or, D., and Tuller, M. 1999. Liquid retention and interfacial area in variably saturated porous media: Upscaling from single-pore to sample-scale model. *Water Resources Research*, **35**(12), 3591–3605.

Øren, P. E., and Bakke, S. 2002. Process based reconstruction of sandstones and prediction of transport properties. *Transport in Porous Media*, **46**(2–3), 311–343.

Øren, P. E., and Bakke, S. 2003. Reconstruction of Berea sandstone and pore-scale modelling of wettability effects. *Journal of Petroleum Science and Engineering*, **39**(3–4), 177–199.

Øren, P. E., and Pinczewski, W. V. 1994. The effect of wettability and spreading coefficients on the recovery of waterflood residual oil by miscible gasflooding. *SPE Formation Evaluation*, **9**(2), 149–156.

Øren, P. E., and Pinczewski, W. V. 1995. Fluid distribution and pore-scale displacement mechanisms in drainage dominated three-phase flow. *Transport in Porous Media*, **20**(1–2), 105–133.

Øren, P. E., Billiotte, J., and Pinczewski, W. V. 1992. Mobilization of waterflood residual oil by gas injection for water-wet conditions. *SPE Formation Evaluation*, **7**(1), 70–78.

Øren, P. E., Billiotte, J., and Pinczewski, W. V. 1994. Pore-Scale Network Modelling of Waterflood Residual Oil Recovery by Immiscible Gas Flooding. SPE 27814, proceedings of the SPE/DOE Improved Oil Recovery Symposium, Tulsa, Oklahoma, 17–20 April.

Øren, P. E., Bakke, S., and Arntzen, O. J. 1998. Extending predictive capabilities to network models. *SPE Journal*, **3**(4), 324–336.

Øren, P. E., Bakke, S., and Held, R. 2007. Direct pore-scale computation of material and transport properties for North Sea reservoir rocks. *Water Resources Research*, **43**(12), W12S04.

Orr, F. M. Jr. 2007. *Theory of Gas Injection Processes*. Tie-Line Publications, Copenhagen.

Ovaysi, S., and Piri, M. 2010. Direct pore-level modeling of incompressible fluid flow in porous media. *Journal of Computational Physics*, **229**(19), 7456–7476.

Ovaysi, S., Wheeler, M. F., and Balhoff, M. 2014. Quantifying the representative size in porous media. *Transport in Porous Media*, **104**(2), 349–362.

Pak, T., Butler, I. B., Geiger, S., van Dijke, M. I. J., and Sorbie, K. S. 2015. Droplet fragmentation: 3D imaging of a previously unidentified pore-scale process during multiphase flow in porous media. *Proceedings of the National Academy of Sciences*, **112**(7), 1947–1952.

Pan, C., Hilpert, M., and Miller, C. T. 2001. Pore-scale modeling of saturated permeabilities in random sphere packings. *Physical Review E*, **64**(6), 066702.

Pan, C., Hilpert, M., and Miller, C. T. 2004. Lattice-Boltzmann simulation of two-phase flow in porous media. *Water Resources Research*, **40**(1), W01501.

Panfilov, M., and Panfilova, I. 2005. Phenomenological meniscus model for two-phase flows in porous media. *Transport in Porous Media*, **58**(1–2), 87–119.

Parker, J. C., Lenhard, R. J., and Kuppusamy, T. 1987. A parametric model for constitutive properties governing multiphase flow in porous media. *Water Resources Research*, **23**(4), 618–624.

Parlange, J. Y., and Hill, D. 1976. Theoretical analysis of wetting front instability in soils. *Soil Science*, **122**(4), 236–239.

Patankar, S. V., and Spalding, D. B. 1972. A calculation procedure for heat, mass and momentum transfer in three-dimensional parabolic flows. *International Journal of Heat and Mass Transfer*, **15**(10), 1787–1806.

Patzek, T. W. 2001. Verification of a complete pore network simulator of drainage and imbibition. *SPE Journal*, **6**(2), 144–156.

Payatakes, A. C. 1982. Dynamics of oil ganglia during immiscible displacement in water-wet porous media. *Annual Review of Fluid Mechanics*, **14**(1), 365–393.

Pentland, C. H. 2011. *Measurements of Non-wetting Phase Trapping in Porous Media*. PhD thesis, Imperial College London.

Pentland, C. H., Tanino, Y., Iglauer, S., and Blunt, M. J. 2010. Capillary Trapping in Water-Wet Sandstones: Coreflooding Experiments and Pore-Network Modeling. SPE 133798, proceedings of the SPE Annual Technical Conference and Exhibition, Florence, 19–22 September.

Pentland, C. H., El-Maghraby, R., Iglauer, S., and Blunt, M. J. 2011. Measurements of the capillary trapping of super-critical carbon dioxide in Berea sandstone. *Geophysical Research Letters*, **38**(6), L06401.

Pereira, G. G. 1999. Numerical pore-scale modeling of three-phase fluid flow: Comparison between simulation and experiment. *Physical Review E*, **59**(4), 4229–4242.

Pereira, G. G., Pinczewski, W. V., Chan, D. Y. C., Paterson, L., and Øren, P. E. 1996. Pore-scale network model for drainage-dominated three-phase flow in porous media. *Transport in Porous Media*, **24**(2), 167–201.

Philip, J. R. 1974. Fifty years progress in soil physics. *Geoderma*, **12**(4), 265–280.

Pickell, J. J., Swanson, B. F., and Hickmann, W. B. 1966. Application of air-mercury and oil-air capillary pressure data in the study of pore structure and fluid distribution. *SPE Journal*, **6**(1), 55–61.

Piller, M., Schena, G., Nolich, M., Favretto, S., Radaelli, F., and Rossi, E. 2009. Analysis of hydraulic permeability in porous media: From high resolution X-ray tomography to direct numerical simulation. *Transport in Porous Media*, **80**(1), 57–78.

Piri, M., and Blunt, M. J. 2005a. Three-dimensional mixed-wet random pore-scale network modeling of two- and three-phase flow in porous media. I. Model description. *Physical Review E*, **71**(2), 026301.

Piri, M., and Blunt, M. J. 2005b. Three-dimensional mixed-wet random pore-scale network modeling of two- and three-phase flow in porous media. II. Results. *Physical Review E*, **71**(2), 026302.

Plug, W. J., and Bruining, J. 2007. Capillary pressure for the sand-CO_2-water system under various pressure conditions. Application to CO_2 sequestration. *Advances in Water Resources*, **30**(11), 2339–2353.

Poiseuille, J. L. 1844. *Recherches expérimentales sur le mouvement des liquides dans les tubes de très-petits diamètres*. Imprimerie Royale, Paris.

Poisson, S. D. 1831. Memoire sur les equations generales de l'equilibre et du mouvement des corps solides elastiques et des fluides. *J. Ecole Polytechnique*, **13**, 1–174.

Popinet, S., and Zaleski, S. 1999. A front-tracking algorithm for accurate representation of surface tension. *International Journal for Numerical Methods in Fluids*, **30**(6), 775–793.

Porter, M, L., Wildenschild, D., Grant, G., and Gerhard, J. I. 2010. Measurement and prediction of the relationship between capillary pressure, saturation, and interfacial area in a NAPL-water-glass bead system. *Water Resources Research*, **46**(8), W08512.

Porter, M. L., Schaap, M. G., and Wildenschild, D. 2009. Lattice-Boltzmann simulations of the capillary pressure-saturation-interfacial area relationship for porous media. *Advances in Water Resources*, **32**(11), 1632–1640.

Princen, H. M. 1969a. Capillary phenomena in assemblies of parallel cylinders: I. Capillary rise between two cylinders. *Journal of Colloid and Interface Science*, **30**(1), 69–75.

Princen, H. M. 1969b. Capillary phenomena in assemblies of parallel cylinders: II. Capillary rise in systems with more than two cylinders. *Journal of Colloid and Interface Science*, **30**(3), 359–371.

Princen, H. M. 1970. Capillary phenomena in assemblies of parallel cylinders: III. Liquid Columns between Horizontal Parallel Cylinders. *Journal of Colloid and Interface Science*, **34**(2), 171–184.

Prodanović, M. 2016. *Digital rocks portal*. https://www.digitalrocksportal.org/.

Prodanović, M., and Bryant, S. L. 2006. A level set method for determining critical curvatures for drainage and imbibition. *Journal of Colloid and Interface Science*, **304**(2), 442–458.

Prodanović, M., Lindquist, W. B., and Seright, R. S. 2006. Porous structure and fluid partitioning in polyethylene cores from 3D X-ray microtomographic imaging. *Journal of Colloid and Interface Science*, **298**(1), 282–297.

Prodanović, M., Lindquist, W. B., and Seright, R. S. 2007. 3D image-based characterization of fluid displacement in a Berea core. *Advances in Water Resources*, **30**(2), 214–226.

Prodanović, M., Bryant, S. L., and Davis, J. S. 2013. Numerical simulation of diagenetic alteration and its effect on residual gas in tight gas sandstones. *Transport in Porous Media*, **96**(1), 39–62.

Prodanović, M., Mehmani, A., and Sheppard, A. P. 2015. Imaged-based multiscale network modelling of microporosity in carbonates. *Geological Society, London, Special Publications*, **406**(1), 95–113.

Rabbani, A., Jamshidi, S., and S., Salehi. 2014. An automated simple algorithm for realistic pore network extraction from micro-tomography images. *Journal of Petroleum Science and Engineering*, **123**, 164–171.

Raeesi, B., and Piri, M. 2009. The effects of wettability and trapping on relationships between interfacial area, capillary pressure and saturation in porous media: A pore-scale network modeling approach. *Journal of Hydrology*, **376**(3–4), 337–352.

Raeini, A. Q., Blunt, M. J., and Bijeljic, B. 2012. Modelling two-phase flow in porous media at the pore scale using the volume-of-fluid method. *Journal of Computational Physics*, **231**(17), 5653–5668.

Raeini, A. Q., Bijeljic, B., and Blunt, M. J. 2014a. Direct simulations of two-phase flow on micro-CT images of porous media and upscaling of pore-scale forces. *Advances in Water Resources*, **231**(17), 5653–5668.

Raeini, A. Q., Bijeljic, B., and Blunt, M. J. 2014b. Numerical modelling of sub-pore scale events in two-phase flow through porous media. *Transport in Porous Media*, **101**(2), 191–213.

Raeini, A. Q., Bijeljic, B., and Blunt, M. J. 2015. Modelling capillary trapping using finite-volume simulation of two-phase flow directly on micro-CT images. *Advances in Water Resources*, **83**, 102–110.

Rahman, T., Lebedev, M., Barifcani, A., and Iglauer, S. 2016. Residual trapping of supercritical CO_2 in oil-wet sandstone. *Journal of Colloid and Interface Science*, **469**, 63–68.

Ramstad, T., Øren, P. E., and Bakke, S. 2010. Simulation of two-phase flow in reservoir rocks using a lattice Boltzmann method. *SPE Journal*, **15**(4), 917–927.

Ramstad, T., Idowu, N., Nardi, C., and Øren, P. E. 2012. Relative permeability calculations from two-phase flow simulations directly on digital images of porous rocks. *Transport in Porous Media*, **94**(2), 487–504.

Ransohoff, T. C, and Radke, C. J. 1988. Laminar flow of a wetting liquid along the corners of a predominantly gas-occupied noncircular pore. *Journal of Colloid and Interface Science*, **121**(2), 392–401.

Ransohoff, T. C., Gauglitz, P. A., and Radke, C. J. 1987. Snap-off of gas bubbles in smoothly constricted noncircular capillaries. *AIChE Journal*, **33**(5), 753–765.

Reeves, P. C., and Celia, M. A. 1996. A functional relationship between capillary pressure, saturation, and interfacial area as revealed by a pore-scale network model. *Water Resources Research*, **32**(8), 2345–2358.

Reynolds, C. A. 2016. *Multiphase Flow Behaviour and Relative Permeability of CO_2-brine and N_2-water in Sandstones*. PhD thesis, Imperial College London.

Reynolds, C. A., and Krevor, S. 2015. Characterising flow behavior for gas injection: Relative permeability of CO_2-brine and N_2-water in heterogeneous rocks. *Water Resources Research*, **51**(12), 9464–9489.

Richards, L. A. 1931. Capillary conduction of liquids through porous mediums. *Journal of Applied Physics*, **1**(5), 318–333.

Robin, M., Rosenberg, E., and Fassi-Fihri, O. 1995. Wettability studies at the pore level: A new approach by the use of cryo-scanning electron microscopy. *SPE Formation Evaluation*, **10**(1), 11–20.

Roof, J. G. 1970. Snap-off of oil droplets in water-wet pores. *SPE Journal*, **10**(1), 85–90.

Roth, S., Hong, Y., Bale, H., Zhao, T., Bhattiprolu, S., Andrew, M., Weichao, C., Gelb, J., and Hornberger, B. 2016. Fully controlled sampling workflow for multi-scale X-ray imaging of complex reservoir rock samples to be used for digital rock physics. GEO 2016-2348618, proceedings of the 12th Middle East Geosciences Conference and Exhibition, 2016, Manama, Bahrain, 4–7 March.

Rothman, D. H. 1988. Cellular-automaton fluids: A model for flow in porous media. *Geophysics*, **53**(4), 509–518.

Rothman, D. H. 1990. Macroscopic laws for immiscible two-phase flow in porous media: Results from numerical experiments. *Journal of Geophysical Research: Solid Earth*, **95**(B6), 8663–8674.

Roux, S., and Guyon, E. 1989. Temporal development of invasion percolation. *Journal of Physics A: Mathematical and General*, **22**(17), 3693–3705.

Rücker, M., Berg, S., Armstrong, R. T., Georgiadis, A., Ott, H., Schwing, A., Neiteler, R., Brussee, N., Makurat, A., Leu, L., Wolf, M., Khan, F., Enzmann, F., and Kersten, M. 2015. From connected pathway flow to ganglion dynamics. *Geophysical Research Letters*, **42**(10), 3888–3894.

Rudman, M. 1997. Volume–tracking methods for interfacial flow calculations. *International Journal for Numerical Methods in Fluids*, **24**(7), 671–691.

Ryazanov, A. V., van Dijke, M. I. J., and Sorbie, K. S. 2009. Two-phase pore-network modelling: Existence of oil layers during water invasion. *Transport in Porous Media*, **80**(1), 79–99.

Sadjadi, Z., Jung, M., Seemann, R., and Rieger, H. 2015. Meniscus arrest during capillary rise in asymmetric microfluidic pore junctions. *Langmuir*, **31**(8), 2600–2608.

Saffman, P. G., and Taylor, G. 1958. The penetration of a fluid into a porous medium or Hele-Shaw cell containing a more viscous liquid. *Proceedings of the Royal Society of London A: Mathematical, Physical and Engineering Sciences*, **245**(1242), 312–329.

Sahimi, M. 1993. Flow phenomena in rocks: From continuum models to fractals, percolation, cellular automata, and simulated annealing. *Reviews of Modern Physics*, **65**(4), 1393–1534.

Sahimi, M. 2011. *Flow and Transport in Porous Media and Fractured Rock: From Classical Methods to Modern Approaches*. John Wiley & Sons, Hoboken, New Jersey.

Sahini, M., and Sahimi, M. 1994. *Applications of Percolation Theory*. CRC Press, Boca Raton, Florida.

Sahni, A., Burger, J., and Blunt, M. 1998. Measurement of Three Phase Relative Permeability during Gravity Drainage using CT. SPE 39655, proceedings of the SPE/DOE Improved Oil Recovery Symposium, Tulsa, Oklahoma, 19–22 April.

Salathiel, R. A. 1973. Oil recovery by surface film drainage in mixed-wettability rocks. *Journal of Petroleum Technology*, **25**(10), 1216–1224.

Saraji, S., Goual, L., and Piri, M. 2010. Adsorption of asphaltenes in porous media under flow conditions. *Energy & Fuels*, **24**(11), 6009–6017.

Saripalli, K. P., Kim, H., Rao, P. S. C., and Annable, M. D. 1997. Measurement of specific fluid-fluid interfacial areas of immiscible fluids in porous media. *Environmental Science & Technology*, **31**(3), 932–936.

Scardovelli, R., and Zaleski, S. 1999. Direct numerical simulation of free-surface and interfacial flow. *Annual Review of Fluid Mechanics*, **31**(1), 567–603.

Schaefer, C. E., DiCarlo, D. A., and Blunt, M. J. 2000. Experimental measurement of air-water interfacial area during gravity drainage and secondary imbibition in porous media. *Water Resources Research*, **36**(4), 885–890.

Schladitz, K. 2011. Quantitative micro-CT. *Journal of Microscopy*, **243**(2), 111–117.

Schlüter, S., Sheppard, A., Brown, K., and Wildenschild, D. 2014. Image processing of multiphase images obtained via X-ray microtomography: A review. *Water Resources Research*, **50**(4), 3615–3639.

Schmatz, J., Urai, J. L., Berg, S., and Ott, H. 2015. Nanoscale imaging of pore-scale fluid-fluid-solid contacts in sandstone. *Geophysical Research Letters*, **42**(7), 2189–2195.

Schmid, K. S., and Geiger, S. 2012. Universal scaling of spontaneous imbibition for water-wet systems. *Water Resources Research*, **48**(3), W03507.

Schmid, K. S., and Geiger, S. 2013. Universal scaling of spontaneous imbibition for arbitrary petrophysical properties: Water-wet and mixed-wet states and Handy's conjecture. *Journal of Petroleum Science and Engineering*, **101**, 44–61.

Schmid, K. S., Geiger, S., and Sorbie, K. S. 2011. Semianalytical solutions for cocurrent and countercurrent imbibition and dispersion of solutes in immiscible two-phase flow. *Water Resources Research*, **47**(2), W02550.

Selker, J. S., Steenhuis, T. S., and Parlange, J. Y. 1992. Wetting front instability in homogeneous sandy soils under continuous infiltration. *Soil Science Society of America Journal*, **56**(5), 1346–1350.

Serra, J. 1986. Introduction to mathematical morphology. *Computer Vision, Graphics, and Image Processing*, **35**(3), 283–305.

Shah, S. M., Gray, F., Crawshaw, J. P., and Boek, E. S. 2016. Micro-computed tomography pore-scale study of flow in porous media: Effect of voxel resolution. *Advances in Water Resources*, **95**, 276–287.

Sharma, M. M., and Wunderlich, R. W. 1987. The alteration of rock properties due to interactions with drilling-fluid components. *Journal of Petroleum Science and Engineering*, **1**(2), 127–143.

Sheppard, A. P., Knackstedt, M. A., Pinczewski, W. V., and Sahimi, M. 1999. Invasion percolation: New algorithms and universality classes. *Journal of Physics A: Mathematical and General*, **32**(49), L521–L529.

Silin, D., and Patzek, T. 2006. Pore space morphology analysis using maximal inscribed spheres. *Physica A: Statistical Mechanics and Its Applications*, **371**(2), 336–360.

Silin, D., Tomutsa, L., Benson, S. M., and Patzek, T. W. 2011. Microtomography and pore-scale modeling of two-phase fluid distribution. *Transport in Porous Media*, **86**(2), 495–515.

Singh, K., Niven, R. K., Senden, T. J., Turner, M. L., Sheppard, A. P., Middleton, J. P., and Knackstedt, M. A. 2011. Remobilization of residual non-aqueous phase liquid in porous media by freeze-thaw cycles. *Environmental Science & Technology*, **45**(8), 3473–3478.

Singh, K., Bijeljic, B., and Blunt, M. J. 2016. Imaging of oil layers, curvature and contact angle in a mixed-wet and a water-wet carbonate rock. *Water Resources Research*, **52**(3), 1716–1728.

Singh, M., and Mohanty, K. K. 2003. Dynamic modeling of drainage through three-dimensional porous materials. *Chemical Engineering Science*, **58**(1), 1–18.

Sivanesapillai, R., Falkner, N., Hartmaier, A., and Steeb, H. 2016. A CSF-SPH method for simulating drainage and imbibition at pore-scale resolution while tracking interfacial areas. *Advances in Water Resources*, **95**, 212–234.

Skauge, A., Sørvik, A., B., Vik., and Spildo, K. 2006. Effect of wettability on oil recovery from carbonate material representing different pore classes. SCA2006-01, proceedings of the Society of Core Analysts Annual Meeting, Trondheim, Norway.

Sohrabi, M., Tehrani, D. H., Danesh, A., and Henderson, G. D. 2004. Visualization of oil recovery by water-alternating-gas injection using high-pressure micromodels. *SPE Journal*, **9**(3), 290–301.

Sok, R. M., Knackstedt, M. A., Sheppard, A. P., Pinczewski, W.V., Lindquist, W. B., Venkatarangan, A., and Paterson, L. 2002. Direct and stochastic generation of network models from tomographic images: Effect of topology on residual saturations. *Transport in Porous Media*, **46**(2–3), 345–371.

Soll, W. E., and Celia, M. A. 1993. A modified percolation approach to simulating three-fluid capillary pressure-saturation relationships. *Advances in Water Resources*, **16**(2), 107–126.

Spalding, D. B., and Patankar, S. V. 1972. A calculation procedure for heat, mass and momentum transfer in three-dimensional parabolic flows. *International Journal of Heat and Mass Transfer*, **15**(10), 1787–1806.

Spanne, P., Thovert, J. F., Jacquin, C. J., Lindquist, W. B., Jones, K. W., and Adler, P. M. 1994. Synchrotron computed microtomograhy of porous media: Topology and transports. *Physical Review Letters*, **73**(14), 2001–2004.

Spelt, P. D. M. 2005. A level-set approach for simulations of flows with multiple moving contact lines with hysteresis. *Journal of Computational Physics*, **207**(2), 389–404.

Spiteri, E. J., Juanes, R., Blunt, M. J., and Orr, F. M. 2008. A new model of trapping and relative permeability hysteresis for all wettability characteristics. *SPE Journal*, **13**(3), 277–288.

Stauffer, D., and Aharony, A. 1994. *Introduction to Percolation Theory*. CRC Press, Boca Raton, Florida.

Steenhuis, T. S., Baver, C. E., Hasanpour, B., Stoof, C. R., DiCarlo, D. A., and Selker, J. S. 2013. Pore scale consideration in unstable gravity driven finger flow. *Water Resources Research*, **49**(11), 7815–7819.

Stokes, G. G. 1845. On the theories of the internal friction of fluids in motion, and of the equilibrium and motion of elastic solids. *Transactions of the Cambridge Philosophical Society*, **8**, 287–305.

Stone, H. L. 1970. Probability model for estimating three-phase relative permeability. *Journal of Petroleum Technology*, **22**(2), 214–218.

Stone, H. L. 1973. Estimation of three-phase relative permeability and residual oil data. *Journal of Canadian Petroleum Technology*, **12**(4), 53–61.

Strebelle, S. 2002. Conditional simulation of complex geological structures using multiple-point statistics. *Mathematical Geology*, **34**(1), 1–21.

Strenski, P. N., Bradley, R. M., and Debierre, J. M. 1991. Scaling behavior of percolation surfaces in three dimensions. *Physical Review Letters*, **66**(10), 1330–1333.

Stüben, K. 2001. An introduction to algebraic multigrid. *ebrary.free.fr*, 413–532.

Suekane, T., Zhou, N., Hosokawa, T., and Matsumoto, T. 2010. Direct observation of trapped gas bubbles by capillarity in sandy porous media. *Transport in Porous Media*, **82**(1), 111–122.

Suicmez, V. S., Piri, M., and Blunt, M. J. 2007. Pore-scale simulation of water alternate gas injection. *Transport in Porous Media*, **66**(3), 259–286.

Suicmez, V. S., Piri, M., and Blunt, M. J. 2008. Effects of wettability and pore-level displacement on hydrocarbon trapping. *Advances in Water Resources*, **31**, 503–512.

Sussman, M., Smereka, P., and Osher, S. 1994. A level set approach for computing solutions to incompressible two-phase flow. *Journal of Computational Physics*, **114**(1), 146–159.

Tallakstad, K. T., Løvoll, G., Knudsen, H. A., Ramstad, T., Flekkøy, E. G., and Måløy, K. J. 2009a. Steady-state, simultaneous two-phase flow in porous media: An experimental study. *Physical Review E*, **80**(3), 036308.

Tallakstad, K. T., Knudsen, H. A., Ramstad, T., Løvoll, G., Måløy, K. J., Toussaint, R., and Flekkøy, E. G. 2009b. Steady-state two-phase flow in porous media: Statistics and transport properties. *Physical Review Letters*, **102**(7), 074502.

Talon, L., Bauer, D., Gland, N., Youssef, S., Auradou, H., and Ginzburg, I. 2012. Assessment of the two relaxation time lattice-Boltzmann scheme to simulate Stokes flow in porous media. *Water Resources Research*, **48**(4), W04526.

Tanino, Y., and Blunt, M. J. 2012. Capillary trapping in sandstones and carbonates: Dependence on pore structure. *Water Resources Research*, **48**, W08525.

Tanino, Y., and Blunt, M. J. 2013. Laboratory investigation of capillary trapping under mixed-wet conditions. *Water Resources Research*, **49**(7), 4311–4319.

Tartakovsky, A. M., and Meakin, P. 2005. A smoothed particle hydrodynamics model for miscible flow in three-dimensional fractures and the two-dimensional Rayleigh-Taylor instability. *Journal of Computational Physics*, **207**(2), 610–624.

Tartakovsky, A. M., and Meakin, P. 2006. Pore scale modeling of immiscible and miscible fluid flows using smoothed particle hydrodynamics. *Advances in Water Resources*, **29**(10), 1464–1478.

Tartakovsky, A. M., Ward, A. L., and Meakin, P. 2007. Pore-scale simulations of drainage of heterogeneous and anisotropic porous media. *Physics of Fluids*, **19**(10), 103301.

Tartakovsky, A. M., Ferris, K. F., and Meakin, P. 2009a. Lagrangian particle model for multiphase flows. *Computer Physics Communications*, **180**(10), 1874–1881.

Tartakovsky, A. M., Meakin, P., and Ward, A. L. 2009b. Smoothed particle hydrodynamics model of non-aqueous phase liquid flow and dissolution. *Transport in Porous Media*, **76**(1), 11–34.

Taylor, H. F., O'Sullivan, C., and Sim, W. W. 2015. A new method to identify void constrictions in micro-CT images of sand. *Computers and Geotechnics*, **69**, 279–290.

Theodoropoulou, M. A., Sygouni, V., Karoutsos, V., and Tsakiroglou, C. D. 2005. Relative permeability and capillary pressure functions of porous media as related to the displacement growth pattern. *International Journal of Multiphase Flow*, **31**(10–11), 1155–1180.

Thompson, K. E, Willson, C. S, White, C. D., Nyman, S., Bhattacharya, J. P, and Reed, A. H. 2008. Application of a new grain-based reconstruction algorithm to microtomography images for quantitative characterization and flow modeling. *SPE Journal*, **13**(2), 164–176.

Thompson, P. A., and Robbins, M. O. 1989. Simulations of contact-line motion: Slip and the dynamic contact angle. *Physical Review Letters*, **63**(7), 766–769.

Thovert, J. F., Salles, J., and Adler, P. M. 1993. Computerized characterization of the geometry of real porous media: Their discretization, analysis and interpretation. *Journal of Microscopy*, **170**(1), 65–79.

Todd, M. R., and Longstaff, W. J. 1972. The development, testing, and application of a numerical simulator for predicting miscible flood performance. *Journal of Petroleum Technology*, **24**(7), 874–882.

Tolman, S., and Meakin, P. 1989. Off-lattice and hypercubic-lattice models for diffusion-limited aggregation in dimensionalities 2–8. *Physical Review A*, **40**(1), 428–437.

Tørå, G., Øren, P. E., and Hansen, A. 2012. A dynamic network model for two-phase flow in porous media. *Transport in Porous Media*, **92**(1), 145–164.

Torsæter, O. 1988. A comparative study of wettability test methods based on experimental results from north sea reservoir rocks. SPE 18281, proceedings of the SPE Annual Technical Conference and Exhibition, Houston, TX, October 2–5.

Treiber, L. E., Archer, D. L., and Owens, W. W. 1972. Laboratory evaluation of the wettability of fifty oil producing reservoirs. *SPE Journal*, **12**(6), 531–540.

Trojer, M., Szulczewski, M. L., and Juanes, R. 2015. Stabilizing fluid-fluid displacements in porous media through wettability alteration. *Physical Review Applied*, **3**(5), 054008.

Tryggvason, G., Bunner, B., Esmaeeli, A., Juric, D., Al-Rawahi, N., Tauber, W., Han, J., Nas, S., and Jan, Y. J. 2001. A front-tracking method for the computations of multiphase flow. *Journal of Computational Physics*, **169**(2), 708–759.

Turner, M. L., Knüfing, L., Arns, C. H., Sakellariou, A., Senden, T. J., Sheppard, A. P., Sok, R. M., Limaye, A., Pinczewski, W. V., and Knackstedt, M. A. 2004. Three-dimensional imaging of multiphase flow in porous media. *Physica A: Statistical Mechanics and Its Applications*, **339**(1–2), 166–172.

Unsal, E., Mason, G., Ruth, D. W., and Morrow, N. R. 2007a. Co- and counter-current spontaneous imbibition into groups of capillary tubes with lateral connections permitting cross-flow. *Journal of Colloid and Interface Science*, **315**(1), 200–209.

Unsal, E., Mason, G., Morrow, N. R., and Ruth, D. W. 2007b. Co-current and counter-current imbibition in independent tubes of non-axisymmetric geometry. *Journal of Colloid and Interface Science*, **306**(1), 105–117.

Unverdi, S., and Tryggvason, G. 1992. A front-tracking method for viscous, incompressible, multifluid flows. *Journal of Computational Physics*, **100**(1), 25–37.

Valavanides, M. S., Constantinides, G. N., and Payatakes, A. C. 1998. Mechanistic model of steady-state two-phase flow in porous media based on ganglion dynamics. *Transport in Porous Media*, **30**(3), 267–299.

Valavanides, M.S., and Payatakes, A.C. 2001. True-to-mechanism model of steady-state two-phase flow in porous media, using decomposition into prototype flows. *Advances in Water Resources*, **24**(3–4), 385–407.

Valvatne, P. H., and Blunt, M. J. 2004. Predictive pore-scale modeling of two-phase flow in mixed wet media. *Water Resources Research*, **40**(7), W07406.

van der Marck, S. C., Matsuura, T., and Glas, J. 1997. Viscous and capillary pressures during drainage: Network simulations and experiments. *Physical Review E*, **56**(5), 5675–5687.

van Dijke, M. I. J., and Sorbie, K. S. 2002a. An analysis of three-phase pore occupancies and relative permeabilities in porous media with variable wettability. *Transport in Porous Media*, **48**(2), 159–185.

van Dijke, M. I. J., and Sorbie, K. S. 2002b. Pore-scale network model for three-phase flow in mixed-wet porous media. *Physical Review E*, **66**(4), 046302.

van Dijke, M. I. J., and Sorbie, K. S. 2002c. The relation between interfacial tensions and wettability in three-phase systems: Consequences for pore occupancy and relative permeability. *Journal of Petroleum Science and Engineering*, **33**(1–3), 39–48.

van Dijke, M. I. J., and Sorbie, K. S. 2003. Pore-scale modelling of three-phase flow in mixed-wet porous media: Multiple displacement chains. *Journal of Petroleum Science and Engineering*, **39**(3–4), 201–216.

van Dijke, M. I. J., and Sorbie, K. S. 2006a. Cusp at the three-fluid contact line in a cylindrical pore. *Journal of Colloid and Interface Science*, **297**(2), 762–771.

van Dijke, M. I. J., and Sorbie, K. S. 2006b. Existence of fluid layers in the corners of a capillary with non-uniform wettability. *Journal of Colloid and Interface Science*, **293**(2), 455–463.

van Dijke, M. I. J., Sorbie, K. S., and McDougall, S. R. 2001a. Saturation dependencies of three-phase relative permeabilities in mixed-wet and fractionally wet systems. *Advances in Water Resources*, **24**(3–4), 365–384.

van Dijke, M. I. J., McDougall, S. R., and Sorbie, K. S. 2001b. Three-phase capillary pressure and relative permeability relationships in mixed-wet systems. *Transport in Porous Media*, **44**(1), 1–32.

van Dijke, M. I. J., Sorbie, K. S., Sohrabi, M., and Danesh, A. 2006. Simulation of WAG floods in an oil-wet micromodel using a 2-D pore-scale network model. *Journal of Petroleum Science and Engineering*, **52**(1–4), 71–86.

van Dijke, M. I. J., Piri, M., Helland, J. O., Sorbie, K. S., Blunt, M. J., and Skjæveland, S. M. 2007. Criteria for three-fluid configurations including layers in a pore with nonuniform wettability. *Water Resources Research*, **43**(12), W12S05.

van Genuchten, M. T. 1980. A closed-form equation for predicting the hydraulic conductivity of unsaturated soils. *Soil Science Society of America Journal*, **44**(5), 892–898.

van Genuchten, M. T., and Nielsen, D. R. 1985. On describing and predicting the hydraulic properties of unsaturated soils. *Annales Geophysicae*, **3**(5), 615–628.

van Kats, F. M., Egberts, P. J. P., and van Kruijsdijk, C. P. J. W. 2001. Three-phase effective contact angle in a model pore. *Transport in Porous Media*, **43**(2), 225–238.

Vicsek, T. 1992. *Fractal Growth Phenomena*. World Scientific, Singapore.

Vizika, O., and Lombard, J. M. 1996. Wettability and spreading: Two key parameters in oil recovery with three-phase gravity drainage. *SPE Reservoir Engineering*, **11**(1), 54–60.

Vizika, O., Avraam, D. G., and Payatakes, A. C. 1994. On the role of the viscosity ratio during low-capillary-number forced imbibition in porous media. *Journal of Colloid and Interface Science*, **165**(2), 386–401.

Vogel, H. J. 2002. *Topological Characterization of Porous Media*. Lecture Notes in Physics, vol. 600. Springer, Berlin & Heidelberg.

Vogel, H. J, and Roth, K. 2001. Quantitative morphology and network representation of soil pore structure. *Advances in Water Resources*, **24**(3–4), 233–242.

Vogel, H. J., Weller, U., and Schlüter, S. 2010. Quantification of soil structure based on Minkowski functions. *Computers & Geosciences*, **36**(10), 1236–1245.

Wang, J., Zhou, Z., Zhang, W., Garoni, T. M., and Deng, Y. 2013. Bond and site percolation in three dimensions. *Physical Review E*, **87**(5), 052107.

Washburn, E. W. 1921. The dynamics of capillary flow. *Physical Review*, **17**(3), 273–283.

Welge, H. J. 1952. A simplified method for computing oil recovery by gas or water drive. *Journal of Petroleum Technology*, **4**(4), 91–98.

Whitaker, S. 1986. Flow in porous media I: A theoretical derivation of Darcy's law. *Transport in Porous Media*, **1**(1), 3–25.

Whitaker, S. 1999. *The Method of Volume Averaging*. Kluwer Academic Publishers, Amsterdam.

Whyman, G., Bormashenko, E., and Stein, T. 2008. The rigorous derivation of Young, Cassie-Baxter and Wenzel equations and the analysis of the contact angle hysteresis phenomenon. *Chemical Physics Letters*, **450**(4–6), 355–359.

Wildenschild, D., and Sheppard, A. P. 2013. X-ray imaging and analysis techniques for quantifying pore-scale structure and processes in subsurface porous medium systems. *Advances in Water Resources*, **51**, 217–246.

Wildenschild, D., Armstrong, R. T., Herring, A. L., Young, I. M., and Carey, J. W. 2011. Exploring capillary trapping efficiency as a function of interfacial tension, viscosity, and flow rate. *Energy Procedia*, **4**, 4945–4952.

Wilkinson, D. 1984. Percolation model of immiscible displacement in the presence of buoyancy forces. *Physical Review A*, **30**(1), 520–531.

Wilkinson, D. 1986. Percolation effects in immiscible displacement. *Physical Review A*, **34**(2), 1380–1391.

Wilkinson, D., and Willemsen, J. F. 1983. Invasion percolation: A new form of percolation theory. *Journal of Physics A: Mathematical and General*, **16**(14), 3365–3376.

Witten, T. A., and Sander, L. M. 1981. Diffusion-limited aggregation, a kinetic critical phenomenon. *Physical Review Letters*, **47**(19), 1400–1403.

Wu, K., Van Dijke, M. I. J., Couples, G. D., Jiang, Z., Ma, J., Sorbie, K. S., Crawford, J., Young, I., and Zhang, X. 2006. 3D stochastic modelling of heterogeneous porous media – applications to reservoir rocks. *Transport in Porous Media*, **65**(3), 443–467.

Wyckoff, R. D., and Botset, H. G. 1936. The flow of gas-liquid mixtures through unconsolidated sands. *Journal of Applied Physics*, **7**(9), 325–345.

Wyckoff, R. D., Botset, H. G., Muskat, M., and Reed, D. W. 1933. The measurement of the permeability of porous media for homogeneous fluids. *Review of Scientific Instruments*, **4**(7), 394–405.

Yan, J. N., Monezes, J. L., and Sharma, M. M. 1993. Wettability alteration caused by oil-based muds and mud components. *SPE Drilling and Completion*, **8**(1), 35–44.

Yang, F., Hinger, F. F., Xiao, X., Liu, Y., Wu1, Z., Benson, S. M., and Toney, M. F. 2015. Extraction of pore-morphology and capillary pressure curves of porous media from synchrotron-based tomography data. *Scientific Reports*, **5**, 10635.

Yang, X., Mehmani, Y., Perkins, W. A., Pasquali, A., Schönherr, M., Kim, K., Perego, M., Parks, M. L., Trask, N., Balhoff, M. T., Richmond, M. C., Geier, M., Krafczyk, M., Luo, L. S., Tartakovsky, A. M., and Scheibe, T. D. 2016. Intercomparison of 3D pore-scale flow and solute transport simulation methods. *Advances in Water Resources*, **95**, 176–189.

Yeong, C. L. Y., and Torquato, S. 1998. Reconstructing random media. II. Three-dimensional media from two-dimensional cuts. *Physical Review E*, **58**(1), 224–233.

Yortsos, Y. C., Xu, B., and Salin, D. 1997. Phase diagram of fully developed drainage in porous media. *Physical Review Letters*, **79**(23), 4581–4584.

Young, T. 1805. An essay on the cohesion of fluids. *Philosophical Transactions of the Royal Society of London*, **95**, 65–87.

Yuan, H. H., and Swanson, B. F. 1989. Resolving pore-space characteristics by rate-controlled porosimetry. *SPE Formation Evaluation*, **4**(1), 17–24.

Zhao, X., Blunt, M. J., and Yao, J. 2010. Pore-scale modeling: Effects of wettability on waterflood oil recovery. *Journal of Petroleum Science and Engineering*, **71**(3–4), 169–178.

Zhou, D., and Blunt, M. 1997. Effect of spreading coefficient on the distribution of light non-aqueous phase liquid in the subsurface. *Journal of Contaminant Hydrology*, **25**(1–2), 1–19.

Zhou, D., and Blunt, M. 1998. Wettability effects in three-phase gravity drainage. *Journal of Petroleum Science and Engineering*, **20**(3–4), 203–211.

Zhou, D., Blunt, M., and Orr, F. M. Jr. 1997. Hydrocarbon drainage along corners of noncircular capillaries. *Journal of Colloid and Interface Science*, **187**(1), 11–21.

Zhou, D., Jia, L., Kamath, J., and Kovscek, A. R. 2002. Scaling of counter-current imbibition processes in low-permeability porous media. *Journal of Petroleum Science and Engineering*, **33**(1–3), 61–74.

Zhou, D., and Orr, F. M. Jr. 1995. The effects of gravity and viscous forces on residual nonwetting-phase saturation. *In Situ*, **19**(3), 249–273.

Zhou, X., R., Morrow N., and S., Ma. 2000. Interrelationship of wettability, initial water saturation, aging time, and oil recovery by spontaneous imbibition and waterflooding. *SPE Journal*, **5**(2), 199–207.

Zhou, Y., Helland, J., and Hatzignatiou, D. G. 2014. Pore-scale modeling of waterflooding in mixed-wet-rock images: Effects of initial saturation and wettability. *SPE Journal*, **19**(1), 88–100.

Zhou, Y., Helland, J. O., and Hatzignatiou, D. G. 2016. Computation of three-phase capillary pressure curves and fluid configurations at mixed-wet conditions in 2D rock images. *SPE Journal*, **21**(1), 152–169.

Zhu, Y., Fox, P. J., and Morris, J. P. 1999. A pore-scale numerical model for flow through porous media. *International Journal for Numerical and Analytical Methods in Geomechanics*, **23**(9), 881–904.

Zimmerman, R. W., Kumar, S., and Bodvarsson, G. S. 1991. Lubrication theory analysis of the permeability of rough-walled fractures. *International Journal of Rock Mechanics and Mining Sciences and Geomechanics Abstracts*, **28**(4), 325–331.

Index

Mount Gambier limestone
  capillary pressure, 88
  displacement statistics, 138, 208
  ganglion size distribution, 174
  image, 21
  J function, 246
  mixed-wet relative permeability, 336
  network, 48
  network properties, 47
  pore-space image, 22
  throat size distribution, 91
Multiphase Darcy law, 254
  assumptions, 255
  for anisotropic media, 256
  including viscous coupling, 255
Multiple displacement, 375
Multiple point statistics, 26
Muskat, Morris, 254

Nano-CT, 19
  images, 24
Napoleon, 16
Navier slip condition, 346
Navier-Stokes equations, 219
  boundary conditions, 221
  fluid boundary conditions, 228
  solution for pipe flow, 221
  solutions of, 247
Network model, 32
  Bentheimer, 37
  Berea, 36, 292
  comparison with direct methods, 249
  computational efficiency, 291
  construction, 35
  dynamic effects, 289
  Estaillades, 48
  generalized approach, 45
  including micro-porosity, 51
  Ketton, 48
  maximal ball construction, 42
  medial axis construction, 40
  Mount Gambier, 48
  process-based construction, 38
  properties, 47
  watershed construction, 42

Oil layers, 194, 197
  conductance, 231, 369
  evidence for, 370
  imaged in Bentheimer, 366
  impact on trapping, 216
  three-phase flow, 364
Oil relative permeability
  hysteresis, 336
  layers in three-phase flow, 368
  oil layers, 327
  predictions for three-phase flow, 392
  three-phase saturation dependence, 386

Oil-wet media
  recovery curves, 419
  relative permeability, 327

Percolation, 95
  conductance, 270
  correlation length, 101, 270
  definition, 96
  flow regimes, 309
  with a gradient, 105
  implications for fluid flow, 109
  implications for oil recovery, 147
  patterns in multiphase flow, 255
  shift in residual saturation, 274
  threshold, 97
  trapping, 169
Permeability
  bundle of tubes, 239
  computation on images, 247
  definition, 237
  example values, 241
  Fontainebleau, 249
  granular packings, 240
  prediction, 251
  range of, 243
  relation to pore size, 239
  units, 238
Phase field model, 298
Pinning
  corner layers, 198
  effect on recovery, 435
  layer conductance, 231
  three-phase contact line, 119
  water layers, 194
  water relative permeability, 328, 330, 335
Piston-like advance
  competition with snap-off, 132
  frontal advance filling pattern, 148
  imbibition, 126
Poiseuille flow, 222
Pore filling
  dynamics of, 141
  imbibition, 128
Pore length scale, 234
Pore shapes, 38
Pore volumes injected, 211, 413
Pore-scale imaging, 18
  contact angle, 70
  examples, 19
  interfacial area, 167
  mixed-wet, 216
  oil layers, 366
  trapped ganglia, 170
  trapping in three-phase flow, 397
Porosity
  definition, 29
  example values, 250, 251
  relationship to permeability, 249